This book has been revised according to the new CCE patter
NCERT syllabus prescribed by the Central Board of Secondary

AWARENESS SCIENCE for Eighth Class

Containing :
Multiple Choice Questions (MCQs) and Questions Based on High Order Thinking Skills (HOTS) (with answers)

Lakhmir Singh
Manjit Kaur

This Book Belongs to :

Name AP Anshuman Behera

Roll No. 1

Class and Section VIII 'C'

School DMS, BBSR

S. CHAND SCHOOL BOOKS
(An imprint of S. Chand Publishing)
A Division of S. Chand & Co. Pvt. Ltd.
7361, Ram Nagar, Qutab Road, New Delhi-110055
Phone: 23672080-81-82, 9899107446, 9911310888; Fax: 91-11-23677446
www.schandpublishing.com; e-mail : helpdesk@schandpublishing.com

Branches :

Ahmedabad : Ph: 27541965, 27542369, ahmedabad@schandgroup.com
Bengaluru : Ph: 22268048, 22354008, bangalore@schandgroup.com
Bhopal : Ph: 4274723, 4209587, bhopal@schandgroup.com
Chandigarh : Ph: 2725443, 2725446, chandigarh@schandgroup.com
Chennai : Ph. 28410027, 28410058, chennai@schandgroup.com
Coimbatore : Ph: 2323620, 4217136, coimbatore@schandgroup.com (Marketing Office)
Cuttack : Ph: 2332580; 2332581, cuttack@schandgroup.com
Dehradun : Ph: 2711101, 2710861, dehradun@schandgroup.com
Guwahati : Ph: 2738811, 2735640, guwahati@schandgroup.com
Haldwani : Mob. 09452294584 (Marketing Office)
Hyderabad : Ph: 27550194, 27550195, hyderabad@schandgroup.com
Jaipur : Ph: 2219175, 2219176, jaipur@schandgroup.com
Jalandhar : Ph: 2401630, 5000630, jalandhar@schandgroup.com
Kochi : Ph: 2378740, 2378207-08, cochin@schandgroup.com
Kolkata : Ph: 22367459, 22373914, kolkata@schandgroup.com
Lucknow : Ph: 4076971, 4026791, 4065646, 4027188, lucknow@schandgroup.com
Mumbai : Ph: 22690881, 22610885, mumbai@schandgroup.com
Nagpur : Ph: 2720523, 2777666, nagpur@schandgroup.com
Patna : Ph: 2300489, 2302100, patna@schandgroup.com
Pune : Ph: 64017298, pune@schandgroup.com
Raipur : Ph: 2443142, Mb. : 09981200834, raipur@schandgroup.com (Marketing Office)
Ranchi : Ph: 2361178, Mob. 09430246440, ranchi@schandgroup.com
Siliguri : Ph. 2520750, siliguri@schandgroup.com (Marketing Office)
Visakhapatnam : Ph. 2782609 (M) 09440100555, visakhapatnam@schandgroup.com (Marketing Office)

First Edition 2002
Revised Edition 2014
Reprints 2005, 2010, 2011, 2012, 2013, 2015

ISBN: 978-81-219-2556-3 Code: 1016E 299

PRINTED IN INDIA
By Vikas Publishing House Pvt. Ltd., Plot 20/4, Site-IV, Industrial Area Sahibabad, Ghaziabad-201010
and Published by S. Chand & Company Pvt. Ltd., 7361, Ram Nagar, New Delhi-110055.

ABOUT THE AUTHORS

LAKHMIR SINGH did his M.Sc. from Delhi University in 1969. Since then he has been teaching in Dyal Singh College of Delhi University, Delhi. He started writing books in 1980. Lakhmir Singh believes that book writing is just like classroom teaching. Though a book can never replace a teacher but it should make the student feel the presence of a teacher. Keeping this in view, he writes books in such a style that students never get bored reading his books. Lakhmir Singh has written more than 15 books so far on all the science subjects: Physics, Chemistry and Biology. He believes in writing quality books. He does not believe in quantity.

MANJIT KAUR did her B.Sc., B.Ed. from Delhi University in 1970. Since then she has been teaching in a reputed school of Directorate of Education, Delhi. Manjit Kaur is such a popular science teacher that all the students want to join those classes which she teaches in the school. She has a vast experience of teaching science to school children, and she knows the problems faced by the children in the study of science. Manjit Kaur has put all her teaching experience into the writing of science books. She has co-authored more than 15 books alongwith her husband, Lakhmir Singh.

It is the team-work of Lakhmir Singh and Manjit Kaur which has given some of the most popular books in the history of science education in India. Lakhmir Singh and Manjit Kaur both write exclusively for the most reputed, respected and largest publishing house of India : S.Chand and Company Pvt. Ltd.

AN OPEN LETTER

Dear Friend,

We would like to talk to you for a few minutes, just to give you an idea of some of the special features of this book. Before we go further, let us tell you that this book has been revised according to the NCERT syllabus prescribed by the Central Board of Secondary Education (CBSE) based on new "Continuous and Comprehensive Evaluation" (CCE) pattern of school education. Just like our earlier books, we have written this book in such a simple style that even the weak students will be able to understand science very easily. Believe us, while writing this book, we have considered ourselves to be the students of Class VIII and tried to make things as simple as possible.

The most important feature of this revised edition of the book is that we have included a large variety of different types of questions as required by CCE for assessing the learning abilities of the students. This book contains :

(*i*) Very short answer type questions (including true-false type questions and fill in the blanks type questions),

(*ii*) Short answer type questions,

(*iii*) Long answer type questions (or Essay type questions),

(*iv*) Multiple choice questions (MCQs),

(*v*) Questions based on high order thinking skills (HOTS), and

(*vi*) Activities.

Please note that answers have also been given for the various types of questions, wherever required. All these features will make this book even more useful to the students as well as the teachers. "A picture can say a thousand words". Keeping this in mind, a large number of coloured pictures and sketches of various scientific processes, procedures, appliances, manufacturing plants and everyday situations involving principles of science have been given in this revised edition of the book. This will help the students to understand the various concepts of science clearly. It will also tell them how science is applied in the real situations in homes, transport and industry.

Other Books by Lakhmir Singh and Manjit Kaur

1. Awareness Science for Sixth Class
2. Awareness Science for Seventh Class
3. Science for Ninth Class (Part 1) PHYSICS
4. Science for Ninth Class (Part 2) CHEMISTRY
5. Science for Tenth Class (Part 1) PHYSICS
6. Science for Tenth Class (Part 2) CHEMISTRY
7. Science for Tenth Class (Part 3) BIOLOGY
8. Rapid Revision in Science
   (A Question-Answer Book for Class X)
9. Science for Ninth Class (J & K Edition)
10. Science for Tenth Class (J & K Edition)
11. Science for Ninth Class (Hindi Edition) :
    PHYSICS and CHEMISTRY
12. Science for Tenth Class (Hindi Edition) :
    PHYSICS, CHEMISTRY and BIOLOGY
13. Saral Vigyan (A Question-Answer Science
    Book in Hindi for Class X)

We are sure you will agree with us that the facts and formulae of science are just the same in all the books, the difference lies in the method of presenting these facts to the students. In this book, the various topics of science have been explained in such a simple way that while reading this book, a student will feel as if a teacher is sitting by his side and explaining the various things to him. We are sure that after reading this book, the students will develop a special interest in science and they would like to study science in higher classes as well.

We think that the real judges of a book are the teachers concerned and the students for whom it is meant. So, we request our teacher friends as well as the students to point out our mistakes, if any, and send their comments and suggestions for the further improvement of this book.

Wishing you a great success,

Yours sincerely,

Lakhmir Singh

Manjit Kaur

396, Nilgiri Apartments,
Alaknanda, New Delhi-110019
E-mail : singhlakhmir@hotmail.com

CONTENTS

CHAPTER

1

Crop Production and Management

All the living organisms like man, animals and plants need food for their growth and survival. The green plants can synthesise their food by the process of photosynthesis by using inorganic substances like carbon dioxide gas and water in the presence of sunlight energy. Man and other animals cannot make food by photosynthesis from carbon dioxide gas and water by using sunlight energy. They need readymade organic food nutrients like carbohydrates, fats, and proteins, etc., for their growth and development. **Man obtains his food from plants as well as animals.** In other words, man has to grow plants and rear animals (bring up animals) to meet his requirements of food.

Many types of plants are grown on a large scale in vast fields because the food grains produced by them are consumed in large amounts. Wheat and rice are two common examples. These are called food grains. In addition to food grains, pulses, vegetables and fruits are also grown on a large scale because they are an important part of our food. The animals such as cow and buffalo are reared to obtain milk whereas goat, fish and hen are reared to get meat and eggs. In this Chapter, we will study the different practices of obtaining food from both, plants as well as animals. Before we go further, we should know the meaning of the word 'crop'. This is discussed below.

Figure 1. The wheat crop in fields.

Crops

When the same kind of plants are grown in the fields on a large scale to obtain foods like cereals (wheat, rice, maize), pulses, vegetables and fruits, etc., it is called a crop. For example, a crop of wheat means that all the plants grown in the fields are that of wheat (see Figure 1). A crop is called '*fasal*' in Hindi. Crops are grown in the soil in the fields by farmers (*kissan*). Some of the examples of crops are given below :

(*i*) Cereal crops (Grain crops) : Wheat, Paddy (Rice), Maize, Millet (*Bajra, Jawar*), Barley
(*ii*) Pulses : Gram (*Chana*), Peas, Beans
(*iii*) Oil seeds : Mustard, Groundnut, Sunflower
(*iv*) Vegetables : Tomato, Cabbage, Spinach
(*v*) Fruits : Banana, Grapes, Guava, Mango, Orange, Apple

Types of Crops

Different crops grow well in different seasons of the year. For example, a crop may grow well in rainy season during summer but it may not grow well in winter season. Similarly, another crop may grow well in winter season but not in rainy season. Based on the seasons (in which they grow well), all the crops are categorised into two main groups :

1. **Kharif crops**, and
2. **Rabi crops.**

The crops which are sown in the rainy season are called kharif crops. The rainy season in India is generally from June to September. The sowing for kharif crops starts in June–July at the beginning of south-west monsoon because these crops (particularly paddy) need substantial amount of water. The kharif crops are harvested at the end of monsoon season during September (or October). **Some of the examples of kharif crops are : Paddy, Maize, Millet, Soyabean, Groundnut and Cotton.** The kharif crops are sometimes also called 'summer crops'. Please note that 'paddy' is 'rice still in the husk'. So, paddy crop gives us rice. In other words, paddy is rice crop. Paddy is grown only in the rainy season because it requires a lot of water. Paddy cannot be grown in the winter season because water available in winter is much less. On the other hand, if wheat is sown in the kharif season, it will not grow well. This is because wheat plants cannot tolerate too much water of the rainy season.

The crops grown in the winter season are called rabi crops. The time period of rabi crops is generally from October to March. The sowing for rabi crops begins at the beginning of winter (October–November) and the crops are harvested by March (or April). **Some of the examples of rabi crops are : Wheat, Gram (*Chana*), Peas, Mustard, and Linseed.**

The people who have no permanent homes and continuously move from one place to another are called 'nomads' (or wanderers). Till about 10000 B.C., people were nomadic. They were continuously moving (or wandering) in groups from place to place in search of food and shelter. These nomadic people ate raw fruits and vegetables found in nature and started hunting animals for food. Later, they settled near the sources of water such as rivers and cultivated land to produce wheat, paddy (rice) and other food crops. This is how agriculture was born. **The growing of plants (or crops) in the fields for obtaining food (like wheat, rice, etc.) is called agriculture.** Agriculture is called '*khetibari*' or '*krishi*' in Hindi. We will now describe the various agricultural practices.

BASIC PRACTICES OF CROP PRODUCTION

In order to raise a crop (or cultivate a crop) successfully and profitably for food production, a farmer has to perform a large number of tasks in a sequence (one after the other). **The various tasks performed by a farmer to produce a good crop are called agricultural practices.** The various agricultural practices which are carried out at various stages of crop production are :

1. **Preparation of soil,**
2. **Sowing,**
3. **Adding manure and fertilisers,**
4. **Irrigation,**

5. Removal of weeds,
6. Harvesting, and
7. Storage of food grains.

In addition to these regular agricultural practices, one more agricultural practice called **'Rotation of crops'** is undertaken sometimes to improve the fertility of soil and increase the crop yield. The various agricultural practices require certain tools or implements which are called agricultural implements. We will now describe all the agricultural practices in detail to know how food is produced on a large scale.

1. PREPARATION OF SOIL

The upper layer of earth is called soil. The crop plants are grown in soil. Soil provides minerals, water, air, humus and anchorage (fixing firmly), to the plants. Preparation of soil is the first step in cultivating a crop for food production. **The soil is prepared for sowing the seeds of the crop by (*i*) ploughing, (*ii*) levelling, and (*iii*) manuring.** Each one of these steps has its own significance. This is described below.

The process of loosening and turning the soil is called ploughing (or tilling). Ploughing (or tilling) of fields is done by using an implement called plough. Ploughs are made of wood or iron, and they have an iron tip for easy penetration into the soil. The ploughs are pulled by a pair of bullocks or by a tractor (see Figures 2 and 3). Actually, the ploughing of small fields is done with the help of animals like bullocks

Figure 2. Ploughing the fields with the help of bullocks.

Figure 3. Ploughing the fields by using tractor.

while large fields are ploughed by using tractors. **The loosening of soil by ploughing is beneficial because of the following reasons :**

(*i*) The loose soil allows the plant roots to penetrate freely and deeper into the soil so that plants are held more firmly to the ground.

(*ii*) The loose soil allows the roots of plants to breathe easily (even when the roots are deep). This is because loose soil can hold a lot of air in its spaces.

(*iii*) The loose soil helps in the growth of worms and microbes present in the soil who are friends of the farmer since they help in further turning and loosening the soil. They also add humus to the soil.

(*iv*) Ploughing also uproots and buries the weeds (unwanted plants) standing in the field and thereby suffocates them to death.

(*v*) The loosening and turning of soil during ploughing brings the nutrient rich soil to the top so that the plants can use these nutrients.

If the soil is very dry, it breaks into large mud 'crumbs' during ploughing. The mud crumbs are then broken down by using a soil plank called 'crumb crusher'.

The ploughed soil is quite loose so it is liable to be carried away by strong winds or washed away by rain water. The removal of top soil by wind and water is called soil erosion. **The ploughed soil is levelled by pressing it with a wooden leveller (or an iron leveller) so that the top soil is not blown away by wind or drained off by water (and soil erosion is prevented).** The levelling of ploughed soil is beneficial because of the following reasons :

(*i*) The levelling of ploughed fields (by pressing) prevents the top fertile soil from being carried away by strong winds or washed away by rain water.

(*ii*) The levelling of ploughed fields helps in the uniform distribution of water in the fields during irrigation.

(*iii*) The levelling helps in preventing the loss of moisture from the ploughed soil.

The levelling of ploughed soil in the fields is done by using an implement called leveller. The soil leveller is a heavy wooden plank or an iron plank. The soil leveller can be pulled by bullocks or by tractor.

'Manuring' means 'adding manure to the soil'. Sometimes, manure is added to the soil before ploughing. Addition of manure to soil before ploughing helps in the proper mixing of manure with the soil. Manure is first transported to the fields. It is then spread out in the fields. When this field is ploughed, the manure gets mixed in the soil properly. Manure contains many nutrients required for the growth of crop plants. So, **manuring is done to increase the fertility of the soil before seeds are sown into it.** Once the soil is ploughed, levelled and manured, it is ready for the sowing of seeds. The soil is watered before sowing.

Agricultural Implements

Before sowing the seeds, it is necessary to loosen and turn the soil in the fields so as to break it to the size of grains. The loosening and turning of soil in the fields is done with the help of various agricultural implements (or tools). The main agricultural implements (or tools) used for loosening and turning the soil are : Plough, Hoe and Cultivator.

(*i*) PLOUGH. Plough is a large agricultural implement which is used for ploughing (or tilling) the soil in the fields. The traditionally used wooden plough is shown in Figure 4. The wooden plough consists of a long log of wood which is called plough shaft (see Figure 4). There is a handle at one end of the ploughshaft. Below the handle is a strong triangular iron strip called ploughshare. The other end of ploughshaft can be attached to a wooden beam which is fixed at right angles to the ploughshaft (see Figure 4). This beam is placed over the neck of two bullocks (or oxen) so as to pull the plough. Thus, the plough is drawn by a pair of bullocks (or other animals such as buffaloes, camels, etc.) (see Figure 2). When the plough is pulled by the bullocks, the farmer holds the handle of the plough and presses down the handle due to which the ploughshare digs into the soil, loosens it and turns it. Nowadays, the traditional wooden plough is increasingly being replaced by the iron plough.

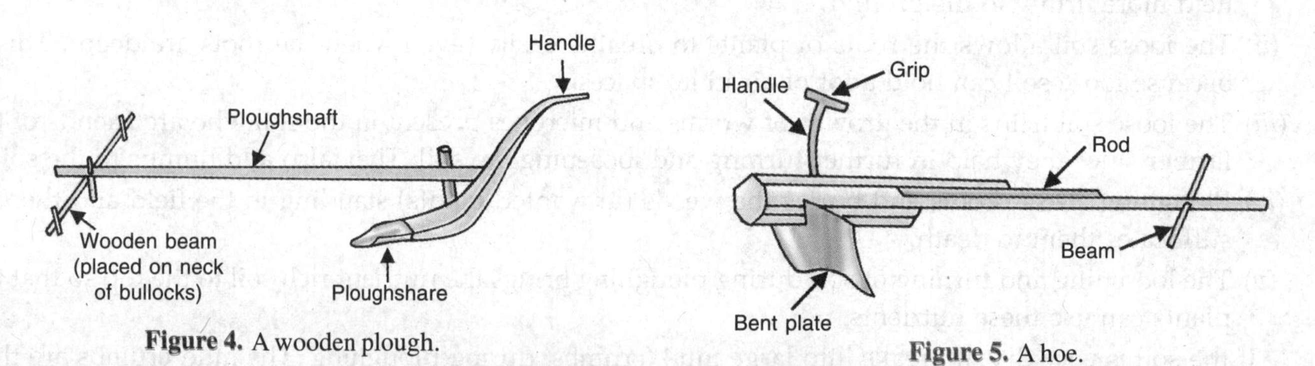

Figure 4. A wooden plough. **Figure 5.** A hoe.

(*ii*) HOE. Hoe is an agricultural implement (or tool) which is used for removing weeds, and loosening and turning the soil (see Figure 5). Hoe consists of a long rod of wood or iron. There is a handle (having grip) at one end of the hoe. A strong, broad and bent plate of iron is fixed below the handle and acts like a blade. The other end of hoe has a beam which is put on the neck of bullocks. Thus, a hoe is also pulled by animals such as a pair of bullocks. The hoe is a kind of modified plough.

***(iii)* CULTIVATOR.** The cultivator is a tractor driven agricultural implement which is used for loosening and turning the soil in the fields quickly (see Figure 6). A cultivator has many ploughshares which can dig into a considerable area of soil at the same time, loosen it and turn it. Due to this, many fields can be ploughed (or tilled) in a short time by using a cultivator. In this way, the use of cultivator saves labour and time. Nowadays, ploughing of large fields is done by using the tractor driven cultivators (see Figure 3).

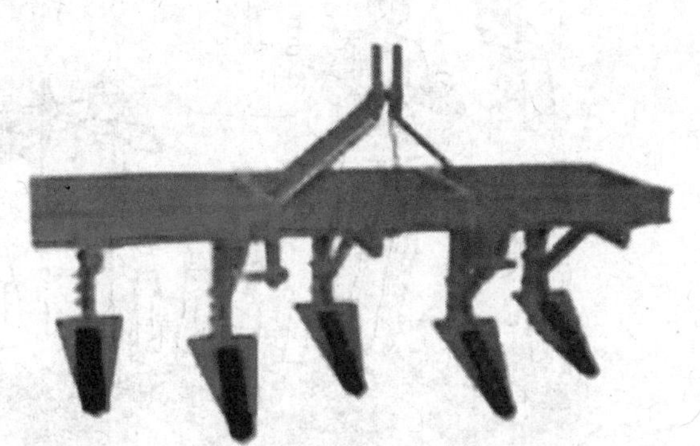

Figure 6. A tractor driven cultivator (or tractor driven plough).

2. SOWING

Once the soil in the field has been prepared by ploughing, levelling and manuring, etc., seeds of the crop can be sown in it. **The process of scattering seeds (or putting seeds) in the ground soil for growing the crop plants is called sowing.** Sowing is the most important part of crop production. Before sowing, good quality seeds are selected. **Good quality seeds are clean and healthy seeds free from infection and diseases.** Farmers prefer to use seeds which give high yield of food grains.

Selection of Seeds

ACTIVITY

We can select good, healthy seeds for sowing as follows : Put all the seeds in a bucket containing water and stir well. Most of the seeds will settle down at the bottom whereas some seeds will float on top. **The seeds which sink at the bottom of the bucket are the healthy seeds.** On the other hand, **the seeds which float on water are the spoiled seeds.** This can be explained as follows : Healthy seeds are heavy, so they sink in water. The seeds which have been partially eaten by pests or damaged by disease become hollow and light, and hence float on water. The seeds may also be treated with fungicide solutions before sowing to prevent the seed-borne diseases of crops.

Methods of Sowing Seeds

Seeds are sown in the soil either by hand or by seed drill. Thus, there are two methods of sowing the seeds in the soil. These are :

(*i*) Sowing by hand, and

(*ii*) Sowing with a seed drill.

The sowing of seeds by hand (or manually) is called broadcasting. In the sowing with hand or manually, the seeds are taken in hand and gradually scattered in the entire ploughed field. This method is, however, not very good because there is no proper spacing or proper depth at which the seeds are sown by hand. Moreover, the seeds scattered on the surface of the soil for sowing can be picked up and eaten by the birds.

The implement used for sowing is a seed drill. A seed drill is a long iron tube having a funnel at the top (see Figure 7). The seed drill is tied to back of the plough and seeds are put into the funnel of the seed drill. And as the plough makes furrows in the soil, the seeds from the seed drill are gradually released and sown into the soil furrows made by the plough. Thus, by using a seed drill for sowing, the seeds are sown at the correct depth and correct intervals (or spacings). The seeds sown with a seed drill are in regular rows. Moreover, when the seeds are sown in furrows by a seed-drill, the seeds get covered by soil. Due to this,

Figure 7. Sowing the seeds by using a traditional seed drill (attached to a plough)

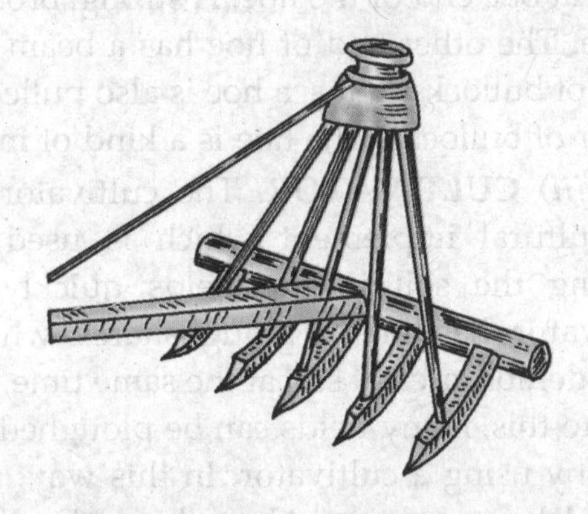

Figure 8. A tractor driven modern seed drill.

these seeds cannot be picked up and eaten by the birds. It is obvious that **the sowing with seed-drill is much better than sowing by hand.** A bullock driven seed-drill has just one long iron tube with a funnel. The tractor driven seed-drill has 5 to 6 iron tubes joined together with a common funnel at the top (see Figure 8). When the seeds are put into the funnel of such a seed drill, the seeds are released through all the tubes and get sown into 5 or 6 furrows of soil simultaneously. By using such tractor driven seed-drills, the sowing of seeds can be completed quickly. Most of the **crops like wheat, gram (*chana*), maize and millet etc, are grown (or cultivated) by sowing the seeds directly into soil.**

Precautions for Sowing Seeds

The following precautions should be taken while sowing seeds in the soil.

(*i*) The Seeds Should be Sown at Right Depth in the Soil Suitable For Germination. If the seeds are just spread on the surface of the soil, then the seeds will be eaten up by the birds. And if the seeds are sown too deep, then they may not germinate because they cannot breathe (cannot get sufficient air) at greater depth. So, the seeds should be sown at right depth in the soil which is suitable for germination. This right depth is learnt by experience.

(*ii*) The Seeds Should be Sown at Right Intervals or Spacings. The seeds should neither be placed too close nor too far apart. This is because if the seeds are sown too close, then plants formed from them will also be too close, and will not get enough sunlight, water, and other nutrients. Thus, an appropriate distance between the seeds is important to avoid overcrowding of plants. This allows the plants to get sufficient sunlight, nutrients and water from the soil. On the other hand, if the seeds are sown too far apart, then it will be a wastage of field space.

(*iii*) The Seeds Should Not be Sown in a Dry Soil. Moisture in the soil is necessary for the germination of seeds. So, if seeds are sown in a dry soil, they may not germinate at all.

(*iv*) The Seeds Should Not be Sown in a Highly Wet Soil. If the seeds are sown in a highly wet soil, then on drying, the soil surface becomes hard and because of this hard surface of soil, the germinating plumule will be unable to come out of ground. Moreover, the seeds are not able to respire properly due to lack of air under these conditions of hard surface of soil.

Advantages of Sowing with a Seed Drill

The sowing of seeds with a seed drill has the following advantages :

(*i*) By using a seed drill for sowing, the seeds are sown at correct depth and correct intervals (or spacings).

(*ii*) The seeds sown with a seed drill are in regular rows.

(*iii*) When the seeds are sown in furrows by a seed drill, the seeds get covered by soil and hence these seeds cannot be picked up and eaten by birds. This prevents damage caused by birds.

(*iv*) Sowing by using a tractor-driven seed drill saves time and labour.

Transplanting (or Transplantation)

Though most of the crops are grown by sowing the seeds directly in the soil but in some crops like paddy (rice) and many vegetables, the seeds are not directly sown in the soil in large fields. In the case of crops like paddy (rice) and vegetables like tomatoes and chillies (*mirch*), the seeds are first sown in a small plot of land or nursery and allowed to grow into **tiny plants called seedlings** by providing them with a good dose of nutrients. After the seeds have grown into tiny plants called seedlings in the seed-bed or nursery, only the healthy and well developed seedlings are then picked out from the nursery bed and transferred or transplanted to the regular field. **The process of transferring the seedlings from the nursery to the main field by hand is called transplantation or transplanting.** During transplantation, proper distance is kept between the various seedlings and also between the various rows of seedlings, to enable each and every plant (formed from seedlings) to get sufficient sunlight, water and other nutrients for normal and healthy growth (see Figure 9). The process of transplantation gives us many advantages over the direct sowing which ultimately leads to an increase in the yield of the crop. The various **advantages of the transplantation** process are given below :

Figure 9. Paddy seedlings being transplanted in the fields.

(*i*) The process of transplantation enables us to select only the better and healthy seedlings for the cultivation of crops. The bad seedlings can be rejected. This selection is, however, not possible when the seeds are directly sown in the soil.

(*ii*) The process of transplantation allows better penetration (deeper penetration) of the roots in the soil.

(*iii*) The process of transplantation promotes better development of the shoot system of plants.

(*iv*) The process of transplantation allows the seedlings to be planted at the right spacings so that the plants may get uniform dose of sunlight, water and nutrients.

The practice of transplantation is used in the cultivation of paddy crop (rice crop) and in the cultivation of many vegetables like tomatoes and chillies. We will now discuss manures and fertilisers.

3. ADDING MANURE AND FERTILISERS

The crop plants need a number of mineral elements for their growth which they get from the soil through their roots. Now, **repeated growing of crops in the same field removes a lot of precious mineral elements, organic matter and other materials from the soil**. Due to this the soil becomes infertile after some time, and the crop yield decreases. So, unless the depleted plant nutrients are put back into the soil from time to time, the growth of crop would be poor. **The deficiency of plant nutrients and organic matter in the soil is made up by adding manures and fertilisers to the soil.**

Manures

Manure is a natural fertiliser. **A manure is a natural substance obtained by the decomposition of animal wastes like cow-dung, human wastes, and plant residues, which supplies essential elements and humus to the soil and makes it more fertile.** Manures are prepared from animal wastes, human wastes and plant residues by the action of micro-organisms. In order to prepare manure, farmers dump animal wastes (animal dung, etc.) and plant wastes (like leaves, etc.) in pits at open places and allow it to decompose slowly. The decomposition is carried out by some micro-organisms. The decomposed animal and plant matter is used as organic manure.

Manures contain a mixture of various nutrient elements and a lot of organic matter (humus) recycled from bio-mass wastes (animal and plant wastes). Though manures are not very rich in plant nutrients like nitrogen, phosphorus and potassium, but they are rich in organic chemical nutrients like humus. Thus, **manures provide a lot of organic matter like humus to the soil** and this humus improves the physical and chemical properties of the soil. **A manure improves the soil texture for better retention of water and aeration.** This is because, being porous, humus can hold more water and air in the soil. In fact, manure makes up the general deficiency of the nutrients in the soil.

A manure is, however, very bulky and voluminous due to which it is inconvenient to store and transport. Moreover, a manure is not "nutrient specific", and hence it is not much helpful when a particular nutrient is required in the soil for a particular crop. A chemical fertiliser, on the other hand, is nutrient specific.

Chemical Fertilisers

Manures are not able to supply the required quantities of the essential plant nutrients like nitrogen, phosphorus and potassium, etc. So, they are to be supplemented with chemical fertilisers. **A chemical fertiliser is a salt or an organic compound containing the necessary plant nutrients like nitrogen, phosphorus or potassium, to make the soil more fertile.** A chemical fertiliser is rich in a particular plant nutrient (such as nitrogen, phosphorus or potassium). Some examples of fertilisers are : Urea, Ammonium sulphate, Superphosphate, Potash and NPK (N = Nitrogen ; P = Phosphorus ; K = Potassium).

The chemical fertilisers are nutrient specific. This means that a chemical fertiliser can provide only nitrogen, only phosphorus or only potassium to the soil, as required. **The chemical fertilisers have plant nutrients in a concentrated form. So, they provide quick replenishment of plant nutrients in the soil and restore its fertility.** Chemical fertilisers have high solubility in water. So, they are easily absorbed by the plants. Chemical fertilisers are made in factories. Chemical fertilisers are easy to transport, store and handle

Figure 10. A bag of chemical fertiliser.

Figure 11. Chemical fertiliser being applied to standing crop in fields.

because they come in bags (see Figure 10). The chemical fertilisers absorb moisture very quickly, so they are packed in air-tight bags.

The chemical fertilisers can be applied before sowing, during irrigation or sprayed on standing crops (see Figure 11). The use of fertilisers has helped the farmers to get better yield of crops such as wheat, paddy (rice) and maize, etc. **The excessive use of fertilisers is harmful** due to the following reasons :

(*i*) The excessive use of fertilisers changes the chemical nature of soil and makes the soil less fertile. For example, the excessive use of fertilisers can make the soil highly acidic or alkaline. The highly acidic or alkaline soil becomes less fertile.

(*ii*) The excessive use of fertilisers causes water pollution in ponds, lakes and rivers, etc.

In order to maintain the fertility of soil, we should substitute some of the fertilisers by organic manure or leave the field fallow (uncultivated) in-between two crops. When a field is kept uncultivated for some time, its fertility is restored naturally.

ACTIVITY TO SHOW THE EFFECT OF MANURE AND FERTILISER ON THE GROWTH OF PLANTS

Take three empty flower pots and mark them *A*, *B* and *C* (see Figure 12). Put some ordinary soil in pot *A*. Add some soil mixed with a little cow-dung manure in pot *B*. And take some soil mixed with a little urea fertiliser in pot *C*. Pour the same amount of water in all the three flower pots. Now take some *moong* or gram seeds and germinate them. Select equal sized seedlings of *moong* or gram. Plant these seedlings in each of the three flower pots. Keep the flower pots in a sunny place and water them daily. Observe the growth of

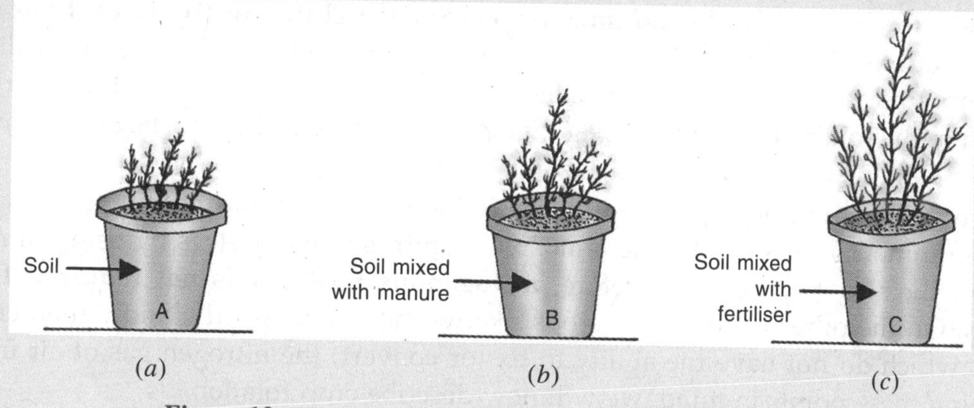

Figure 12. Effect of manure and fertiliser on the growth of plants.

seedlings in the three flower pots after 7 to 10 days. We will find that the seedlings planted in ordinary soil in pot *A* show the minimum growth [see Figure 12(*a*)]. The seedlings planted in soil containing manure in pot *B* show better growth [see Figure 12(*b*)]. But the seedings planted in soil containing fertiliser in pot *C* show the maximum growth as well as the fastest growth [see Figure 12(*c*)]. From this activity we conclude that manure and fertilisers help the plants to grow better and faster.

Before we end this discussion, we would like to give the main differences between manures and fertilisers in tabular form.

Differences between Manures and Fertilisers

Manure	*Fertiliser*
1. A manure is a natural substance obtained by the decomposition of animal wastes like cow dung, human waste, and plant residues.	1. A fertiliser is a salt or an organic compound.
2. A manure is not very rich in essential plant nutrients like nitrogen, phosphorus and potassium.	2. Fertilisers are very rich in plant nutrients like nitrogen, phosphorus and potassium.
3. A manure provides a lot of organic matter like humus to the soil.	3. A fertiliser does not provide any humus to the soil.
4. A manure is absorbed slowly by the plants because it is not much soluble in water.	4. Being soluble in water, a fertiliser is readily absorbed by the plants.
5. Manure can be prepared in the fields.	5. Fertilisers are prepared in factories.

Advantages of Manure

The manure is an organic material. **The organic manure is considered better than fertilisers because of the follwing reasons :**

(*i*) Manure enhances the water-holding capacity of the soil.

(*ii*) Manure makes the soil porous due to which the exchange of gases becomes easy.

(*iii*) Manure increases the number of useful microbes in the soil.

(*iv*) Manure improves the texture of the soil.

Another method of replenishing the soil with nutrients (such as nitrogen) is through crop rotation. Before we discuss crop rotation as a means of improving the fertility of soil, we should know something about leguminous plants or leguminous crops. **The pulses, peas, beans, groundnut, gram (*chana*) and clover (*berseem*) are leguminous crops.** The root nodules of leguminous plants have nitrogen-fixing bacteria (called *Rhizobium* bacteria) which can directly fix (or convert) the nitrogen gas present in air to form nitrogen compounds. In other words, **leguminous crops have the ability to fix atmospheric nitrogen to form nitrogen compounds**. These nitrogen compounds go into the soil and improve its fertility. Some of these nitrogen compounds are used by the leguminous crop for its own growth and the rest of nitrogen compounds are left in the soil. Thus, planting a leguminous crop like pulses, peas, beans, groundnut, gram and clover, etc., in a field results in nitrogen-rich soil.

The planting of a leguminous crop in a field has the same effect as adding nitrogenous fertiliser in the field. Since leguminous crops can fix the atmospheric nitrogen themselves by using nitrogen-fixing bacteria present in their root nodules, therefore, **nitrogenous fertiliser is not required for growing leguminous crops**. On the other hand, cereal crops like wheat, maize, paddy, and millet, etc., are non-leguminous crops which do not have the ability to fix (or convert) the nitrogen gas of air into nitrogen compounds. Keeping these points in mind, we will now describe crop rotation.

Crop Rotation

The fertility of soil can be improved by crop rotation. **The practice in which different types of crops (leguminous crops and non-leguminous crops) are grown alternately in the same field or soil is called crop rotation.** In crop rotation, the cereal crops like wheat, maize, paddy and millet are grown alternately with leguminous crops like pulses, peas, beans, groundnut and clover, etc., in the same field. For example, when a cereal crop like maize crop is grown first, it takes away a lot of nitrogen from the soil for its growth and makes the soil nitrogen deficient. And next, when the leguminous crop like pulses or groundnut is grown in the same field, then the leguminous crop with its nitrogen fixing bacteria, enriches the soil with nitrogen compounds and increases its fertility. And when another cereal crop like wheat is grown after that, then wheat can utilise this extra nitrogen from the soil for its growth and produce a bumper crop. In this way, rotating different crops (leguminous and non-leguminous crops) in the same field replenishes the soil with nitrogen naturally and leads to increase in the crop production. **Rotation of crops has the following advantages :**

(*i*) Rotation of crops improves the fertility of the soil by replenishing it with nitrogen and hence brings about an increase in the production of food grains.

(*ii*) Rotation of crops saves a lot of nitrogenous fertiliser. This is because the leguminous crops grown during the rotation of crops can fix atmospheric nitrogen with the help of their nitrogen fixing bacteria, and there is no need to add nitrogenous fertiliser to the soil.

3. IRRIGATION

All the crop plants need water for their growth. The crop plants absorb water from the soil. The amount of water in the soil is not constant throughout the year. Water from the soil is lost constantly by evaporation and percolation to lower depths of the ground. It is, therefore, necessary to supply water to the crop plants in the fields, periodically. **The process of supplying water to crop plants in the fields is called**

irrigation. Just as we cannot survive without water for a long time, in the same way, plants also cannot survive without water for a long time. For example, if we stop watering plants grown in our home for a considerable time, the plants become pale, wilt and ultimately die. Water is absorbed by the roots of the plants. Alongwith water, minerals and fertilisers are also absorbed by crop plants. Plants contain nearly 90 per cent water.

Why is Irrigation Necessary

(*i*) Irrigation before ploughing the fields makes the soil soft due to which the ploughing of fields becomes easier.

(*ii*) Irrigation is necessary to provide moisture for the germination of seeds. This is because seeds do not grow in dry soil.

(*iii*) Irrigation is necessary to maintain the moisture of soil for healthy crop growth so as to get good yield.

(*iv*) Irrigation is necessary for the absorption of nutrient elements by the plants from the soil. The irrigation water dissolves the nutrients present in the soil to form a solution. This solution of nutrients is then absorbed by the roots for the development of plants.

(*v*) Water supplied to the crops during irrigation protects the crop plants from hot air currents as well as frost.

Factors Affecting Irrigation Requirements of Crops

The irrigation requirements (or water requirements) of crops depend on three factors :

(*i*) **Nature of the crop,**

(*ii*) **Nature of the soil,** and

(*iii*) **Season.**

Each crop needs a specific amount of water during the various stages of its growth and ripening. Some crops need more water whereas others require less water. For example, **paddy crop (rice crop) is transplanted in standing water and requires continuous irrigation whereas other crops like wheat, gram (*chana*) and cotton, etc., do not require so much water.** For cereal crops like wheat, irrigation is needed at only three stages : before ploughing the field ; at the time of flowering ; and at the time of development of grain.

There are two important types of soils in which the crops are grown ; Sandy soil and Clayey soil. **The crops grown in a sandy soil need irrigation more frequently whereas the frequency of irrigation for the crops grown in a clayey soil is comparatively less.** This can be explained as follows : Sandy soil is highly porous having high permeability. So, when we irrigate the crops standing in a sandy soil, then water quickly percolates down the soil and the crop plants are not able to absorb adequate amount of water. So, **due to the poor water retaining capacity of the sandy soil, the crops cultivated in sandy soil need more frequent irrigation.** On the other hand, clayey soil is much less permeable than sandy soil due to which it can retain water for a much longer time. So, when the crop grown in a clayey soil is irrigated, the water remains in the soil for a longer time, and hence the plants can absorb this water in adequate amount. So, **due to better water retaining capacity of the clayey soil, the crops cultivated in clayey soil need irrigation less frequently.**

Figure 13. Water in canal is used for irrigation.

The frequency of irrigation of crops also varies from season to season. For example, **the frequency of irrigation (or watering) of the crops is higher in summer season**. This is because during the hot days of summer, the rate of evaporation of water from the soil and the leaves of crop plants is increased. On the other hand, the frequency of irrigation (or watering) of the crops is comparatively lower in the colder winter season.

Sources of Irrigation

Crops are supplied water for irrigation from different sources like : **Rivers, Canals, Wells, Tube-wells, Dams (Reservoirs), Ponds and Lakes**. Even **rain** is a source of irrigation of crops. The water available in wells, lakes and canals is lifted up by different methods in different regions for taking it to the fields.

Traditional Methods of Irrigation

The various traditional methods of irrigation are :

(*i*) Moat (Pulley system),

(*ii*) Chain pump,

(*iii*) *Dhekli*, and

(*iv*) *Rahat* (Lever system).

MOAT. In the moat system of irrigation, water is drawn out from a well by using a big container tied to a long rope which moves over a pulley fixed at one edge of the well (see Figure 14). The rope tied to container is usually pulled by animals such as bullock (buffalo or camel). When the rope is pulled at the free end, the container filled with water (tied at the other end of rope) comes out of the well (see Figure 14).

Figure 14. Moat.

The farmer pours out water from the container into the fields and lowers empty container back into the well to get it refilled. The water-filled container again comes out of the well when the bullock pulls the rope and this process is repeated to get a continuous supply of water for irrigation.

CHAIN PUMP. Chain pump is an arrangement to lift water from a source of water like a stream, pond or lake (which is at a lower level than the fields) so as to provide irrigation in the fields. A chain pump consists of two large wheels, one fixed at the lower level of water source and the other fixed at the higher level above the fields (see Figure 15). The two wheels are connected by a chain passing over them. On the chain are hung small buckets. Below the bottom wheel is the source of water (like a stream, pond or lake) from which the water is to be lifted up to the fields. A handle is attached to the axle of the upper wheel. When the handle attached to the upper wheel is rotated by the farmer, the wheels connected by the chain start turning. When the lower wheel turns, the buckets attached to chain dip in the stream and get filled with water (see Figure 15). The moving chain then lifts these buckets filled with water up to the upper

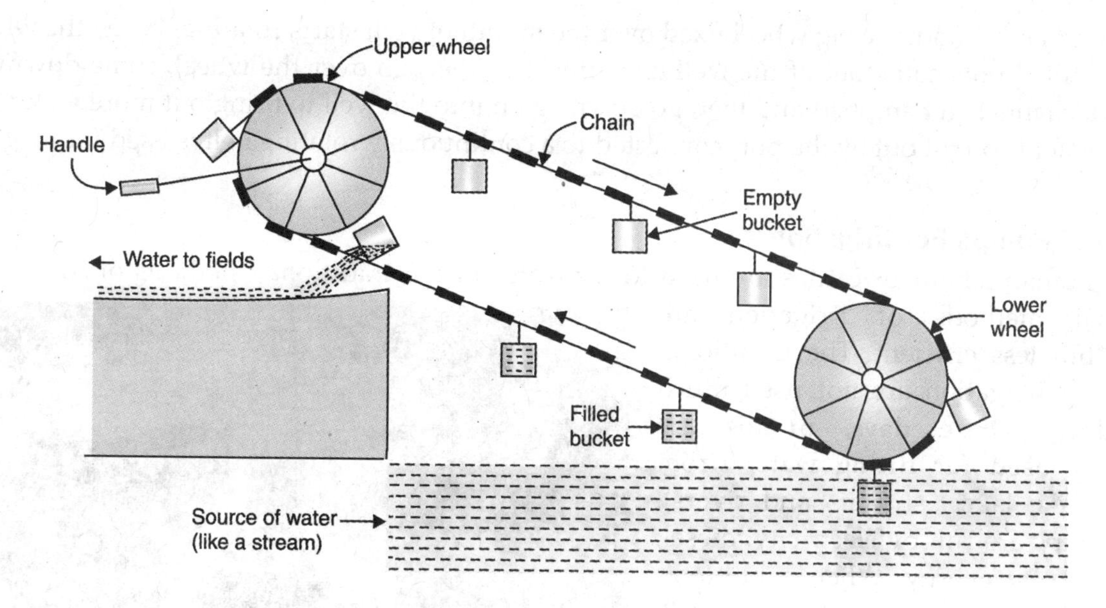

Figure 15. Chain pump.

wheel where the buckets tilt, and get emptied in the fields to provide water for irrigation. The moving chain then carries the empty buckets down to the lower wheel to be filled with water again. This process is repeated due to which the water from stream (pond or lake) is continuously lifted up into the fields.

DHEKLI. *Dhekli* is an arrangement to lift water from shallow wells by using the principle of simple lever (The word *'dhekli'* means *'lever-beam of well'*). In *dhekli*, a long wooden beam is supported over a forked vertical support fixed in the ground near the well in such a way that its longer arm is towards the well and shorter arm away from it (see Figure 16). A bucket is tied to the end of longer lever arm with a rope in such a way that it hangs over the mouth of the well. A weight is tied to the end of shorter lever arm. In order to lift water from the well, the end of longer lever arm carrying the bucket is pulled down by the rope. When the bucket is lowered into the well, it gets filled with water. On releasing the rope, the weight attached to the end of shorter lever arm comes down and lifts the water-filled bucket out of the well. The farmer then gets hold of the bucket, tilts it and pours out its water into the field. This process is repeated so as to get a large amount of water from the well required for irrigation.

Figure 16. *Dhekli.*

RAHAT **(LEVER SYSTEM).** In the *rahat* system of irrigation, water is drawn out from a well. In this method, there is a large wheel fixed on an axle above the mouth of the well (see Figure 17). A long belt with many, many small metal pots is put over the circumference of big wheel which can move over the wheel when the wheel turns. The lower end of the long belt of pots dips in the water of the well. The big wheel is turned by using a lever system driven by the force of bullocks (or other animals such as buffalo or camel). When the bullocks move the long, horizontal handle of the lever-system by

Figure 17. *Rahat.*

going round and round, the big wheel fixed over the mouth of well starts rotating. When the wheel rotates, the water filled pots come out of the well one after the other, go over the wheel, come downward, pour water in a channel, get emptied and then go down again into the well to bring out more water (see Figure 17). The water brought out by the pots connected to a continuously rotating belt is used for irrigation in the fields.

The Use of Pumps For Irrigation

The human labour or cattles are used to lift water in the traditional methods of irrigation. So, the traditional methods of irrigation are cheaper but less efficient. The traditional methods of irrigation are not used much these days. **These days, pumps are commonly used for lifting water** (from wells, ponds, lakes, streams and rivers). These pumps are run by electricity, diesel, biogas or solar energy. When a pump is used to draw out water from a narrow well, it is called a tube-well. Nowadays, tube-wells are being used increasingly for lifting underground water to be used for irrigation in agriculture (see Figure 18). We will now describe some modern methods of irrigation in which though the water is pumped out through tube-wells but used in such a way that its wastage is prevented.

Figure 18. Tube-well.

Modern Methods of Irrigation

The modern methods of irrigation help us to use water economically (by preventing its wastage). The two main modern methods of irrigation are :

(*i*) Sprinkler system, and

(*ii*) Drip system.

SPRINKLER SYSTEM. In the sprinkler system of irrigation, a main pipeline is laid in the fields. Perpendicular pipes having rotating nozzles at the top are joined to the main pipeline at regular intervals. When water from a tube-well is allowed to flow through the main pipeline under pressure with the help of a pump, it escapes from the rotating nozzles (see Figure 19). This water gets sprinkled on the crop plants as if it is raining. The sprinkler system of irrigation is more useful for the **uneven land** where sufficient water is not available. The sprinkler system is also very useful for **sandy soil.**

Figure 19. Sprinkler system of irrigation.

Figure 20. Drip irrigation system.

DRIP SYSTEM. In the drip irrigation system, there is a network of narrow pipes (or tubes) with small holes, in the fields (see Figure 20). When water flows through the narrow pipes, it falls drop by drop at the position of roots of the plants. This water is absorbed by the soil in the root zone of the plants and utilised by the plants. There is no run-off (or wastage) of irrigation water. Drip system is the best technique for watering (or irrigating) **fruit plants, trees** and **gardens**. Drip irrigation system has the following advantages :

(*i*) Drip system provides water to plants drop by drop. So, water is not wasted at all.

(*ii*) Drip system minimises the use of water in agriculture. So, drip system of irrigation is very useful in those regions where the availability of water is poor.

We will now discuss the removal of weeds from crop fields.

5. REMOVING THE WEEDS (OR WEEDING)

When we grow a food crop in the field, then in addition to the crop plants, many small, unwanted plants also germinate and grow in the field naturally. **The unwanted plants (or wild plants) which grow alongwith a cultivated crop are called weeds** (see Figure 21). The growth of weeds in the fields is harmful because they consume a lot of fertiliser, water, sunlight and space, meant for the crop plants and reduce the crop yield, and lower the quality of food grains. Since the presence of weeds in the fields will reduce the crop-yield, therefore, it is necessary to remove them from time to time. Though most of the weeds get uprooted during the ploughing of fields but they reappear when the crop grows. The weeds multiply and spread very fast because they produce a large quantity of seeds.

Figure 21. The small plants in this picture are weeds.

The type of weeds vary from field to field, from crop to crop, and also from season to season. Some of the common weeds (unwanted plants) found in wheat and rice fields are :

(*i*) Wild oat (*Javi*)

(*ii*) Grass (*Ghass*)

(*iii*) Amaranthus (*Chaulai*)

(*iv*) Chenopodium (*Bathua*)

Since weeds are harmful to the crops, they must be removed from the fields. **The process of removing weeds (unwanted plants) from a crop field is called weeding.** Weeding is necessary because weeds compete with crop plants for water, nutrients, light and space, and hence affect the growth of the crop. Some weeds are poisonous for human beings and animals whereas some weeds interfere in harvesting. **The best time for the removal of weeds is before they produce flowers and seeds.** Weeding is done by hand or with the help of implements like trowel (*khurpa*). Weeds can also be destroyed (or controlled) by spraying special chemicals called weedicides in the crop fields. Thus, the various methods of weeding (controlling weeds or eradicating weeds) are as follows :

1. Removal of Weeds by Pulling Them Out With Hand. Weeds can be removed from the crop fields just by pulling them up with hands (see Figure 22). When we pull the weeds, they get uprooted from the field. These uprooted weeds can then be thrown away.

2. Removal of Weeds by Using a Trowel (*Khurpa*). Weeds can be removed by digging or cutting them close to the ground from time to time with the help of an implement called trowel (or *khurpa*). A trowel (or *khurpa*) is shown in Figure 23.

Figure 22. Removing weeds by pulling them up.

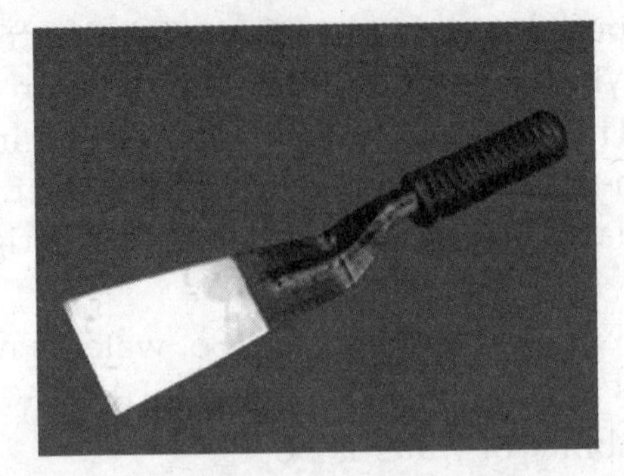

Figure 23. This is a trowel (*khurpa*). It is used to remove weeds from crop fields.

Figure 24. Weeds being destroyed by spraying weedicide.

3. Destroying the Weeds by Spraying Special Chemicals Called Weedicides. The poisonous chemicals which are used to kill weeds (unwanted plants) in the fields are called weedicides. Some of the common weedicides are : **2,4-D, MCPA and Butachlor**. A solution of the weedicide in water is sprayed on the standing crops in the fields with a sprayer (see Figure 24). The weedicides kill (destroy) the weeds (unwanted plants) but do not damage the main crop. The weedicides are sprayed in the crop fields before the flowering and seed formation in the weeds takes place. Since weedicides are poisonous chemicals, therefore, spraying of weedicides may affect the health of the person who handles the weedicide sprayer. The weedicides should be sprayed on the standing crops very carefully. During the spraying of weedicides, the person should cover his nose and mouth properly with a piece of cloth (so as to prevent the inhaling of poisonous weedicide).

6. HARVESTING

It normally takes about three or four months for a food crop to mature. Lush green wheat fields and paddy fields turn to golden yellow at the end of this period (see Figure 25). This change in colour from

Figure 25. Mature wheat crop standing in the fields.

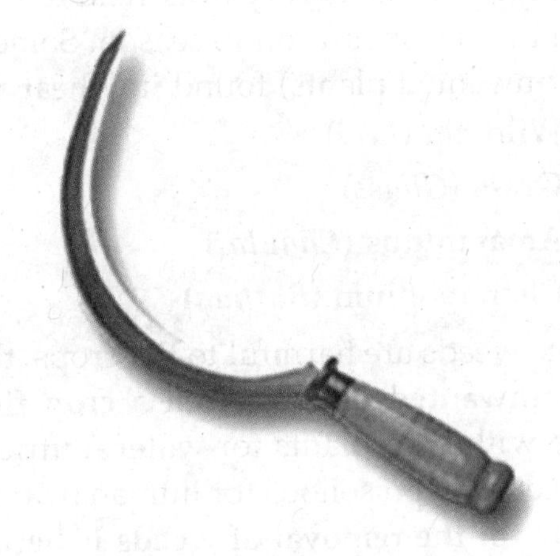

Figure 26. Sickle.

green to golden is due to the maturity of crop or ripening of crop. Once the crop has matured then it is ready for cutting and gathering. **The cutting and gathering of the matured food crop is called harvesting.** In harvesting, the crops like wheat or rice are cut close to the ground by hand using a cutting tool called sickle (see Figure 26). This is called manual harvesting. In large fields, wheat and paddy crops are cut by a motorised machine called harvester.

After harvesting the crop, the next step is threshing. **The process of beating out the grains from the harvested crop plants is called threshing.** Threshing is done to take out the grain from its outer covering called chaff. In the traditional method of threshing, the harvested crop is spread on the ground in a small area and various cattle like bullocks, buffaloes and camels are made to walk over it again and again in a circle. The cattle's feet crush the harvested crop plants due to which the chaff breaks up and the grain comes out. During threshing, the leaves and stems of the crop plants are converted into very small pieces called hay which is used as a fodder for animals. In larger farms, **a motorised machine called thresher is also used for the threshing process.**

Though the process of threshing brings out grains from the cut and dried crop plants, but this grain is mixed with chaff (outer inedible covering of grain) and hay, and has to be cleaned by separating from chaff and hay, before it can be used. This is done by the process of winnowing. **The process of separating grain from chaff and hay with the help of wind is called winnowing.** When the grain mixed with chaff and hay is made to fall from a height in blowing wind, the grain, being heavy, falls straight to the ground, whereas the chaff and hay, being much lighter, are carried some distance away by the wind. In this way, the grains form a separate heap and can be collected and packed in gunny bags.

Figure 27. This is a 'combine'. The 'combine' is a combined 'harvester and thresher'.

These days 'combines' (also called combine harvesters) are being used in large farms for harvesting related operations (see Figure 27). **A combine is a huge machine which cuts the standing cereal crop (like wheat) in the fields, threshes it and separates the chaff from grain in one operation.** This grain is clean and can be directly filled in gunny bags (and there is no need of winnowing). One of the disadvantages of using the machines like 'combines' is that it reduces the yield of hay (*bhoosa*) which is used as a fodder for cattle. This is because the combines cut only the upper part of the standing crop, and not from near the ground.

The stubs of crop plants left in the fields after harvesting are sometimes burnt by the farmers. The burning of stubs is harmful due to two reasons :

(*i*) The burning of stubs of crop plants in the fields causes air pollution.

(*ii*) The burning of stubs in the fields may cause accidental fire to the harvested crop lying in the fields and damage it.

Harvest Festivals. When the crops mature and become ready to be harvested, the fields turn golden yellow. The sight of golden fields of standing crops, laden with grains, fills the hearts of farmers with joy. The period of harvest is of great joy and happiness in all the parts of India. Men and women celebrate the harvest season in the form of festivals. The special festivals in India associated with the harvest seasons are : Pongal, Baisakhi, Holi, Nabanya, and Bihu. We will now discuss the storage of food grains.

7. STORAGE OF FOOD GRAINS

The fresh food grains (like wheat) obtained by the harvesting of crops contain more moisture than required for their safe storage. So, **the food grains (like wheat) obtained by harvesting the crops are dried in the sunshine before storing, to reduce their moisture.** It is necessary to reduce the moisture content of grains before storing to prevent their spoilage during storage. This is because **the higher moisture content in food grains promotes the growth of fungus and moulds on the stored grains which damages them** (and makes them lose their germination capacity). The farmers store the dried food grains at home in metal bins (metal drums) and jute bags (called gunny bags). Dried *neem* leaves are used for storing food grains at home. For example, when wheat is stored at home in iron drums,

Figure 28. These are wheat grains (a kind of food grains) obtained by harvesting and threshing wheat crop.

then some dried *neem* leaves are put in it. Dry *neem* leaves protect the stored food grains from pests such as insects and micro-organisms.

The Government Agencies like Food Corporation of India (FCI) buy grains from farmers on large scale and store it in big godowns so that it can be supplied throughout the country, round the year. **The large scale storage of food grains (like wheat and rice) is done in two ways :**

(*i*) in gunny bags, and

(*ii*) in grain silos.

The most common method of storing food grains on a large scale is to fill them in gunny bags, stitch the mouth of gunny bags tightly, and keep these gunny bags one over the other in big godowns (see Figure 29). Pesticide solutions are sprayed on the stacked gunny bags in the godown from time to time to protect

Figure 29. Food grains stored in gunny bags in granary.

Figure 30. Grain-silos for storing foodgrains.

the grains from damage by pests during storage. The population of rats in the godown is also controlled by killing them with rat poison from time to time. Though a gunny bag is not an ideal container for food grains but its greatest advantage is that food grains filled in gunny bags can be easily transported and distributed at various places.

In addition to the gunny-bag method of storing food grains, the grain-silos are also used for storing food grains on large scale (or bulk storage of food grains). The grain-silos are specially designed big and tall cylindrical structures (see Figure 30). The grain-silos have inbuilt arrangements for the protection of stored food grains from pests (like insects) and micro-organisms.

FOOD FROM ANIMALS

Though we get most of our food from crop plants, animals also provide us food. **The food provided by animals consists of Milk, Eggs and Meat.** The food obtained from animals is very rich in proteins. In fact, animal food provides certain proteins which are not present in plant foods. Most of the food obtained from animals also contains a good amount of fat but it contains very little of carbohydrates. Animal food, however, contains minerals and vitamins. The food obtained from animals is more expensive than that obtained from plant sources. The animals which provide us food are mainly of two types :

1. Milk yielding animals (or Milch animals), and
2. Meat and Egg yielding animals.

The examples of milk yielding animals (or milch animals) are : Cow, Buffalo and Goat. Milk is a perfect natural diet. Milk and its products (called dairy products) like Butter, *Ghee*, Curd and Cheese are highly nutritious foods. **The examples of meat and egg yielding animals are : Goat, Sheep, Fish, and**

Poultry (Chicken, Hen and Duck). Out of these animals, goat, sheep and fish give us meat. **Poultry gives us meat as well as eggs.** Honey is another nutritious food obtained from animals. It is obtained from insects called 'bees' (or honeybees).

Animal Husbandry

Just as each crop has its own requirements of proper soil, irrigation, manures and fertilisers and weedicides, in the same way, each domestic animal has its own needs of food, shelter, and health care. **The branch of agriculture which deals with the feeding, shelter, health and breeding of domestic animals is called animal husbandry**. The various practices necessary for raising animals for food and other purposes (or the elements of animal husbandry) are :

1. Proper feeding of animals,
2. Proper shelter for animals,
3. Prevention and cure of animal diseases, and
4. Proper breeding of animals.

Milk giving animals (milch animals or milch cattle) like cows and buffaloes are reared on small scale in rural homes. On a large scale, they are reared in big dairy farms.

Figure 31. Fish is an important source of animal food.

Fish as Food

Fish is an important source of animal food. Many people living in the coastal areas (sea-side areas) consume fish as a major part of their diet. Fish is rich in proteins. It is a highly nutritious and easily digestible food. Fish liver oil is rich in vitamin A and vitamin D. For example, Cod liver oil (or Cod fish liver oil) is rich in vitamin A and vitamin D. We are now in a position to **answer the following questions :**

Very Short Answer Type Questions

1. Which agricultural practice is carried out with the help of a sickle ?
2. What name is given to the cutting and gathering of a food crop like wheat or paddy ?
3. Name the tool (or implement) used in the traditional harvesting of crops.
4. Name the process of beating out the grains from harvested crop.
5. Name the machine used in recovering the grain from already cut crop.
6. Name the machine which does the cutting of standing crops and recovers the grain too.
7. Name the process in which grains are separated from chaff and hay with the help of wind.
8. Name three food materials obtained from animals.
9. Name two domestic animals which are used to obtain milk.
10. Name one meat yielding animal and one egg yielding animal.
11. Name an animal food obtained from insects.
12. What name is given to that branch of agriculture which deals with feeding, shelter, health and breeding of domestic animals ?
13. Name the major food nutrient provided by fish.
14. Name the vitamin/vitamins present in cod liver oil.
15. Name one Government Agency which is involved in procuring food grains (like wheat and rice) from farmers and storing them properly.
16. What type of organisms grow on stored food grains having higher moisture content ?
17. Which crop is generally grown between two cereal crops in crop rotation to restore the fertility of soil ?
18. State one advantage of growing a leguminous crop between two cereal crops.
19. Name the nitrogen-fixing bacteria present in root nodules of leguminous plants.
20. Which agricultural practice comes first : harvesting or weeding ?
21. Which is the first step in the cultivation of a crop ?
22. For what purpose is a hoe used ?

23. Name the implement used in sowing.
24. Name the practice used for cultivating paddy.
25. Name the two types of substances which are added to the fields by the farmers to maintain the fertility of soil.
26. Some grass is growing in a wheat field. What will it be known as ?
27. Name one crop which can tolerate standing water (water-logging) in the field and one which cannot.
28. Which is the best time for the removal of weeds ?
29. Name two methods of irrigation which conserve water.
30. Fill in the following blanks with suitable words :
    (*a*) The same kind of plants grown and cultivated on a large scale at a place is called ....
    (*b*) The first step before growing crops is ....of soil.
    (*c*) For growing a crop, sufficient sunlight, ..., and .... from the soil are essential.
    (*d*) Damaged seeds would .... on top of water.
    (*e*) Crop rotation helps in the replenishment of soil with ....
    (*f*) The supply of water to crops at different intervals is called ....
    (*g*) The unwanted plants present in a crop field are called ....
    (*h*) Dried.............leaves are used for storing food grains at home.
    (*i*) Many people living in the ....areas consume fish as a major part of their diet.

Short Answer Type Questions

31. (*a*) Why is it is necessary to dry the harvested food grains before storage ?
    (*b*) What are the two ways in which farmers store food grains ?
32. Out of drip system and sprinkler system of irrigation, which one is more suitable :
    (*a*) for uneven land ? (*b*) for sandy soil ?
    (*c*) for watering fruit plants ? (*d*) where availability of water is poor ?
33. (*a*) What are weeds ? Name any one weed found in a crop field.
    (*b*) How do weeds affect the growth of crops ?
34. Explain how, the irrigation requirements depend on the nature of the crop.
35. Explain how, the irrigation requirements of a crop depend on the nature of soil in which the crop is grown.
36. Describe the sprinkler system of irrigation. State its advantages.
37. Explain the drip system of irrigation. State two advantages of the drip system of irrigation.
38. How do the irrigation requirements of a wheat crop differ from that of a paddy crop ?
39. Explain why, the frequency of irrigation of crops is higher in summer season.
40. How are weeds removed from the crop fields ? Name one implement used for weeding.
41. If wheat is sown in the kharif season, what would happen ? Discuss.
42. Which of the following are kharif crops and which are rabi crops ?
    Wheat, Paddy, Gram, Maize, Mustard, Cotton, Soyabean, Linseed, Peas, Groundnut
43. What is a crop ? Give two examples of crops.
44. What are the two types of crops based on seasons ? Give one example of each type.
45. Name the various agricultural practices in the right sequence in which they are undertaken by the farmers.
46. Describe briefly, how soil is prepared for sowing the seeds.
47. Why do farmers carry out levelling of the ploughed fields ?
48. What are the advantages of sowing seeds with a seed drill ?
49. Explain why, the seeds should be sown at right spacings.
50. What is ploughing (or tilling) ? Name any two implements used for tilling the fields.
51. State two beneficial effects of ploughing the fields (or loosening and turning the soil).
52. (*a*) State the function of Food Corporation of India.
    (*b*) What is done to protect the grains stored in gunny bags in big godowns from damage ?
53. Define manure. What are the advantages of manure ?
54. What is a fertiliser ? Name any two fertilisers. State two harmful effects caused by the excessive use of fertilisers.
55. Explain how, soil gets affected by the repeated growing of crops in the same fields. How does use of fertilisers help the farmers ?

56. What is weeding ? Why is weeding necessary ?
57. What are weedicides ? Name one weedicide.
58. What precaution should be taken while spraying weedicides ? Why ?
59. Give any four differences between manures and fertilisers.
60. Define the terms : (*i*) harvesting, (*ii*) threshing, and (*iii*) winnowing.
61. (*a*) What are the two ways in which food grains are stored on a large scale ?
(*b*) What is the advantage of storing food grains in gunny bags ?
62. Name two traditional methods of irrigation and two modern methods of irrigation.
63. What is a 'combine' which is used in agriculture ? State its functions.
64. What is 'animal husbandry' ?
65. What are the various practices necessary for raising animals for food and other purposes ?

Long Answer Type Questions

66. (*a*) What is meant by kharif crops ? Give two examples of kharif crops.
(*b*) What is meant by rabi crops ? Give two examples of rabi crops.
67. (*a*) What is meant by 'sowing' ? What are the various methods of sowing the seeds ?
(*b*) What precautions should be taken in sowing the seeds ?
68. What are good quality seeds ? You are given a sample of wheat seeds. How will you select good, healthy seeds for sowing ?
69. (*a*) What is the process of 'transplantation' in agriculture ? Give examples of two crops which are usually grown by this process.
(*b*) State two advantages of the process of transplantation in growing crops.
70. (*a*) What is irrigation ? Why is irrigation necessary ?
(*b*) Name the various sources of irrigation in our country.

Multiple Choice Questions (MCQs)

71. Which of the following crops would enrich the soil with nitrogen ?
(*a*) apple (*b*) pea (*c*) paddy (*d*) potato
72. Which of the following is not a kharif crop ?
(*a*) paddy (*b*) mustard (*c*) maize (*d*) groundnut
73. In agriculture, broadcasting is used for :
(*a*) ploughing the fields (*b*) rotating the crops (*c*) removing the weeds (*d*) sowing the seeds
74. Fish liver oil is rich in :
A. Vitamin A B. Vitamin B C. Vitamin C D. Vitamin D
(*a*) A and B (*b*) B and C (*c*) A and D (*d*) only D
75. Which of the following is not grown by transplantation ?
(*a*) chillies (*b*) tomatoes (*c*) peas (*d*) paddy
76. Which of the following is not a rabi crop ?
(*a*) soyabean (*b*) peas (*c*) wheat (*d*) linseed
77. One of the following crop is not cultivated by sowing its seeds directly into soil. This one is :
(*a*) wheat (*b*) gram (*chana*) (*c*) paddy (*d*) maize (*makka*)
78. Tomatoes are cultivated by the practice called :
(*a*) transpiration (*b*) translocation (*c*) transportation (*d*) transplantation
79. Which of the following cannot be provided to the soil by a chemical fertiliser ?
(*a*) nitrogen (*b*) humus (*c*) potassium (*d*) phosphorus
80. Which of the following is not grown by transplantation ?
(*a*) chillies (*b*) tomatoes (*c*) paddy (*d*) papaya
81. The *Rhizobium* bacteria present in the root nodules of pea plants can fix one of the following from the atmosphere. This one is :
(*a*) hydrogen (*b*) oxygen (*c*) nitrogen (*d*) halogen
82. The process of beating out grains from the harvested wheat crop is called :
(*a*) beating (*b*) crushing (*c*) threshing (*d*) weeding
83. The food obtained from animals is very rich in :
(*a*) fats (*b*) carbohydrates (*c*) minerals (*d*) proteins

84. The Government Agency responsible for purchasing grains from the farmers, safe storage and distribution is :
(*a*) CBI (*b*) FBI (*c*) FCI (*d*) FDI

85. The process of removing unwanted plants from a crop field is called :
(*a*) breeding (*b*) weeding (*c*) transplanting (*d*) harvesting

86. Poultry gives us :
(*a*) eggs (*b*) meat (*c*) meat as well as eggs (*d*) honey

87. Which of the following is not a correct statement for sowing seeds ?
(*a*) seeds should be sown at right intervals
(*b*) seeds should be sown at right depth
(*c*) seeds should be sown in dry soil
(*d*) seeds should not be sown in highly wet soil

88. Which of the following system of irrigation is preferred for the uneven land ?
(*a*) chain pump irrigation system (*b*) drip irrigation system
(*c*) sprinkler irrigation system (*d*) river irrigation system

89. The two crops which are not grown by sowing their seeds directly into the soil in large fields are :
A. Peas B. Tomatoes C. Chillies D. Maize
(*a*) A and B (*b*) B and C (*c*) A and C (*d*) only C

90. The best technique of watering the fruit plants and trees is :
(*a*) chain pump system (*b*) sprinkler system (*c*) moat (pulley system) (*d*) drip system

Questions Based on High Order Thinking Skills (HOTS)

91. Arrange the following practices in the correct order as they appear in the sugarcane crop production :
Sending crop to sugar factory ; Irrigation ; Harvesting ; Sowing ; Preparation of soil ; Ploughing the field ; Manuring. **(NCERT Book Question)**

92. Match items in column *A* with those in column *B* :

A	B
(*i*) Kharif crops	(*a*) Food for cattle
(*ii*) Rabi crops	(*b*) Urea and superphosphate
(*iii*) Chemical fertilisers	(*c*) Animal excreta, cow-dung, and plant waste
(*iv*) Organic manure	(*d*) Wheat, gram, pea
	(*e*) Paddy and maize

93. Name two crops which are cultivated :
(*a*) by sowing seeds directly into fields.
(*b*) by transplanting.

94. Farmers in Northern India grow legumes as fodder in one season and wheat in the next season.
(*a*) What is this practice known as ?
(*b*) How does this practice help in the replenishment of soil ?

95. A student lists the following agricultural practices for crop production :
Irrigation ; Removal of weeds ; Preparation of soil ; Storage of food grains ; Sowing ; Adding manure and fertilisers
Which agricultural practice is missing from the above list ?

ANSWERS

20. Weeding **26.** Weed **30.** (*a*) crop (*b*) preparation (*c*) nutrients ; water (*d*) float (*e*) nitrogen (*f*) irrigation (*g*) weeds (*h*) *neem* (*i*) coastal **32.** (*a*) Sprinkler system (*b*) Sprinkler system (*c*) Drip system (*d*) Drip system **71.** (*b*) **72.** (*b*) **73.** (*d*) **74.** (*c*) **75.** (*c*) **76.** (*a*) **77.** (*c*) **78.** (*d*) **79.** (*b*) **80.** (*d*) **81.** (*c*) **82.** (*c*) **83.** (*d*) **84.** (*c*) **85.** (*b*) **86.** (*c*) **87.** (*c*) **88.** (*c*) **89.** (*b*) **90.** (*d*) **91.** Ploughing the field ; Preparation of soil, Sowing; Manuring ; Irrigation ; Harvesting ; Sending crop to sugar factory **92.** (*i*) *e* (*ii*) *d* (*iii*) *b* (*iv*) *c* **93.** (*a*) Wheat ; Gram (*b*) Paddy ; Tomatoes **94.** (*a*) Crop rotation (*b*) The *Rhizobium* bacteria present in the root nodules of legumes fix the nitrogen gas of the atmosphere to form nitrogen compounds. Some of these nitrogen compounds go into the soil and replenish it **95.** Harvesting

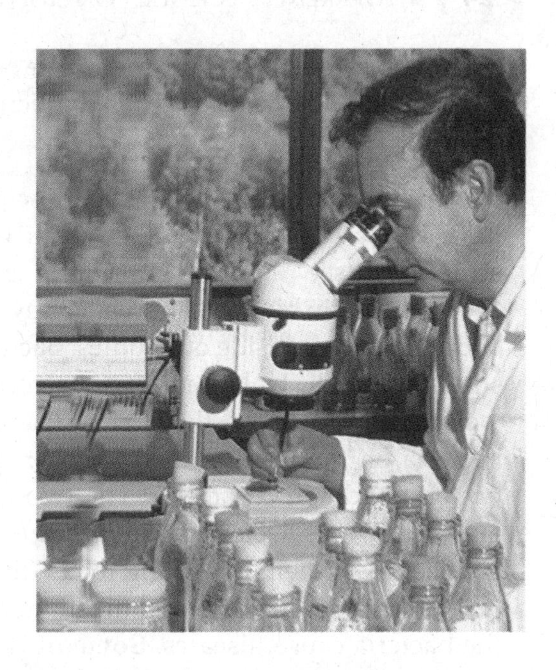

Micro-Organisms : Friend and Foe

Many living organisms are present in soil, water, and air around us. Some of these organisms are so small that we cannot see them with naked eyes. We need a magnifying instrument called microscope to see these extremely small organisms. These extremely small organisms are known as micro-organisms (micro = extremely small). We can now say that : **Those organisms which are too small to be seen without a microscope are called micro-organisms.** Thus, micro-organisms cannot be seen with the naked eye. Micro-organisms can be seen only with the help of a microscope.

Though we cannot see the micro-organisms around us, we become aware of the presence of micro-organisms through their actions like spoiling our food and causing diseases. Thus, **some of the micro-organisms are harmful to us.** The micro-organisms like certain bacteria and fungi make our food go bad. The micro-organisms also cause diseases in humans, other animals and plants. The diseases like common cold, malaria, skin infections, typhoid, tuberculosis, tetanus, cholera, measles, chickenpox, smallpox and AIDS, etc., are all caused by the action of various types of micro-organisms. Some of the micro-organisms grow on our food and cause food poisoning.

Some of the micro-organisms are also useful to us. For example, the micro-organisms like certain bacteria help in making food products such as curd and cheese. Micro-organisms are also useful in making bread, cakes, pastries, alcohol, acetic acid (vinegar) and medicines called antibiotics. Some micro-organisms decompose the organic waste of dead plants and animals into simple substances and clean up the environment. They also help in recycling the materials (like carbon and nitrogen) in nature.

Major Groups of Micro-Organisms

Micro-organisms are classified into five major groups. These groups are : Bacteria, Viruses, Protozoa, and some Fungi and Algae. Micro-organisms may be unicellular (single celled) or multicellular (many-celled). Please note that the singular of bacteria is bacterium; the singular of viruses is virus ; the singular of protozoa is protozoan ; the singular of fungi is fungus ; and the singular of algae is alga. We will now describe the various types of micro-organisms very briefly.

1. Bacteria

Bacteria are very small, single-celled micro-organisms which have cell walls but do not have an organised nucleus and other structures (see Figure 1). Bacteria are found in large numbers everywhere : in air ; soil and water ; every surface around us ; on our bodies and even inside our bodies. Bacteria are larger than viruses but still very small. Unlike viruses, bacteria feed, move and respire, as well as reproduce on their own. There are mainly three groups of bacteria on the basis of their shape : spherical bacteria, rod-shaped bacteria and spiral bacteria. The two common examples of bacteria are *Lactobacillus* bacteria and *Rhizobium* bacteria. Some of the bacteria are useful and help in making foods (like curd), nitrogen fixation and decomposition of waste organic matter. On the other hand, some of the bacteria cause diseases. **Some of the human diseases caused by bacteria are cholera, typhoid, tuberculosis (TB), diphtheria, whooping cough and food poisoning.**

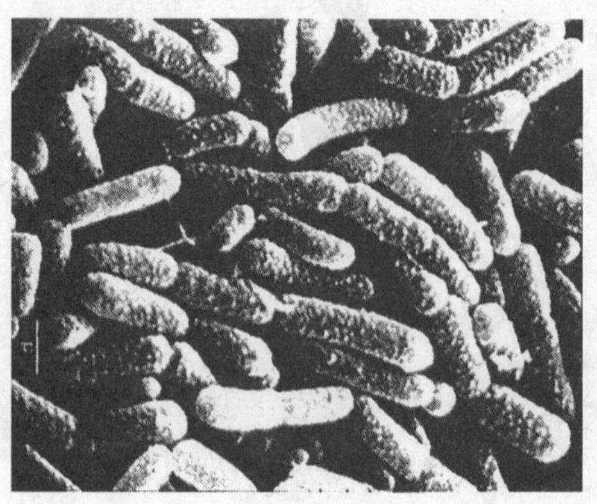

Figure 1. Bacteria (These are actually rod-shaped bacteria).

2. Viruses

Viruses are the smallest micro-organisms which can develop only inside the cells of the host organisms (which may be animal, plant or bacterium). Viruses are much smaller than bacteria (see Figure 2). **Viruses do not show most of the characteristics of living things.** For example, viruses do not respire, feed, grow, excrete, or move on their own. They just reproduce. Viruses are able to reproduce if they enter a living cell. That is, **viruses can reproduce and multiply only inside the cells of other organisms (such as animal cells, plant cells or bacteria cells).** Thus, as long as viruses are outside the living cells, they behave as non-living things. But as soon as the viruses enter the living cells of other organisms, they start behaving as living things by carrying out the process of reproduction. Due to this reason, viruses are said to lie on the border line dividing the living things from non-living things. **Viruses are the agents of disease.** Viruses cause a variety of diseases in human beings, other animals and plants. **The human diseases such as common cold, influenza (flu), measles, polio, chickenpox, and smallpox are all caused by viruses.** The two examples of viruses are 'common cold virus' and 'Human Immunodeficiency Virus' (HIV). Common cold virus causes common cold disease whereas HIV causes AIDS disease (AIDS stands for Acquired Immune Deficiency Syndrome). Diseases caused by virus (or viral diseases) cannot be treated with antibiotics.

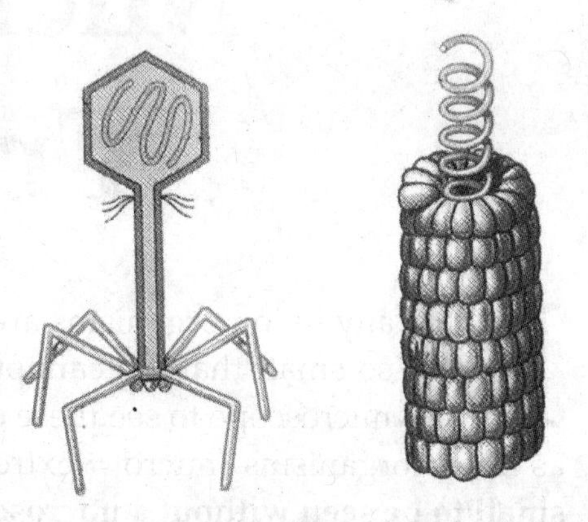

Figure 2. Viruses.

3. Protozoa

Protozoa are a group of single-celled micro-organisms which are classified as animals. Protozoa are animal like just as algae are plant like. Protozoa are found in ponds, lakes, dirty water drains, rivers, sea-water and damp soil. Some common examples of protozoa are : *Amoeba*, *Paramecium*, *Entamoeba* and *Plasmodium*. A few protozoa are shown in Figure 3. Many protozoa are parasites and cause diseases. **Diseases like dysentery and malaria are caused by protozoa.** For example, *Entamoeba* is a protozoan which causes a disease known as amoebic dysentery.

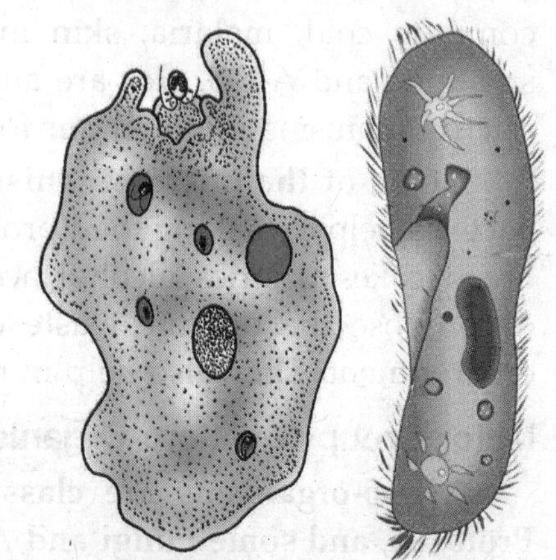

(*a*) *Amoeba* (*b*) *Paramecium*

Figure 3. Protozoa.

And *Plasmodium* is a protozoan which causes a disease called malaria in humans. *Plasmodium* is commonly known as 'Malarial Parasite' (MP).

4. Algae

Algae is a large group of simple, plant-like organisms. They contain chlorophyll and produce food by photosynthesis just like plants. Algae, however differ from plants because they do not have proper roots, stems and leaves. Some of the examples of algae are : *Chlamydomonas*, *Spirogyra*, Blue-green algae ; Diatoms and Seaweeds (see Figure 4). Only some of the algae are unicellular. Most of the algae are multicellular. For example, *Chlamydomonas* and diatoms are single-celled algae whereas blue-green algae and *Spirogyra* are multicellular algae. The blue-green algae have the ability to fix nitrogen gas of atmosphere.

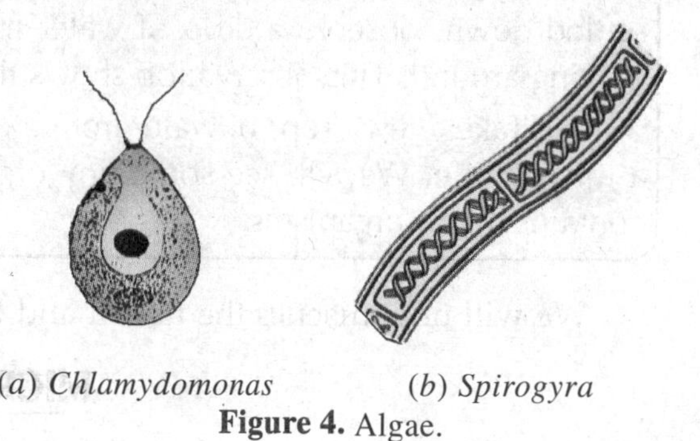

(*a*) *Chlamydomonas* (*b*) *Spirogyra*

Figure 4. Algae.

5. Fungi

Fungi are a large group of organisms which do not have chlorophyll and do not photosythesise. Some examples of fungi are : Yeast, Moulds (such as Bread mould, *Penicillium* and *Aspergillus*), Mushrooms, Toadstools and Puffballs (see Figure 5). All fungi (except yeast) are made up of fine threads called hyphae

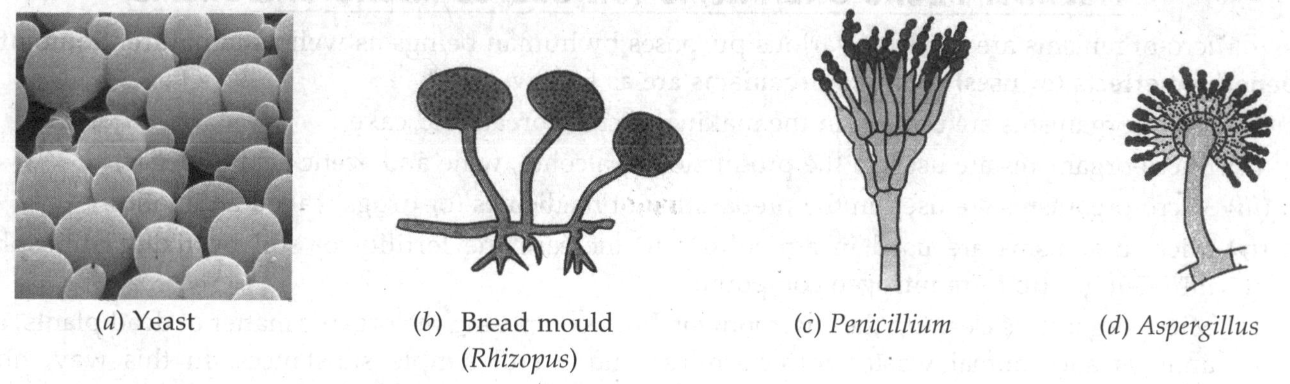

(*a*) Yeast (*b*) Bread mould (*Rhizopus*) (*c*) *Penicillium* (*d*) *Aspergillus*

Figure 5. Fungi.

(pronounced as hi-fee). Some fungi look like plants but they cannot make their own food like the plants do. Fungi need moist and warm conditions to grow. Most of the fungi are saprophytes which feed on dead things like remains of dead plants and animals. Some of the fungi are parasites. They feed on living things and cause diseases. A mould is a fur-like growth of minute fungi occurring on organic matter in moist and warm conditions. Some of the examples of mould fungi are : *Rhizopus* (Bread mould), *Penicillium* and *Aspergillus*. The fungi like yeast and moulds are very small in size and can be seen clearly only with a microscope. Thus, yeast and moulds are the fungi which can be considered to be micro-organisms. The fungi such as mushrooms, toadstools and puffballs are bigger in size. **Some of the human diseases caused by fungi are ringworm and athlete's foot.**

Where Do Micro-Organisms Live

Micro-organisms are found practically everywhere in all types of habitats. Micro-organisms are found in air, soil and water bodies (like ponds, lakes, wells, rivers and sea). Micro-organisms can live and survive in almost all kinds of environment like hot springs, ice-cold waters, saline water (salty water), desert soil or marshy land. They also occur in dead and decomposed organic matter (plant and animal matter). Micro-organisms are present inside the human body and that of other animals. The micro-organisms also live as parasites on other living things, including us.

ACTIVITY

We can show the presence of micro-organisms in soil and water by performing the following activities :

(*i*) Collect some moist soil from the field in a beaker and add water to it. After the soil particles have settled down, observe a drop of water from the beaker under a microscope. We will see tiny organisms moving around. This observation shows that soil contains micro-organisms.

(*ii*) Take a few drops of water from a pond. Spread this water on a clean glass slide and observe through a microscope. We will see some tiny organisms moving around. This observation shows that pond water contains micro-organisms.

We will now discuss the useful and harmful micro-organisms in detail.

MICRO-ORGANISMS AND US

Initially it was thought that all the micro-organisms are harmful and cause diseases. Later on scientists discovered that only a handful of micro-organisms are harmful and cause diseases. Most of the micro-organisms are harmless and some of the micro-organisms are even beneficial to us (or useful to us). Micro-organisms play an important role in our lives. We will now describe the beneficial effects and harmful effects of micro-organisms in detail, one by one.

FRIENDLY MICRO-ORGANISMS (OR USEFUL MICRO-ORGANISMS)

Micro-organisms are used for various purposes by human beings as well as in nature. **Some of the beneficial effects (or uses) of micro-organisms are as follows :**

(*i*) Micro-organisms are utilised in the making of curd, bread and cake.

(*ii*) Micro-organisms are used in the production of alcohol, wine and acetic acid (vinegar).

(*iii*) Micro-organisms are used in the preparation of medicines (or drugs) called antibiotics.

(*iv*) Micro-organisms are used in agriculture to increase the fertility of soil by fixing atmospheric nitrogen gas (to form nitrogen compounds).

(*v*) Micro-organisms clean up the environment by decomposing the organic matter of dead plants, dead animals and animal wastes into harmless and useful simple substances. In this way, micro-organisms help in the recycling of materials in nature.

We will now study the beneficial effects (or uses) of micro-organisms in somewhat detail.

1. Making of Curd

Milk is turned into curd by bacteria. In order to make curd, a little pre-made curd is added to warm milk and set aside for some time. Curd contains several micro-organisms including *Lactobacillus* bacterium (Plural of *Lactobacillus* is *Lactobacilli*). *Lactobacilli* bacteria promote the formation of curd from milk. When a little of pre-made curd is added to warm milk, then *Lactobacilli* bacteria present in curd multiply in milk and convert it into curd. This happens as follows : Milk contains a sugar called lactose. *Lactobacilli* bacteria convert the lactose sugar into lactic acid. This lactic acid then converts milk into curd. We can now say that the micro-organism utilised in making curd from milk is bacterium (or bacteria). The name of this bacterium is *Lactobacillus*. An important ingredient of *idlis* and *bhaturas* is curd. Curd is added in making *idlis* and *bhaturas* to make them soft and spongy. Bacteria are also involved in the making of cheese, pickles, and many other food items.

Figure 6. The micro-organism 'yeast' is used to make bread.

2. Making of Bread

Yeast is used in the baking industry for making bread. When yeast is mixed in dough for making bread, the yeast reproduces rapidly and gives out carbon dioxide gas during respiration. The bubbles of carbon dioxide gas fill the dough and increase its volume. This makes the bread 'rise'. The holes in the bread are due to the bubbles of carbon dioxide given off during the baking process (see Figure 6). This makes the bread light, soft and spongy. For the same reason, yeast is also used in making cakes and pastries. We can perform an activity to demonstrate the action of yeast in the making of bread as follows.

ACTIVITY

Take half a kilogram of white flour (*maida*), add some sugar and mix with warm water. Then add a small amount of yeast powder and knead the mixture of white flour, sugar, yeast powder and water to make a soft

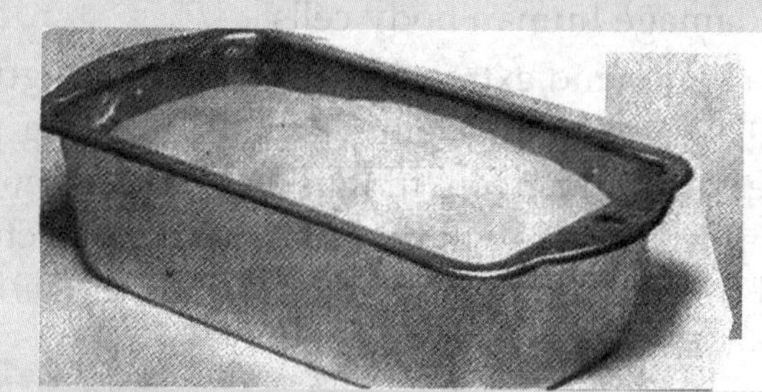

(*a*) This is freshly prepared dough containing yeast powder

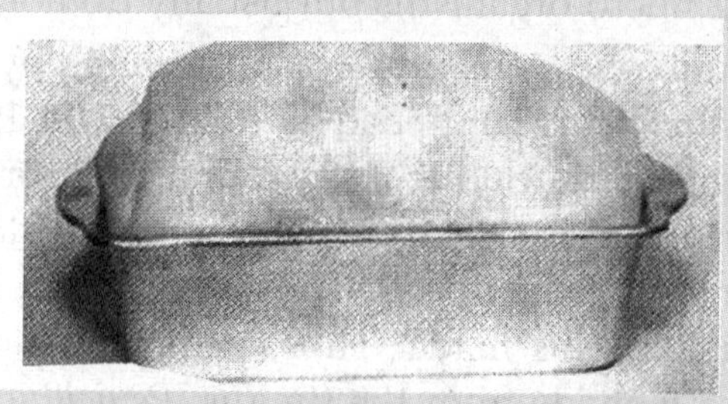

(*b*) Carbon dioxide produced by yeast has caused the dough to expand and rise

Figure 7.

dough [see Figure 7(*a*)]. Keep this dough aside for about 2 hours. We will find that the volume of dough has increased [see Figure 7(*b*)]. We say that the dough rises. From this discussion we conclude that yeast is the micro-organism used in making bread, cakes and pastries. Yeast is a fungus.

3. Commercial Use of Micro-Organisms

Micro-organisms are used for the large scale production of alcohol and acetic acid (vinegar). Yeast is the micro-organism which is used for the large scale production of alcohol. This alcohol is then used in making wine, beer and whisky as well as industrial spirit. So, we can now say that yeast is used to produce alcoholic drinks (such as wine, beer, whisky, etc.) and industrial spirit. Yeast is capable of converting sugar into alcohol (and carbon dioxide). The sugar for making alcohol comes from substances such as cane juice and fruit juices, or from substances such as barley, maize, rice, etc. (that contain starch which gets converted into sugar). **The process of conversion of sugar into alcohol by the action of yeast is called fermentation.** Fermentation was discovered by Louis Pasteur in 1857. In order to make alcohol, yeast is grown on natural sugars present in grains like barley, maize, rice, cane juice and fruit juices, etc. Yeast converts sugar into alcohol. This alcohol is then used for various purposes. We can carry out the process of alcoholic fermentation in the laboratory as follows.

ACTIVITY

Take a 500 mL beaker and fill it three-fourths with water. Dissolve 2 or 3 teaspoonfuls of sugar in it. Add half a teaspoonful of yeast powder to the sugar solution. Cover the beaker and allow the mixture of sugar solution and yeast powder to stand in a warm place for 4 to 5 hours. Now, smell the solution. We will get a characteristic pleasant smell coming from the beaker. This is the smell of alcohol (because sugar has been converted into alcohol by yeast). Now taste the solution from the beaker. We will get a 'burning taste'. This burning taste is due to the formation of alcohol.

A dilute solution of acetic acid is called vinegar. **Bacteria can turn alcohol into acetic acid (or vinegar).** In order to produce acetic acid (or vinegar) on a large scale, first alcohol is made by using yeast. The *Acetobacter* bacteria are then added to alcohol and air is bubbled through it. In the presence of oxygen (of air), *Acetobacter* bacteria convert alcohol into acetic acid (or vinegar).

4. Medicinal Use of Micro-Organisms

A medicine which stops the growth of, or kills the disease-causing micro-organisms is called an antibiotic. **The source of antibiotic medicines are micro-organisms.** The antibiotics are manufactured by growing specific micro-organisms (and used to cure a number of diseases). These days, a large number of antibiotics are being produced from micro-organisms such as fungi and bacteria. **Some of the common antibiotics which are made from fungi and bacteria are : Penicillin, Streptomycein, Erythromycein and Tetracycline.** Nowadays, many antibiotics are also being made synthetically. Many different antibiotics are now available to treat a wide variety of diseases caused by pathogenic micro-organisms. Antibiotics kill the disease-causing micro-organisms but usually do not damage human body cells.

The first antibiotic 'penicillin' was discovered by chance and extracted from the tiny fungus (a mould) called '*Penicillium*'. This happened as follows : In 1929, Alexander Fleming was cultivating a culture of disease-causing bacteria. Suddenly he found the spores of a little green 'fungus' in one of his culture plates. He noticed that the presence of tiny green fungus stopped the growth of disease-causing bacteria. In fact, it also killed many of the disease-causing bacteria. From this fungus (or mould) called '*Penicillium*', the antibiotic penicillin was made. Thus, penicillin is an antibiotic medicine which is made from the fungus '*Penicillium*'. Penicillin controls bacterial and fungal infections.

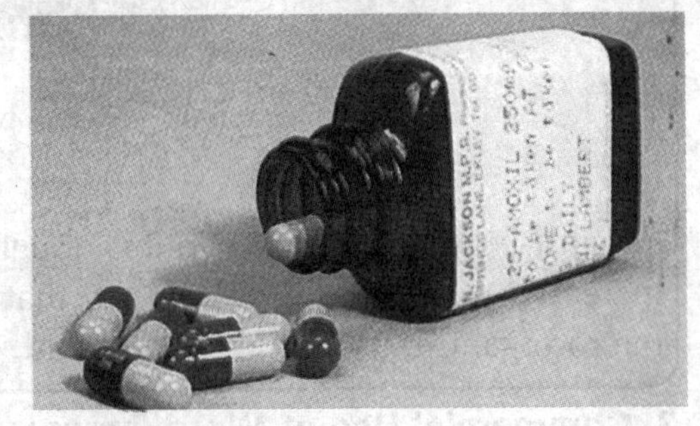

Figure 8. These capsules are of 'penicillin' antibiotic.

Antibiotics are used to treat many diseases in humans. Whenever we fall ill, the doctor may give us some antibiotic tablets, capsules or injections (such as penicillin) (see Figure 8). Antibiotics are very effective in curing diseases caused by micro-organisms such as bacteria and fungi. **Antibiotics are, however, not effective against diseases caused by viruses.** For example, antibiotics cannot be used to cure diseases like 'common cold', 'flu', and 'viral fever' because these are caused by viruses. **Some of the precautions to be observed in the use of antibiotics are as follows :**

(*i*) Antibiotics should be taken only on the advice of a qualified doctor.

(*ii*) A person must complete the 'full course' of antibiotics prescribed by the doctor.

(*iii*) The antibiotics should be taken in proper doses as advised by the doctor. If a person takes antibiotics in wrong doses (or when not needed), it may make the antibiotics less effective when the person might need it in future.

(*iv*) Antibiotics should not be taken unnecessarily. Antibiotics taken unnecessarily may kill the useful bacteria in the body and harm us.

Antibiotics can be used to treat many diseases in animals. Antibiotics are even mixed with the feed of live stock (cattle like cows, buffaloes, etc.) and poultry birds to control microbial diseases in animals. **Antibiotics are also used to control many plant diseases.**

Vaccine

We have just studied that micro-organisms are used to make medicines called antibiotics which can cure many diseases in human beings, animals and plants. Mirco-organisms are also used to make vaccines. **A vaccine is a special kind of preparation (or medicine) which provides immunity (or protection) against a particular disease.** Vaccines are given to healthy persons so that they may not get certain diseases throughout their life (even if they are exposed to the pathogens of these diseases later on in life). These days vaccines are made on a large scale from micro-organisms to protect human beings and other animals from several diseases. A vaccine works as follows :

A vaccine contains the dead or weakened but alive micro-organisms of a disease (which are harmless and do not actually give a diseasc). When the vaccine containing dead or alive micro-organisms is introduced into the body of a healthy person orally (by mouth) or by injection, the body of that person responds by producing some substances called 'antibiotics' in its blood. These antibodies kill any 'alive' disease-causing micro-organisms present in the vaccine. Some of the antibodies remain in the blood of the person for a very long time and fight against the same micro-organisms and kill them if they happen to enter the body naturally at a later date (when the person is exposed to disease). So, due to the presence of antibodies in the blood, a person remains protected from that particular disease. Thus, **a vaccine develops the immunity from a disease.**

A number of diseases can be prevented by vaccination. **Vaccination is the process of giving a vaccine orally (by mouth) or by injection which provides protection against a particular disease.** For keeping good health, we must prevent the diseases by vaccination at the proper time. Since children are more susceptible to diseases, so all the children should be vaccinated at proper ages to provide them immunity from certain diseases. **The diseases which can be prevented by vaccination of children at proper age are : Polio, Smallpox, Cholera, Typhoid, Hepatitis, Tuberculosis (TB), Tetanus, Measles, Rabies, Diphtheria and Pertussis (Whooping cough).** To develop the fighting capability in the body to a disease in called 'immunisation'. After getting vaccinated, the child becomes 'immune' to a particular disease. This means that the child becomes protected against that disease. He will never get that disease. **Edward Jenner discovered the vaccine for smallpox in 1798.** A worldwide campaign against smallpox has finally led to its eradication from most parts of the world. Under the National Health Programme in our country the vaccines for several diseases are given free of cost at all the Government Health Centres. Polio disease is prevented by giving Oral Polio Vaccine (OPV) (see Figure 9). Many times we see advertisements on TV and in newspapers to protect the children from polio under the Pulse Polio Programme by giving them polio drops. The polio drops given to children are actually a vaccine.

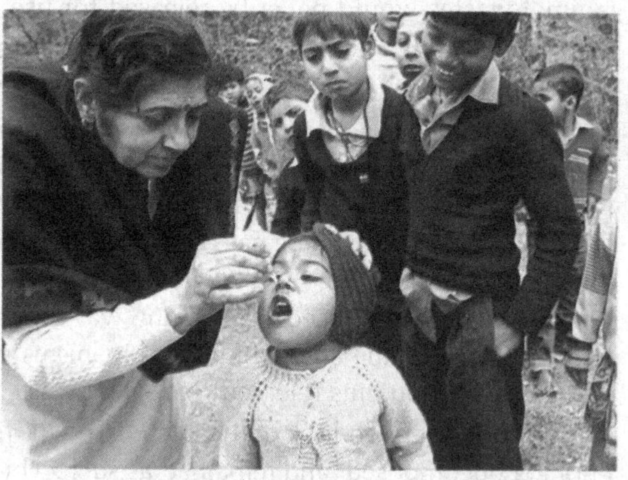

Figure 9. This picture shows a child being given oral vaccine against polio disease.

5. Increasing Soil Fertility

Some of the micro-organisms present in the soil can fix nitrogen gas from the atmosphere to form nitrogen compounds. These nitrogen compounds mix with the soil and increase the fertility of soil. For example, **some bacteria and blue-green algae are able to 'fix' nitrogen gas from the atmosphere to enrich the soil with nitrogen compounds and increase its fertility.** The nitrogen-fixing bacteria and blue-green algae are called **biological nitrogen fixers.** The nitrogen-fixing blue-green algae are shown in Figure 10. Since blue-green algae store in them nitrogen compounds made by nitrogen-fixation, they are used as fertiliser in agriculture. The addition of blue-green algae to barren fields increases the nitrogen content of the soil and makes it fertile. *Rhizobium* bacteria present in the root nodules of leguminous plants (like peas, beans, etc.) also fix atmospheric nitrogen and increase soil fertility.

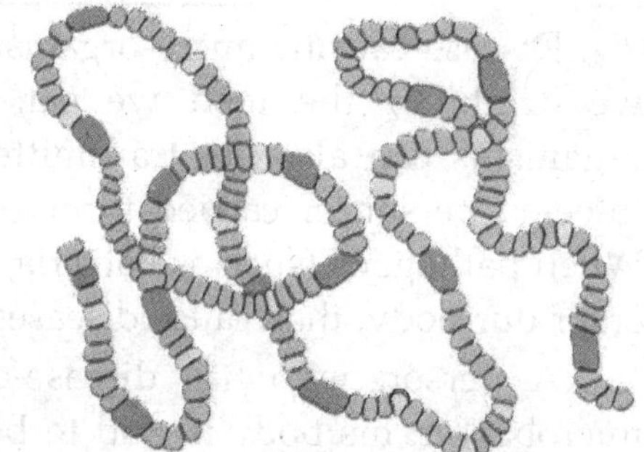

Figure 10. The nitrogen-fixing blue-green algae.

6. Cleaning the Environment

Some micro-organisms (like certain bacteria and fungi) decompose the organic matter present in dead plants, dead animals and animal wastes, and convert them into simple substances which mix up with the soil. These simple substances contain plant nutrients which are again used by new plants for their growth. **Since micro-organisms decompose the harmful and smelly dead remains of plants and animals, and**

animal wastes (like faeces, dung, urine, etc.) into harmless materials, they clean the environment. If, however, there were no micro-organisms (called decomposers) in the soil, then the dead plants, dead animals and animal wastes would keep on piling up in the environment and make it dirty. ***In addition to cleaning the environment, the micro-organisms also help in recycling the nutrients (present in dead plants, dead animals and animal wastes) in nature which can then be used as food by green plants.*** If there were no micro-organisms (called decomposers), then the nutrients present in dead plants, dead animals and animal wastes would never be released for use by new plants.

ACTIVITY

We will now describe on activity to show that some micro-organisms decompose waste plant materials and convert them into useful manure. Take two flower pots and mark them *A* and *B*. Fill each flower pot half with soil. Take some plant wastes such as fruit and vegetable peels, fallen leaves, etc., and bury them in soil in pot *A*. Bury a polythene bag, an empty glass bottle and a broken plastic toy in the soil in pot *B*. Keep both the pots aside for 3 to 4 weeks. If we now observe the pot *A*, we will find that the plant wastes buried in it have been decomposed. The plant wastes (fruit and vegetable peels, fallen leaves, etc.) have been decomposed by the action of micro-organisms present in the soil and converted into manure. This manure contains the nutrients released from plant wastes. These nutrients can be used for growing new plants.

If, however, we look at the pot *B*, we will find that the polythene bag, glass bottle, and plastic toy did not get decomposed, and remained as such. This is because the micro-organims present in soil are not able to decompose polythene bag, glass and plastic and convert them into manure. The micro-organisms present in soil can decompose only the organic matter present in dead plants, dead animals and animal wastes, etc.

HARMFUL MICRO-ORGANISMS

Micro-organisms can be harmful in many ways. For example, some of the micro-organisms cause diseases in human beings, other animals and plants. **Those micro-organisms which cause diseases are called pathogens.** Thus, pathogens are disease-causing micro-organisms. Pathogens can be bacteria, viruses, protozoa or fungi, etc. Micro-organisms cause diseases such as tuberculosis (TB), tetanus, diphtheria, whooping cough, cholera, typhoid, AIDS, food poisoning, malaria, smallpox and chickenpox, etc. Some micro-organisms spoil food, clothing and leather objects. We will now study some of the harmful activities of micro-organisms in detail.

DISEASE-CAUSING MICRO-ORGANISMS IN HUMANS

Disease-causing micro-organisms (or pathogens) enter our body through the air we breathe, the water we drink, or the food we eat. The disease-causing micro-organisms can also get transmitted by direct contact with an infected person or carried through an insect (or other animal). When pathogens (such as bacteria, viruses, protozoa, fungi, etc.) enter our body, they cause diseases.

Figure 11. Sneezing by this woman having common cold is putting thousands of 'common cold viruses' into the air.

A person who has disease-causing micro-organisms (or microbes) in his body is said to be an 'infected person'. **Those microbial diseases which can spread from an infected person to a healthy person through air, water, food or physical contact, etc., are called communicable diseases.** In communicable diseases, the disease-causing germs (or infection) get transmitted from a human being, an animal or the environment to another human being. **Some of the examples of communicable diseases are : Common cold, Cholera, Chickenpox, Tuberculosis (TB), Malaria, and AIDS.** For example, the disease called 'common cold' spreads by breathing air containing micro-organisms. This happens as follows : The disease 'common cold' is caused by a virus. A person suffering from common cold is infected with 'common cold virus'. When the person suffering from common cold sneezes, fine droplets of moisture carrying thousands of common cold viruses are spread in the air around him (or her) (see Figure 11). When a healthy person breathes in this contaminated air containing common cold virus, the virus enters his body and he also gets 'common cold' disease. **The communicable diseases can occur and spread in the following ways** :

(*i*) by breathing of air containing micro-organisms,
(*ii*) by taking infected food or water,
(*iii*) through insect bites (such as mosquito bites),
(*iv*) by sharing infected injection needles, and
(*v*) by physical contact with an infected person (or by using articles of an infected person such as towel, clothes, bed, utensils, etc.).

For example, the common cold disease spreads by breathing air containing micro-organisms ; the disease called cholera spreads by taking infected food or water ; the disease called malaria spreads through insect bites (mosquito bites) ; and the disease called AIDS spreads by sharing infected injection needles or through physical contact (sexual contact) with an infected person.

Prevention of Communicable Diseases

Some of the methods for preventing the occurrence and spreading of communicable diseases are given below :

(*i*) A person suffering from common cold should always cover his mouth and nose with a handkerchief while sneezing, so that micro-organisms do not get into the air. We should also keep a safe distance from a person having common cold.
(*ii*) We should keep our food covered to protect it from getting infected by flies. We should also drink clean and safe water.
(*iii*) We should protect ourselves from mosquito bites by using mosquito nets over our beds while sleeping, by putting fine wire mesh on doors and windows, or by using mosquito repellant creams and devices.
(*iv*) We should make sure that only disposable syringes and needles are used for giving us injections.
(*v*) We should avoid physical contact with an infected person, and not use his towel, clothes or bed. The towel, clothes and utensils used by an infected person should be washed and cleaned separately with soap and hot water.
(*vi*) Some of the diseases can be prevented by vaccination at proper time.

Carriers of Disease-Causing Micro-Organisms

There are some insects in our environment which transfer disease-causing microbes into our body (either by contaminating our food or by biting into our body), and spread diseases. The two most common insects which carry disease-causing micro-organisms (microbes or pathogens) are the housefly and mosquito. **The insect (or other animal) which transmits disease-causing micro-organisms to humans (without itself suffering from them) is called a 'carrier'**. We can now say that there are some insects which act as carriers of disease-causing micro-organisms. The two most common carriers of disease-causing micro-organisms (or microbes) are :

(*i*) Housefly, and
(*ii*) Mosquito.

We will now describe the role of houseflies and mosquitoes in spreading diseases. Please note that houseflies breed and feed in filthy places (insanitary places) which contain a lot of disease-causing micro-organisms (microbes or pathogens). And mosquitoes breed in pools of stagnant water.

The Role of Housefly in Spreading Diseases

The houseflies lay eggs on garbage dumps. So, they breed on filth and refuse (*kachra*). The houseflies feed on garbage, animal excreta, dead organic matter and exposed human food (uncovered human food). The body and legs of housefly bear a lot of fine hair (see Figure 12). When the housefly sits on a garbage heap, human excreta or other filth and refuse, then millions of disease-causing micro-organisms (like bacteria) present in them stick to the hairy legs and other body parts of the housefly [see Figure 13(*a*)]. And when this housefly now sits on our uncovered food, then the micro-organisms sticking to the hair on its legs and other body parts are transferred to food [see Figure 13(*b*)]. In this way our food gets contaminated with disease-causing micro-organisms. When this contaminated food is consumed by a person, then the disease causing micro-organisms enter into his body and cause various diseases. The person gets sick. Thus, the housefly carries disease-causing micro-organisms (or germs) on the hair of its legs and other body parts. It is the habit of housefly of sitting on garbage and human food alternately which is responsible for the transmission of micro-organisms to our body and causing diseases. **Some of the dangerous diseases spread by houseflies are : Cholera, Tuberculosis (TB), Typhoid and Diarrhoea.**

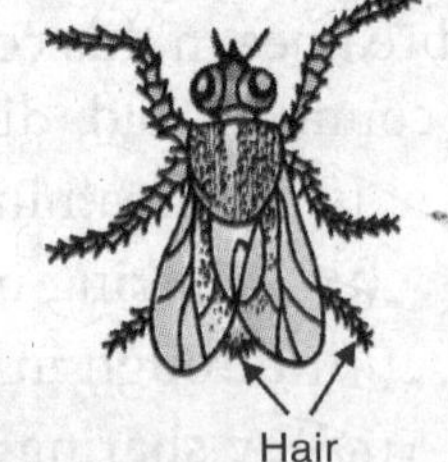

Figure 12. Housefly.

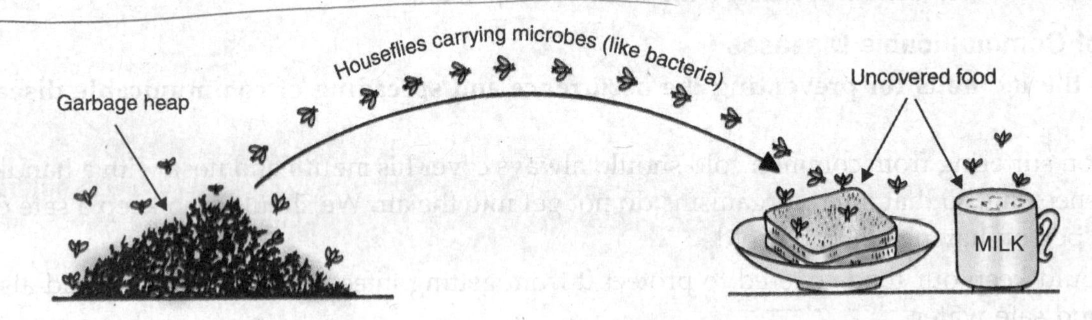

(*a*) Garbage heap (excreta, etc.): This contains microbes like bacteria

(*b*) Uncovered food: The houseflies drop microbes like bacteria in this food

Figure 13. Diagram to show how houseflies transfer microbes (like bacteria) from dirty places like garbage heaps to uncovered food and cause diseases.

Prevention of Diseases Spread by Houseflies

The spreading of diseases by houseflies can be prevented in the following ways :

(*i*) We should not leave household garbage here and there. The garbage should be put in the garbage bins which should be kept covered. This will prevent the houseflies from breeding because they will not be able to lay their eggs on garbage.

(*ii*) The food should always be kept covered so that flies cannot sit on it.

(*iii*) We should avoid eating uncovered food items from the road-side stalls.

(*iv*) The flies should be killed by using insecticide spray and baits.

(*v*) Some of the diseases spread by houseflies can be prevented by vaccination.

Role of Mosquitoes in Spreading Diseases

Mosquito is another insect which spreads diseases by transmitting disease-causing micro-organisms (or microbes) (see Figure 14). Mosquito acts as a carrier of disease-causing micro-organisms and spreads diseases from one person to another. Please note that housefly carries the disease-causing microbes on the hair (outside its body) but the mosquito carries microbes inside its body. Mosquitoes breed in stagnant water of ponds, dirty drains, pools, ditches, and shallow lakes, etc.

Figure 14. Mosquito.

The most common disease spread by mosquitoes is 'malaria'. Actually, it is the female *Anopheles* mosquito which carries the parasite of malaria. The malarial parasite (called *Plasmodium*) causes malaria disease. It is called a parasite because it lives on the blood of a person as its food on entering his body. We will now describe how female *Anopheles* mosquito spreads malaria disease.

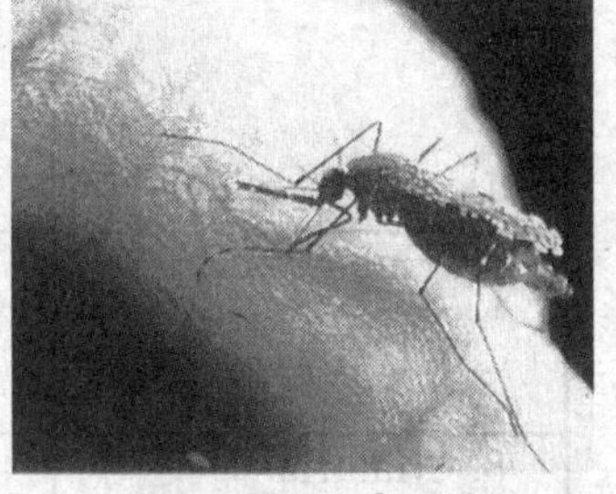

Figure 15. A female *Anopheles* mosquito sucking blood from a person's arm.

(*i*) When a female *Anopheles* mosquito bites a person suffering from malaria disease, it sucks the blood of that person which contains the malarial parasite microbes (see Figure 15) (Malarial parasite is a protozoan called *Plasmodium*).

(*ii*) And when this infected *Anopheles* mosquito now bites a healthy person to suck his blood, it transfers the malarial parasite microbes into his blood stream alongwith saliva.

(*iii*) By receiving malarial parasite microbes in blood, the healthy person also gets malaria disease.

Please note that only the female *Anopheles* mosquito sucks blood, so only the female *Anopheles* mosquito carries malarial parasite and spreads malaria. The male *Anopheles* mosquito does not suck blood of a person and hence does not spread malaria disease. **Another disease spread by mosquitoes is 'dengue'.** Dengue is caused by a virus. The female *Aedes* mosquito acts as a carrier of dengue virus and spreads the dengue disease from person to person.

Prevention of Diseases Spread by Mosquitoes

We can prevent (or control) the spreading of diseases like malaria and dengue caused by mosquitoes in the following ways :

(*i*) All the mosquitoes breed in water. So, the pools of stagnant water around the houses should be drained out so that mosquitoes may not breed in them. We should not let water collect in coolers, tyres, flower pots, etc. By keeping our surroundings clean and dry, we can prevent mosquitoes from breeding. And when there are no mosquitoes, there will be no malaria or dengue.

(*ii*) The windows and doors of the house should have fine iron wire mesh so that mosquitoes cannot enter the house.

(*iii*) Insecticides should be sprayed in houses periodically to kill mosquitoes.

(*iv*) Oil should be sprayed on the surface of water in dirty water drains to kill the larvae of mosquitoes.

(*v*) Mosquito repellant cream should be applied on the exposed parts of the body before sleeping at night. Mosquito repellant devices can also be used.

(*vi*) Mosquito net should be used over beds while sleeping to prevent mosquito bites.

Please note that **the micro-organism which causes a disease is known as causative micro-organism of that disease.** For example, tuberculosis disease is caused by bacteria, so the causative micro-organisms of tuberculosis are bacteria. Some of the common human diseases, their causative micro-organisms, modes of transmission and general methods of prevention are given below.

Some Common Human Diseases Caused by Micro-Organisms

Human disease	*Causative micro-organism*	*Mode of transmission*	*General preventive measures*
1. Tuberculosis 2. Measles 3. Chickenpox 4. Polio	Bacteria Virus Virus Virus	Air Air Air/Contact Air/Water	(*i*) The patient should be kept in complete isolation. (*ii*) The personal belongings of the patient should be kept away from those of others. (*iii*) Vaccination should be done at suitable age.
5. Cholera 6. Typhoid	Bacteria Bacteria	Water/Food Water	(*i*) Maintain personal hygiene and good sanitary habits.

			(*ii*) Eat properly cooked food. (*iii*) Drink clean and boiled drinking water. (*iv*) Vaccination should be done.
7. Hepatitis B	Virus	Water	(*i*) Drink clean and boiled drinking water. (*ii*) Vaccination should be done.
8. Malaria	Protozoan	Mosquito	(*i*) Do not allow water to collect in surroundings to prevent breeding of mosquitoes.
9. Dengue	Virus	Mosquito	(*ii*) Spray insecticides in homes to kill mosquitoes. (*iii*) Put fine wire mesh on doors and windows to prevent mosquitoes from entering into house. (*iv*) Use mosquito nets over beds for sleeping. (*v*) Use mosquito repellant cream and devices.

Disease-Causing Micro-Organisms in Animals

Just like human beings, several micro-organisms cause diseases in other animals (such as cow, buffalo, sheep, goat and poultry birds). **Some of the examples of diseases caused in animals by the micro-organisms (or microbes) are : Foot and mouth disease ; Anthrax and Aspergillosis.**

(*i*) Foot and mouth disease of animals (like cattle) is caused by a virus. Thus, foot and mouth disease is a viral disease of animals. The causative micro-organism of foot and mouth disease is **virus**. The cattle suffering from this disease get blisters on feet and mouth.

(*ii*) Anthrax is a dangerous disease of animals (like cattle) which is caused by a bacterium. Thus, anthrax disease is a bacterial disease of animals. The causative micro-organism of anthrax disease is a **bacterium** (known as *Bacillus anthracis*). The bacterium *Bacillus anthracis* which causes anthrax disease in animals was discovered by Robert Koch in 1876.

(*iii*) Aspergillosis is a disease of animals (like poultry birds) which is caused by a fungus. Thus, aspergillosis is a fungal disease of animals. The causative micro-organism of aspergillosis disease is a **fungus**.

Disease-Causing Micro-Organisms in Plants

Just like human beings and other animals, micro-organisms also cause diseases in plants. For example, several micro-organisms (or microbes) cause diseases in plants like wheat, rice, potato, sugarcane, orange, apple, and others. **Some of the common plant diseases caused by micro-organisms are : Rust of wheat ; Citrus canker and Yellow vein mosaic of *bhindi* (Okra).**

(*i*) The plant disease called 'rust of wheat' is caused by fungi. Thus, the causative micro-organism of 'rust of wheat' disease is **fungus**. The rust of wheat disease is transmitted through air and seeds. This means that the modes of transmission of the 'rust of wheat' disease are : **Air** and **Seeds**. As the name suggests, the rust of wheat disease occurs in wheat plants.

(*ii*) The plant disease called 'citrus canker' is caused by bacteria. So, the causative micro-organisms of 'citrus canker' disease are **bacteria**. The 'citrus canker' disease is transmitted through air. This means that the mode of transmission of citrus canker disease of plants is : **Air.** As the name suggests, the citrus canker disease occurs in citrus trees such as those of lemon, lime, orange, etc.

(*iii*) The plant disease called 'yellow vein mosaic of *bhindi* (or Okra)' is caused by a virus. So, the causative micro-organism of this plant disease is **virus**. The 'yellow vein mosaic of *bhindi*' disease is transmitted through insects. Thus, the mode of transmission of yellow vein mosaic of *bhindi* disease is : **Insect.**

The diseases of plants caused by micro-organisms reduce the yield and quality of various crops. The plant diseases can be controlled by the use of certain chemicals which kill the disease-causing micro-organisms (or insects). We will now discuss food poisoning and preservation of food.

FOOD POISONING

If the food is not covered properly, stored properly or preserved properly, then it gets spoiled by the action of micro-organisms (such as bacteria and fungi) on it. Micro-organisms that grow on our food sometimes produce toxic substances (poisonous substances). The food spoiled in this manner starts giving foul smell and bad taste. Its colour may also change (see Figure 16). If such a food is eaten, it will lead to food poisoning. **The disease caused due to the presence of a large number of micro-organisms (like bacteria and fungi) in the food, or due to the presence of toxic substances in food formed by the action of micro-organisms, is called food poisoning.** Thus, food poisoning occurs due to the consumption of food spoilt by some micro-organisms. **The major symptoms of food poisoning are : Vomiting, Diarrhoea (Loose motions), Pain in abdomen, Headache and Fever.** Food poisoning can cause serious illness and even death. The micro-organisms (like bacteria and fungi) which cause food poisoning come into food from the air, dirty hands, unclean food containers, flies, cockroaches, insects, rats, or sick farm animals. The two most common examples of bacteria which cause food poisoning are bacteria *Salmonella* and bacteria *Clostridium botulinum*. An example of fungus which causes food poisoning is *Aspergillus*. The spoiling of food is a chemical change. **Food is a very precious material. It should not be allowed to get spoiled, become useless, and cause food poisoning, etc.** We 'preserve' food to prevent it from getting spoiled. We will now discuss the preservation of food.

Figure 16. A slice of bread spoiled by the growth of fungus on it.

PRESERVATION OF FOOD

The food materials like milk, fruits, vegetables, meat, fish and cooked food, etc., get spoiled easily. This is because they contain a lot of water due to which the food-spoiling micro-organisms can grow in them easily. We can prevent the spoilage and contamination of easily spoilable food materials like milk, fruits, vegetables, meat, fish and cooked food, etc., by proper methods (or techniques) of food preservation. **The process in which the food materials are given a suitable physical or chemical treatment to prevent their spoilage is called food preservation.** Some of the methods for preserving foods are : (*i*) Sun-drying (or Dehydration) (*ii*) Heating (*iii*) Cooling (or Refrigeration) (*iv*) Deep freezing (*v*) Addition of common salt (*vi*) Addition of sugar (*vii*) Addition of mustard oil and vinegar (*viii*) Use of special chemical preservatives (such as sodium metabisulphite, sodium benzoate and citric acid) (*ix*) Pasteurisation, and (*x*) Packing food in air-tight packets. The method to be used for preserving a particular food material depends on the nature of the food material. Different types of foods are preserved by using different methods of preservation. The preservation of food by different methods will become clear from the following examples.

1. Preservation of Food by Sun-Drying (or Dehydration)

Drying (or dehydration) means removal of water from food materials which are to be preserved. Sun-drying (or dehydration) reduces the water content (or moisture content) of food materials and makes them dry. In the absence of moisture, the food does not get spoiled because the micro-organisms which spoil food do not grow in dry food. **The vegetables like Spinach (*Palak, Saag*), *Methi* leaves, Cauliflower and Peas (*Mutter*) are preserved in our homes by the sun-drying method.** These dried vegetables can be stored safely for a long time and used whenever required. Grapes are preserved by drying to make raisins (*kishmish*).

2. Preservation of Food by Heating

Heating kills many micro-organisms and prevents the food from spoilage. So, some foods can be preserved just by heating. For example, **we boil milk to prevent it from spoilage.** When we heat the milk during boiling, then the food-spoiling bacteria present in it get killed. So, the boiled milk remains good for a longer time. It does not get sour quickly.

3. Preservation of Food by Cooling (or Refrigeration)

Low temperature inhibits the growth of micro-organisms. So, the food-spoiling bacteria do not grow and multiply in cold conditions (having low temperature). Thus, when food is kept in a cold place (like that in a refrigerator), then the food does not get spoiled easily. It remains fresh for a much longer period. **The food materials like milk, kneaded flour (dough), cooked food (like cooked vegetables and pulses), and fresh fruits and vegetables are kept in a cool place like refrigerator to prevent their spoilage.**

4. Preservation of Food by Deep Freezing

Preservation of food by deep freezing means preservation of food by excessive cold (at temperatures much below 0°C). Deep freezing of food can be done by placing it in the 'freezer compartment' of our refrigerator or in special 'deep freeze refrigerators'. When the food is kept in a deep freezer (whose temperature is much below 0°C), the food gets frozen. At the very low temperature in deep freezer, the growth of food-spoiling micro-organisms is prevented completely. Due to this, the frozen food remains unspoiled and fresh for long periods. Thus, food can be frozen in deep freezers and kept fresh even for months. This frozen food can be cooked and eaten whenever needed. **Deep freezing method is used for the preservation of foods like meat, fish and their products ; fruits and vegetables.**

5. Preservation of Food by Common Salt

Common salt prevents the growth of food-spoiling micro-organisms due to which it is used to preserve a number of food materials. **Common salt has been used to preserve meat and fish for ages.** Meat and fish are covered with dry salt to prevent the growth of bacteria. Such meat and fish do not get spoiled easily. They remain good for a long time. **Common salt is also used to preserve fruits such as raw mangoes, lemon and *amla* (in the form of their pickles) and tamarind (in the form of *chutney*)**. Common salt does not allow bacteria or fungus to grow on fruits and vegetables preserved in it. For example, if ripe mangoes are kept as such for some time, they rot and get spoiled but the raw mangoes preserved by using common salt in the form of pickle do not get spoiled for a long time.

6. Preservation of Food by Sugar

Sugar is used as a preservative in making jams and jellies from fruits (see Figure 17). Sugar reduces the moisture content from the fruits which inhibits the growth of micro-organisms like bacteria which spoil the fruits, etc. **The fruits which are preserved in the form of jams and jellies by using sugar as preservative are : Apple, Ripe mango, Orange, Strawberry, Pineapple and Guava, etc.** The fruits preserved in the form of food materials like jams and jellies can be stored safely for a considerable time and used later.

Figure 17. Jams.

Figure 18. Pickles.

7. Preservation of Food by Mustard Oil and Vinegar

Mustard oil and vinegar (*sirka*) are widely used as preservatives for the preservation of fruits and vegetables in the form of pickles (*achar*) (see Figure 18). The use of mustard oil or vinegar prevents the spoilage of fruits and vegetables because food-spoiling bacteria cannot live in such an environment. Some of the fruits and vegetables which can be spoiled easily are preserved in the form of their pickles by using mustard oil or vinegar. **Mustard oil and vinegar are used as preservatives for preserving fruits such as raw mango, *amla* and lemon, etc., in the form of their pickles.**

8. Preservation of Food by Using Special Chemicals as Preservatives

The three special chemicals which are used as preservatives in the preservation of food are : Sodium metabisulphite, Sodium benzoate and Citric acid. **Sodium metabisulphite and sodium benzoate are used to preserve foods such as jams, jellies, juices and squashes** so as to save them from spoilage. And **citric acid is used as a preservative in confectionary (sweets).** These special chemicals kill the food-spoiling bacteria but they do not harm us.

9. Preservation of Food by Pasteurisation

A French scientist named Louis Pasteur has given an excellent method of preserving food. This method is called 'pasteurisation' after his name. **The method of pasteurisation is used for the preservation of milk in big milk dairies,** and it involves the process of heating, followed by quick cooling. Milk is preserved by the method of pasteurisation as follows : First the milk is heated to about 70°C for 15 to 30 seconds to kill most of the bacteria present in it. Next, this hot milk is cooled very quickly to a low temperature to prevent any remaining bacteria from growing further. And then this milk is stored in cold (in refrigerators). Pasteurised milk can be consumed without boiling because it is free from harmful micro-organisms. The milk that comes in packets also does not get spoiled for a fairly long time. This is because it is pasteurised milk.

10. Preservation of Food by Packing in Air-Tight Packets

These days, dry fruits and even vegetables are sold in sealed, air-tight packets to prevent the attack of micro-organisms on them. This helps the dry fruits and vegetables to remain unspoiled for a longer time.

NITROGEN FIXATION

Our atmosphere contains a lot of nitrogen gas. In fact, our atmosphere has 78 per cent nitrogen gas. The atmospheric nitrogen gas cannot be utilised directly by the plants or animals. In order that the nitrogen gas of atmosphere can be utilised by plants for their growth, it has first to be converted into nitrogen compounds (which can be absorbed by the roots of the plants). **The process of converting nitrogen gas of atmosphere (or air) into compounds of nitrogen (which can be used by the plants) is called nitrogen fixation.** The '*nitrogen gas*' is the '*free nitrogen*' whereas '*nitrogen compounds*' (like nitrates) are said to be '*fixed nitrogen*'. The nitrogen gas of atmosphere (or air) can be 'fixed' (converted into nitrogen compounds) (*i*) by certain nitrogen-fixing bacteria present in the soil, (*ii*) by *Rhizobium* bacteria present in the root nodules of leguminous plants, (*iii*) by blue-green algae, and (*iv*) by lightning. The nitrogen fixing *Rhizobium* bacteria live in the root nodules of leguminous plants (such as peas, beans, etc.) (see Figure 19). *Rhizobium* bacteria have symbiotic relationship with leguminous plants. Thus, some nitrogen-fixing bacteria live freely in the soil whereas other nitrogen-fixing bacteria (*Rhizobium* bacteria) live in the root nodules of leguminous plants. Nitrogen gas of atmosphere also

Figure 19. The nodules on the roots of this pea plant (a leguminous plant) contain nitrogen-fixing '*Rhizobium*' bacteria which convert nitrogen gas of air into nitrogen compounds.

gets fixed through the action of lightning in the sky. This happens as follows : When lightning takes place in the sky during thunderstorm, a high temperature is produced in the atmosphere. At this high temperature, nitrogen gas of air combines with oxygen gas of air to form nitrogen compounds. These nitrogen compounds dissolve in rain water, fall to earth with rain water and go into the soil. Nitrogen of atmosphere can also be fixed by artificial methods. We will study the artificial fixation of nitrogen in higher classes.

THE NITROGEN CYCLE

Nitrogen is required by both, plants and animals for their growth and development. Nitrogen is an essential component of proteins which make up the bodies of plants and animals. Nitrogen is also present in chlorophyll, nucleic acids and vitamins. The same nitrogen element is circulated again and again through living things (like plants and animals) and non-living things (like air, soil and water). **The circulation of nitrogen element through living things (plants and animals) and non-living environment (air, soil and water) is called nitrogen cycle in nature.** A labelled diagram of nitrogen cycle in nature is

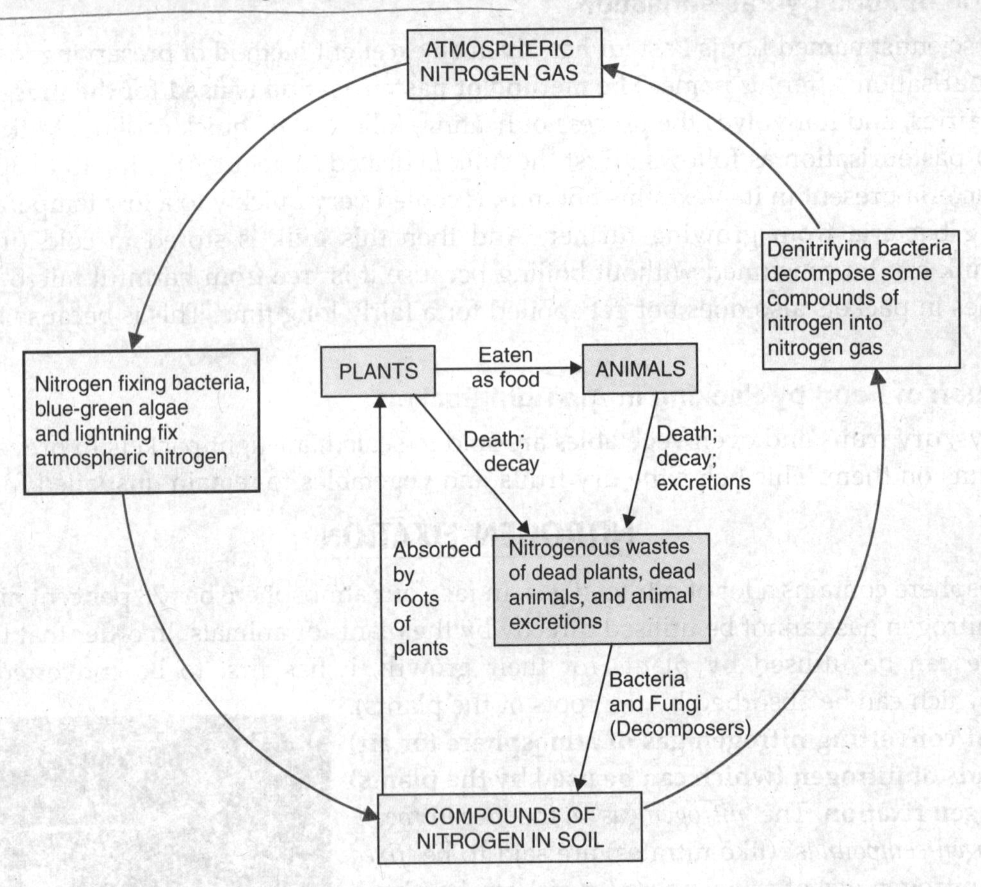

Figure 20. Nitrogen cycle in nature.

given in Figure 20. We will now describe the nitrogen cycle in nature. The main steps in the nitrogen cycle in nature are as follows :

(*i*) The atmosphere (or air) contains nitrogen gas. The nitrogen-fixing bacteria (present in the soil and in the root nodules of leguminous plants), blue-green algae and lightning in the sky fix nitrogen gas from the atmosphere and convert it into compounds of nitrogen which go into soil.

(*ii*) The plants take compounds of nitrogen from the soil for their growth. The plants absorb the nitrogen compounds from the soil through their roots. The plants convert the compounds of nitrogen into plant proteins and other organic compounds which make up the body of plants.

(*iii*) The plants are eaten up by animals as food. Animals convert plant proteins into animal proteins and other organic compounds which make up their body. Some animals also eat other animals to obtain nitrogen compounds. Thus, animals obtain nitrogen compounds by eating plants as well as other animals.

(*iv*) When plants and animals die, the complex nitrogen compounds (like proteins, etc.) present in their dead bodies are decomposed and converted into simple compounds of nitrogen by certain bacteria and fungi present in the soil. Animal excretions (urine, etc.) are also converted into simple compounds of nitrogen. All the simple compounds of nitrogen formed in this way go into the soil. In this way, the compounds of nitrogen which were taken by the plants from the soil during their growth are returned to the soil. From the soil, these nitrogen compounds are again absorbed by the new plants for their growth and this part of nitrogen cycle is repeated endlessly.

(*v*) Some of the compounds of nitrogen (formed from the decay of dead plants and animals) are decomposed by denitrifying bacteria present in the soil to form nitrogen gas. This nitrogen gas goes back into the atmosphere (from where it initially came) (This process is the reverse of fixation of nitrogen). In this way, the nitrogen gas which was removed from the atmosphere during fixation is put back into the atmosphere.

From the atmosphere, nitrogen gas is used again during nitrogen fixation and the nitrogen cycle is repeated in nature again and again. **As a result of nitrogen cycle in nature, the percentage of nitrogen gas in the atmosphere (or air) remains constant.** We are now in a position to **answer the following questions :**

Very Short Answer Type Questions

1. Name the instrument (or device) which is needed to see the micro-organisms.
2. What is the name of micro-organisms which reproduce only inside the living cells of other organisms ?
3. What are the major groups of micro-organisms ?
4. Name any two human diseases caused by bacteria.
5. Name any two human diseases caused by viruses.
6. Name any two human diseases caused by protozoa.
7. Name any two human diseases caused by fungi.
8. Which micro-organism is utilised in making curd from milk ?
9. Name the micro-organism which is used for the large scale production of alcohol.
10. Name any two antibiotics.
11. Name an antibiotic extracted from fungus (mould). Name the fungus.
12. Name any four diseases which can be prevented by vaccination.
13. Name the scientist who discovered the vaccine for smallpox.
14. Name the scientist who discovered 'penicillin'.
15. State an important function performed by blue-green algae.
16. Name one 'biological nitrogen-fixer'.
17. Name two common insects which act as carriers of disease-causing micro-organisms (or disease-causing microbes).
18. Name any two diseases spread by housefly.
19. Name the insect which is the carrier of parasite of malaria.
20. Name the insect which carries dengue virus.
21. Which of the two spreads dengue : mosquito or housefly ?
22. Name two diseases spread by mosquitoes.
23. Name the microbe which causes malaria disease.
24. Name one disease which spreads by breathing in air containing micro-organisms.
25. Name one disease which spreads through insect bites.
26. Name one disease which spreads through infected food or water.
27. Name the causative micro-organisms of the following animal diseases :
    (*a*) Foot and mouth disease (*b*) Anthrax
28. Name two food materials which are preserved by sun-drying method in our homes.
29. Name two food materials which are preserved by using common salt.

30. Name two food materials which can be preserved by using sugar.
31. Name two food materials which are usually preserved by deep freezing.
32. Name some of the preservatives which are used in the preservation of fruits as jams and jellies.
33. Name some of the preservatives which are used in the preservation of fruits and vegetables as pickles.
34. Name two food materials which can be preserved by using oil or vinegar.
35. Name any two special chemicals which are used as food preservatives.
36. Name the micro-organisms which can fix atmospheric nitrogen in the soil.
37. What type of plants can fix nitrogen gas of the air into compounds of nitrogen ?
38. Name the micro-organisms present in the soil and in the root nodules of leguminous plants which can fix atmospheric nitrogen.
39. Name two leguminous plants which can fix nitrogen.
40. Fill in the following blanks with suitable words :
(*a*) Alcohol is produced with the help of ....
(*b*) Blue-green algae fix............directly from air to enhance fertility of soil.
(*c*) Micro-organisms can be seen with help of a...............
(*d*) Cholera is caused by...............
(*e*) Common salt has been used to preserve.......... and .... for ages.
(*f*) The food material which is preserved by pasteurisation is ....
(*g*) As a result of nitrogen cycle, the percentage of nitrogen in the atmosphere remains more or less...............

Short Answer Type Questions

41. How do viruses differ from other micro-organisms such as bacteria ?
42. What are micro-organisms ? Give any two examples of micro-organisms.
43. Can micro-organisms be seen with the naked eye ? If not, how can they be seen ?
44. (*a*) How do houseflies carry disease-causing microbes (or pathogens) ?
(*b*) State any two ways of preventing diseases spread by houseflies.
45. (*a*) How do mosquitoes carry disease-causing micro-organisms and spread diseases ?
(*b*) Mention any three ways of preventing diseases spread by mosquitoes.
46. (*a*) What is meant by fermentation ? Name the scientist who discovered fermentation.
(*b*) Which micro-organism converts sugar into alcohol during fermentation ?
47. (*a*) How do micro-organisms help in increasing soil fertility ?
(*b*) How do micro-organisms help in cleaning the environment ?
48. What are antibiotics ? What precautions must be taken while taking antibiotics ?
49. Why are antibiotics not effective against 'common cold' and 'flu' ?
50. What is the full form of HIV ? Name the disease caused by HIV.
51. Describe how, curd is made from milk. Name the bacterium which converts milk into curd.
52. Name the micro-organism used in bread-making which makes the bread-dough rise. How does it make the dough rise ?
53. What is food poisoning ? How is food poisoning caused ?
54. (*a*) What is meant by food preservation ? Name any five methods of preserving food.
(*b*) How do you preserve cooked food at home ?
55. (*a*) Why should we not let water collect anywhere in the neighbourhood ?
(*b*) Name one animal disease each caused : (*i*) by virus (*ii*) by bacteria (*iii*) by fungus.
56. Where do *Rhizobium* bacteria live ? What is their function ?
57. Name any two (*a*) bacteria (*b*) viruses (*c*) protozoa (*d*) algae, and (*e*) fungi.
58. State the beneficial effects (or usefulness) of micro-organisms in our lives.
59. Describe the method of pasteurisation for the preservation of milk.
60. Name one plant disease each caused : (*a*) by fungi (*b*) by virus (*c*) by bacteria.
61. Which disease is spread by :
(*a*) female *Anopheles* mosquito ?
(*b*) female *Aedes* mosquito ?

62. Name two fruits which are preserved :
    (*a*) in the form of pickles.
    (*b*) in the form of jams.
63. What is the mode of transmission of the following diseases ?
    (*a*) Rust of wheat (*b*) Citrus canker (*c*) Yellow vein mosaic of *bhindi* (Okra)
64. Name any two animal diseases and two plant diseases caused by micro-organisms.
65. State the causative micro-organisms and modes of transmission of the following human diseases :
    (*i*) Tuberculosis (*ii*) Measles (*iii*) Chickenpox (*iv*) Polio (*v*) Cholera
    (*vi*) Typhoid (*vii*) Hepatitis B (*viii*) Malaria (*ix*) Dengue

Long Answer Type Questions

66. (*a*) What is meant by communicable diseases ? Name any two communicable diseases.
    (*b*) What are the various ways in which communicable diseases can occur and spread ?
67. (*a*) Name any five human diseases caused by micro-organisms. Also name the causative micro-organisms and mode of transmission for each of these diseases.
    (*b*) State the various ways of preventing the occurrence and spreading of communicable diseases.
68. (*a*) What is a vaccine ? How does a vaccine work ?
    (*b*) Why are children given vaccination ?
69. What is meant by 'nitrogen fixation' ? State two ways in which nitrogen gas of the atmosphere can be 'fixed' in nature to get nitrogen compounds in the soil.
70. Draw a neat, labelled diagram of nitrogen cycle in nature. Which natural phenomenon occurring in the sky is responsible for nitrogen fixation ?

Multiple Choice Questions (MCQs)

71. The bread dough rises because of :
    (*a*) heat (*b*) grinding (*c*) growth of yeast cells (*d*) kneading
72. Yeast is used in the production of :
    (*a*) sugar (*b*) alcohol (*c*) hydrochloric acid (*d*) oxygen
73. The process of conversion of sugar into alcohol is called :
    (*a*) nitrogen fixation (*b*) moulding (*c*) fermentation (*d*) infection
74. Which of the following is an antibiotic ?
    (*a*) sodium bicarbonate (*b*) streptomycein (*c*) alcohol (*d*) yeast
75. The most common carrier of communicable diseases is :
    (*a*) ant (*b*) housefly (*c*) dragonfly (*d*) spider
76. The carrier of malaria-causing protozoan is :
    (*a*) female *Anopheles* mosquito (*b*) cockroach
    (*c*) housefly (*d*) female Aedes mosquito
77. The vaccine for smallpox was discovered by :
    (*a*) Alexander Fleming (*b*) Edward Jenner (*c*) Louis Pasteur (*d*) Rober Koch
78. Alcohol can be converted into vinegar by the action of micro-organisms called :
    (*a*) viruses (*b*) yeast (*c*) protozoa (*d*) bacteria
79. The first antibiotic called penicillin was extracted from :
    (*a*) a bacterium (*b*) a protozoan (*c*) a fungus (*d*) an alga
80. Which of the following is not a communicable disease ?
    (*a*) cholera (*b*) cancer (*c*) chickenpox (*d*) malaria
81. Which of the following increase the fertility of soil ?
    A. *Lactobacillus* bacteria B. *Rhizobium* bacteria C. *Spirogyra* algae D. Blue-green algae
    (*a*) A and B (*b*) B and C (*c*) A and D (*d*) B and D
82. Which of the following cannot be used as a food preservative ?
    (*a*) sodium metabisulphite (*b*) sodium hydroxide (*c*) sodium benzoate (*d*) citric acid
83. Which of the following disease is not caused by bacteria ?
    (*a*) cholera (*b*) typhoid (*c*) tuberculosis (*d*) measles

84. The micro-organisms which can reproduce and multiply only inside the cells of other organisms are :
(*a*) protozoa (*b*) fungi (*c*) bacteria (*d*) viruses

85. The dengue disease spread by *Aedes* mosquito is caused by :
(*a*) bacteria (*b*) virus (*c*) protozoan (*d*) fungus

86. Which of the following disease is not caused by viruses ?
(*a*) measles (*b*) smallpox (*c*) cholera (*d*) polio

87. The micro-organism which is capable of converting sugar into alcohol and carbon dioxide is :
(*a*) bacterium (*b*) fungus (*c*) alga (*d*) protozoan

88. Which of the following is not a use of micro-organisms ?
(*a*) preparation of medicines (or drugs) (*b*) preparation of food by photosynthesis
(*c*) recycling of materials in nature (*d*) increasing the fertility of soil

89. The malaria disease is caused by a :
(*a*) virus (*b*) protozoan (*c*) bacterium (*d*) fungus

90. The parasite called *Plasmodium* causes a disease known as :
(*a*) measles (*b*) polio (*c*) malaria (*d*) dengue

Questions Based on High Order Thinking Skills (HOTS)

91. After consuming a dish of mutton, a person complained of nausea, vomiting, diarrhoea, and pain in the abdomen.
(*a*) What type of disease is he suffering from ?
(*b*) What causes this disease ?

92. Match the micro-organisms in column *A* with their action in column *B* :

A	B
(*i*) Bacteria	(*a*) Fixing nitrogen
(*ii*) *Rhizobium*	(*b*) Setting of curd
(*iii*) *Lactobacillus*	(*c*) Baking of bread
(*iv*) Yeast	(*d*) Causing malaria
(*v*) A protozoan	(*e*) Causing cholera
(*vi*) A virus	(*f*) Causing AIDS
(*vii*) *Penicillium*	(*g*) Producing antibiotics

93. To which category of micro-organisms do the following belong ?
Amoeba, Lactobacillus, Chlamydomonas, Penicillium, Yeast, HIV

94. Name the causative micro-organisms of the following plant diseases :
(*a*) Rust of wheat (*b*) Citrus canker (*c*) Yellow vein mosaic of *bhindi* (Okra)

95. The mosquito P is a carrier of virus and spreads a disease Q. Another mosquito R is the carrier of protozoan S and spreads a disease called T.
(*a*) Name (*i*) mosquito P, and (*ii*) disease Q.
(*b*) Name (*i*) mosquito R (*ii*) protozoan S, and (*iii*) disease T.
(*c*) What is the sex of mosquito P ?
(*d*) What is the sex of mosquito R ?

ANSWERS

40. (*a*) yeast (*b*) nitrogen (*c*) microscope (*d*) bacteria (*e*) meat ; fish (*f*) milk (*g*) constant **71.** (*c*) **72.** (*b*) **73.** (*c*) **74.** (*b*) **75.** (*b*) **76.** (*a*) **77.** (*b*) **78.** (*d*) **79.** (*c*) **80.** (*b*) **81.** (*d*) **82.** (*b*) **83.** (*d*) **84.** (*d*) **85.** (*b*) **86.** (*c*) **87.** (*b*) **88.** (*b*) **89.** (*b*) **90.** (*c*) **91.** (*a*) Food poisoning (*b*) Micro-organisms (like bacteria and fungi) present in spoilt dish of mutton **92.** (*i*) *e* (*ii*) *a* (*iii*) *b* (*iv*) *c* (*v*) *d* (*vi*) *f* (*vii*) *g* **93.** *Amoeba* : Protozoa ; *Lactobacillus* : Bacteria ; *Chlamydomonas* : Algae ; *Penicillium* : Fungi ; Yeast : Fungi ; HIV : Viruses **94.** (*a*) Fungus (*b*) Bacteria (*c*) Virus **95.** (*a*) (*i*) *Aedes* mosquito (*ii*) Dengue (*b*) (*i*) *Anopheles* mosquito (*ii*) *Plasmodium* (*iii*) Malaria (*c*) Female (*d*) Female

CHAPTER

3

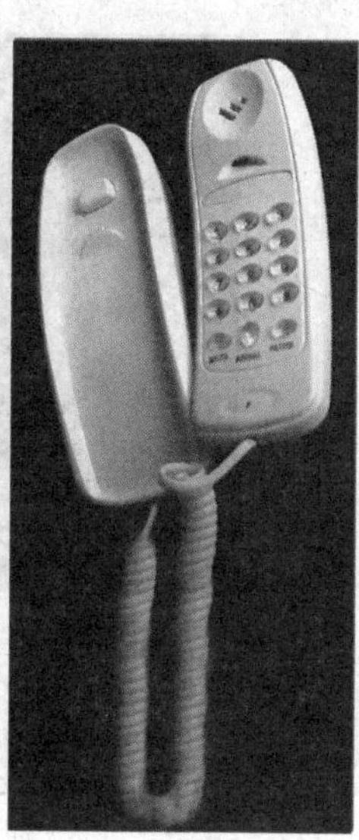

Synthetic Fibres and Plastics

A very thin, thread-like strand from which cloth is made, is called a fibre. Fabric means cloth. Fabric is made by weaving or knitting long, twisted threads called 'yarn' made from fibres. The clothes which we wear are made of fabrics. Fabrics are made from fibres obtained from 'natural' or 'artificial' sources (synthetic sources). Thus, all the fibres can be divided into two groups :

(*i*) **Natural fibres,** and

(*ii*) **Synthetic fibres.**

The fibres obtained from plants and animals are called natural fibres. Cotton, flax, jute, wool and silk are natural fibres. Cotton, flax and jute fibres come from plants whereas wool and silk come from animals. **The synthetic fibres are made by human beings.** Rayon, nylon, polyester and acrylic are synthetic fibres. We have studied the natural fibres in Classes VI and VII. In this Class we will study synthetic fibres. Before we go further and discuss synthetic fibres in detail, we should know the meaning of the term 'polymer'. This is described below.

A polymer is a 'very big molecule' formed by the combination of a large number of small molecules. The small molecules (of chemical compounds) which join together to form a polymer are called **'monomers'**. The monomers which make a polymer may all be of the 'same compound' or of 'two different compounds' (see Figures 1 and 2). The word 'polymer' comes from two Greek words 'poly' meaning 'many' and 'mer' meaning 'units'. So, ***a polymer is made of many small 'repeating units' (of chemical compounds) called monomers.***

Polymers are of two types : Natural polymers and Synthetic polymers. Cotton, wool and silk are natural polymers. For example, cotton fibre is made of a natural polymer called cellulose. Cellulose is a polymer which is made up of a large number of small glucose molecules (or glucose units) joined one after the other (see Figure 1). The walls of all the plant cells are made up of cellulose. So, wood contains a large

Figure 1. A cellulose polymer chain is made from only one type of molecules arranged one after the other like the 'same coloured' beads in the above picture.

Figure 2. A nylon polymer chain is made from two types of molecules arranged alternately just like the 'two different coloured' beads in the above picture.

amount of cellulose polymer. Thus, polymers occur in nature too. Nylon, polyester, acrylic, polythene, polyvinyl chloride (PVC), bakelite, and melamine are synthetic polymers (or man-made polymers). For example, nylon fibre is made up of nylon polymer in which two different types of molecules (or monomer units) are combined alternately to form long chains (see Figure 2). Please note that **the term 'synthetic' means made by humans in an industrial process (and not occurring naturally).** We will now discuss synthetic fibres.

SYNTHETIC FIBRES

The man-made fibres produced from chemical substances are called synthetic fibres. Synthetic fibres are made in industry by the chemical process called 'polymerisation'. *A synthetic fibre is a long chain of small units joined together.* Each small unit is a chemical compound (called organic compound). Many, many such small units join together one after the other to form a very large single unit called polymer. It is this man-made polymer which forms synthetic fibres. Thus, **a synthetic fibre is a polymer made from the molecules of a monomer (or sometimes two monomers) joined together to form very long chains** (see Figure 2). Synthetic fibres are also known as man-made fibres or artificial fibres.

Types of Synthetic Fibres

Depending upon the type of chemicals used for manufacturing synthetic fibres, **there are four major types of synthetic fibres (or man-made fibres).** These are :

(i) **Rayon,**
(ii) **Nylon,**
(iii) **Polyester,** and
(iv) **Acrylic.**

Rayon is a man-made fibre made from a *natural* material called cellulose (obtained from wood pulp). Nylon, polyester and acrylic are *fully* synthetic fibres which do not require a natural material (like cellulose) for their manufacture. These fully synthetic fibres are prepared by a number of processes by using raw materials (or chemical compounds) of petroleum origin, called **petrochemicals.** We will now study all these synthetic fibres, one by one. Let us start with rayon. Before we do that, we should know the meaning of the term 'wood pulp'. **Wood pulp is a soft, wet mass of fibres obtained from wood.** Wood pulp contains a large amount of natural polymer called **'cellulose'.**

1. RAYON

We have studied in Class VII that silk is a natural fibre obtained from silkworms. The fabric (or cloth) made from natural silk fibres is very costly. But the beautiful texture (feel, appearance, shine) of natural silk fabrics fascinated everyone. So, attempts were made to make silk artificially which would be cheaper than natural silk. Towards the end of 19th century, scientists were successful in obtaining fibres having properties similar to that of silk. This fibre was called rayon. **Rayon is often regarded as artificial silk.**

Rayon is a man-made fibre prepared from a natural raw material (called cellulose) by chemical treatment. The cellulose required for making rayon is obtained from 'wood pulp'. So, we can also say that *rayon is obtained by the chemical treatment of wood pulp* (*which contains cellulose*). Rayon is produced as follows :

(*i*) Wood pulp is dissolved in an alkaline solution (sodium hydroxide solution) to form a sticky liquid called 'viscose'.

(*ii*) Viscose is forced to pass through the tiny holes of a metal cylinder (called spinneret) into a solution of sulphuric acid when a silk like thread of rayon is formed.

Since rayon is made from naturally occurring polymer (cellulose) present in wood pulp, therefore, rayon is neither a fully synthetic fibre nor a fully natural fibre. It is a semi-synthetic fibre. **Rayon is different from truly synthetic fibres because it is obtained from a natural material (wood pulp).** Although rayon is obtained from a natural resource called wood pulp, yet it is said to be a man-made fibre. This is because it is obtained by the chemical treatment of wood pulp in factories. Rayon fibre is chemically identical to cotton but it has shine like silk. Since rayon resembles silk in appearance, therefore, ***rayon is also called artificial silk.*** Rayon is cheaper than natural silk and can be woven like silk fibres. Rayon can also be dyed in a variety of colours.

Uses of Rayon

(*i*) Rayon is used in textile industry for making clothing like sarees, blouses, dresses, socks, etc.

(*ii*) Rayon (mixed with cotton) is used to make furnishings such as bed-sheets, curtains, blankets, etc.

(*iii*) Rayon (mixed with wool) is used to make carpets.

(*iv*) Rayon is used in medical field for making bandages and surgical dressings.

(*v*) Rayon is used in tyre industry for the manufacture of tyre cord.

2. NYLON

Nylon is a synthetic fibre. In fact, nylon is the first *fully* synthetic fibre made by man without using any natural raw materials (from plants or animals). It was made in the year 1931. The chemical compounds (or monomers) used in making nylon are now obtained from petroleum products called petrochemicals. Actually, nylon is made up of the repeating units of a chemical called an 'amide'. So, **nylon is a polyamide** (which is a polymer). The name **NYLON** comes from the fact that it was developed in New York (**NY**) and London (**LON**). Nylon is a thermoplastic polymer (which can be melted by heating). Molten nylon is forced through the tiny holes in a spinneret to make nylon fibres (or nylon threads), or cast into desired shapes.

Some of the important properties of nylon fibres are as follows : Nylon fibres are very strong, fairly elastic, lightweight and lustrous. Nylon fibres absorb very little water, so clothes made of nylon are easy to wash and dry. Nylon is wrinkle resistant. Nylon fibres have high abrasion resistance (high wear and tear resistance), so they are very durable (long lasting). Nylon is not attacked by moths and ordinary chemicals. Due to all these properties, nylon fibres have become very popular for making clothes.

Uses of Nylon

(*i*) Nylon is used for making textiles (fabrics) like sarees, shirts, neck-ties, tights, socks and other garments.

(*ii*) Nylon is used in making curtains, sleeping bags and tents.

(*iii*) Nylon is used in making ropes, car seat-belts, fishing nets, tyre cord, strings for sports rackets and musical instruments, bristles for toothbrushes and paint brushes. Nylon is used for making parachutes and ropes for rock climbing. All these uses of nylon are due to the high strength of nylon fibres. A nylon thread is actually stronger than a steel wire of similar thickness.

(*iv*) Nylon is used as a plastic for making machine parts.

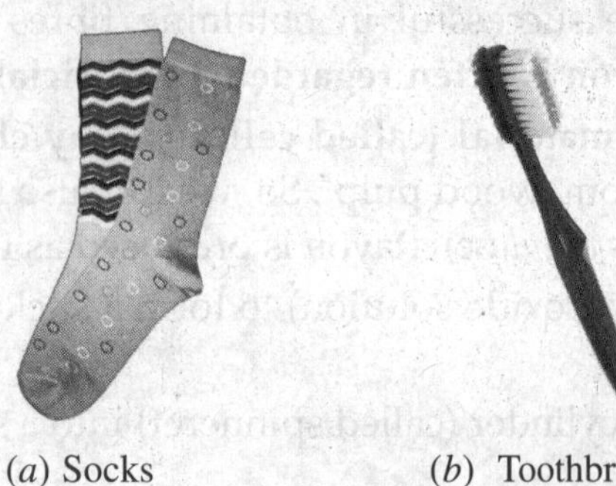

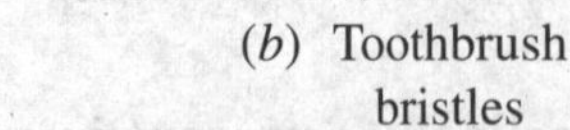

(*a*) Socks (*b*) Toothbrush bristles (*c*) Parachute (*d*) Climbing ropes

Figure 3. Some of the uses of nylon fibres.

POLYESTER

Polyester is another synthetic fibre. Actually, polyester is the general name of synthetic fibres which contain many ester groups. Polyester (poly + ester) is made up of the repeating units of a chemical called an 'ester' (Esters are the organic chemical substances which give fruits their sweet smell). We can now say that *polyester is a synthetic fibre in which the polymer units are linked by ester groups*. **Terylene is a popular polyester fibre.** The chemical compounds (or monomers) used in making polyester fibres are made from petroleum products called petrochemicals. Like nylon, **polyester is also a thermoplastic polymer.** When molten polyester is forced through the tiny holes of a spinneret, then thin polyester fibres (or polyester threads) are formed. The polyester yarn can be woven to make fabrics.

Most of the properties of polyester fibres (like terylene) are similar to those of nylon. Polyester fibres are, however, *stronger* than nylon fibres. Polyester fibres are also *softer* than nylon fibres. Since polyester fabric is strong, wrinkle resistant, easy to wash and dry, not attacked by moths and ordinary chemicals, and has high abrasion resistance, it is quite suitable for making dress materials. This is why we see many people around us wearing polyester shirts and other dresses. Sometimes, natural fibres (such as cotton or wool) are mixed with polyester (or terylene) to make blended fabrics. Blended fabrics are sold by the names like polycot (or terrycot) and polywool (or terrywool), etc. As the name suggests, these fabrics are made by mixing (or blending) two types of fibres. For example, polycot is a mixture of polyester and cotton. Similarly, polywool is a mixture of polyester and wool.

Uses of Polyester Fibres

(*i*) The most important use of polyester (like terylene) is in making fabrics for sarees, dress materials and curtains. Polyester mixed with cotton (called polycot or terrycot) is used for making shirts, trousers and other dresses. Polyester mixed with wool (called polywool or terrywool) is used for making suits.

(*ii*) Polyester is used for making sails of sail-boats. The polyester sails are light, strong, do not stretch and do not rot in contact with water.

(*iii*) Polyester is used for making water hoses for fire-fighting operations.

(*iv*) Polyester is used for making conveyer belts.

From the above discussion we conclude that synthetic fibres have become very popular. **The synthetic fibres have become very popular** because they are strong and elastic, and have low water absorption. Synthetic fibres are lightweight, long lasting and extremely fine. They are wrinkle resistant, chemically unreactive and not attacked by moths or common chemicals. Due to these properties, synthetic fibres are *much more superior* to natural fibres like cotton, wool and silk.

PET

PET is a very familiar form of polyester. PET is the abbreviation of the synthetic polymer called 'Poly-Ethylene Terephthalate'. PET can be made into a fibre or a plastic. In discussing synthetic fibres, PET is generally referred to as 'polyester' while the term PET is usually used for the plastic form. PET as a plastic is very lightweight. It is naturally colourless with high transparency. PET is strong and impact-resistant. As a plastic, **PET is replacing materials like glass.** Unlike glass, PET is shatterproof. PET is used for making bottles, jars and utensils. For example, PET bottles are used for fizzy drinks and PET jars are used for storing sugar, salt, spices and rice, etc., in our homes (see Figure 4). PET is also used for making thin films and many other useful products. It is clear that polyester fibres and PET bottles and jars are made of the same material.

Figure 4. All these containers are made of PET.

ACRYLIC

Acrylic is a synthetic fibre. Acrylic fibre is made from a chemical called 'acrylonitrile' by the process of polymerisation. Acrylic is lightweight, soft and warm with a wool-like feel. Acrylic retains its shape, resists shrinkage and wrinkles. It can be dyed very well. Acrylic fibres are strong and durable. Acrylic absorbs very little water so it has 'quick-dry' quality. Acrylic fibres are resistant to moths and most chemicals.

Due to its wool-like feel, acrylic fibre is often used as a substitute for wool. The wool obtained from natural sources (like sheep) is quite expensive. Acrylic offers a less expensive alternative to natural wool. So, the clothes made from acrylic are relatively cheaper but more durable than those made from natural wool. Many of the sweaters which the people wear in winter, and the shawls and blankets which people use, are actually not made from natural wool, though they appear to be made from wool. They are made from synthetic fibre called acrylic. Acrylic fibre is used for making sweaters, shawls, blankets, jackets, sportswear, socks, furnishing fabrics, carpets and as lining for boots and gloves.

Characteristics of Synthetic Fibres

Synthetic fibres have unique characteristics which make them popular dress materials. The important characteristics (or properties) of synthetic fibres are given below :

1. Synthetic fibres are very strong. On the other hand, natural fibres like cotton, wool and silk have low strength.

2. Synthetic fibres are more durable. Synthetic fibres have high resistance to abrasion (wear and tear). Due to this, the clothes made of synthetic fibres are very durable (long lasting). On the other hand, natural fibres like cotton, wool and silk have low abrasion resistance due to which the clothes made of natural fibres are not much durable. They do not last long.

3. Synthetic fibres absorb very little water. Due to this, the clothes made of synthetic fibres dry up quickly. On the other hand, natural fibres like cotton, wool, and silk absorb a lot of water. So, the clothes made of natural fibres do not dry up quickly.

4. Synthetic fibres are wrinkle resistant. Due to this, the clothes made of synthetic fibres do not get crumpled easily during washing or wear. They keep permanent creases. On the other hand, natural fibres like cotton, wool and silk are not wrinkle resistant. So, the clothes made of natural fibres get crumpled easily during washing and wear.

5. Synthetic fibres are quite lightweight. On the other hand, natural fibres are comparatively heavy.

6. Synthetic fibres are extremely fine. So, the fabrics made from synthetic fibres have a very smooth texture. On the other hand, natural fibres are not so fine. Due to this, the fabrics made from natural fibres do not have a very smooth texture.

7. Synthetic fibres are not attacked by moths. On the other hand, natural fibres are damaged by moths.

8. Synthetic fibres do not shrink. So, the clothes made of synthetic fibres retain their original size even after washing. On the other hand, natural fibres shrink after washing.

9. Synthetic fibres are less expensive and readily available as compared to natural fibres.

10. Clothes made from synthetic fibres are easier to maintain as compared to those made from natural fibres.

We will now describe **an activity to compare the strengths of synthetic fibres with those of natural fibres.** For this activity, we require threads of natural fibres (such as wool, cotton and silk) and of synthetic fibres (like nylon and polyester) which are of nearly the same thickness and of the same length.

ACTIVITY

Take an iron stand with a clamp. Take a woollen thread of about 50 cm length. Tie one end of the woollen thread to the clamp so that it hangs freely. Tie a pan to the lower end of the woollen thread (as shown in Figure 5) so that weights can be placed in it. Put a weight in the pan. Go on adding more and more weights in the pan till the woollen thread breaks. Note down the total weight required to break the woollen thread. This weight indicates the strength of the woollen thread. Repeat this activity by using similar threads of cotton, silk, nylon and polyester. Note down the weights required to break all these threads, one by one. We will find that :

(*i*) Minimum weight is required to break the woollen thread showing that woollen thread has the minimum strength.

(*ii*) More weight is required to break the cotton thread showing that the strength of cotton thread is greater than that of woollen thread.

(*iii*) Still more weight is needed to break the silk thread, indicating that silk thread is stronger than the cotton thread.

(*iv*) Much more weight is required to break the nylon thread showing that nylon thread has a greater strength than the silk thread.

(*v*) Maximum weight is needed to break the polyester thread indicating that the polyester thread is even stronger than the nylon thread.

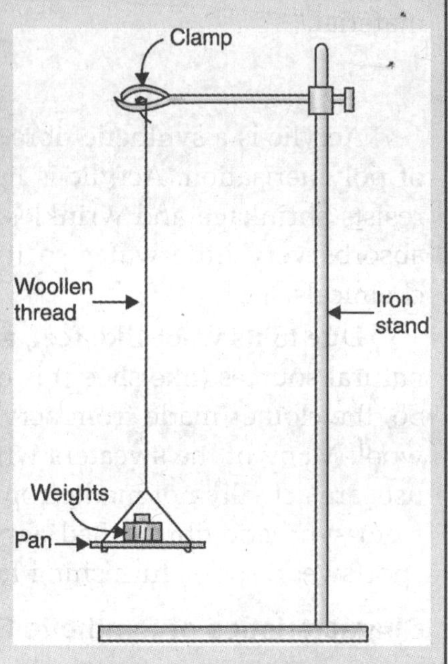

Figure 5. Arrangement to determine the relative strengths of natural fibres and synthetic fibres.

Based on this activity, we can now arrange the natural fibres and synthetic fibres in the order of their increasing strengths as : Wool, Cotton, Silk, Nylon, Polyester. This activity tells us that the synthetic fibres (like nylon and polyester) are stronger than the natural fibres (like wool, cotton and silk).

We will now describe **an activity to demonstrate the difference in the 'water absorption' property of synthetic fibres and natural fibres.**

ACTIVITY

(*i*) Take a piece of cloth made of a synthetic fibre (like polyester cloth) of about half a square metre size. Also take another piece of cloth made of a natural fibre (like cotton cloth) of exactly the same size.

(*ii*)Take two mugs and put 500 mL of water in each mug. Soak the piece of polyester cloth in water taken in one mug for about 5 minutes. Soak the piece of cotton cloth in water taken in the other mug for an equal time.

(*iii*) Take out the soaked pieces of polyester cloth and cotton cloth from the two mugs and compare the volume of water which remains behind in each mug. We will find that *more* water is left behind in the mug in which polyester cloth was soaked. This shows that *polyester cloth absorbs less water.* Much *less* water is left behind in the mug in which cotton cloth was soaked, showing that *cotton cloth absorbs much more water.* From these observations we conclude that **synthetic fibres (like polyester) absorb much less water than natural fibres (like cotton).**

(*iv*) Spread the wet piece of polyester cloth and the wet piece of cotton cloth in sunshine so as to dry them. We will find that the wet piece of polyester cloth dries up rapidly but the wet piece of cotton cloth takes much longer time to get dried. From these observations we conclude that **the wet synthetic fibres (like polyester) dry up quickly but the wet natural fibres (like cotton) do not dry up quickly.**

So far we have described the advantages of synthetic fibres. We will now describe some disadvantages of synthetic fibres (over the natural fibres).

A disadvantage of synthetic fibres is that they melt on heating. If a person is wearing clothes made of synthetic fibres and his clothes catch fire accidently, then the synthetic fibres of clothes melt and stick to the body of the person causing severe burns. This can be disastrous for the person concerned. We should, therefore, not wear synthetic clothes (made of nylon, polyester, etc.) while working in the kitchen or in a science laboratory. The natural fibres (like cotton, wool, etc.) do not melt on heating. So, it is quite safe to wear clothes made of natural fibres while working in the kitchen or in a science laboratory.

Another disadvantage of synthetic fibres is that the clothes made of synthetic fibres are not suitable for wearing during hot summer weather. This can be explained as follows : Synthetic fibres are extremely fine, so the clothes made of synthetic fibres do not have sufficient pores for the sweat to come out, evaporate and cool our body. Due to this, *clothes made of synthetic fibres make us feel hot and uncomfortable during summer*. Clothes made of natural fibres (like cotton) are more comfortable during summer. This is because the large pores of cotton clothes allow the body sweat to come out through them, evaporate and make us feel cool and comfortable. So, if we want to buy shirts for summer, we should buy cotton shirts and not the shirts made from synthetic materials (like polyester).

The manufacturing of fully synthetic fibres (like nylon, polyester and acrylic, etc.) is helping in the conservation of forests. This is because the fully synthetic fibres are manufactured from petrochemicals (obtained from crude oil 'petroleum'), so no trees have to be cut down for making them. On the other hand, semi-synthetic fibres like rayon are made from wood pulp which requires cutting down of forest trees. We will now discuss plastics.

PLASTICS

We use a large number of articles (or things) made of plastics in our everyday life. Some of the articles made of plastics which are used by us in our everyday life are plastic bags (polythene bags), water bottles, buckets, mugs, water tanks, water pipes, ballpoint pens, combs, toothbrushes, toys, shoes, tea-strainers,

Figure 6. Some of the articles (or things) made of plastics which are used in our everyday life.

cups, plates, chairs, tables, insulation of electric wires, covers of electric switches, plugs, sockets and bulb-holders, etc. (see Figure 6). Some of the parts of radio, television, refrigerator, cars, buses, trucks, scooters, trains, aeroplanes, ships and spacecrafts are also made of various types of plastics. The list of things made of plastics and used in homes, transport and industry is endless. Plastic articles are available in all possible shapes and sizes.

A plastic is a synthetic material which can be moulded (or set) into desired shape when soft and then hardened to produce a durable article (the term *'plastic'* means *'easy to mould'*). Like synthetic fibres, **plastics are also polymers.** This means that plastics consist of very long molecules made by joining many small molecules together. The starting materials for plastics are obtained from petroleum products called 'petrochemicals'. Some of the examples of plastics are : Polythene, Poly-Vinyl Chloride (PVC), Bakelite, Melamine and Teflon. Nylon is also a plastic. The major properties and uses of some of the plastics are given below.

Polythene (poly + ethene = polythene). Polythene is a plastic obtained by the polymerisation of a chemical compound known as ethene. Polythene is tough and durable. Polythene is used in making polythene bags (plastic bags), waterproof plastic sheets, bottles, buckets and dustbins. Polythene is also used for packaging.

Polyvinyl chloride (commonly known as PVC) is a strong and hard plastic. It is not as flexible as polythene. PVC is used for making insulation for electric wires, pipes, garden hoses, raincoats, seat covers, etc.

Bakelite is a very hard and tough plastic. Bakelite is a poor conductor of heat and electricity. Bakelite is used for making the handles of various cooking utensils (such as frying pans and pressure cookers, etc.). Bakelite is used for making handles of cooking utensils because (*i*) it is a poor conductor of heat, and (*ii*) it does not become soft on getting heated (It is a thermosetting plastic). Bakelite is also used for making electrical fittings such as electric switches, plugs and sockets, etc. Bakelite is used for making electric switches, plugs and sockets, etc., because (*i*) it does not conduct electricity, and (*ii*) it does not become soft on getting heated.

Melamine is a plastic which can tolerate heat better than other plastics and resists fire. Melamine is used for making floor tiles, unbreakable kitchenware (cups, plates, etc.), ashtrays and fire-resistant fabrics. **Melamine is a fire-resistant plastic.** The uniforms of fire-men have a coating of melamine plastic to make them fire-resistant. Special plastic cookware made of melamine is used in microwave ovens for cooking food. In microwave ovens, the heat cooks the food but does not affect the plastic vessel.

Teflon is a special plastic on which oil and water do not stick. Oil and water do not stick on teflon plastic because it has a slippery surface. Teflon also withstands high temperatures. **Teflon is used for giving non-stick coating on cookwares (like non-stick frying pans).** Teflon is also used for making soles (bottoms) of electric irons.

Plastics are used extensively in the health care industry. For example, plastics are used in the packing of tablets, for making syringes, doctor's gloves, threads for stitching wounds, and a number of medical instruments.

Types of Plastics

Plastics are of two types : Thermoplastics and Thermosetting plastics. We will now discuss these two types of plastics in detail, one by one. Let us start with thermoplastics.

1. Thermoplastics

Some plastics get soft (or melt) when heated, and hard again when they are cooled. Such plastics can be made soft and hard again and again. **A plastic which can be softened repeatedly by heating and can be moulded into different shapes again and again, is called a thermoplastic.** Thermoplastics are flexible so they can be bent easily (without breaking). Thermoplastics are also known as 'thermosoftening' plastics. **Some of the examples of thermoplastics are : Polythene and Polyvinyl chloride (PVC).**

If we take a plastic bottle (polythene bottle) and add quite hot water in it, the plastic bottle gets deformed—its shape changes and becomes irregular. This happens because the bottle is made of a thermoplastic (like polythene) which becomes soft on getting heated by hot water and changes shape. This activity shows that the articles made of thermoplastics become soft on heating. Let us now take a plastic bottle and press it by applying the force of our hands. We will find that the plastic bottle bends easily. This shows that the articles made of thermoplastics bend easily. In other words, we can say that thermoplastics are flexible. In fact, ***thermloplastics are used for making those articles which do not get too hot, and are flexible.*** Thermoplastics are used for making insulation of electric wires and cables, various types of plastic containers (plastic bottles, plastic jars, etc.), combs, toys, plastic bags, raincoats, seat covers, bristles of

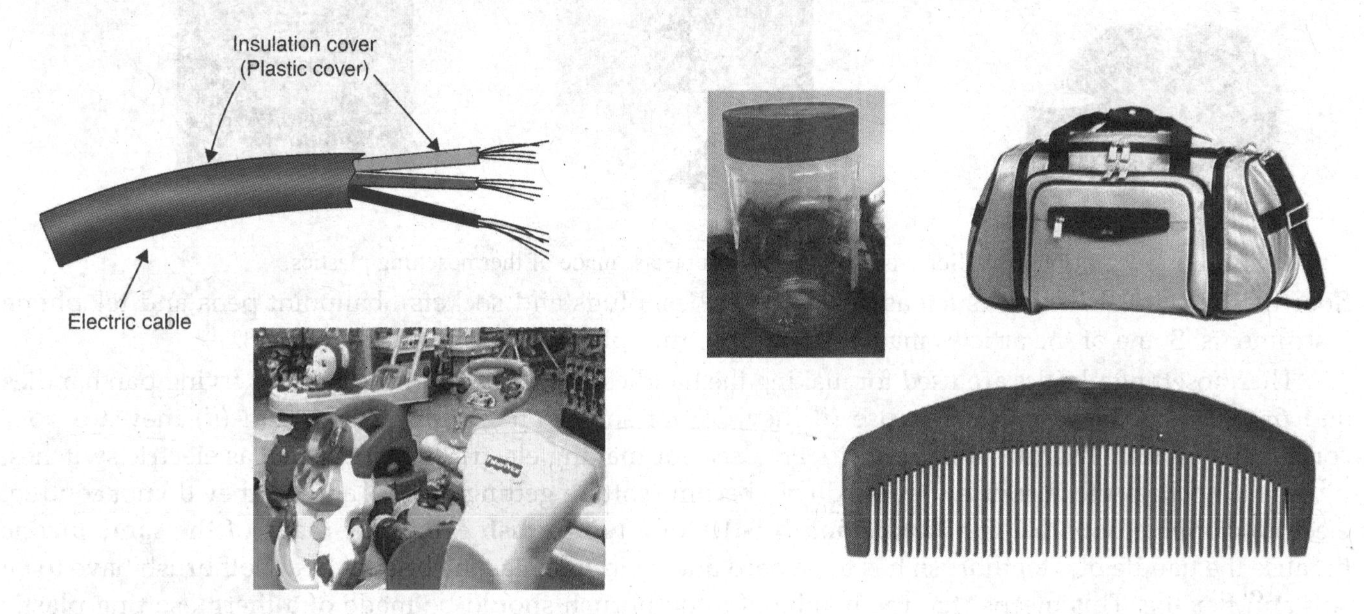

Figure 7. Some of the articles (or objects) made of thermoplastics (or thermosoftening plastics).

brushes, packaging materials and chairs. Thermoplastics are used for making the insulation of electric wires because (*i*) they do not conduct electricity, and (*ii*) they are flexible. Some of the articles (or objects) made of thermoplastics are shown in Figure 7.

2. Thermosetting Plastics

There are some plastics which get soft only once—the first time they are heated after being made. When such plastics are heated for the first time, they become soft (or melt) and can be moulded to make an article of any desired shape. On cooling, this article becomes very hard and rigid. When this plastic article is heated again, it does not become soft at all. **A plastic which once set, does not become soft on heating and cannot be moulded a second time, is called a thermosetting plastic.** Once set in a given shape and solidified, a thermosetting plastic cannot be re-softened or re-moulded. Thus, an article (or object) made of thermosetting plastic will retain its original shape permanently, even on heating. Thermosetting plastics are also known as thermosets. **Some of the examples of thermosetting plastics are : Bakelite and Melamine.** Thermosetting plastics are hard and rigid. Thermosetting plastics are not flexible. Due to this, thermosetting plastics cannot bend. When an article made of thermosetting plastic is forced to bend, it breaks.

If we take a discarded electric switch and put it in hot water for some time, we will find that the electric switch does not become soft. This is because an electric switch is made of a thermosetting plastic (called bakelite). This activity shows that thermosetting plastics do not become soft on heating. If we try to bend an electric switch by applying the force of our hands, we find that it does not bend at all. This shows that the articles made of thermosetting plastics are hard and rigid. The articles made of thermosetting plastics are not flexible. They do not bend at all. ***Thermosetting plastics are used for making those articles which***

may get too hot during use and are hard and rigid (so that they do not bend at all). Thermosetting plastics are used for making handles of cooking utensils (such as frying pans, pressure cookers, etc.), plates, cups,

Figure 8. Some of the articles (or objects) made of thermosetting plastics.

floor tiles, electrical fittings (such as electric switches, plugs and sockets), ballpoint pens and telephone instruments. Some of the articles made of thermosetting plastics are shown in Figure 8.

Thermosetting plastics are used for making the handles of cooking utensils (such as frying pan handles and pressure cooker handles) because (*i*) they do not soften on getting heated, and (*ii*) they are poor conductors of heat. Thermosetting plastics are used for making electrical fittings such as electric switches, plugs and sockets, etc., because (*i*) they do not become soft on getting heated, and (*ii*) they do not conduct electricity. Please note that **the handle and bristles of a toothbrush cannot be made of the same plastic** because the handle of a toothbrush has to be hard and rigid whereas the bristles of a toothbrush have to be soft and flexible. This means that the handle of a toothbrush should be made of a thermosetting plastic whereas its bristles should be made of a thermoplastic.

We can tell whether a given plastic is thermoplastic or thermosetting plastic from the way it behaves on heating. If a given plastic article softens on heating, then it will be a thermoplastic. On the other hand, if the given plastic article does not become soft on heating, then it will be a thermosetting plastic. Moreover, thermoplastics are flexible and can be bent whereas thermosetting plastics are very hard and rigid which cannot be bent at all. Please note that **the articles made of thermoplastics can be recycled whereas the articles made of thermosetting plastics cannot be recycled.**

Thermoplastics and Thermosetting Plastics Differ in Structure

We will now discuss why thermoplastics can be softened by heat but thermosetting plastics cannot be softened by heat. This is due to the difference in their structure. It can be explained as follows : Both, thermoplastics and thermosetting plastics are made up of long chain molecules called polymers. **In thermoplastics, the long polymer chains are not cross-linked with one another** [see Figure 9(*a*)]. Due to

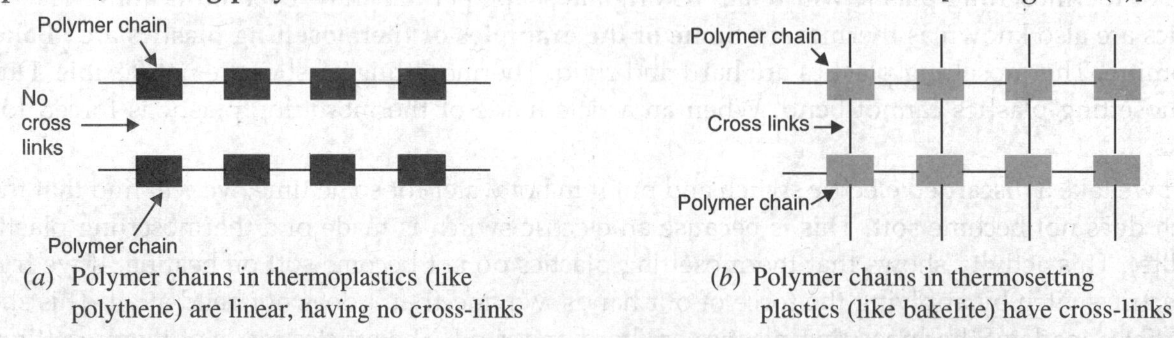

(*a*) Polymer chains in thermoplastics (like polythene) are linear, having no cross-links

(*b*) Polymer chains in thermosetting plastics (like bakelite) have cross-links

Figure 9. Arrangement of polymer chains in thermoplastics and thermosetting plastics.

this, on heating, the individual polymer chains can slide over one another and thermoplastic material becomes soft and ultimately melts. On the other hand, **in thermosetting plastics, the long polymer chains are cross linked with one another** [see Figure 9(*b*)]. These cross-links prevent the displacement (or sliding) of individual polymer chains on being heated. Due to this, thermosetting plastics do not become soft on heating (or change their shape on heating) once they have been set into a particular shape. For example, polythene is a thermoplastic having linear polymer chains with no cross-linkages, so it becomes soft on heating. On the other hand, bakelite is a thermosetting plastic having long polymer chains connected through cross-links (or held strongly through cross-links), due to which it does not become soft on heating.

USEFUL PROPERTIES OF PLASTICS

Plastics have many useful properties which make them materials of choice for all sorts of uses. Due to these special properties, plastics have many advantages over the traditional materials like metals, wood, etc., for making various articles. The important properties of plastics which make them very useful materials are given below.

(*i*) Plastic is Chemically Unreactive

We know that metals like iron get rusted (or corroded) when left exposed to air and water (moisture). This is because metals are chemically reactive. Plastics are chemically unreactive. Plastics do not react with air and water. Due to this, ***plastics are resistant to corrosion***. In other words, plastics are not affected by the weather. Plastics are also often unaffected by various chemicals (including acids and bases). Since plastics are unreactive and resist corrosion, the plastic containers are used to store various kinds of materials, including many chemicals.

(*ii*) Plastics are Poor Conductors of Heat and Electricity

Plastics do not conduct heat or electricity, so they can be used as 'insulators'. Plastics, being poor conductors of heat, are used where heat is to be kept away from reaching our hands. For example, the handles of cooking utensils (like frying pans and pressure cookers) are made of plastic so that we can hold the hot cooking utensil safely (without getting our hands burnt). Since plastics are poor conductors of electricity, they are used as electrical insulators. For example, electric wires have plastic covering as insulation so as to protect us from electric current passing through them. The handles of screw drivers used by electricians are also made of plastic because it is an electrical insulator. Electric switches, plugs and sockets also have plastic covers.

(*iii*) Plastics Can be Moulded into Different Shapes

Since plastics can be easily moulded, they are used to make a large variety of articles (or objects) having different shapes and sizes such as buckets, mugs, furniture (chairs, tables, etc.), bags, sheets, slippers, electrical fittings, toys, combs, toothbrushes, etc. The list is endless.

(*iv*) Plastics are Quite Cheap and Easily Made

Plastics are generally cheaper than metals. Plastics can also be made much more easily than metals. Due to these properties, plastics are now widely used for making many of the household and industrial articles which were earlier made from metals. For example, the buckets used in our homes were earlier made from iron metal sheet but these days most of the buckets are made of plastics.

(*v*) Plastic is Light, Strong and Durable

Plastics have low density, so they are lighter than metals. Plastics also have good strength and they are durable (long lasting). It is because of the lower price, easy availability, lightweight, good strength, durability and corrosion-resistance of plastic that the plastic containers are preferred for storing food, water, milk, jams, juices, pickles, squashes and soft drinks, etc. Being lighter than metals, plastics are also used in cars, aircrafts and spacecrafts.

BIODEGRADABLE AND NON-BIODEGRADABLE MATERIALS

A material which gets decomposed through natural processes (such as the action of bateria) is called biodegradable. Plant wastes (such as peels of vegetables and fruits, fallen leaves, left-over food stuffs, etc.), animal wastes, paper, cotton cloth, woollen cloth, jute and wood, are all biodegradable materials. Biodegradable materials rot away with time and hence do not cause pollution in the environment. Thus, ***biodegradable materials are environment friendly.***

A material which is not easily decomposed by natural processes (such as the action of bacteria) is called non-biodegradable. Plastics, glass, tin, aluminium cans, and other metal objects are non-biodegradable. Non-biodegradable materials do not rot away on their own and hence cause pollution in the environment. So, ***non-biodegradable materials are not environment friendly.***

From the above discussion we conclude that two types of waste materials are produced in our day to day life : biodegradable wastes and non-biodegradable wastes. The biodegradable wastes and non-biodegradable wastes in our homes should be collected separately and disposed of separately.

PLASTICS AND THE ENVIRONMENT

When we go to the market, we usually get things put in plastic bags (polythene bags), or wrapped in plastic sheets or packed in plastic cartons. After we reach home, the plastic bags (polythene bags), plastic sheets and plastic cartons are no longer needed and become a waste. This is just one way in which the plastic wastes keep on getting accumulated in our homes. Ultimately the plastic wastes are dumped alongwith the household garbage. **The use of plastics has a bad effect on the environment.** The use of plastic materials affects the environment because of the following reasons :

(*i*) Plastic is non-biodegradable. So, the articles made of plastics (such as plastic bags, bottles and cartons) do not rot when they are thrown away after use. The waste plastic articles keep on accumulating in the surroundings and pollute the environment (see Figure 10). Thus, plastics are not environment friendly.

Figure 10. Plastic is a non-biodegradable waste.

(*ii*) The waste plastic articles (like polythene bags, etc.) thrown here and there carelessly get into dirty water drains and sewers, and clog them (block them). This makes the dirty drain water (or sewer water) to flow over the streets and roads causing unhygienic conditions.

(*iii*) Sometimes the animals (like cows) eat up the used polythene bags or plastic wrappers alongwith the left-over food and vegetable wastes thrown on garbage dumps. The plastic wastes can choke the respiratory system of these animals or form a plastic lining in their stomach. This can cause the death of these animals.

(*iv*) When the plastic waste materials are burnt, they produce poisonous gases which pollute the air.

The Disposal of Plastic Wastes is a Major Problem. This is because of the following two big disadvantages of plastics :

(*i*) The articles made of plastics are non-biodegradable. They do not decompose (or rot) easily. This causes a great problem in the disposal of plastic wastes. So, plastic wastes cannot be disposed of easily.

(*ii*) The burning of plastic wastes gives out harmful gases which pollute the air. So, it is not advisable to dispose of the used plastic articles by burning.

How to Save the Environment From Excessive Plastic Wastes

Since the use of plastic articles is not good for the environment, we should take some steps to save the environment from the harmful effects of excessive use of plastics. Please note that plastics are very useful materials for us, so it is not possible to stop the use of plastic articles altogether. We can take steps only to minimise the use of plastics, wherever possible. **Some of the steps which can be taken to save the environment from plastic wastes are as follows :**

(*i*) We should try to reduce (or minimise) the use of plastics by using other materials in their place. For example, we should use bags made of cotton cloth or jute for shopping instead of polythene bags (plastic bags). Paper bags can also be used. Similarly, a stainless steel lunch box can be used instead of a plastic lunch box.

(*ii*) We should not throw polythene bags (plastic bags), wrappers of chips, biscuits and other eatables in water bodies, on the roads, in parks or picnic places. The used plastic materials should be put in the dustbins provided at various public places. This will keep our surroundings clean and also prevent the blockage of dirty water drains and sewers.

(*iii*) We should reuse the plastic containers which come with jams, pickles, oils and other packed food materials for storing salt, spices, tea-leaves, and sugar, etc., in the kitchen. We can also reuse the plastic carry bags for shopping purposes instead of throwing them as a waste.

(*iv*) Plastic wastes should be recycled. All the plastic wastes in the homes, shops and industry should be collected and sent for recycling to plastic making factories. In plastic factory, the waste plastic articles are melted and used to make new plastic articles. During recycling of used plastic articles, certain colours are added. This is to tell the buyers that it is a recycled plastic product, and to avoid its use for the storage of food. Most of the articles made of thermoplastics can be recycled. The articles made of thermosetting plastics cannot be recycled.

We should remember the 3R's to save the environment from the harmful effects of the excessive use of plastics. **The three R's stand for : Reduce, Reuse and Recycle**. This means that we should reduce the use of plastic articles by using articles made of other suitable materials ; we should reuse plastic articles wherever possible ; and we should recycle old and discarded plastic articles, if possible. We are now in a position to **answer the following questions :**

Very Short Answer Type Questions

1. Name the units of which cellulose polymer is made.
2. Name the man-made fibre prepared from natural materials.
3. Name the man-made fibre which is regarded as artificial silk.
4. Name the fibre obtained by the chemical treatment of wood pulp (or cellulose).
5. Name the first fully synthetic fibre.
6. Name the fibre used for making parachutes and rock climbing ropes.
7. Which synthetic fibre contains the organic group similar to those which give fruits their sweet smell ?
8. Which synthetic fibre feels like wool and used as a substitute for wool ?
9. To which kind of synthetic fibres does terylene belong ?
10. State one disadvantage of using synthetic fibres for making clothes.
11. Name the form of polyester which is replacing materials like glass and used for making bottles and jars.
12. Name four different plastics.
13. Give one use of teflon.
14. Which of the two is a thermosetting plastic : PVC or bakelite ?
15. Fill in the following blanks with suitable words :
    (*a*) Synthetic fibres are also called..........or..........fibres.
    (*b*) Synthetic fibres are made from raw materials called............
    (*c*) Like synthetic fibres, plastic is also a ....
    (*d*) The use of plastics can be reduced by using bags made of ....or ....instead of polythene bags.

Short Answer Type Questions

16. What is a polymer ? Name the natural polymer of which cotton is made.
17. State the characteristics of synthetic fibres.

18. What is nylon ? State the important properties of nylon.
19. Give the important uses of nylon.
20. What is polyester ? Name a popular polyester.
21. Arrange the following fibres in the order of increasing strength (keeping the fibre of least strength first) :
    Nylon, Cotton, Wool, Polyester, Silk
22. What is PET ? State the uses of PET.
23. What is acrylic ? State one important property of acrylic.
24. Write the uses of acrylic fibres.
25. Why should we not wear clothes made of synthetic fibres (like nylon or polyester) while working in the kitchen ?
26. What type of shirts should we buy for summer : cotton shirts or shirts made from synthetic materials (like polyester) ? Give reason for your answer.
27. Explain how, manufacturing of synthetic fibres is actually helping in the conservation of forests.
28. What are plastics ? Name any five commonly used articles made of plastics.
29. What are the various types of plastics ? Give two examples of each type of plastics.
30. Why are thermoplastics not used for making frying pan handles ?
31. Explain why, frying pan handles are made of thermosetting plastics.
32. Why are electric switches, plugs and sockets made of thermosetting plastics ?
33. Explain the difference between thermoplastics and thermosetting plastics.
34. Should the handle and bristles of a toothbrush be made of the same type of plastic material ? Explain your answer.
35. Explain why, plastic containers are preferred for storing food.
36. Choose the thermoplastics and thermosetting plastics from the following :
    Melamine, Polythene, Bakelite, Polyvinyl chloride
37. State two uses of polythene.
38. Write the full form of PVC. Is it thermoplastic or thermosetting plastic ?
39. Write two uses of bakelite.
40. State two uses of melamine.
41. Give two uses of PVC.
42. Write some of the uses of plastics in healthcare industry.
43. Classify the following as biodegradable and non-biodegradable materials :
    Woollen clothes, Polythene bags, Paper, Aluminium cans, Toothbrush, Peels of vegetables and fruits, Cotton cloth, Jute bag, Electric switch, Frying pan handle
44. State whether plastic is biodegradable or non-biodegradable ? Give reasons for your answer.
45. Explain how, the use of plastics has a bad effect on the environment.
46. Explain why, the disposal of plastic wastes is a major problem. Give two reasons only.
47. What are the various ways to save the environment from excessive plastic wastes ?
48. How do carelessly thrown plastic bags (polythene bags) affect :
    (*a*) dirty water drains and sewers ?
    (*b*) animals (such as cows) ?
49. What is meant by the 3R's principle in the context of use of plastics ?
50. State the various ways in which we can avoid (or minimise) the use of plastics.

Long Answer Type Questions

51. (*a*) What is rayon ? How is rayon made ?
    (*b*) Give any two uses of rayon.
52. (*a*) What are synthetic fibres ? Name any two synthetic fibres.
    (*b*) Why have synthetic fibres become more popular than natural fibres ?
53. (*a*) What are thermoplastics ? Give two examples of thermoplastics.
    (*b*) What are thermosetting plastics ? Give two examples of thermosetting plastics.
54. Explain why, thermoplastics become soft on heating but thermosetting plastics do not become soft on heating. Draw labelled diagrams to illustrate your answer.
55. What is meant by biodegradable and non-biodegradable materials ? Give examples of two biodegradable and two non-biodegradable materials.

Multiple Choice Questions (MCQs)

56. Rayon is different from truly synthetic fibres because :
(*a*) it has a silk-like appearance.
(*b*) it is obtained from wood pulp.
(*c*) its fibres can be woven like those of natural fibres.
(*d*) it can be dyed in wide variety of colours.

57. The synthetic material which can be used for making fabrics as well as shatterproof bottles and jars is :
(*a*) nylon (*b*) rayon (*c*) polyester (*d*) acrylic

58. Which of the following has cross-linked polymer chains ?
(*a*) bakelite (*b*) polyester (*c*) PVC (*d*) nylon

59. The man-made fibre made from the cellulose polymer is :
(*a*) nylon (*b*) acrylic (*c*) rayon (*d*) polyester

60. Which of the following is not a thermoplastic polymer ?
(*a*) polyester (*b*) melamine (*c*) nylon (*d*) polyvinyl chloride

61. The synthetic polymer which can be used as a substitute for wool for making sweaters and shawls, etc., is :
(*a*) nylon (*b*) polyester (*c*) terylene (*d*) acrylic

62. Which of the following is not a synthetic fibre ?
(*a*) nylon (*b*) flax (*c*) acrylic (*d*) polyester

63. The synthetic fibre which contains the organic groups similar to those which give fruits their 'sweet smell' is :
(*a*) nylon (*b*) acrylic (*c*) terylene (*d*) rayon

64. The man-made fibre rayon is chemically identical to :
(*a*) wool (*b*) silk (*c*) jute (*d*) cotton

65. One of the following man-made fibre is not prepared from raw materials obtained from petrochemicals. This one is :
(*a*) polyester (*b*) nylon (*c*) rayon (*d*) acrylic

66. Which of the following plastics do not have cross-links between their polymer chains ?
A. Nylon B. Melamine C. Terylene D. Bakelite
(*a*) A and B (*b*) B and C (*c*) A and C (*d*) C and D

67. The clothes of a person working in the kitchen catch fire accidently causing severe burns. The person is most likely wearing clothes made of :
(*a*) flax (*b*) rayon (*c*) terylene (*d*) cotton

68. The plastic which is coated on the uniforms of firemen to make them fire-resistant is :
(*a*) bakelite (*b*) polythene (*c*) teflon (*d*) melamine

69. Which of the following is a man-made fibre prepared from wood-pulp ?
(*a*) flax (*b*) nylon (*c*) acrylic (*d*) rayon

70. The manufacture of one of the following artificial fibres contributes to deforestation. This fibre is :
(*a*) nylon (*b*) rayon (*c*) terylene (*d*) acrylic

71. The non-stick coating on frying pans is that of a plastic called :
(*a*) polyvinyl chloride (*b*) melamine (*c*) bakelite (*d*) teflon

72. Which of the following plastics is used for making electric switches ?
(*a*) teflon (*b*) melamine (*c*) PET (*d*) bakelite

73. Which of the following are thermosetting polymers ?
A. Melamine B. Terylene C. Polythene D. Bakelite
(*a*) A and B (*b*) B and C (*c*) A and D (*d*) B and D

74. The similarity between artificial silk and cotton is that :
(*a*) both are non-biodegradable (*b*) both melt on heating
(*c*) both are amide polymers (*d*) both are cellulose polymers

75. Which of the following plastic objects can be recycled ?
A. Electric socket B. Polythene bag C. PVC pipe D. Ashtray
(*a*) A and B (*b*) B and C (*c*) A and D (*d*) C and D

Questions Based on High Order Thinking Skills (HOTS)

76. Match the terms of column *A* correctly with the phrases given in column *B* :

A	B
(*i*) Polyester	(*a*) Prepared by using wood pulp
(*ii*) Teflon	(*b*) Used for making parachutes
(*iii*) Rayon	(*c*) Used to make non-stick cookware
(*iv*) Nylon	(*d*) Fabrics do not wrinkle easily

77. Which plastic is used :
(*a*) for making uniforms of fire-men fire resistant ?
(*b*) for giving non-stick.coating on frying pans ?
(*c*) for making handles of frying pans ?
(*d*) for making insulation (covering) of electric wires ?
(*e*) for making electric switches ?
(*f*) for making flexible water bottles ?

78. Which of the following articles made of plastics 'can be recycled' and which 'cannot be recycled' ? Give reasons for your choice.
Telephone instruments, Plastic toys, Cooker handles, Plastic covering on electrical wires, Electric switches, Ballpoint pens, Carry bags, Plastic bottles, Plastic chairs

79. Out of the following materials :
Cotton, Nylon, Terylene, Wool, PET, Acrylic
(*a*) Which materials are polyesters ?
(*b*) Which material is a polyamide ?
(*c*) Which material is used as a substitute for wool ?
(*d*) Which material is used as a substitue for glass ?

80. The synthetic fibre A is chemically a polyamide whereas the synthetic fibre B contains a large number of ester groups. Another synthetic fibre C is made of a polymer D which consists of a number of glucose units joined one after the other.
(*a*) Which fibre could be (*i*) terylene (*ii*) rayon, and (*iii*) nylon ?
(*b*) Name the polymer D.
(*c*) Which fibre (A, B or C) is prepared from a natural raw material ?
(*d*) Which fibre (A, B or C) contains the same type of groups as those in a PET jar ?

ANSWERS

7. Polyester **9.** Polyester **11.** PET (Poly-Ethylene Terephthalate) **14.** Bakelite **15.** (*a*) man-made ; artificial (*b*) petrochemicals (*c*) polymer (*d*) cotton; jute **56.** (*b*) **57.** (*c*) **58.** (*a*) **59.** (*c*) **60.** (*b*) **61.** (*d*) **62.** (*b*) **63.** (*c*) **64.** (*d*) **65.** (*c*) **66.** (*c*) **67.** (*c*) **68.** (*d*) **69.** (*d*) **70.** (*b*) **71.** (*d*) **72.** (*d*) **73.** (*c*) **74.** (*d*) **75.** (*b*) **76.** (*i*) *d* (*ii*) *c* (*iii*) *a* (*iv*) *b* **77.** (*a*) Melamine (*b*) Teflon (*c*) Bakelite (*d*) PVC (Poly-Vinyl Chloride) (*e*) Bakelite (*f*) Polythene **78.** Can be recycled : Plastic toys, Plastic covering on electrical wires, Carry bags, Plastic bottles, Plastic chairs — *All these are made of thermoplastics* ; Cannot be recycled : Telephone instruments, Cooker handles, Electric switches, Ballpoint pens — *All these are made of thermosetting plastics* **79.** (*a*) Terylene ; PET (*b*) Nylon (*c*) Acrylic (*d*) PET **80.** (*a*) (*i*) B (*ii*) C (*iii*) A (*b*) Cellulose (*c*) C (Rayon) (*d*) B (Terylene)

CHAPTER

4

Materials : Metals and Non-Metals

A **substance which cannot be broken down into two (or more) simpler substances by chemical reactions (by applying heat, light or electricity) is called an element**. For example, iron is an element because it cannot be broken down into two or more simpler substances by the usual methods of carrying out chemical reactions such as by applying heat, light or electricity. Please note that the elements themselves are the simplest substances. That is why they cannot be broken down (or split up) into any more simpler substances. Some of the common elements are : Hydrogen, Helium, Carbon, Nitrogen, Oxygen, Sulphur, Phosphorus, Silicon, Chlorine, Bromine, Iodine, Sodium, Potassium, Magnesium, Calcium, Aluminium, Iron, Zinc, Copper, Silver, Gold and Mercury. Every element is represented by a 'symbol'. All the elements have separate symbols. No two elements can have the same symbol. A symbol is the short way to write an element. For example, the symbol of Hydrogen is H whereas the symbol of Magnesium is Mg. The symbol of an element also represents 'one atom' of that element. We will discuss the symbols of elements in detail in higher classes.

The smallest particle of an element is called 'atom'. A sample of an element contains only one kind of atoms. This gives us another definition of element which can be written as : **An element is a substance which is made up of only one kind of atoms**. There are as many type of atoms as are elements. So, different elements are made up of different kinds of atoms. For example, sulphur element is made up of only sulphur atoms whereas iron element is made up of only iron atoms. The atoms of an element remain unaffected by the physical changes in the element. For example, an atom of liquid sulphur (molten sulphur) or vapour forms of sulphur would be exactly the same as that of solid sulphur. Although there is an enormous variety of substances in the universe but the number of elements forming these substances is limited. **There are only 92 naturally occurring elements known to us at present**. An important classification of elements is in terms of metals and non-metals. This is discussed on the next page.

METALS AND NON-METALS

On the basis of their properties, all the elements can be divided into two main groups : metals and non-metals. Iron, copper and aluminium are examples of metals whereas carbon, oxygen and sulphur are examples of non-metals. All the metals have similar properties. All the non-metals have also similar properties. But **the properties of non-metals are opposite to those of metals**. Both, metals as well as non-metals are used in our daily life. We also use a large number of compounds of metals and non-metals.

Before we go further and give the characteristics (identifying properties) of metals and non-metals, we should know the meaning of some new terms such as malleable, ductile, brittle and lustrous. **Malleable** means which can be beaten with a hammer to form thin sheets (without breaking). **Ductile** means which can be stretched (or drawn) to form thin wires. **Brittle** means which breaks into pieces on hammering or stretching. **Lustrous** means shiny (*chamakdar*). Keeping these points in mind, we will now give the characteristics of metals and non-metals.

Characteristics of Metals

The important characteristics of metals are as follows. **Metals are malleable and ductile elements which are good conductors of heat and electricity**. Metals are lustrous or shiny. Metals are usually hard and strong. They cannot be cut easily. All the metals are solids except mercury which is a liquid metal. Metals have high melting points and boiling points. Metals have high densities which means they are heavy. Metals are sonorous which means that metals make a ringing sound when we strike them with a hard object.

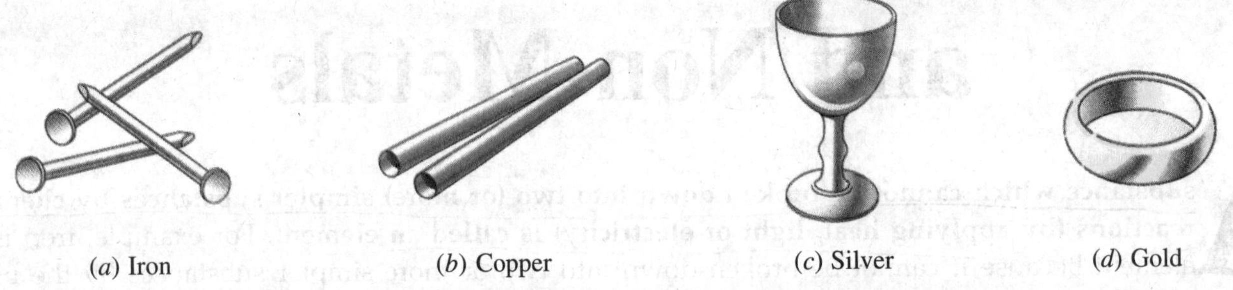

(*a*) Iron (*b*) Copper (*c*) Silver (*d*) Gold

Figure 1. Iron, copper, silver and gold are metals.

Some of the examples of metals are : Iron, Copper, Aluminium, Zinc, Silver, Gold, Platinum, Chromium, Sodium, Potassium, Magnesium, Nickel, Cobalt, Tin, Calcium, Lead, Cadmium, Mercury, Antimony, Tungsten, Manganese and Uranium. Out of 92 naturally occurring elements, 70 elements are metals. All these metals are solids except one metal mercury which is a liquid.

Characteristics of Non-Metals

The important characteristics of non-metals are as follows. **Non-metals are the elements which are neither malleable nor ductile, they are brittle. Non-metals do not conduct heat and electricity.** Non-metals are not lustrous or shiny, they are dull in appearance. Solid non-metals are usually neither hard nor strong. They can be cut easily. Non-metals can be solid, liquid or gases at the room temperature. Non-metals have usually low melting points and boiling points. Non-metals have low densities which means they are light. Non-metals are not sonorous, which means non-metals do not make a ringing sound when we strike them with a hard object.

Some of the examples of non-metals are : Carbon, Sulphur, Phosphorus, Hydrogen, Oxygen, Nitrogen, Fluorine, Chlorine, Bromine, Iodine, Helium, Neon, Argon, Krypton and Xenon. Of the 92 naturally occurring elements, 22 elements are non-metals. Out of these, 10 non-metals are solids, 1 non-metal (bromine) is a liquid whereas the remaining 11 non-metals are gases.

(*a*) Carbon

(*b*) Sulphur

Figure 2. Carbon and sulphur are non-metals.

Metalloids

There are some elements which show some properties of metals and the other properties of non-metals. **The elements whose properties are intermediate between those of metals and non-metals are known as metalloids**. For example, metals are good conductors of electricity whereas non-metals do not conduct electricity at all but metalloids conduct electricity to a small extent. Thus, metalloids are semiconductors. The examples of metalloids are : **Silicon, Germanium, Arsenic and Tellurium**. We will learn more about metalloids in higher classes. We will now discuss the properties of metals and non-metals in detail.

PHYSICAL PROPERTIES OF METALS AND NON-METALS

Metals and non-metals show different physical properties. The important physical properties of metals and non-metals are given below.

1. Malleability

(*i*) Metals are Malleable. This means that metals can be beaten into thin sheets with a hammer.

If we take a piece of aluminium metal, place it on a block of iron and beat it with a hammer, we will find that the piece of aluminium metal turns into a thin aluminium sheet, without breaking. And we say that aluminium metal is malleable or it shows malleability. **The property which allows the metals to be hammered into thin sheets is called malleability**. Malleability is an important characteristic property of metals. Most of the metals are malleable. Gold and silver are the best malleable metals and can be hammered into very fine sheets or foils. Aluminium and copper are also highly malleable metals. For example, aluminium metal can be hammered to form thin aluminium foils. Copper metal can also be hammered to form copper sheets. Iron is also a quite malleable metal which can be hammered to form iron sheets. **It is due to the property of malleability that metals can be bent to form objects of different shapes by beating with a hammer**. For example, it is because of the property of malleability of iron that an ironsmith can change the shape of a block of iron metal by hammering to make different iron objects such as an axe, a spade or a shovel, etc.

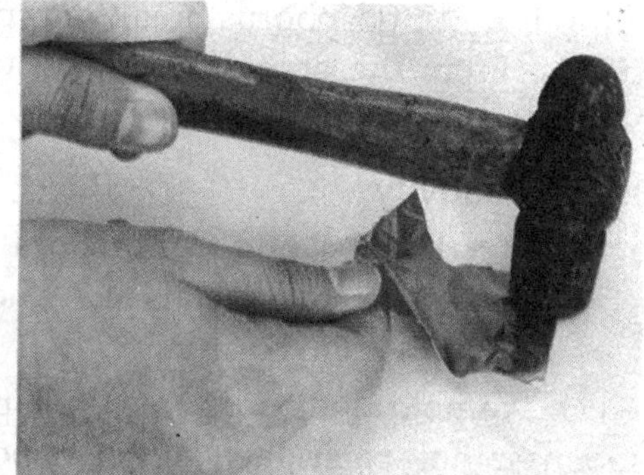

Figure 3. Metals can be hammered into sheets. They are malleable.

(*ii*) Non-Metals are Not Malleable. Non-Metals are Brittle. This means that non-metals cannot be beaten into thin sheets with a hammer. Non-metals break into small pieces when hammered.

Carbon is a non-metal. Carbon is found in many forms such as charcoal, coke, graphite and diamond, etc. Coal is also mainly carbon. The pencil lead is a form of carbon called graphite (see Figure 4). If we take a piece of carbon (say, a pencil lead or charcoal) and beat it with a hammer, it will break into pieces. We cannot hammer carbon (without breaking) to obtain thin sheets of carbon. Thus, carbon is a non-metal which is not malleable. Carbon is brittle. Sulphur is also a non-metal. If we hammer a piece of sulphur, it will break into smaller pieces. We cannot hammer sulphur to obtain thin sheets of sulphur. Thus, sulphur is a non-metal which is not malleable, it is brittle. From this discussion we conclude that we cannot obtain thin sheets by beating non-metals. When beaten with a hammer, solid non-metals break into pieces. **The property due to which non-metals break on hammering is called brittleness.** Brittleness is a characteristic property of solid non-metals.

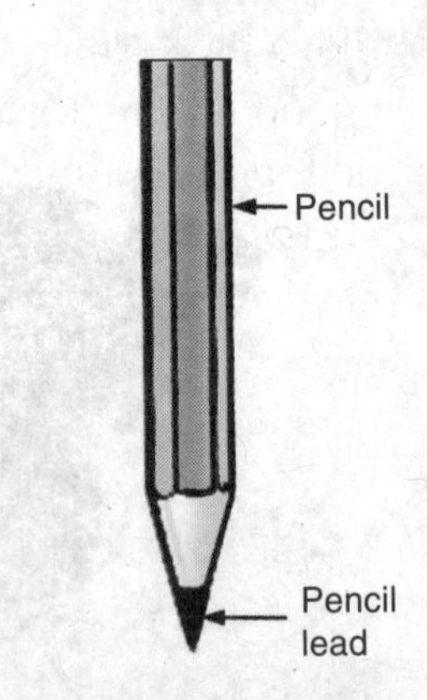

Figure 4. Pencil lead is made of graphite (which is a form of non-metal carbon). Pencil lead breaks easily, it is brittle.

2. Ductility

(*i*) **Metals are Ductile.** This means that metals can be drawn (or stretched) into thin wires.

Most of the metals are ductile. Gold and silver are among the best ductile metals. Copper and aluminium metals are also very ductile and can be drawn into thin copper wires and aluminium wires (which are used as electric wires). Iron, magnesium, and tungsten metals are also quite ductile and can be drawn into wires. Iron wires are used for making wire gauzes. Magnesium wires are used in science experiments in the laboratory. And thin wires of tungsten metal are used for making the filaments of electric bulbs. **The property which allows the metals to be drawn into wires is called ductility**. Ductility is another characteristic property of metals. From the above discussion we conclude that : **Generally, metals are malleable and ductile.**

Figure 5. Metals can be drawn into wires. They are ductile.

(*ii*) **Non-Metals are Not Ductile.** This means that non-metals cannot be drawn into wires. They are easily snapped on stretching.

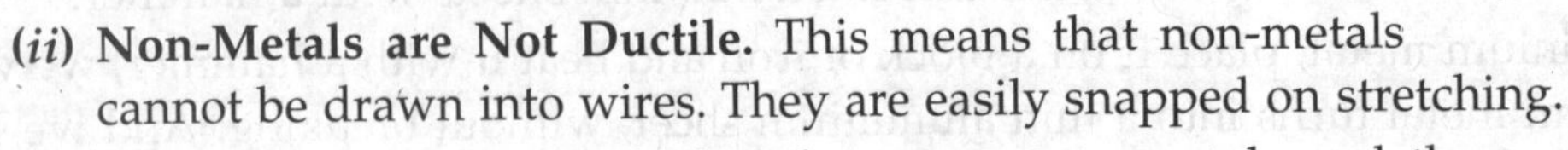

For example, sulphur and phosphorus are non-metals and they are not ductile. When stretched, sulphur and phosphorus break into pieces and do not form wires. Thus, we cannot get wires from non-metals. From the above discussion we conclude that : **Non-metals are neither malleable nor ductile. Non-metals are brittle.**

3. Conductivity

(*i*) **Metals are Good Conductors of Heat and Electricity.** This means that metals allow heat and electricity to pass through them easily.

If we hold one end of a metal spoon (like an aluminium spoon) in hot water, then its other end becomes hot very soon. This is because the metal of spoon conducts heat (or carries heat) from one end to the other end quickly. And we say that the metal spoon is a good conductor of heat. **Copper, silver, gold, aluminium and iron metals are good conductors of heat.** Though all the metals are good conductors of heat, **silver metal is the best conductor of heat.** Copper metal is a better conductor of heat than aluminium metal. The cooking utensils

Figure 6. Metals conduct heat well. That is why this frying pan is made of metal.

(like a frying pan, etc.) are made of metals because metals are good conductors of heat (see Figure 6). Being a good conductor of heat, the metallic bottom of cooking utensil transfers the heat of gas stove quickly to the food kept inside it. We cannot hold a hot metal pan directly because it will conduct the heat quickly to our hand causing burns. We have to hold a hot metal pan from its handle made of plastic or wood (because plastic and wood do not conduct heat).

ACTIVITY

Let us now show the conduction of electricity by metals. We take a cell, a torch bulb fitted in a holder and some connecting wires (copper wires) with crocodile clips, and connect them to make an electric circuit as shown in Figure 7. Let us insert a piece of aluminium foil between the ends of crocodile clips *A* and *B*. We will see that the torch bulb lights up at once (see Figure 7). This means that aluminium foil allows electric current to pass through it. In other words, aluminium metal is a good conductor of electricity. Let us now remove the aluminium foil and insert an iron nail between the two ends of crocodile clips *A* and *B*. The bulb will light up again showing that iron metal is also a good conductor of electricity. The connecting wires used in making the circuit shown in Figure 7 are made of copper metal. Since the copper connecting wires allow electric current to pass through them, therefore, copper metal is also a good conductor of electricity. From this activity we conclude that metals are good conductors of electricity. Copper wires are used in household electric wiring because copper metal is a very good conductor of electricity. Copper metal is a better conductor of electricity than aluminium. Silver metal is the best conductor of electricity.

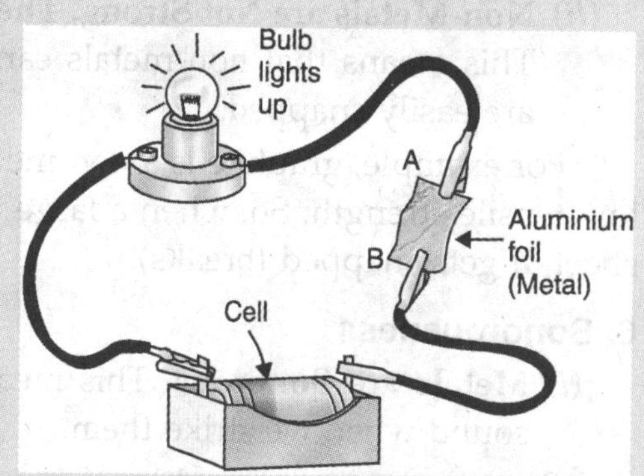

Figure 7. Metals conduct electricity. Here an aluminium foil is conducting electricity.

(*ii*) **Non-Metals are Poor Conductors of Heat and Electricity.** This means that non-metals do not allow heat and electricity to pass through them.

For example, sulphur is a non-metal which does not conduct heat or electricity. Similarly, a piece of coal (which is mainly carbon non-metal) also does not conduct heat or electricity. Many of the non-metals are, in fact, very good insulators. There are, however, some exceptions. A form of the carbon element, **diamond is a non-metal which is a good conductor of heat.** And another form of carbon element, **graphite is a non-metal which is a good conductor of electricity.** Being a good conductor of electricity, graphite is used for making electrodes (as that in dry cells).

4. Lustre

(*i*) **Metals are Lustrous (or Shiny).** This means that metals have a shiny appearance.

If we observe the freshly cut surfaces of metals, we will find that they have a shiny appearance. This is called metallic lustre (or *chamak*). The shiny appearance of metals makes them useful in making jewellery and decoration pieces. For example, gold and silver are used for making jewellery because they are bright and shiny.

(*ii*) **Non-Metals are Not Lustrous. They are Dull.** This means that non-metals are not shiny, they are dull in appearance.

For example, sulphur and phosphorus are solid non-metals which do not have lustre (or shine). They are dull in appearance. There is, however, one exception. **Iodine is a non-metal having lustre** (or *chamak*). Iodine has a shining surface like that of metals.

Figure 8. Metals are lustrous (or shiny). These bangles are made of gold because it is a highly lustrous (or shiny) metal.

5. Strength

(*i*) Metals are Usually Strong. They have High Tensile Strength. This means that metals can hold large weights without snapping (without breaking).

For example, iron metal (in the form of steel) is very strong having a high tensile strength. Due to this iron metal is used in the construction of bridges, buildings, railway lines, girders, machines, vehicles and chains, etc. Though most of the metals are strong but some of the metals are not strong. For example, sodium and potassium metals are not strong. They have low tensile strength.

Figure 9. Iron metal (as steel) is very strong. It is used to build bridges.

(*ii*) Non-Metals are Not Strong. They have Low Tensile Strength. This means that non-metals cannot hold large weights. They are easily snapped.

For example, graphite is a non-metal which is not strong. It has a low tensile strength. So, when a large weight is placed on a graphite sheet, it gets snapped (breaks).

6. Sonorousness

(*i*) Metals are Sonorous. This means that metals make a ringing sound when we strike them.

Sonorous means capable of producing a ringing sound. If we drop a metal coin or a metal utensil on the floor of our house, we hear a ringing sound. And when the clapper (or hammer) of an electric bell strikes the metal gong, even then a ringing sound is produced. So, the metal objects make a ringing sound when we drop them on a hard floor or strike them with a hard object. We say that the metals are sonorous (or capable of producing a ringing sound). Suppose we have two boxes, one box made of metal and the other box made of wood, which are similar in appearance. We can tell which box is made of metal by striking them with a small hammer. The box which produces a ringing sound on being struck by the hammer will be the one made of metal. Metal sheets are used for making bells (like the bicycle bells and temple bells). The use of metals for making bells is based on their property of being sonorous.

Figure 10. Metals are sonorous so they are used to make bells.

(*ii*) Non-Metals are Not Sonorous. This means that solid non-metals do not make a ringing sound when we strike them.

If we drop a piece of carbon (say, a piece of charcoal) or a lump of sulphur on the floor or strike them with a hammer, we do not hear any ringing sound. This means that carbon and sulphur non-metals are not sonorous. They are not capable of producing a ringing sound when struck.

7. Hardness

(*i*) Metals are Generally Hard. This means that most of the metals cannot be cut easily.

Though most of the metals are hard but their hardness varies from one metal to another. For example, if we try to cut a thin sheet of iron metal with a pair of scissors, we will find that it is very, very difficult to cut the sheet of iron. This is because iron

Figure 11. Sodium metal is so soft that it can be cut with a knife.

metal is very hard. On the other hand, a thin sheet of aluminium metal can be cut easily by using scissors. This means that aluminium metal is less hard. There are, however, some exceptions to this property of hardness of metals. **Sodium and potassium metals are soft and can be easily cut with a knife.** For example, if we try to cut a lump of sodium metal with a dry knife, we will find that it can be easily cut into small pieces (just like wax). This shows that sodium metal is soft. Magnesium metal can also be cut easily.

(*ii*) Most of the Solid Non-Metals are Quite Soft. This means that most of the solid non-metals can be cut easily.

For example, sulphur and phosphorus are soft non-metals which can be easily cut into pieces with a knife. Only one non-metal, diamond, is very hard. In fact, diamond is the hardest natural substance known.

Comparison Between the Physical Properties of Metals and Non-Metals

We will now compare the physical properties of metals and non-metals in tabular form. These physical properties can be used to distinguish between metals and non-metals.

Differences in Physical Properties of Metals and Non-Metals

Metals	*Non-Metals*
1. Metals are malleable and ductile.	1. Non-metals are neither malleable nor ductile. They are brittle.
2. Metals are good conductors of heat and electricity.	2. Non-metals are poor conductors of heat and electricity (*except graphite which is a good conductor of electricity*).
3. Metals are lustrous (or shiny).	3. Non-metals are not lustrous (or shiny). They are dull.
4. Metals are strong. They have high tensile strength (*except sodium and potassium which are not strong and have low tensile strength*).	4. Non-metals are not strong. They have a low tensile strength.
5. Metals are sonorous. They make a ringing sound when struck.	5. Non-metals are not sonorous. They do not make a ringing sound when struck.
6. Metals are generally hard (*except sodium and potassium which are soft metals*).	6. Solid non-metals are quite soft (*except diamond which is extremely hard*).

An element can be identified as being a metal or a non-metal by comparing its properties with the general properties of metals and non-metals. While doing so we should, however, keep the various exceptions to the general properties of metals and non-metals in mind. We will now answer one question based on metals and non-metals.

Sample Problem. State two reasons for believing that copper is a metal and sulphur is a non-metal.

Answer. The two properties which tell us that copper is a metal and sulphur is a non-metal are given below.

Copper	*Sulphur*
1. Copper is malleable and ductile. It can be hammered into thin sheets and drawn into wires.	1. Sulphur is neither malleable nor ductile. It is brittle. Sulphur breaks into pieces when hammered or stretched.
2. Copper is a good conductor of heat and electricity.	2. Sulphur is a poor conductor of heat and electricity.

CHEMICAL PROPERTIES OF METALS AND NON-METALS

Metals and non-metals show different chemical properties. Some of the important chemical properties of metals and non-metals are given below.

1. Reaction with Oxygen

(*i*) **Metals react with oxygen to form metal oxides. Metal oxides are basic in nature.**

Metal	+	Oxygen	⟶	Metal oxide
		(From air)		(*Basic oxide*)

Thus, on burning, metals react with oxygen to form basic oxides. **The basic metal oxides turn red litmus to blue**. This property will become clear from the following activity.

ACTIVITY

Magnesium is a metal. We take a magnesium wire, hold it with a pair of tongs and heat it over a flame. Magnesium wire burns vigorously producing a bright white light to form an ash (which is magnesium oxide). We put this magnesium oxide ash in a boiling tube, add a little water in the boiling tube and shake it. We will find that magnesium oxide dissolves in water partially. We now add some red litmus solution to the boiling tube and observe the change in colour. We will see that red litmus solution turns blue. This shows that magnesium oxide is basic in nature (because only basic substances turn red litmus to blue). From this activity we conclude that **magnesium is a metal which forms a basic oxide (magnesium oxide) on burning in air.** This basic magnesium oxide turns red litmus blue.

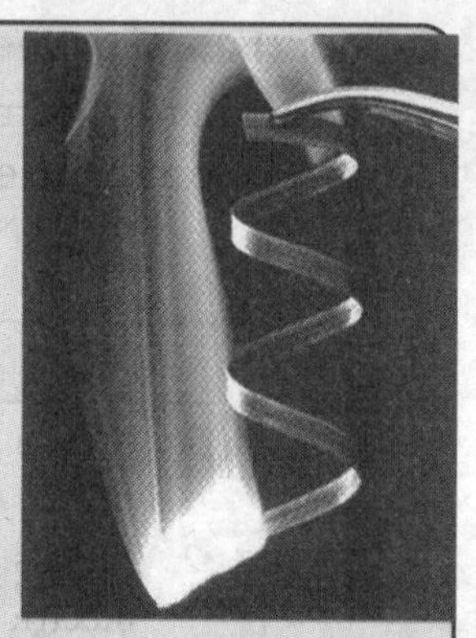

Figure 12. Magnesium burns brightly in air to form magnesium oxide.

The chemical reactions involved in the above activity are given below :

(*a*) When magnesium burns in air, it combines with the oxygen of air to form magnesium oxide (which is a basic oxide) :

Magnesium	+	Oxygen	⟶	Magnesium oxide
(Mg)		(O_2)		(MgO)
(A metal)				(A basic oxide)

(*b*) Magnesium oxide dissolves partially in water to form magnesium hydroxide solution :

Magnesium oxide	+	Water	⟶	Magnesium hydroxide
(MgO)		(H_2O)		[$Mg(OH)_2$]
(A basic oxide)				(A base)

Magnesium hydroxide turns red litmus to blue showing that it is a base and that magnesium oxide is a basic oxide.

Similarly, sodium is a metal which forms a basic oxide, sodium oxide (Na_2O). A solution of sodium oxide in water turns red litmus blue. Please note that when we want to react a metal with oxygen, we usually burn the metal in air. It is the oxygen present in air which combines with the metal on burning to form a metal oxide.

We will now discuss the reaction of iron metal with oxygen of air which takes place in nature. We have studied the rusting of iron in Class VII. During the rusting of iron, iron metal combines slowly with the oxygen of air in the presence of water (moisture) to form a compound called 'iron oxide'. This iron oxide is rust. The reaction of iron metal with oxygen in the presence of water can be written as follows :

Iron	+	Oxygen	+	Water	⟶	Iron oxide
(Fe)		(O_2)		(H_2O)		(Fe_2O_3)
		From damp air				Rust
						(A basic oxide)

ACTIVITY

We will now describe an activity to show that iron oxide (or rust) is basic in nature. Take a spoonful of rust (from any rusted iron object) in a test-tube, add a little of water and shake it well. In this way, we will get a suspension of iron oxide in water. Test this suspension of iron oxide (or rust) with blue litmus paper and red litmus paper, one by one. We will find that the red litmus paper turns blue. This shows that iron oxide suspension is basic in nature (because only basic substances turn red litmus paper to blue). From this activity we conclude that iron metal forms a basic oxide (iron oxide) on reaction with oxygen of air. In other words, rust is basic in nature.

We will now describe the reaction of copper metal with moist air which takes place in nature. When a copper object is exposed to moist air for a long time, then copper reacts with water, carbon dioxide and oxygen present in moist air to form a green coating on the copper object. The green coating (or green material) is a mixture of copper hydroxide [$Cu(OH)_2$] and copper carbonate ($CuCO_3$) which is formed by the action of moist air on copper object. This reaction can be written as :

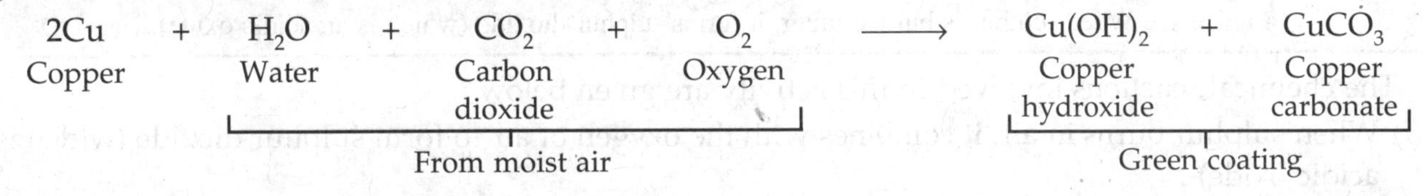

Thus, when a copper vessel is exposed to moist air for a long time, it acquires a green coating on its surface. The mixture of copper hydroxide and copper carbonate which forms the green coating is commonly known as 'basic copper carbonate' (because it is basic in nature). If we make a suspension of a little of green coating (from a copper vessel) in water and test it with litmus papers, we will find that it turns red litmus paper to blue. This shows that the green coating formed on a copper vessel (or any other copper object) is basic in nature. **The formation of green coating of basic copper carbonate on the surface of copper objects on exposure to moist air is called corrosion of copper.** While iron rusts, other metals corrode.

(*ii*) Non-metals react with oxygen to form non-metal oxides. Non-metal oxides are acidic in nature.

$$\underset{}{\text{Non-metal}} + \underset{\text{(From air)}}{\text{Oxygen}} \longrightarrow \underset{(\textit{Acidic oxide})}{\text{Non-metal oxide}}$$

Thus, non-metals react with oxygen to form acidic oxides. **The acidic non-metal oxides turn blue litmus to red.** This property will become more clear from the following activity.

ACTIVITY

Sulphur is a non-metal. We take a small amount of sulphur powder in a deflagrating spoon and heat it over a flame [Figure 13(*a*)]. As soon as sulphur starts burning with a blue flame, we introduce the deflagrating spoon in a gas jar and allow the sulphur to burn inside the gas jar [Figure 13(*b*)]. Cover the gas jar with a lid to prevent the gas being formed from escaping. Sulphur burns in the air of gas jar to form a pungent smelling gas (sulphur dioxide) [Figure 13(*c*)]. Remove the deflagrating spoon from the gas jar. We now put some water in the gas jar, cover it with a lid and shake it to dissolve sulphur dioxide gas. Add some blue litmus solution to the gas jar. We will see that the blue litmus solution turns red. This shows that sulphur dioxide gas is acidic in nature (because only acidic substances turn blue litmus to red). From this activity we conclude that **sulphur is a non-metal which forms an acidic oxide (sulphur dioxide) on burning in air.**

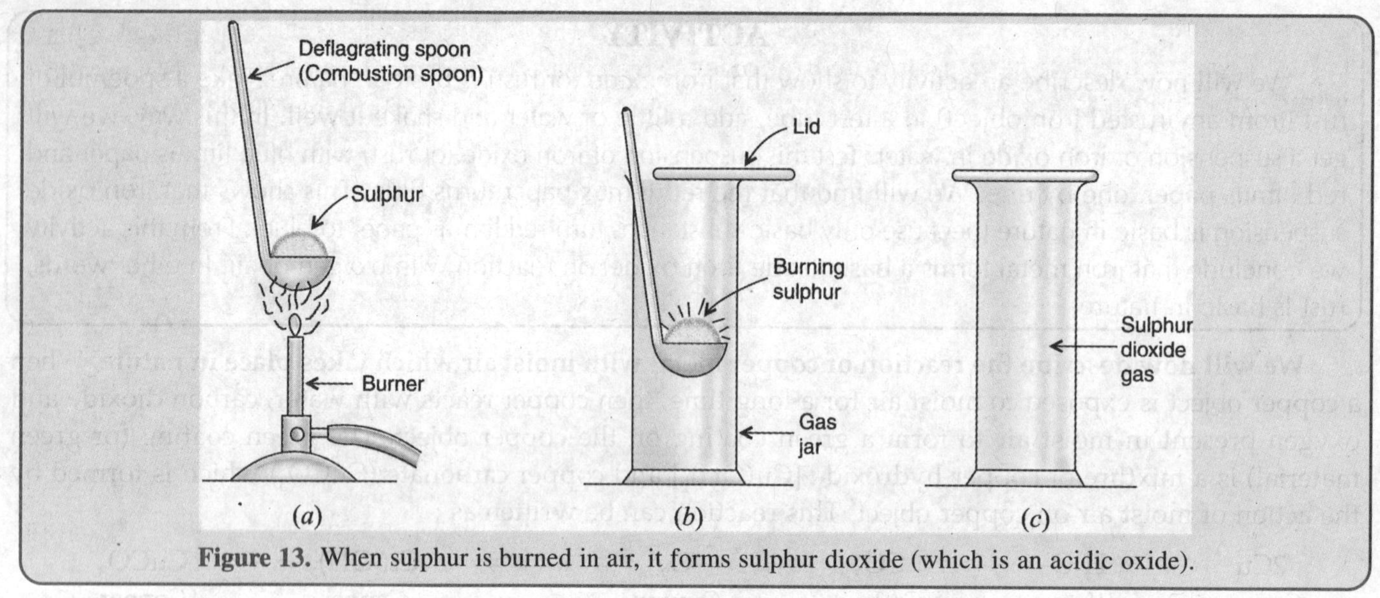

Figure 13. When sulphur is burned in air, it forms sulphur dioxide (which is an acidic oxide).

The chemical reactions involved in this activity are given below :

(*i*) When sulphur burns in air, it combines with the oxygen of air to form sulphur dioxide (which is an acidic oxide) :

Sulphur	+	Oxygen	⟶	Sulphur dioxide
(S)		(O_2)		(SO_2)
(A non-metal)				(An acidic oxide)

(*ii*) Sulphur dioxide dissolves in water to form sulphurous acid solution :

Sulphur dioxide	+	Water	⟶	Sulphurous acid
(SO_2)		(H_2O)		(H_2SO_3)
(An acidic oxide)				(An acid)

This sulphurous acid turns blue litmus to red showing that it is an acid and that sulphur dioxide is acidic in nature. Similarly, carbon is a non-metal which forms an acidic oxide, carbon dioxide (CO_2). A solution of carbon dioxide in water turns blue litmus red. From the above discussions we conclude that :

(*i*) Metals form metal oxides on burning in air. **Metal oxides are basic in nature and turn red litmus to blue.**

(*ii*) Non-metals form non-metal oxides on burning in air. **Non-metal oxides are acidic in nature and turn blue litmus to red.**

This property of the nature of oxides can be used to identify whether a given element is a metal or a non-metal. This is because :

(*a*) If an element forms a basic oxide (which turns red litmus blue), **then the element will be a metal.**

(*b*) If an element forms an acidic oxide (which turns blue litmus red), **then the element will be a non-metal.**

We have studied acids and bases in Class VII. We can now understand that **metal oxides are basic in nature and form bases on dissolving in water**. On the other hand, **non-metal oxides are acidic and form acids on dissolving in water**. Some of the bases and acids, and the metals and non-metals present in them (from whose oxides they are formed) are given below :

Metals in Bases

Name of base	*Name of metal*
1. Magnesium hydroxide	Magnesium
2. Calcium hydroxide	Calcium
3. Sodium hydroxide	Sodium
4. Potassium hydroxide	Potassium

Non-Metals in Acids

Name of acid	*Name of non-metal*
1. Sulphurous acid	Sulphur
2. Sulphuric acid	Sulphur
3. Nitric acid	Nitrogen
4. Carbonic acid	Carbon

Please note that most of the non-metals form acidic oxides but there are some exceptions. This is because some of the non-metals form neutral oxides (which are neither acidic nor basic). For example, hydrogen is a non-metal which forms a neutral oxide H_2O (which is commonly known as water).

2. Reaction with Water

(*i*) When a metal reacts with water, then a metal hydroxide and hydrogen gas are formed.

$$\text{Metal} + \text{Water} \longrightarrow \text{Metal hydroxide} + \text{Hydrogen}$$

The vigour (or intensity) of reaction of a metal with water depends on its chemical reactivity. Some metals react vigorously even with cold water, some metals react with hot water, some metals react with steam whereas some metals do not react even with steam. For example, sodium is a very reactive metal, therefore, sodium metal reacts violently even with cold water. Magnesium is a comparatively less reactive metal so it reacts slowly with cold water, it reacts rapidly only with hot boiling water or steam. The metals like zinc and iron are less reactive which react slowly even with steam. And the metals like copper, silver and gold are so unreactive that they do not react with water or even with steam. The reaction of sodium metal with water is described below.

Sodium metal reacts violently (explosively) with cold water forming sodium hydroxide solution and hydrogen gas :

$$\underset{(Na)}{\text{Sodium}} + \underset{(H_2O)}{\text{Water}} \longrightarrow \underset{\substack{(NaOH)\\(\text{A base})}}{\text{Sodium hydroxide}} + \underset{(H_2)}{\text{Hydrogen}}$$

Thus, **the gas liberated when sodium metal (or any other metal) reacts with water is hydrogen.** The reaction of sodium metal with water can be studied as follows.

ACTIVITY

We cut a small piece of sodium metal carefully and dry it by using a filter paper. This piece of sodium metal is placed in water filled in a beaker. We will find that the piece of sodium metal starts moving in water making a hissing sound due to formation of bubbles of a gas and reacts with water causing little explosions. Soon the piece of sodium metal catches fire (see Figure 14). When the reaction stops, touch the beaker. We will feel the beaker to be somewhat hot. This is because heat is produced in this reaction. If we test the solution in the beaker with red and blue litmus papers one by one, we will find that it turns red litmus paper blue. This shows that the solution formed by the reaction of sodium and water is basic in nature. These observations can be explained as follows : Sodium metal reacts with water to form sodium hydroxide and hydrogen gas. A lot of heat is also produced in this reaction. This heat burns the hydrogen gas as well as sodium metal. The burning of hydrogen gas causes little explosions. The formation of sodium hydroxide makes the solution basic. And this basic solution turns red litmus paper blue.

Figure 14. Reaction of sodium metal with water.

Sodium is a very reactive metal. It reacts with the moisture (water), oxygen and other gases present in air. So, if sodium metal is kept exposed to air, it will react with the various components of air and get spoiled. In order to prevent its reaction with the moisture and other gases of air, **sodium metal is always stored under kerosene.** Potassium metal is also highly reactive. So, potassium metal is also stored in kerosene.

(*ii*) Non-metals do not react with water.

Sulphur is a non-metal. Sulphur does not react with water. In fact, some of the reactive non-metals are kept under water to protect them from the action of air. For example, phosphorus is a very reactive non-metal element. If phosphorus is kept open in the air, it reacts with the oxygen of air and catches fire. So, **in order to protect phosphorus from atmospheric air, it is stored in a bottle containing water.**

3. Reaction with Acids

(*i*) Most of the metals react with dilute acids to form salts and hydrogen gas.

$$\text{Metal} + \text{Acid} \longrightarrow \text{Salt} + \text{Hydrogen}$$

Only the less reactive metals like copper, silver and gold do not react with dilute acids. The vigour of reaction of a metal with dilute acid depends on the chemical reactivity of the metal. Depending on their reactivity, some metals react violently (explosively) with dilute acids, some metals react rapidly with dilute acids, some metals react with dilute acids only on heating whereas some metals do not react with dilute acids at all. The reaction of magnesium metal with dilute hydrochloric acid is given below.

Magnesium metal reacts with *dilute* hydrochloric acid to form magnesium chloride and hydrogen gas. This reaction can be written as :

$$\underset{(Mg)}{\text{Magnesium}} + \underset{(HCl)}{\text{Hydrochloric acid}} \longrightarrow \underset{(MgCl_2)}{\text{Magnesium chloride}} + \underset{(H_2)}{\text{Hydrogen}}$$

Aluminium, iron and zinc metals also react with dilute hydrochloric acid to form the corresponding metal chlorides and hydrogen gas. This hydrogen gas burns with a 'pop' sound when a lighted matchstick is brought near the mouth of test-tube (containing metal and dilute hydrochloric acid). **The less reactive metals like copper, silver and gold do not react with dilute acids** (like dilute hydrochloric acid or dilute sulphuric acid). Thus, **copper, silver and gold do not produce hydrogen gas with dilute acids.**

ACTIVITY

We will now describe a simple activity to show that **when a metal reacts with a dilute acid, then hydrogen gas is produced**. We take four test-tubes. Put a small piece of magnesium ribbon in the first test-tube, a piece of aluminium foil in the second test-tube, some iron filings in the third test-tube and a piece of uncovered copper wire in the fourth test-tube. Add 10 mL of dilute hydrochloric acid in each test-tube and warm them gently. Test the gas produced in each test-tube by bringing a lighted matchstick (or burning matchstick) near the mouth of each test-tube.

(*i*) When we bring a lighted matchstick near the mouth of the first test-tube containing a piece of magnesium ribbon and dilute hydrochloric acid, the gas produced burns with a 'pop' sound, showing that it is hydrogen gas.

(*ii*) When we bring a lighted matchstick near the mouth of the second test-tube containing a piece of aluminium foil and dilute hydrochloric acid, the gas burns with a 'pop' sound showing that it is hydrogen gas.

(*iii*) When we bring a lighted matchstick near the mouth of the third test-tube containing iron filings and dilute hydrochloric acid, the gas burns with a 'pop' sound showing that it is hydrogen gas.

(*iv*) When we bring a lighted matchstick near the mouth of the fourth test-tube containing a piece of copper wire and dilute hydrochloric acid, nothing happens showing that no hydrogen gas is produced in this case.

This activity shows that though magnesium, aluminium and iron metals react with dilute hydrochloric acid to produce hydrogen gas but copper metal does not react with dilute hydrochloric acid to form hydrogen gas. Dilute sulphuric acid reacts with these metals in a similar way.

Copper metal also does not react with dilute sulphuric acid to produce hydrogen gas. Copper metal, however, reacts with hot and concentrated sulphuric acid but no hydrogen gas is produced. We will study these reactions in higher classes.

The reactions of metals with acids have some important implications in our daily life. Certain foodstuffs like citrus fruit juices (say, orange juice), pickles, *chutney* and curd, etc., contain acids. **When foodstuffs containing acids are kept in iron, aluminium or copper containers, the acids present in them**

react with the metal of the container slowly to form toxic salts (or poisonous salts). And these toxic salts can make us sick and damage our health. So, iron, aluminium and copper containers (or utensils) should not be used to store acidic foods like citrus fruit juices (such as orange juice), pickles, *chutney* and curd, etc. For example, we cannot store lemon pickle in an aluminium vessel. This is because the acid present in lemons will react with aluminium metal of vessel to form toxic salts which can make us sick and damage our health.

(*ii*) Non-metals do not react with dilute acids.

Non-metals do not react with dilute acids to form salts and hydrogen gas. For example, if we take some sulphur powder (or charcoal powder) in a test-tube and add dilute hydrochloric acid, then no reaction takes place even on heating. This shows that sulphur and carbon non-metals do not react with dilute acids and hence no hydrogen gas is produced. Some of the non-metals, however, react with hot and concentrated sulphuric acid and nitric acid but no hydrogen gas is produced in such cases. We will study these reactions in higher classes.

4. Reaction with Bases

(*i*) Some metals react with bases to form salts and hydrogen gas.

$$\text{Metal} + \text{Base} \longrightarrow \text{Salt} + \text{Hydrogen}$$

Aluminium is a metal and sodium hydroxide is a base. When aluminium is heated with sodium hydroxide solution, then sodium aluminate (salt) and hydrogen gas are formed :

$$\underset{(NaOH)}{\text{Sodium hydroxide}} + \underset{(Al)}{\text{Aluminium}} \longrightarrow \underset{(NaAlO_2)}{\text{Sodium aluminate}} + \underset{H_2}{\text{Hydrogen}}$$

ACTIVITY

We can demonstrate the formation of hydrogen gas in the reaction of aluminium metal with sodium hydroxide as follows : Take 5 mL of freshly prepared sodium hydroxide solution in a boiling tube. Drop a piece of aluminium foil in the sodium hydroxide solution and heat the boiling tube over a burner. Bring a lighted matchstick near the mouth of boiling tube. We will find that the gas produced burns with a 'pop' sound showing that it is hydrogen gas.

Zinc metal also reacts with sodium hydroxide solution to form hydrogen gas. Thus, **aluminium and zinc are the two common metals which react with bases (like sodium hydroxide) to produce hydrogen gas.** In general we can say that : Some metals react with sodium hydroxide to produce hydrogen gas. Please note that **all the metals do not react with bases (like sodium hydroxide) to produce hydrogen gas**.

(*ii*) Some non-metals react with bases (like sodium hydroxide) but no hydrogen gas is produced. The reactions of non-metals with bases are complex. We will study these reactions in higher classes.

Before we go further and study the displacement reactions of metals, we should know the reactivity series of metals. This is described below.

REACTIVITY SERIES OF METALS. Some metals are chemically very reactive whereas other metals are less reactive or unreactive. On the basis of vigour of reactions of various metals with oxygen, water and acids, as well as the displacement reactions, the metals have been arranged in a group (or series) according to their chemical reactivities. **The arrangement of metals in a vertical column in the order of decreasing reactivities is called the reactivity series of metals** (or activity series of metals). In reactivity series, the most reactive metal is placed at the top whereas the least reactive metal is placed at the bottom. The reactivity series of some common metals is given on the next page.

Reactivity Series of Metals

Potassium	(K)	..........Most reactive metal
Sodium	(Na)	
Calcium	(Ca)	
Magnesium	(Mg)	
Aluminium	(Al)	Decreasing
Zinc	(Zn)	chemical
Iron	(Fe)	reactivity
Lead	(Pb)	
Copper	(Cu)	
Silver	(Ag)	↓
Gold	(Au)	.........Least reactive metal

Please note that **potassium is the most reactive metal, so it has been placed at the top of the reactivity series.** As we come down in the reactivity series, the chemical reactivity of metals decreases gradually. For example, sodium is less reactive than potassium ; calcium is less reactive than sodium ; magnesium is less reactive than calcium ; aluminium is less reactive than magnesium ; zinc is less reactive than aluminium ; iron is less reactive than zinc, copper is less reactive than iron and silver is less reactive than copper. **Gold being the least reactive metal here has been placed at the bottom of the reactivity series.** We should remember the reactivity series of metals because it will help us in understanding the displacement reactions of metals.

5. Displacement Reactions

(*i*) A more reactive metal displaces a less reactive metal from its salt solution. This means that when a more reactive metal is placed in the salt solution of a less reactive metal, then the more reactive metal displaces (pushes out) the less reactive metal from its salt solution. The more reactive metal becomes a part of the salt whereas the less reactive metal is set free. Let us take some examples to make this point more clear.

(*a*) Reaction of Iron Metal with Copper Sulphate Solution. When a strip of iron metal (or an iron nail) is placed in copper sulphate solution for some time, then the blue colour of copper sulphate solution fades and a red-brown coating of copper metal is deposited on the iron strip (or iron nail). This reaction can be written as :

Copper sulphate	+	Iron	⟶	Iron sulphate	+	Copper
($CuSO_4$)		(Fe)		($FeSO_4$)		(Cu)
(Blue solution)		(Grey)		(Greenish solution)		(Red-brown)

In this case the solution turns greenish due to the formation of iron sulphate. We know that iron metal is more reactive than copper metal. So, **in this reaction, a more reactive metal 'iron' is displacing a less reactive metal 'copper' from its salt solution, copper sulphate solution.** The products of this displacement reaction are 'iron sulphate solution' and 'copper metal'. Please note that the blue colour of copper sulphate solution changes to greenish due to the formation of iron sulphate (which is green in colour). The copper metal produced in this displacement reaction forms a red-brown coating over the iron strip (or iron nail). In the above displacement reaction, **iron metal displaces copper from copper sulphate solution.** This displacement reaction takes place because iron is more reactive than copper. The displacement reaction between iron metal and copper sulphate solution can be performed as follows.

ACTIVITY

We take about 50 mL of water in a beaker and dissolve 5 grams of copper sulphate in it to obtain copper sulphate solution (which is blue in colour). Put a clean iron nail in copper sulphate solution in the beaker and keep the beaker undisturbed for some time. We will find that the blue colour of copper sulphate solution starts fading gradually. And the iron nail gets covered with a red-brown layer of copper metal. This change takes place because iron metal displaces copper metal from its compound copper sulphate. It is the copper metal set free from its compound which forms a red-brown layer on the surface of iron nail.

Figure 15. This iron nail has been kept in copper sulphate solution. A red-brown layer of displaced copper metal has formed on the nail.

We will now discuss the reverse case in which a copper strip is placed in iron sulphate solution.

A less reactive metal cannot displace a more reactive metal from its salt solution. For example, if we place a strip of copper metal in iron sulphate solution for some time, then no displacement reaction takes place. That is :

Iron sulphate	+	Copper	⟶	No displacement reaction
($FeSO_4$)		(Cu)		

This displacement reaction does not occur because copper metal is less reactive than iron metal. So, a less reactive metal 'copper' cannot displace a more reactive metal 'iron' from its salt solution, iron sulphate solution. Thus, **copper metal cannot displace iron from iron sulphate solution.**

(*b*) Reaction of Zinc Metal with Copper Sulphate Solution. When a strip of zinc metal is placed in copper sulphate solution for some time, then the blue colour of copper sulphate solution fades gradually and red-brown copper metal is deposited on the zinc strip. This reaction can be written as :

Copper sulphate	+	Zinc	⟶	Zinc sulphate	+	Copper
($CuSO_4$)		(Zn)		($ZnSO_4$)		(Cu)
(Blue solution)		(Silvery white)		(Colourless solution)		(Red-brown)

We know that zinc metal is more reactive than copper metal. So, **in this reaction, a more reactive metal 'zinc' is displacing a less reactive metal 'copper' from its salt solution, copper sulphate solution.** The products of this displacement reaction are 'zinc sulphate solution' and 'copper metal'. Please note that the blue colour of the copper sulphate solution gradually disappears due to the formation of colourless zinc sulphate solution. The copper metal which is formed in this displacement reaction deposits on the zinc strip in the form of a red-brown coating. In the above displacement reaction, **zinc metal displaces copper metal from copper sulphate solution.** This displacement reaction between zinc and copper sulphate solution occurs because zinc is more reactive than copper. Let us now discuss the reverse case in which a copper strip is placed in zinc sulphate solution.

Figure 16. Zinc displaces copper from copper sulphate solution.

If we place a strip of copper metal in zinc sulphate solution, then no displacement reaction will take place. That is :

Zinc sulphate	+	Copper	⟶	No displacement reaction
($ZnSO_4$)		(Cu)		

This displacement reaction does not take place because copper metal is less reactive than zinc metal. So, a less reactive metal 'copper' cannot displace a more reactive metal 'zinc' from its salt solution, zinc sulphate solution. Thus, **copper cannot displace zinc from zinc sulphate solution.**

(c) Reaction of Zinc Metal with Iron Sulphate Solution. When a strip of zinc metal is placed in iron sulphate solution, then a displacement reaction takes place to form zinc sulphate solution and iron metal. This reaction can be written as :

Iron sulphate	+	Zinc	⟶	Zinc sulphate	+	Iron
($FeSO_4$)		(Zn)		($ZnSO_4$)		(Fe)
(Greenish solution)		(Silvery white)		(Colourless solution)		(Grey)

In this reaction, zinc metal displaces iron metal from iron sulphate solution. This displacement reaction takes place because zinc is more reactive than iron. Let us now discuss the reverse case in which an iron nail is placed in zinc sulphate solution.

If we place an iron nail in zinc sulphate solution, then no displacement reaction takes place. That is :

Zinc sulphate	+	Iron	⟶	No displacement reaction
($ZnSO_4$)		(Fe)		

This displacement reaction does not occur because iron metal is less reactive than zinc metal. So, a less reactive metal iron cannot displace a more reactive metal zinc from zinc sulphate solution.

Please note that magnesium metal is more reactive than zinc, iron and copper. So, magnesium metal can displace zinc, iron and copper metals from their salt solutions.

***(ii)* A more reactive non-metal displaces a less reactive non-metal from its salt solution.** We will study these displacement reactions in higher classes.

Comparison Between the Chemical Properties of Metals and Non-Metals

We will now compare the important chemical properties of metals and non-metals in tabular form. Alongwith physical properties, the chemical properties can be used to distinguish between metals and non-metals.

Differences in Chemical Properties of Metals and Non-Metals

Metals	*Non-Metals*
1. Metals form basic oxides.	1. Non-metals form acidic oxides.
2. Metals react with water (or steam) to produce hydrogen gas (*except copper, silver and gold which do not react with water or steam*).	2. Non-metals do not react with water (or steam).
3. Metals react with dilute acids to produce hydrogen gas (*except copper, silver and gold which do not react with dilute acids*).	3. Non-metals do not react with dilute acids.

We will now answer some questions based on the properties of metals and non-metals.

Sample Problem 1. An element reacts with oxygen to form an oxide. An aqueous solution of this oxide turns red litmus paper blue. Is the element a metal or a non-metal ? Give reason for your answer.

Answer. We know that basic substances turn red litmus to blue. Since an aqueous solution of this oxide turns red litmus to blue, it means that the oxide is basic in nature. Now, basic oxides are formed by metal elements. So, the given element is a metal.

Sample Problem 2. Why is iron not deposited over a copper plate when the copper plate is dipped in iron sulphate solution ?

Answer. Copper is less reactive than iron, so copper is not able to displace iron from iron sulphate solution to form free iron metal. Since no iron metal is formed, it is not deposited over copper plate.

Sample Problem 3. Consider the following displacement reactions :

(*i*) Copper sulphate (solution) + Iron ⟶ Iron sulphate (solution) + Copper

(*ii*) Iron sulphate (solution) + Zinc ⟶ Zinc sulphate (solution) + Iron

On the basis of these two displacement reactions, find out which is the most reactive metal and which is the least reactive metal out of copper, iron and zinc.

Answer. (*i*) In the first reaction, iron metal displaces copper metal from copper sulphate solution, therefore, *iron is more reactive than copper.*

(*ii*) In the second reaction, zinc metal displaces iron metal from iron sulphate solution, so *zinc is more reactive than iron.*

Now, since zinc is more reactive than iron, and iron is more reactive than copper, therefore, in this case : *zinc is the most reactive metal* whereas *copper is the least reactive metal.*

Sample Problem 4 : Saloni took a piece of burning charcoal and collected the gas evolved in a test-tube.

(*a*) How will she find the nature of the gas ?

(*b*) Write down word equations of all the reactions taking place in this process.

(NCERT Book Question)

Answer. Charcoal is a form of carbon (which is a non-metal). When carbon (charcoal) burns in air, it forms an acidic oxide called carbon dioxide (which is a gas).

(*a*) To find the nature of carbon dioxide gas, dissolve it in water. Test the aqueous solution of carbon dioxide gas with blue litmus paper and red litmus paper. The aqueous solution of carbon dioxide gas will turn blue litmus paper to red showing that carbon dioxide is acidic in nature.

(*b*) When carbon (or charcoal) burns in air, it combines with the oxygen of air to form carbon dioxide. The word equation for this reaction is :

Carbon (Charcoal) + Oxygen ⟶ Carbon dioxide

Carbon dioxide dissolves in water to form carbonic acid (which turns blue litmus paper to red). The word equation for this reaction is :

Carbon dioxide + Water ⟶ Carbonic acid

USES OF METALS

Metals are used in our everyday life for a large number of purposes. Metals are used in making nails, screws, utensils, water boilers, electric wires, office furniture, cars, buses, trains, aeroplanes, satellites, various types of machines, and many, more things used by us. **Some of the important uses of metals are given below :**

1. Iron, copper and aluminium metals are used to make cooking utensils, and water boilers for factories.

2. Copper metal is used for making electric wires for household wiring, electric motors, armature of dynamos, and many other electrical appliances. Aluminium metal is used for making electric cables (thick wires) and over-head electric transmission lines.

3. Aluminium foils are used for packaging medicines, chocolates, food items and many other materials.

4. Aluminium metal (in the form of alloys) is used to make aeroplanes.

5. Iron metal (in the form of steel) is used to make nails, screws, nut-bolts, pipes, railings, gas cylinders, stoves, water tanks, office furniture, industrial tools and machines, buildings and bridges, railway lines, transport vehicles such as cars, buses and trains, household goods and agricultural implements.

6. Zinc metal is used for galvanising iron to protect it from rusting.
7. Silver and gold metals are used to make jewellery.
8. Mercury metal is used in making thermometers.

In our body, iron is present in the pigment of red blood cells called 'haemoglobin' (which transports oxygen from the lungs to all the tissues of our body). If there is deficiency of iron in our body, then there will be less haemoglobin in our blood. Lower level of haemoglobin in the blood will cause shortage of oxygen in the body leading to tiredness and weakness.

USES OF NON-METALS

All the living things around us like plants and animals (including human beings) are made up mainly of compounds of non-metals like carbon, hydrogen, oxygen, nitrogen, sulphur and phosphorus, etc. Non-metals are used in our day to day life for a large number of purposes. **Some of the important uses of the common non-metals like oxygen, nitrogen, chlorine, iodine, sulphur, phosphorus and carbon are given below :**

1. Oxygen is a non-metal which is used by plants and animals (including human beings) for breathing. Thus, oxygen non-metal is essential for maintaining our life. Oxygen non-metal is also used in the process of burning (or combustion) of fuels in homes, factories and transport vehicles.

2. Nitrogen is a non-metal which is used in making fertilisers to enhance the growth of plants. Nitrogen gas (being inert) is used in food packaging instead of air, to keep the food fresh.

3. Chlorine is a non-metal which is used in the water purification process. Chlorine has the ability to kill germs, so chlorine makes the drinking water supply germ-free.

4. Iodine is a non-metal which is used to make purple-coloured solution called 'tincture iodine' which is applied on cuts and wounds as an antiseptic.

5. Sulphur and phosphorus are the non-metals which are used in fireworks (crackers, etc.)

6. Sulphur is a non-metal which is used in the vulcanisation of rubber (or hardening of rubber).

7. Carbon is a non-metal which is used as a fuel. The forms of carbon used as a fuel are charcoal, coke and coal.

We are now in a position to **answer the following questions :**

Very Short Answer Type Questions

1. What is the general name of the elements whose properties are intermediate between those of metals and non-metals ?
2. Name one metal and one non-metal which exist in liquid state at room temperature.
3. Name the property :
   (*a*) which allows metals to be hammered into thin sheets.
   (*b*) which enables metals to be drawn into wires.
4. Name two metals which are soft and can be easily cut with a knife.
5. If a metal coin is dropped on hard floor, it produces a ringing sound. What is this property of metals known as ?
6. Name the property of iron metal due to which it can be hammered to make objects of different shapes such as an axe, a spade or a shovel.
7. Name a non-metal which is very hard.
8. Name a non-metal which is a good conductor of electricity.
9. State one chemical property which can be used to distinguish a metal from a non-metal.
10. How do metal oxides differ from non-metal oxides ?
11. An element forms an oxide which is acidic in nature. Is the element a metal or a non-metal ?
12. An element forms an oxide which is basic in nature. State whether the element is a metal or a non-metal ?
13. Write a word equation for the reaction of magnesium with oxygen.

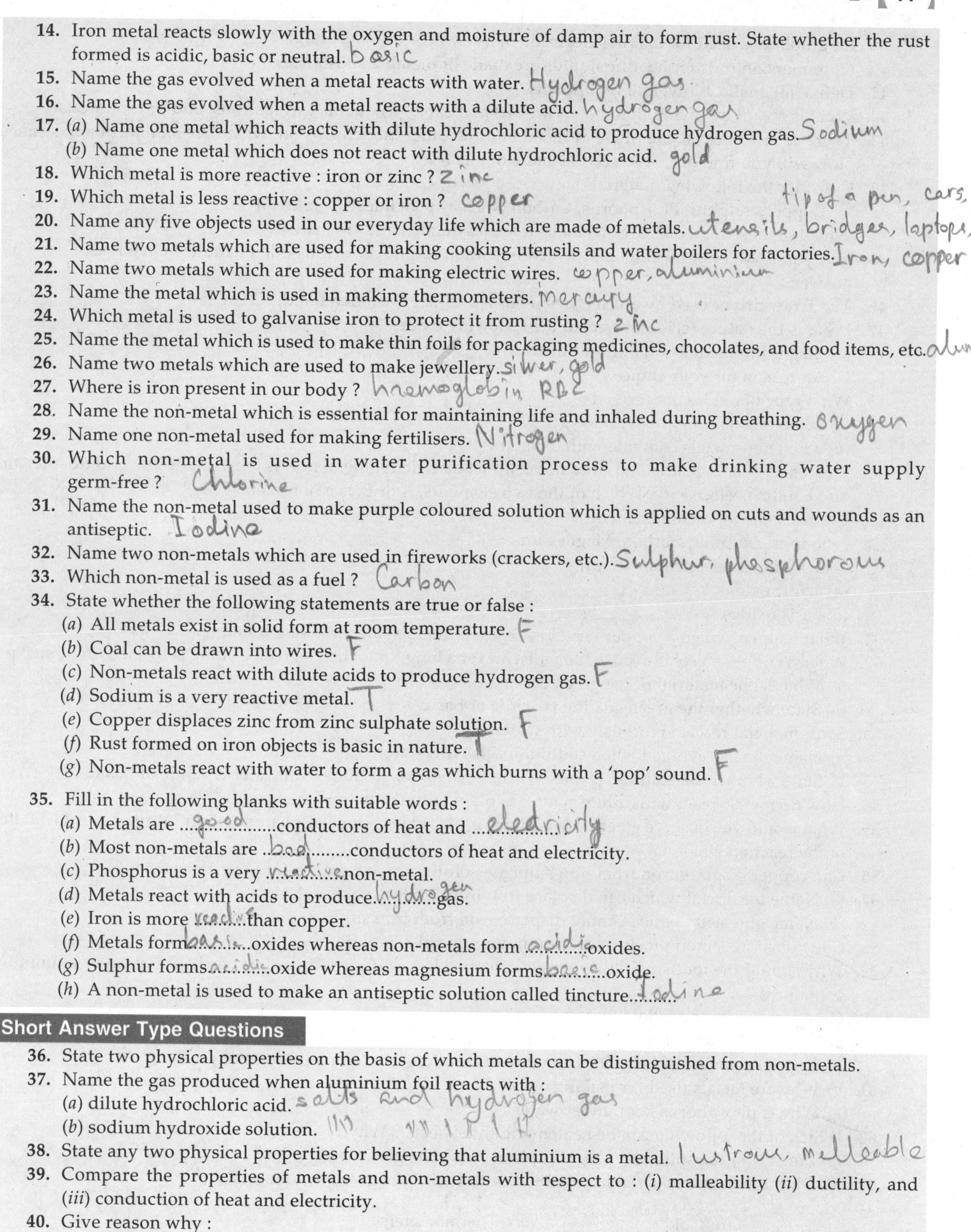

14. Iron metal reacts slowly with the oxygen and moisture of damp air to form rust. State whether the rust formed is acidic, basic or neutral.
15. Name the gas evolved when a metal reacts with water.
16. Name the gas evolved when a metal reacts with a dilute acid.
17. (*a*) Name one metal which reacts with dilute hydrochloric acid to produce hydrogen gas.
    (*b*) Name one metal which does not react with dilute hydrochloric acid.
18. Which metal is more reactive : iron or zinc ?
19. Which metal is less reactive : copper or iron ?
20. Name any five objects used in our everyday life which are made of metals.
21. Name two metals which are used for making cooking utensils and water boilers for factories.
22. Name two metals which are used for making electric wires.
23. Name the metal which is used in making thermometers.
24. Which metal is used to galvanise iron to protect it from rusting ?
25. Name the metal which is used to make thin foils for packaging medicines, chocolates, and food items, etc.
26. Name two metals which are used to make jewellery.
27. Where is iron present in our body ?
28. Name the non-metal which is essential for maintaining life and inhaled during breathing.
29. Name one non-metal used for making fertilisers.
30. Which non-metal is used in water purification process to make drinking water supply germ-free ?
31. Name the non-metal used to make purple coloured solution which is applied on cuts and wounds as an antiseptic.
32. Name two non-metals which are used in fireworks (crackers, etc.).
33. Which non-metal is used as a fuel ?
34. State whether the following statements are true or false :
    (*a*) All metals exist in solid form at room temperature.
    (*b*) Coal can be drawn into wires.
    (*c*) Non-metals react with dilute acids to produce hydrogen gas.
    (*d*) Sodium is a very reactive metal.
    (*e*) Copper displaces zinc from zinc sulphate solution.
    (*f*) Rust formed on iron objects is basic in nature.
    (*g*) Non-metals react with water to form a gas which burns with a 'pop' sound.
35. Fill in the following blanks with suitable words :
    (*a*) Metals are ....conductors of heat and ....
    (*b*) Most non-metals are ....conductors of heat and electricity.
    (*c*) Phosphorus is a very ....non-metal.
    (*d*) Metals react with acids to produce..................gas.
    (*e*) Iron is more ....than copper.
    (*f*) Metals form..................oxides whereas non-metals form ....oxides.
    (*g*) Sulphur forms..................oxide whereas magnesium forms..................oxide.
    (*h*) A non-metal is used to make an antiseptic solution called tincture..................

Short Answer Type Questions

36. State two physical properties on the basis of which metals can be distinguished from non-metals.
37. Name the gas produced when aluminium foil reacts with :
    (*a*) dilute hydrochloric acid.
    (*b*) sodium hydroxide solution.
38. State any two physical properties for believing that aluminium is a metal.
39. Compare the properties of metals and non-metals with respect to : (*i*) malleability (*ii*) ductility, and (*iii*) conduction of heat and electricity.
40. Give reason why :
    (*a*) copper metal is used for making electric wires.

(*b*) graphite is used for making electrode in a cell.
(*c*) immersion rods for heating liquids are made of metallic substances.

41. Define (*a*) malleability, and (*b*) ductility.
42. What is meant by saying that metals are : (*i*) malleable (*ii*) ductile (*iii*) lustrous, and (*iv*) sonorous ?
43. There are two boxes, one made of metal and the other made of wood, which are similar in appearance. How will you find out which box is made of metal ?
44. Consider the following materials :
    Copper, Sulphur, Phosphorus, Carbon (such as pencil lead), Gold, Silver.
    Which of these materials are : (*i*) malleable and ductile, and (*ii*) brittle ?
45. Can you hold a hot metallic pan which is without a plastic or a wooden handle ? Give reason for your answer.
46. The screw driver used by an electrician has a plastic or wooden handle. Why ?
47. What is the nature (acidic/basic) of the following oxides ?
    (*a*) Magnesium oxide (*b*) Sulphur dioxide
    Given reason for your choice.
48. What type of oxides are formed :
    (*a*) when metals combine with oxygen ?
    (*b*) when non-metals combine with oxygen ?
49. Element *A* is soft, brittle and does not conduct electricity. Element *B* is hard, malleable and ductile, and also conducts electricity. Which of the two elements, *A* or *B*, is a non-metal ?
50. Consider the following elements :
    Sodium, Sulphur, Carbon, Magnesium
    Which of these elements will form :
    (*a*) acidic oxides ?
    (*b*) basic oxides ?
51. What happens when a copper vessel is exposed to moist air for a long time ?
52. When a copper object is exposed to moist air for a long time, then a green coating is formed on its surface.
    (*a*) What is the material of the green coating ?
    (*b*) State whether the green coating is acidic or basic.
53. Sodium metal reacts vigorously with water.
    (*a*) Name the gas evolved when sodium reacts with water.
    (*b*) State whether the solution formed by the reaction of sodium with water is acidic or basic.
54. How do metals react with dilute acids ? Explain with the help of an example.
55. What would you observe when a strip of zinc is placed in copper sulphate solution ? Write a word equation of the reaction which takes place.
56. Can copper displace iron from iron sulphate solution ? Give reason for your answer.
57. (*a*) Name one metal which can displace iron from iron sulphate solution.
    (*b*) Name one metal which cannot displace iron from iron sulphate solution.
58. Can you store lemon pickle in an aluminium utensil ? Explain.
59. Why should the foodstuffs like orange juice, pickles, *chutney* and curd not be kept in iron or aluminium containers ?
60. Give reasons for the following :
    (*a*) Sodium and potassium are stored in kerosene.
    (*b*) Copper cannot displace zinc from its salt solution (zinc sulphate solution).
61. (*a*) Why are metals used for making bells ?
    (*b*) Why is phosphorus kept under water ?
62. Which of the following can be beaten into thin sheets ? Why ?
    (*a*) Zinc (*b*) Phosphorus (*c*) Sulphur (*d*) Oxygen
63. Match the substances given in column *A* with their uses given in column *B* :

A	*B*
(*i*) Gold	(*a*) Thermometers
(*ii*) Iron	(*b*) Electric wires
(*iii*) Aluminium	(*c*) Wrapping food

(*iv*) Carbon (*d*) Jewellery
(*v*) Copper (*e*) Machinery
(*vi*) Mercury (*f*) Fuel **(NCERT Book Question)**

64. Give one use each of the following metals :
(*a*) Iron (*b*) Copper (*c*) Aluminium (*d*) Zinc (*e*) Mercury

65. State one use each of the following non-metals :
(*a*) Oxygen (*b*) Nitrogen (*c*) Sulphur (*d*) Chlorine (*e*) Iodine

Long Answer Type Questions

66. (*a*) What are metals ? Name five metals.
(*b*) What are non-metals ? Name five non-metals.

67. (*a*) What are metalloids ? Name two metalloids.
(*b*) Classify the following elements into metals, non-metals and metalloids :
Copper, Sulphur, Aluminium, Oxygen, Silicon, Nitrogen, Germanium, Mercury, Chlorine, Sodium.

68. (*a*) What happens when sulphur dioxide is dissolved in water ? Write a word equation for the reaction which takes place.
(*b*) What happens when an iron nail is placed in copper sulphate solution ? Write word equation of the reaction involved.

69. (*a*) State five characteristics of metals and five characteristics of non-metals.
(*b*) State five uses of metals and five uses of non-metals.

70. Compare the chemical properties of metals and non-metals in tabular form.

Multiple Choice Questions (MCQs)

71. An element is soft and can be cut easily with a knife. It is very reactive and cannot be kept open in the air. It reacts vigorously with water. This element is most likely to be :
(*a*) magnesium (*b*) potassium (*c*) phosphorus (*d*) aluminium

72. Which one of the following four metals would be displaced from the solution of its salt by the other three metals ?
(*a*) zinc (*b*) silver (*c*) copper (*d*) magnesium

73. Sulphur element is said to be :
(*a*) ductile (*b*) hard (*c*) malleable (*d*) brittle

74. An element X reacts with water to from a solution which turns phenolphthalein indicator pink. The element X is most likely to be :
(*a*) sulphur (*b*) sodium (*c*) carbon (*d*) silicon

75. The non-metal which exists in the liquid state at room temperature is :
(*a*) fluorine (*b*) chlorine (*c*) bromine (*d*) iodine

76. A basic oxide will be formed by the element :
(*a*) sulphur (*b*) phosphorus (*c*) potassium (*d*) carbon

77. "Is malleable and ductile". This best describes :
(*a*) a metal (*b*) a compound (*c*) a non-metal (*d*) a mixture

78. The metal which will not produce hydrogen gas on reacting with dilute sulphuric acid is :
(*a*) sodium (*b*) silver (*c*) iron (*d*) zinc

79. The element which is stored under kerosene is :
(*a*) sulphur (*b*) phosphorus (*c*) sodium (*d*) silicon

80. Which of the following pairs cannot undergo displacement reaction ?
(*a*) iron sulphate solution and magnesium
(*b*) zinc sulphate solution and iron
(*c*) zinc sulphate solution and calcium
(*d*) silver nitrate solution and copper

81. Which of the following metal exists in the liquid state at room temperature ?
(*a*) magnesium (*b*) manganese (*c*) mercury (*d*) sodium

82. The element Z burns in air to form an oxide. The aqueous solution of this oxide turns blue litmus to red. The element Z is most likely to be.
(*a*) carbon (*b*) calcium (*c*) iron (*d*) magnesium

83. Which of the following elements is a metalloid ?
(*a*) sodium (*b*) sulphur (*c*) silicon (*d*) silver

84. Which of the following elements will produce an oxide that will dissolve in water to form an acid ?
(*a*) carbon (*b*) calcium (*c*) chromium (*d*) copper

85. The least reactive metal among the following is :
(*a*) magnesium (*b*) lead (*c*) silver (*d*) sodium

86. You are given a solution of iron sulphate. Which of the following do you think cannot displace iron from iron sulphate ?
(*a*) magnesium (*b*) calcium (*c*) copper (*d*) zinc

87. When a vessel is exposed to moist air for a long time, then a green coating is formed on its surface. The vessel must be made of :
(*a*) zinc (*b*) magnesium (*c*) iron (*d*) copper

88. Which among the following is the most reactive metal ?
(*a*) copper (*b*) calcium (*c*) iron (*d*) magnesium

89. The element whose oxide will turn red litmus solution to blue will be :
(*a*) sodium (*b*) sulphur (*c*) carbon (*d*) phosphorus

90. Which of the following is not a characteristic property of iron ?
(*a*) malleability (*b*) brittleness (*c*) ductility (*d*) sonorousness

Questions Based on High Order Thinking Skills (HOTS)

91. Which of the following reactions will not occur ? Why not ?

(*a*) Zinc sulphate (solution) + Copper ⟶ Copper sulphate (solution) + Zinc

(*b*) Copper sulphate (solution) + Iron ⟶ Iron sulphate (solution) + Copper

92. One day Reeta went to a jeweller's shop with her mother. Her mother gave an old gold jewellery to goldsmith to polish. Next day when they brought the jewellery back, they found that there was a slight loss in its weight. Can you suggest a reason for the loss in weight ? **(NCERT Book Question)**

93. An element burns in air to form an oxide. The aqueous solution of this oxide turns blue litmus paper red. State whether the element is a metal or a non-metal. Name one such element.

94. An element burns in air to form an oxide. The aqueous solution of this oxide turns turmeric paper red. State whether the element is a metal or non-metal. Name one such element.

95. The metal X reacts with dilute hydrochloric acid to form a gas Y. The metal X also reacts with sodium hydroxide solution (on heating) to form the same gas Y. When a lighted matchstick is applied, this gas burns by producing a 'pop' sound.
(*a*) Name two metals which could behave like X.
(*b*) Name the gas Y.

ANSWERS

1. Metalloids **3.** (*a*) Malleability (*b*) Ductility **5.** Sonorousness **6.** Malleability **11.** Non-metal **12.** Metal **14.** Basic **17.** (*a*) Zinc (*b*) Copper **18.** Zinc **19.** Copper **34.** (*a*) False (*b*) False (*c*) False (*d*) True (*e*) False (*f*) True (*g*) False **35.** (*a*) good ; electricity (*b*) poor (*c*) reactive (*d*) hydrogen (*e*) reactive (*f*) basic ; acidic. (*g*) acidic ; basic (*h*) iodine **37.** (*a*) Hydrogen (*b*) Hydrogen **49.** Element A **57.** (*a*) Zinc (*b*) Copper **63.** (*i*) *d* (*ii*) *e* (*iii*) *c* (*iv*) *f* (*v*) *b* (*vi*) *a* **71.** (*b*) **72.** (*b*) **73.** (*d*) **74.** (*b*) **75.** (*c*) **76.** (*c*) **77.** (*a*) **78.** (*b*) **79.** (*c*) **80.** (*b*) **81.** (*c*) **82.** (*a*) **83.** (*c*) **84.** (*a*) **85.** (*c*) **86.** (*c*) **87.** (*d*) **88.** (*b*) **89.** (*a*) **90.** (*b*) **91.** Reaction (*a*) will not occur because copper is less reactive than zinc **92.** While polishing old gold jewellery, a thin, outer dull layer of gold is removed by treatment with a chemical. This leads to a slight loss in the weight of jewellery **93.** Non-metal ; Sulphur **94.** Metal ; Magnesium **95.** (*a*) Aluminium ; Zinc (*b*) Hydrogen

Coal and Petroleum

In our everyday life, we use a large number of materials for our basic needs. Some of these materials are found in nature whereas others are man-made. Air, water, soil, sunlight, coal, petroleum, natural gas and minerals are obtained from nature, so they are called natural resources. On the other hand, plastics, synthetic fibres, paints, drugs, explosives, etc., are all man-made materials.

Inexhaustible and Exhaustible Natural Resources

Anything in the environment *'which can be used'* is called a *'resource'*. All the natural resources can be classified into two main groups :

(*i*) **Inexhaustible natural resources,** and

(*ii*) **Exhaustible natural resources.**

The term 'inexhaustible' means something 'which cannot be used up completely'. **Those natural resources which are present in unlimited quantity in nature and are not likely to be exhausted by human activities are called inexhaustible natural resources**. The examples of inexhaustible natural resources are : Sunlight, Air and Water. There is a never ending supply of inexhaustible resources in nature. The inexhaustible resources can be used again and again. They last forever.

The term 'exhaustible' means 'something which can be used up completely' (so that nothing is left behind). **Those resources which are present in a limited quantity in nature and can be exhausted by human activities, are called exhaustible natural resources**. The examples of exhaustible natural resources are : Coal, Petroleum, Natural gas, Minerals, Forests and Wildlife, etc. The exhaustible natural resources do not last forever.

In this Chapter we will study some exhaustible sources of energy like coal, petroleum and natural gas. **Coal, petroleum and natural gas are also called non-renewable sources of energy.** This is because when all the coal, petroleum and natural gas present under the earth will get used up (or exhausted), no more supply of these fuels will be available in the near future. Before we go further, we should know the meaning of the term 'fossil'. ***Fossils are the remains of the pre-historic plants or animals, buried under the earth millions of years ago.***

FOSSIL FUELS

The natural fuels formed from the remains of living organisms buried under the earth long, long ago, are called fossil fuels. Coal, petroleum and natural gas are fossil fuels. Coal, petroleum and natural gas are called fossil fuels because they were formed by the decomposition of the remains of pre-historic plants and animals (fossils) buried under the earth long, long ago. Fossil fuels are exhaustible natural resources because once all the fossil fuels are used up, they will be gone forever.

How Fossil Fuels were Formed

Fossil fuels were formed from the dead remains of living organisms (plants and animals) buried under the earth millions of years ago. This happened as follows : The plants and animals which died millions of years ago, were gradually buried deep in the earth and got covered with sediments like mud and sand, away from the reach of air. In the absence of air, the chemical effects of heat, pressure and bacteria, converted the buried remains of plants and animals into fossil fuels like coal, petroleum and natural gas. Please note that the buried remains of large land plants were converted into coal whereas those of tiny marine plants and animals were converted into petroleum and natural gas.

COAL

Coal is a hard, black combustible mineral that consists mainly of carbon (see Figure 1). Coal is found in deep coal mines under the surface of the earth. In India, coal is found mainly in Bihar, West Bengal, Orissa and Madhya Pradesh. Coal is found in abundance in our country and it is the most important source of energy in our country.

Figure 1. Coal.

How Coal was Formed

Coal was formed by the decomposition of large land plants and trees buried under the earth about 300 million years ago. This happened as follows : About 300 million years ago, the earth had dense forests in low-lying wet land areas. Due to natural processes like earthquakes, volcanoes and floods, etc., these forests were buried under the surface of earth. As more soil deposited over them, they were compressed. The temperature also rose as they sank deeper and deeper. Due to high pressure and high temperature inside the earth, and in the absence of air, the wood of buried forest plants and trees was slowly converted into coal. ***The slow process by which the dead plants buried deep under the earth have become coal is called carbonisation.*** Since coal was formed from the remains of plants, therefore, coal is called a fossil fuel.

Coal is a Source of Energy

Coal is mainly carbon. When heated in air, coal burns and produces mainly carbon dioxide gas. A lot of heat energy is also produced during the burning of coal. This can be written as :

$$\underset{\text{(Coal)}}{\text{Carbon}} + \underset{\text{(From air)}}{\text{Oxygen}} \longrightarrow \text{Carbon dioxide} + \text{Heat}$$

Coal is important because it can be used as a source of heat energy as such (just by burning it), or it can be converted into other forms of energy such as coal gas, coke or electricity. The real source of energy of coal is the solar energy (or sun's energy). This is because the plants and trees which decomposed to form coal grew on the earth by absorbing sunlight energy during the process of photosynthesis.

Uses of Coal

(*i*) Coal is used as a fuel in homes and industry.
(*ii*) Coal is used as a fuel at Thermal Power Plants for generating electricity.
(*iii*) Coal is used to make coal gas which is an important industrial fuel.

(*iv*) Coal is used to make coke.

(*v*) Earlier, coal was used as a fuel to make 'steam' to run steam engines of trains.

(*vi*) Coal was also used as a source of organic chemicals.

Products of Coal

When coal is heated strongly in closed retorts in the absence of air, a number of useful products are obtained. The various useful products obtained by processing the coal by heating in the absence of air are :

(*i*) **Coal gas,**

(*ii*) **Coal tar**, and

(*iii*) **Coke.**

Coal gas, coal tar and coke are called products of coal. Please note that these products are obtained when coal is heated in the absence of air. This is because if coal is heated in the presence of air, then coal burns to produce mainly carbon dioxide gas and no other useful products are obtained. ***The strong heating of coal in the absence of air is called destructive distillation of coal.*** We will now describe the various products of coal in somewhat detail.

Coal Gas

Coal gas is a gaseous fuel which is obtained by the strong heating of coal in the absence of air during the processing of coal to get coke. Coal gas is mainly a mixture of methane and hydrogen, with some carbon monoxide. All the gases present in coal gas can burn to produce heat, due to which coal gas is an excellent fuel (having high calorific value). Coal gas is used as a fuel in industries (which are situated near the coal processing plants). When coal gas burns, it also produces a good amount of light. So, in the past, coal gas has also been used for lighting purposes (or illumination purposes). Coal gas was used for street lighting for the first time in London in the year 1810. It was used for street lighting in New York around 1820. These days, however, coal gas is used as a source of heat rather than light.

Coal Tar

Coal tar is a thick, black liquid having an unpleasant smell which is obtained by heating coal in the absence of air (see Figure 2). Coal tar is not a single compound. Coal tar is a mixture of about 200 carbon compounds (or organic compounds). The useful carbon compounds (or organic compounds) present in coal tar include benzene, toluene, naphthalene, anthracene, phenol and aniline. Thus, the naphthalene balls used to repel moths and other insects (in stored clothes, etc.) are obtained from coal tar. The various compounds present in coal tar are separated by the process of fractional distillation. The compounds (or products) obtained from coal tar are used as starting materials for manufacturing a large number of substances used in everyday life and industry. For example, **the products of coal tar are used to make synthetic fibres, drugs (medicines), plastics, synthetic dyes, perfumes, paints, varnishes, pesticides, photographic materials, roofing materials and explosives, etc.** Coal tar has been traditionally used for metalling the roads. These days, however, bitumen (a petroleum product) is being used increasingly for metalling the road surfaces (in place of coal tar).

Figure 2. Coal tar.

Coke

Coke is a tough and porous black solid substance (see Figure 3). Coke is prepared by heating coal in the absence of air. When coal is heated in the absence of air, then coal gas and coal tar are eliminated, and coke is left behind as a black residue. Thus, *coal minus volatile constituents is coke*. Coke is an almost pure form of carbon. It is 98 per cent carbon.

Figure 3. Coke.

Coke is mainly used as a reducing agent in the extraction of metals (like iron, zinc, etc.) Coke is used in the manufacture of steel. Coke is also used as a fuel. Coke is a better fuel than coal because it produces more heat on burning than an equal amount of coal. Moreover, coke burns without producing any smoke whereas coal produces a lot of smoke on burning.

PETROLEUM

Petroleum is a dark coloured, thick crude oil found deep below the ground in certain areas. It has an unpleasant odour. The name 'petroleum' means 'rock oil' (*petra* = rock ; *oleum* = oil). It is called petroleum because it is found under the crust of earth trapped in rocks. Petroleum is not a single chemical compound. Petroleum is a complex mixture of compounds known as hydrocarbons (Hydrocarbons are compounds which are made up of only two elements : carbon and hydrogen). Petroleum is insoluble in water. Petroleum is a natural resource obtained from deep oil wells which are dug in certain areas of the earth. Just like coal, petroleum is also a fossil fuel. Please note that petroleum is also called 'crude oil' or 'mineral oil'.

Figure 4. Petroleum (Crude oil).

How Petroleum was Formed

Petroleum (oil) was formed by the decomposition of the remains of tiny plants and animals buried under the sea millions of years ago. It is believed that millions of years ago, the tiny plants and animals which lived in the sea, died. Their dead bodies sank to the bottom of sea and were soon covered with mud and sand. Due to high pressure, heat, action of bacteria, and in the absence of air, the dead remains of tiny plants and animals were slowly converted into petroleum. The petroleum thus formed got trapped between two layers of impervious rocks (non-porous rocks), forming an oil deposit.

Occurrence and Extraction of Petroleum

Petroleum occurs deep under the surface of earth between two layers of impervious rocks (see Figure 5). Petroleum is lighter than water, so it floats over water. Petroleum oil deposits are usually found mixed with water, salt and earth particles (sand, etc.). Petroleum does not occur in all the places of earth. It is found in only certain areas of the earth. Natural gas occurs above the petroleum oil trapped under the rocks (see Figure 5).

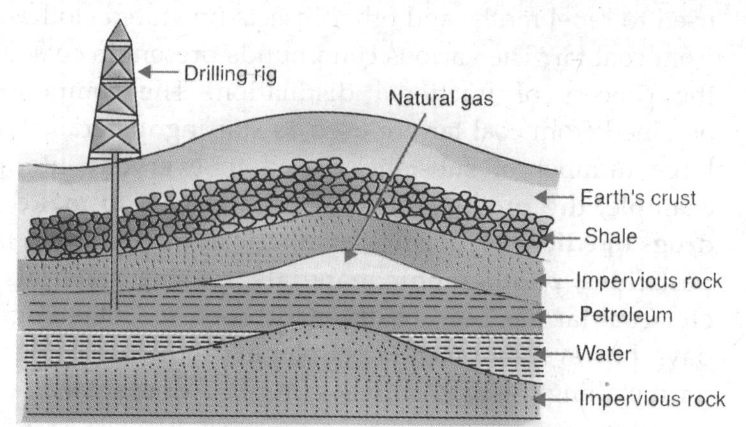

Figure 5. Petroleum and natural gas deposits under the surface of earth.

Petroleum is extracted by drilling holes (called oil wells) in the earth's crust, where the presence of oil has been predicted by survey. The oil wells are drilled by using 'drilling rigs'(A drilling rig is a large structure with equipment for drilling an oil well). When an oil well is drilled through the rocks, natural gas comes out first with a great pressure and for a time, the crude petroleum oil comes out by itself due to gas pressure. After the gas pressure has subsided, petroleum is pumped out of the oil well. Some wells dug into the earth yield both petroleum and natural gas but some wells yield only natural gas but no oil. **Early drilling of oil wells for getting petroleum was done only on land.** Later on, oil wells were also drilled under the sea-bed by using new techniques. Thus, **some of the oil wells are now drilled under the sea for the extraction of petroleum.**

The world's first oil well was drilled in Pennsylvania (USA) in 1859. Eight years later in 1867, oil was struck at Makum in Assam. In India, petroleum (oil) is found in Assam, Gujarat, Mumbai High (off-shore area), and near the basins of Godavari and Krishna rivers. The off-shore oil bearing area called Mumbai High is located in high seas at a distance of about 150 kilometres west of Mumbai city. The oil deposits of Mumbai High are buried at a depth of about 1000 metres below the sea-bed. A special platform of steel has been erected in the Mumbai High sea to pump out petroleum from under the sea-bed.

Refining of Petroleum

The crude petroleum oil is a complex mixture of solid, liquid and gaseous hydrocarbons. It is not very useful to us as such. So, before petroleum can be used for specific purposes, it has to be refined (or purified). **The process of separating crude petroleum oil into more useful fractions is called refining**. The refining of petroleum (or separation of petroleum) into different fractions is based on the fact that the different fractions of petroleum have different boiling point ranges. **The refining of petroleum is carried out in an oil refinery** (see Figure 6). The crude petroleum oil extracted from oil wells is taken to the 'oil refinery' through pipes. In the oil refinery, crude petroleum oil is refined (or separated) into different useful fractions. **The separation of petroleum into different fractions is done by the process of 'fractional distillation'**. Fractional distillation is a process in which fractions of petroleum having different boiling point ranges are collected separately. **The various useful fractions obtained by the refining of petroleum are : Petroleum gas, Petrol, Kerosene, Diesel, Lubricating oil, Paraffin wax and Bitumen.**

Figure 6. An oil refinery.

The Various Fractions of Petroleum and Their Uses

The refining of petroleum gives the fractions (or products) such as petroleum gas, petrol, kerosene, diesel, lubricating oil, paraffin wax and bitumen. The important uses of the various fractions of petroleum are given below.

(*i*) PETROLEUM GAS. Petroleum gas is used as a fuel in homes and industry. Petroleum gas is used as a fuel as such or in the form of Liquefied Petroleum Gas (LPG).

(*ii*) PETROL. Petrol is used as a fuel in light motor vehicles (such as cars, motorcycles, and scooters, etc.). Petrol is also used as a solvent for drycleaning.

(*iii*) KEROSENE. Kerosene is used as a fuel in wick stoves and pressure stoves to cook food. Kerosene is also used in lanterns for lighting purposes (see Figure 7). A special grade of kerosene oil is used as aviation fuel in jet aeroplanes.

(*iv*) DIESEL. Diesel is used as a fuel in heavy motor vehicles (such as buses, trucks, tractors, and diesel train engines). Diesel is also used to run pump sets for irrigation in agriculture and in electric generators (to produce electricity on a small scale).

(*v*) LUBRICATING OIL. Lubricating oil is used for lubrication in machines and engines (like car engines).

Figure 7. Kerosene is used as a fuel in lanterns.

(vi) **PARAFFIN WAX**. Paraffin wax is used for making candles, vaseline, ointments, wax paper, and grease.

(vii) **BITUMEN**. Bitumen is used for road surfacing. It is also used for water-proofing the roofs of buildings. Bitumen is used in making black paints.

Please note that the fuels such as petroleum gas, petrol, kerosene, and diesel are also fossil fuels (because they are obtained from a major fossil fuel called petroleum).

The most common fuel used in homes is liquefied petroleum gas (or LPG). **The petroleum gas which has been liquefied under pressure is called liquefied petroleum gas**. The liquefied petroleum gas (or LPG) consists mainly of butane (C_4H_{10}) (which has been liquefied by applying pressure). Thus, the domestic gas cylinders like 'Indane' contain mainly 'butane' (see Figure 8). *The gas used for domestic cooking is called liquefied petroleum gas because it is obtained from petroleum and it is liquefied by compression before filling into the gas cylinders.* When we turn on the knob of the gas cylinder, the pressure is released, due to which the highly volatile LPG is converted into gas. This gas goes into the burner of LPG stove. When a lighted matchstick is applied to the burner, the gas burns with a blue flame producing a lot of heat. This heat is used for cooking food. **Liquefied petroleum gas (LPG) is a good fuel because of its following advantages :**

Figure 8. These gas cylinders contain LPG.

(i) LPG burns easily.
(ii) LPG has a high calorific value. Due to this, a given amount of LPG produces a lot of heat.
(iii) LPG burns with a smokeless flame and hence does not cause air pollution.
(iv) LPG does not produce any poisonous gases on burning.
(v) LPG does not leave behind any solid residue on burning.

We will now discuss natural gas.

NATURAL GAS

Natural gas consists mainly of methane with small quantities of ethane and propane. In fact, natural gas contains about 95% methane, the remaining being ethane and propane. Natural gas occurs deep under the crust of earth either alone or alongwith oil above the petroleum deposits. Thus, some wells dug into the earth produce only natural gas whereas others produce natural gas as well as petroleum oil. Natural gas is formed under the earth by the decomposition of vegetable matter lying under water. This decomposition is carried out by anaerobic bacteria in the absence of air. Just like coal and petroleum, natural gas is also a fossil fuel.

India has vast reserves of natural gas. In India, natural gas has been found in Tripura, Rajasthan, Maharashtra, and in Krishna-Godavari delta. **When natural gas is compressed by applying pressure, it is called Compressed Natural Gas (which is written in short form as CNG)**. In fact, natural gas is stored under high pressure as compressed natural gas (or CNG). It becomes easier to store, transport and use natural gas in the form of CNG. Natural gas is called a clean fuel because it burns without producing any smoke and does not cause air pollution.

Advantages of Using Natural Gas (or Compressed Natural Gas, CNG)

1. Natural gas (or CNG) is a good fuel because it burns easily and produces a lot of heat. Moreover, natural gas burns with a smokeless flame and causes no air pollution. It also does not produce any poisonous gases on burning. Natural gas does not leave behind any solid residue on burning. Natural gas is, therefore, a clean fuel (as compared to other fossil fuels).

2. Natural gas (or CNG) is a complete fuel in itself and can be used directly for heating purposes in homes and industry. There is no need to add anything else to it.

3. A great advantage of natural gas is that it can be supplied to homes and factories through a network of underground pipes and this eliminates the need for additional storage and transport. Such a network of pipelines for the supply of natural gas exists in Vadodara (in Gujarat), in some parts of Delhi and a few other places.

Uses of Natural Gas (or CNG)

1. Natural gas is used as a domestic and industrial fuel.
2. Natural gas is used as a fuel in Thermal Power Stations for generating electricity.
3. Compressed natural gas (CNG) is being used increasingly as a fuel in transport vehicles (like cars, buses, etc.) in place of petrol and diesel. CNG is a good alternative to petrol and diesel in vehicles because it is a cleaner fuel and does not cause much air pollution. In fact, CNG is being used in many vehicles these days to reduce air pollution in cities. CNG which is used in vehicles is filled in cylinders. These cylinders can be refilled at CNG Filling Stations (see Figure 9).

Figure 9. CNG being filled in a car at a Filling Station.

4. Natural gas is used as a source of hydrogen gas needed to manufacture fertilisers. When natural gas is heated strongly, the methane present in it decomposes to form carbon and hydrogen. This hydrogen is then used to manufacture fertilisers.
5. Natural gas is used as a starting material for the manufacture of a number of chemicals (which are called petrochemicals).

PETROCHEMICALS

Many useful chemicals (or substances) are obtained from petroleum and natural gas. **Those chemicals which are obtained from petroleum and natural gas are called petrochemicals**. Some examples of petrochemicals are : methyl alcohol, ethyl alcohol, formaldehyde, acetone, acetic acid, ethylene, benzene, toluene, vinyl chloride and hydrogen. **Petrochemicals are very important because they are used to manufacture a wide range of useful materials** such as : Detergents, Synthetic fibres (like Polyester, Nylon, Acrylic, etc.), Plastics (such as Polythene, Polyvinyl chloride, Bakelite, etc.), Synthetic rubber, Drugs, Dyes, Perfumes, Fertilisers, Insecticides and Explosives, etc. Hydrogen gas is obtained as a petrochemical from natural gas. Hydrogen gas obtained from natural gas is used in the manufacture of fertilisers (such as ammonium nitrate and urea). Thus, petroleum is not only a source of fuels but also provides raw materials (in the form of petrochemicals) to manufacture a large number of useful substances. Due to its great commercial importance, petroleum is also called "black gold".

Energy Resources of Earth are Limited

Most of the energy that we use today comes mainly from the three exhaustible resources of the earth : coal, petroleum and natural gas **The amount of coal, petroleum and natural gas present in the earth is limited.** The known reserves of coal, petroleum and natural gas will last only for about 100 years. Once the present stock of coal, petroleum and natural gas present in the earth gets exhausted, no new supplies of these fossil fuels will be available to us in the near future (because it takes millions of years to convert the dead organisms into fossil fuels in nature). So, fossil fuels should be used with care and caution, and not wasted at all so that the existing reserves of fossil fuels can be used over as long a period as possible. Moreover, **the burning of fossil fuels is a major source of air pollution. The use of fossil fuels is also linked to global warming** (because they produce a lot of greenhouse gas 'carbon dioxide' on burning). So, the use of lesser fossil fuels will lead to cleaner environment and smaller risk of global warming. From the above discussion we conclude that **we should use fossil fuels only when absolutely necessary because :**

(*i*) it will ensure the availability of fossil fuels for a longer period of time.
(*ii*) it will reduce air pollution and lead to a cleaner environment.
(*iii*) it will reduce the risk of global warming.

Please note that **the fossil fuels such as coal, petroleum and natural gas cannot be prepared in the laboratory** from dead organisms (dead plants and animals). This is because the formation of fossil fuels is a very, very slow process and the conditions for their formation cannot be created in the laboratory.

How to Save Petrol and Diesel

Petrol and diesel are the two main fuels which are used for driving vehicles. We should make every effort to avoid the wastage of these precious fuels. In India, the **Petroleum Conservation Research Association (PCRA)** advises people on how to save petrol (or diesel) while driving vehicles. The various tips for minimising the wastage of petrol and diesel while driving vehicles are as follows :

(*i*) Drive the vehicle at a constant and moderate speed as far as possible.
(*ii*) Switch off the vehicle's engine at traffic lights or at a place where a person has to wait.
(*iii*) Ensure correct air pressure in the tyres of the vehicle. Low tyre pressure consumes more fuel.
(*iv*) Ensure regular maintenance of the vehicle (including engine tuning).

We are now in a position to **answer the following questions :**

Very Short Answer Type Questions

1. Name three useful products of coal.
2. Which product of coal is used as a reducing agent in the extraction of metals ?
3. Name the process by which plant material (or vegetation) buried deep under the earth was slowly converted into coal.
4. Name the product of coal which is thick black liquid having an unpleasant smell.
5. Name any five substances used in everyday life which are manufactured starting from the products of coal tar.
6. Name an important source from which naphthalene balls are obtained.
7. Which substance is used for metalling the roads these days in place of coal tar ?
8. Name the most common fuel used in light motor vehicles.
9. Name the fuel which is used in jet aircraft engines.
10. Name the petroleum product used to drive heavy vehicles.
11. Name the petroleum product which is commonly used for electric generators.
12. What is the full form of LPG ?
13. Is it possible to extract petroleum from under the sea-bed ?
14. What is the full form of CNG ?
15. Name the major component of natural gas.
16. Name any two places in India where natural gas is found.
17. Name a fossil fuel other than coal and petroleum.
18. Name two places in India where coal is found.
19. Name the petroleum product used for surfacing of roads.
20. Name any four places in India where petroleum is found.
21. Write the full form of PCRA.
22. State whether the following statements are true or false :
    (*a*) Coke is almost pure form of carbon.
    (*b*) Coal tar is a mixture of various substances.
    (*c*) Kerosene is not a fossil fuel.
    (*d*) CNG is more polluting than petrol.
    (*e*) Fossil fuels can be made in the laboratory.
23. Fill in the following blanks with suitable words :
    (*a*) Fossil fuels are ...., ... and ....
    (*b*) Coal contains mainly........

(*c*) The slow process of conversion of dead vegetation into coal is called ....

(*d*) The process of separation of different constituents from petroleum is called ....

(*e*) The least polluting fuel for vehicles is ....

(*f*) The burning of fossil fuels causes air.............and also leads to global............

Short Answer Type Questions

24. Explain why, fossil fuels are exhaustible natural resources.

25. Describe how coal was formed. What is this process called ?

26. What happens when coal is heated in air ? State the uses of coal.

27. State the uses of coke.

28. What are the constituents of coal gas ? State one use of coal gas.

29. What are the major products (or fractions) of petroleum refining ? Give one use of each petroleum product.

30. What are the advantages of using natural gas (or CNG) as a fuel ?

31. State the various uses of natural gas.

32. What is CNG ? State its one use.

33. Where is natural gas found ? Why is natural gas called a clean fuel ?

34. What are the advantages of using LPG as fuel ?

35. Name any five useful substances which are manufactured from petrochemicals.

36. Which material is called 'black gold' ? Why ?

37. (*a*) Where and when was the world's first oil well drilled ?

(*b*) Where and when was oil first struck in India ?

38. State one use each of the following products of petroleum :

(*a*) Petroleum gas (*b*) Petrol (*c*) Diesel (*d*) Lubricating oil (*e*) Bitumen

39. What is the major cause of air pollution ? Write the various tips for minimising the wastage of petrol/ diesel while driving vehicles.

40. Why should we use fossil fuels only when absolutely necessary ?

41. State (*a*) two uses of kerosene, and (*b*) two uses of paraffin wax.

Long Answer Type Questions

42. (*a*) What is meant by inexhaustible natural resources ? Name two inexhaustible natural resources.

(*b*) What is meant by exhaustible natural resources ? Name any two exhaustible natural resources.

43. (*a*) What are fossil fuels ? Name three fossil fuels.

(*b*) Describe how, fossil fuels were formed.

44. (*a*) What is petroleum ? Where does petroleum occur ?

(*b*) Describe the process of formation of petroleum.

45. (*a*) What are petrochemicals ? Name any two petrochemicals.

(*b*) Why are petrochemicals so important ?

Multiple Choice Questions (MCQs)

46. Which one of the following is not a fossil fuel ?

(*a*) petrol (*b*) coke (*c*) charcoal (*d*) coal

47. The major component of LPG is :

(*a*) hydrogen (*b*) carbon monoxide (*c*) methane (*d*) butane

48. Which is the major component of CNG ?

(*a*) ethane (*b*) propane (*c*) methane (*d*) butane

49. The gas which occurs above the petroleum oil trapped under the rocks is called :

(*a*) biogas (*b*) petroleum gas (*c*) natural gas (*d*) coal gas

50. Which of the following is being used as a source of hydrogen gas needed to manufacture fertilisers ?

(*a*) biogas (*b*) natural gas (*c*) coal gas (*d*) petroleum gas

51. One of the following is not an exhaustible source of energy. This one is :

(*a*) natural gas (*b*) petroleum gas (*c*) coal gas (*d*) biogas

52. The slow process by which the large land plants and trees buried deep under the earth have become coal is called :
(*a*) carbonation (*b*) carburation (*c*) carbonisation (*d*) carbocation

53. Which of the following is used as a reducing agent in the extraction of iron metal ?
(*a*) coal (*b*) bitumen (*c*) charcoal (*d*) coke

54. Which of the following is usually referred to as 'black gold' ?
(*a*) coke (*b*) coal tar (*c*) petroleum (*d*) coal

55. The various compounds present in coal tar are separated by the process of :
(*a*) simple distillation (*b*) destructive distillation
(*c*) fractional distillation (*d*) fractional crystallisation

56. Which of the following is not obtained as a fraction during the refining of petroleum ?
(*a*) kerosene (*b*) natural gas (*c*) lubricating oil (*d*) bitumen

57. Which one of the following is an inexhaustible natural resource ?
(*a*) coal (*b*) petroleum (*c*) water (*d*) forests

Questions Based on High Order Thinking Skills (HOTS)

58. The substance W is a fossil fuel. It occurs deep below the ground in certain areas of the earth. Another fossil fuel X is found trapped above the deposits of W. When W is subjected to a process called Y, then a number of different products are collected at different temperature ranges which are put to different uses. A special grade of product Z obtained in this way is used as aviation fuel in jet aeroplanes.
(*a*) What are (*i*) W, and (*ii*) X ?
(*b*) What is the physical state of (*i*) W, and (*ii*) X ?
(*c*) Name the process Y.
(*d*) Name the product Z.

59. The material A is a fossil fuel which is extracted from the earth. It is said to be formed from the buried, large land plants by a very slow process B. When A is heated in the absence of air in a process called C, then it gives three products D, E and F. The product D is used as a reducing agent in the extraction of metals, the product E is used as an industrial fuel whereas the product F has been traditionally used for metalling the roads.
(*a*) What could material A be ?
(*b*) What is (*i*) physical state, and (*ii*) colour, of A ?
(*c*) Name the processes (*i*) B, and (*ii*) C.
(*d*) What are (*i*) D, (*ii*) E, and (*iii*) F ?

60. The fossil fuel P is formed under the earth by the decomposition of vegetable matter lying under water by the action of anaerobic bacteria. The major component of fuel P is Q. The fossil fuel P is used as a source of gas R needed to manufacture nitrogenous fertilisers. When P is filled in metal cylinders and used as a fuel in motor vehicles, it is called S. What are P, Q, R and S ?

ANSWERS

2. Coke **22.** (*a*) True (*b*) True (*c*) False (*d*) False (*e*) False **23.** (*a*) coal ; petroleum ; natural gas (*b*) carbon (*c*) carbonisation (*d*) refining (*e*) CNG (*f*) pollution ; warming **46.** (*c*) **47.** (*d*) **48.** (*c*) **49.** (*c*) **50.** (*b*) **51.** (*d*) **52.** (*c*) **53.** (*d*) **54.** (*c*) **55.** (*c*) **56.** (*b*) **57.** (*c*) **58.** (*a*) (*i*) Petroleum (*ii*) Natural gas (*b*) (*i*) Thick liquid (*ii*) Gas (*c*) Fractional distillation (*d*) Kerosene **59.** (*a*) Coal (*b*) (*i*) Solid (*ii*) Black (*c*) (*i*) Carbonisation (*ii*) Destructive distillation (*d*) (*i*) Coke (*ii*) Coal gas (*iii*) Coal tar **60.** P : Natural gas ; Q : Methane ; R : Hydrogen ; S : Compressed natural gas (CNG).

CHAPTER 6

Combustion and Flame

The burning of a substance in the oxygen of air in which heat and light are produced, is called combustion. So, in most simple words, '*combustion*' means '*burning*'. In this Chapter we will study the chemical process of combustion (or burning) and the types of flames produced during this process.

COMBUSTION

A chemical process in which a substance reacts with the oxygen (of air) to give heat and light is called combustion. The light which is given off during combustion can be in the form of a 'flame' or as a 'glow'. For example, wood burns by producing a flame (see Figure 1). But charcoal burns by producing light in the form of glow. The substance which undergoes combustion is said to be combustible. It is also called a fuel. We will now give some examples of combustion.

Figure 1. This picture shows the combustion (or burning) of wood in which a flame is produced.

(*i*) Combustion of Magnesium. If a magnesium ribbon is heated, it starts burning (or undergoes combustion). When a magnesium ribbon burns, it combines with the oxygen of air to form magnesium oxide, and liberates heat and light. The combustion of magnesium can be written as follows :

$$\text{Magnesium} + \underset{\text{(From air)}}{\text{Oxygen}} \xrightarrow{\text{Combustion}} \text{Magnesium oxide} + \text{Heat} + \text{Light}$$

Thus, the burning of magnesium in air to produce heat and light is a combustion process. In this reaction, magnesium is a combustible substance.

(*ii*) Combustion of Charcoal. Charcoal is mainly carbon. If we hold a piece of charcoal with a pair of tongs and heat it on the flame of a burner, it starts burning (or undergoes combustion). When charcoal burns, then the carbon of charcoal combines with the oxygen of air to form carbon dioxide. A lot of heat is produced in this combustion reaction but only a little light is produced (which makes the charcoal glow). The combustion of charcoal can be written as follows :

$$\underset{\text{(Charcoal)}}{\text{Carbon}} + \underset{\text{(From air)}}{\text{Oxygen}} \xrightarrow{\text{Combustion}} \text{Carbon dioxide} + \text{Heat} + \text{Light}$$

Coal also contains a lot of carbon. So, coal also burns in air producing carbon dioxide, heat and light. Thus, charcoal and coal are combustible substances.

In both the examples of combustion given above, we find that **oxygen is necessary for combustion (or burning) to take place**. Actually, **oxygen is a supporter of combustion**. In most of the cases of combustion, oxygen is provided by the air around us. So, in a way, **air is also a supporter of combustion. This is because air contains oxygen**. So, whether we write oxygen as supporter of combustion or air as supporter of combustion, it will mean the same thing.

Food is a fuel for our body. During respiration, the digested food (like glucose) is broken down by reaction with oxygen in the body cells to produce carbon dioxide, water and heat energy. This heat energy is utilised by our body. Thus, **respiration is a kind of slow combustion of food which takes place in the body to produce heat energy.**

We have all seen the brown rust present on iron nails and other iron objects. The rust is formed when iron slowly combines with the oxygen present in air (in the presence of moisture) to form iron oxide. **The process of rusting of iron is an example of slow combustion**. The rusting liberates very little heat but no light.

The sun produces heat and light. The heat and light produced in the sun are not due to ordinary combustion (which takes place in the presence of oxygen of air). **In the sun, heat and light are produced due to nuclear reactions** (in which hydrogen is converted into helium with the release of heat and light).

Combustible and Non-Combustible Substances

Take a piece of paper. Apply a lighted matchstick to this piece of paper. We will find that the piece of paper starts burning. We say that the piece of paper burns (or the piece of paper undergoes combustion). We now take a piece of stone and apply a lighted matchstick to it. We will find that the piece of stone does not burn. We say that the piece of stone does not undergo combustion. This means that all the substances around us do not burn (or do not undergo combustion). So, there are two types of substances around us :

(*i*) **Combustible substances,** and

(*ii*) **Non-combustible substances.**

Those substances which can burn are called combustible substances. In other words, *those substances which can undergo combustion are called combustible substances.* Some of the combustible substances are : Paper, Cloth (Fabrics), Straw (Dry grass), Cooking gas (LPG), CNG, Kerosene oil, Wood, Charcoal, Coal, Cow-dung cakes, Petrol, Diesel, Alcohol, Matchstick and Magnesium ribbon, etc. ***A combustible substance is also called a fuel.***

Those substances which do not burn are called non-combustible substances. In other words, *those substances which do not undergo combustion are called non-combustible substances.* Some of the non-combustible substances are : Stone, Glass, Cement, Bricks, Soil, Sand, Water, Iron nails, Copper objects and Asbestos, etc. From this discussion we conclude that some of the substances around us are combustible whereas others are non-combustible.

We will now learn the various conditions which are necessary for combustion (or burning) of a substance to take place. These can also be considered to be the conditions necessary to start a fire.

CONDITIONS NECESSARY FOR COMBUSTION

There are three conditions which are necessary for combustion to take place. These are :

1. **Presence of a combustible substance (A substance which can burn)**
2. **Presence of a supporter of combustion (like air or oxygen)**
3. **Heating the combustible substance to its ignition temperature**

We will now discuss these three conditions required for combustion (or burning) of substances in detail.

1. Combustible Substance

The presence of a combustible substance is necessary for combustion to take place. So, when fire starts in a room, we remove all the combustible substances like wooden furniture, clothes, books and papers, etc., quickly from the room so that the fire may not spread due to the presence of a large number of combustible substances. *A combustible substance is actually the 'food for fire'.*

2. Supporter of Combustion

The most common supporter of combustion which we have around us is air. So, we can say that **air is necessary for combustion**. We can demonstrate that air is necessary for combustion by performing a simple activity as follows.

ACTIVITY

Light a candle with a burning matchstick and fix it on a table. We will see that this candle keeps on burning [see Figure 2(*a*)]. This uncovered candle keeps burning because it is getting continuous supply of fresh air from the surroundings. Let us now cover the burning candle with an inverted gas jar. We will see that the candle stops burning after some time. In other words, the candle gets extinguished [see Figure 2(*b*)]. The candle stops burning (or gets extinguished) because the supply of fresh air to the burning candle is cut off by the gas jar cover. Since no fresh air is available to the burning candle, it stops burning (or gets extinguished). This observation shows that air is necessary for combustion (or burning) to take place.

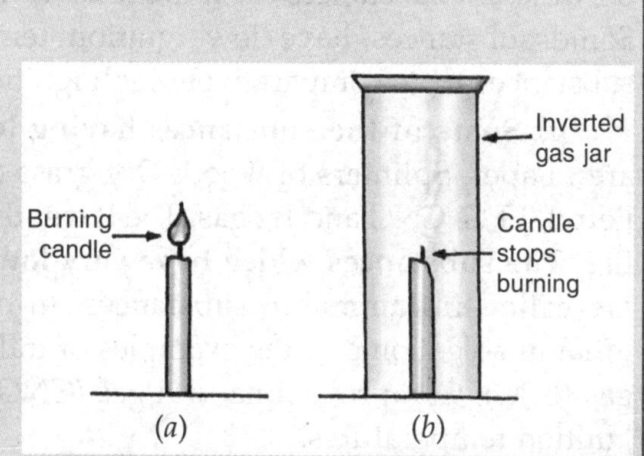

Figure 2. Activity to show that air is necessary for combustion (or burning).

We will now give some examples from our everyday life which will tell us that **a supporter of combustion like air is necessary for combustion to take place.**

(*i*) If burning charcoal is covered with a vessel, it stops burning after some time, that is, the charcoal fire gets extinguished after some time. Actually, when we cover the burning charcoal with a vessel, the supply of supporter of combustion (air) to the burning charcoal is cut off and hence the charcoal fire stops.

(*ii*) When the clothes of a person catch fire, the person is covered with a blanket to extinguish the fire. This is because when the burning clothes of a person are covered with a blanket, the supply of air to the burning clothes is cut off due to which the clothes stop burning – the fire gets extinguished.

From all the above examples we find that when the supply of supporter of combustion (air) to a burning substance is cut off, then the process of burning (or combustion) also stops. Thus, *a supporter of combustion (like air) is necessary for combustion to take place.* If, however, there is no supporter of combustion (like air), then a combustible substance cannot burn even if it is heated to its ignition temperature.

3. Ignition Temperature

Before a combustible substance can catch fire and burn, it must be heated to a certain *minimum* temperature by supplying heat from outside. **The lowest temperature at which a substance catches fire and starts burning, is called its ignition temperature**. It is necessary to heat a combustible substance to its ignition temperature so that it may undergo combustion (or burn). The ignition temperature of paper is 233°C. This means that a piece of paper has to be heated at least to a temperature of 233°C so that it may catch fire and start burning. A combustible substance cannot catch fire (or burn) as long as its temperature is lower than its ignition temperature.

We usually apply a burning matchstick (or lighter) to make a substance burn. This burning matchstick supplies heat to raise the temperature of the substance to its ignition temperature and make it burn. This will become more clear from the following example : A piece of paper does not catch fire (or does not burn) at the room temperature because the ignition temperature of paper is much higher than the room temperature. When we apply a burning matchstick to the piece of paper, it starts burning (see Figure 3). This is because the heat produced by burning matchstick heats the piece of paper to its ignition temperature and makes it burn (or undergo combustion).

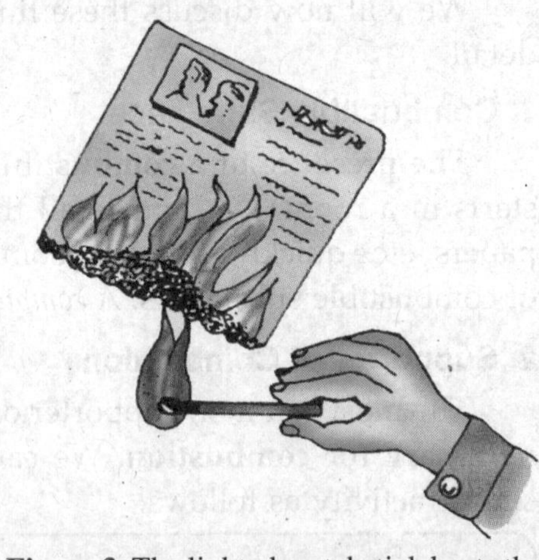

Figure 3. The lighted matchstick heats the paper to its ignition temperature due to which the paper catches fire and starts burning.

The ignition temperatures of different substances are different. So, different substances catch fire and burn at different temperatures. Some substances have low ignition temperatures whereas other substances have comparatively high ignition temperatures.

(*i*) Some of the substances having low ignition temperatures are : Paper, Splinters of wood, Dry grass (Straw), White phosphorus, Cloth (Fabrics), Alcohol, Kerosene, Petrol, LPG, CNG and Biogas. The lower the ignition temperature, the more easily the substance will catch fire. **The substances which have very low ignition temperatures and can easily catch fire with a flame are called inflammable substances.** In other words, a substance which is easily set on fire is called inflammable. Some of the examples of inflammable substances are : Petrol, Alcohol, Liquefied petroleum gas (LPG), Compressed natural gas (CNG) and Biogas. All these inflammable substances have very low ignition temperatures.

The fuels having very low ignition temperatures are very dangerous to use. For example, **petrol has a much lower ignition temperature than that of kerosene**. Due to its very low ignition temperature, a can full of petrol catches fire very easily on being lighted with a matchstick and burns explosively. This is why, petrol is not used in stoves. On the other hand, kerosene has a comparatively higher ignition temperature due to which it burns smoothly in a kerosene stove.

(*ii*) Some of the substances having high ignition temperatures are : Coal, Charcoal, Log of wood and Cow-dung cakes (*Uple*). The substances having high ignition temperatures catch fire with difficulty. The substances having high ignition temperatures burn only on strong heating. They cannot be burnt directly by a lighted matchstick.

We will now describe **how a matchstick is lighted**. A matchstick does not catch fire and burn on its own at room temperature because the ignition temperature of matchstick is much higher than the room temperature. A matchstick is lighted by rubbing it on the rough surface provided on the side of the matchbox (see Figure 4). The heat produced by friction raises the temperature of the chemicals present on the matchstick head to their ignition temperature. Due to this, the chemicals present on the head of the matchstick catch fire and the matchstick starts burning (see Figure 4). Thus, **a matchstick starts burning on rubbing it on the side of the matchbox because the heat produced by friction heats the chemicals at the head of the matchstick to their ignition temperature and make it catch fire.**

Figure 4. Lighting of a matchstick.

Kerosene oil and wood do not catch fire on their own at room temperature. This is because the ignition temperatures of kerosene oil and wood are higher than the room temperature. Now, if kerosene is heated a little (say, by a burning matchstick), it will catch fire easily. This is because kerosene oil has a comparatively low ignition temperature which is reached even on little heating. But if wood is heated a little, it does not catch fire. This is because wood has a much higher ignition temperature which is not reached by the little heat being supplied to it by a matchstick. From this we conclude that the ignition temperature of kerosene oil is much lower than that of wood. Since kerosene oil has a low ignition temperature and it can catch fire easily, we have to take special care in storing kerosene oil. **Sometimes we see cooking oil in a frying pan catching fire when the frying pan is kept on the burning gas stove for a long time.** This happens because the cooking oil gets heated to its ignition temperature when kept over a burning stove for a long time.

We know that **a matchstick can light a tiny splinter of wood but not a big log of wood.** This can be explained as follows : A splinter of wood has a low ignition temperature. Now, a burning matchstick can produce sufficient heat to reach the ignition temperature of the splinter of wood (which is low), therefore, a matchstick can light (or burn) a splinter of wood directly. The ignition temperature of a log of wood is high which cannot be reached by the small heat produced by a burning matchstick. So, a matchstick cannot light (or burn) a log of wood directly. In order to burn a log of wood (as in a *chulha*), a small fire is first started by burning straw (or dry grass) with a matchstick, and then the log of wood is placed over this fire. The considerable heat of this fire then heats the log of wood to its ignition temperature due to which the log of wood starts burning.

Coal has a high ignition temperature, so a coal fire cannot be started by using a lighted matchstick directly. This is because the small heat produced by burning matchstick is not sufficient to heat the coal to its ignition temperature (which is high). **A coal fire is started indirectly as follows :** A piece of cloth is dipped in kerosene oil and some pieces of wood are arranged over it. When the kerosene soaked piece of cloth is ignited by a lighted matchstick, it starts burning. The heat produced by the burning of kerosene soaked cloth makes the pieces of wood to burn. The coal pieces are then placed over the burning wood pieces. The large heat produced by the burning wood pieces heats the coal to its ignition temperature due to which the coal also starts burning. This starts the coal fire.

Figure 5. A forest fire.

Sometimes we hear of 'forest fires' which occur on their own. **The forest fires occur during the hottest summer days.** This happens as follows : During extreme heat of summer, sometimes the ignition temperature of dry grass in the forest is reached, which makes the dry grass catch fire. From the burning grass, the fire spreads to bushes and trees, and very soon the whole forest is on fire (see Figure 5). It is very difficult to control such forest fires. We will now describe an activity which shows that it is essential for a substance to reach its ignition temperature to 'catch fire' and 'start burning'.

ACTIVITY

Make two paper cups by folding two round sheets of paper. Keep one paper cup empty but put about 50 mL of water in the other paper cup. Heat both the paper cups separately over a candle flame (see Figure 6). We will see that the empty paper cup catches fire easily and starts burning [see Figure 6(*a*)]. On the other hand, the paper cup containing water does not catch fire. The water in this paper cup becomes hot gradually [see Figure 6(*b*)]. If we continue heating this paper cup, we can even boil the water in it (without the paper cup catching fire). These observations can be explained as follows :

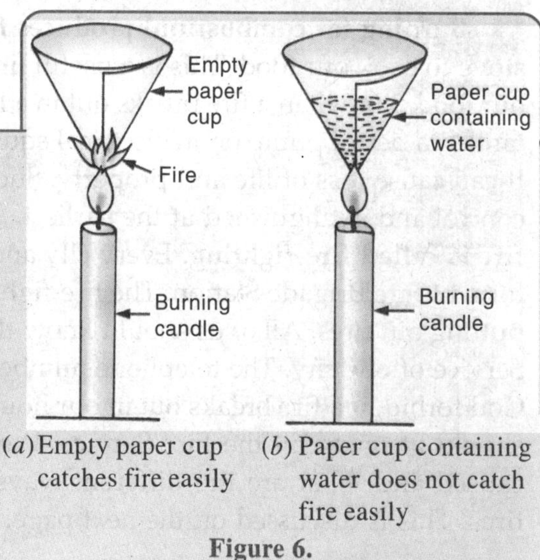

(*a*) Empty paper cup catches fire easily

(*b*) Paper cup containing water does not catch fire easily

Figure 6.

(*i*) When we heat the empty paper cup over the candle flame, then the ignition temperature of paper is reached quickly. Due to this, the empty paper cup catches fire quickly and starts burning.

(*ii*) When we heat the paper cup containing water, then the heat supplied to the paper cup is transferred to water inside it by conduction. Due to the continuous transfer of heat from paper cup to water, the paper cup does not get heated too much and its ignition temperature is not reached. So, in the presence of water, the ignition temperature of paper cup is not reached, and hence the paper cup does not catch fire (or does not burn).

We know that **it is difficult to burn a heap of green leaves but dry leaves catch fire easily**. This can be explained as follows : The green leaves contain a lot of water. This water does not allow the green leaves to get heated to their ignition temperature easily and makes the burning of green leaves difficult. On the other hand, since dry leaves do not contain water, they get heated to their ignition temperature easily and hence catch fire easily.

The History of Matchstick

A short, thin piece of wood having chemicals coated at one end, which is used to light a fire by rubbing against a rough surface, is called matchstick. The history of matchstick is very old. About 5000 years ago, small, thin pieces of pinewood dipped in sulphur at one end, were used as matchsticks in ancient Egypt. The modern safety match (or matchstick) was developed only about 200 years ago.

Earlier, a mixture of antimony trisulphide, potassium chlorate and white phosphorus with some glue and starch was applied to the head of a matchstick made of suitable wood. When the head of this matchstick was rubbed against a rough surface, white phosphorus got ignited due to the heat of friction. This started the combustion (or burning) of matchstick. White phosphorus is poisonous. So, the use of white phosphorus in making matchsticks proved to be dangerous for the workers engaged in making matchsticks as well as for the users.

These days, the head of matchstick (or safety match) contains only antimony trisulphide and potassium chlorate. The rough rubbing surface on the side of the matchbox has a coating of powdered glass and a little red phosphorus (Red phosphorus is much less dangerous than white phosphorus). When the matchstick is rubbed against the rough surface of matchbox, some red phosphorus is converted into white phosphorus. This white phosphorus immediately reacts with potassium chlorate in the matchstick head to produce sufficient heat to ignite antimony trisulphide and make the matchstick head burn. We will now describe how to control unwanted fires.

HOW DO WE CONTROL FIRE

Burning (or combustion) produces fire. Fire is useful as well as harmful. We make small fire in a gas stove to cook our food. This is a useful fire. Without fire, we cannot cook our food. But when a fire breaks out in a house, an office, a factory, an oil tanker, a petrol pump or an electrical equipment, then the fire is harmful. It can cause loss of life and property. Such fires must be brought under control and extinguished at the earliest. **The process of extinguishing a fire is called fire-fighting.** Every city and town has a Fire Service in the form of Fire Brigade Station. The fire-fighters of Fire Brigade specialise in putting out fires. All of us should know the telephone number of the Fire Service of our city. The telephone number of Fire Service in Delhi is 101. God forbid, if a fire breaks out in our house or in our neighbourhood, we should at once call the Fire Service. We should also know how fire is caused and what are the different ways of fire-fighting or putting off fires. This is discussed on the next page.

Figure 7. A building on fire.

Any fire needs three things to be present : Fuel (Combustible substance), Air (or Oxygen) and Heat. If any one of these three things is removed, then the burning will stop and fire will be extinguished. Thus, a fire can be extinguished in three ways :

1. By removing the fuel (combustible substances)
2. By removing the heat (by cooling with water)
3. By cutting off the air supply to the burning substances (with carbon dioxide, etc.)

We will now discuss these three ways of extinguishing fire in detail, one by one.

1. Remove the Fuel (or Combustible Substances)

A fuel (or combustible substance) is a food for fire. So, when fire starts in a room, all the combustible substances like furniture, clothes, and books, etc., (which can burn easily) should be removed at once so that fire may not spread. If possible, cooking gas cylinder should be removed and electricity should be switched off. If the fire is near a pile of wood that could provide fuel to keep the fire going, the pile of wood should be removed from there as soon as possible. It is, however, usually not possible to remove all the combustible materials from the place of fire.

2. Remove the Heat

Water is used to remove heat from a burning substance and to make it too cool to burn further. Water is the most common fire extinguisher for ordinary fires. Water extinguishes fire by cooling the burning substances. ***When water is thrown on a burning substance, it gets cooled below its ignition temperature and stops burning***. The fire gets extinguished. For example, when fire brigade man throws a strong stream of water on a building on fire, the burning materials get cooled to below their ignition temperatures and fire is extinguished. The water vapour produced by the action of heat of fire on water surround the burning material and help in cutting off the supply of air. This also helps in extinguishing fire. Thus, fire-men extinguish the fire by throwing water under pressure on the burning things such as houses, factories or other buildings. Water works as a fire extinguisher only when things like wood and paper, etc., are on fire. The fires caused by burning oil (or petrol) or those caused by electricity, however, cannot be extinguished by using water. This is explained below.

The fire produced by burning oil and petrol (like fire in frying pan, oil tanks, petrol pumps and airports, etc.) cannot be extinguished by using water. This is because of the following reason : Water is heavier than oil and petrol. So, when water is thrown over burning oil (or petrol), it (water) settles down. The oil (or petrol) floats on water and continues to burn. Thus, fires caused by burning oil (or petrol) cannot be extinguished by pouring water over it.

The fires caused by electrical short-circuit in an electrical appliance or in electric wiring should not be extinguished by throwing water. This is because of the following reason : Ordinary water conducts electricity to some extent. So, when water is thrown over the burning electrical appliance (or burning electric wires), it can give electric shock to the persons involved in fire-fighting. Thus, water cannot be used to extinguish fires caused by electricity.

3. Cut Off the Air Supply

Many fires can be extinguished by cutting off air supply to the burning substances. The air supply to a burning substance can be cut off in a number of ways such as covering the burning substances with carbon dioxide, sand (or soil), a blanket or a damp cloth, etc.

The electrical fires are extinguished by using carbon dioxide gas fire extinguisher. Carbon dioxide gas is denser than air and forms a layer around the burning substances. Carbon dioxide layer covers the fire like a blanket due to which fresh air cannot reach the burning substances. The burning substance does not get oxygen of air and hence stops burning. In this way, the fire gets extinguished. Please note that carbon dioxide gas neither burns itself nor supports

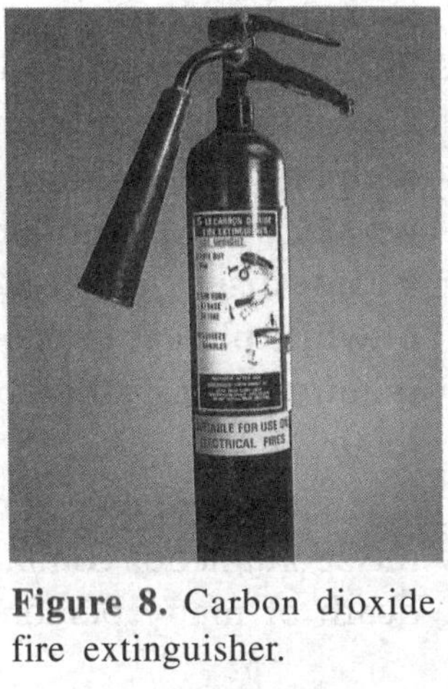

Figure 8. Carbon dioxide fire extinguisher.

burning (or combustion). An added advantage of carbon dioxide is that it does not harm the electrical equipment. The fires caused by the burning of inflammable materials like oil or petrol are also extinguished by using carbon dioxide fire extinguishers.

Carbon dioxide used for extinguishing fire can be stored as a liquid at high pressure in cylinders (called fire extinguishers). When released from the cylinder, carbon dioxide expands enormously in volume and cools down. In this way, carbon dioxide not only forms a blanket around the burning substance, it also cools down the burning substance. This makes carbon dioxide an excellent fire extinguisher. Another way to obtain carbon dioxide for extinguishing a fire is to release a lot of dry powder of chemicals like sodium bicarbonate (baking soda) or potassium bicarbonate over the fire. The heat of fire decomposes these chemicals to produce carbon dioxide gas. And this carbon dioxide then extinguishes the fire.

A small fire can be extinguished by throwing sand (or soil) over it. For example, when sand is thrown over burning kerosene oil, the sand covers it like a blanket. The sand cuts off the air supply to the burning kerosene oil due to which the fire gets extinguished. **The cooking oil fire in a frying pan in the kitchen can be extinguished by covering the pan with a fire blanket or a damp cloth**. When the frying pan is covered with a fire blanket or a damp cloth, the supply of air to the burning cooking oil is cut off and hence the fire gets extinguished. **If the clothes of a person working in the kitchen catch fire, the person is immediately covered with a blanket.** When the burning clothes of a person are covered with a blanket, the supply of air to the burning clothes is cut off and hence the burning (or fire) stops.

TYPES OF COMBUSTION

There are various types of combustion. The three important types of combustion are :

(*i*) **Rapid combustion,**

(*ii*) **Spontaneous combustion,** and

(*iii*) **Explosive combustion (or Explosion).**

We will now describe these three types of combustion reactions in a little more detail, one by one.

1. Rapid Combustion

The combustion reaction in which a large amount of heat and light are produced in a short time is called rapid combustion. When we bring a lighted matchstick or a lighter near the burner of a gas stove in the kitchen, the cooking gas starts burning at once producing a lot of heat and some light. So, **the immediate burning of cooking gas (LPG) in a gas stove to give heat and light, is an example of rapid combustion**. The burning of kerosene oil in a kerosene stove and the burning of wax in a candle are also examples of rapid combustion.

2. Spontaneous Combustion

The combustion reaction which occurs on its own (without the help of any external heat), is called spontaneous combustion. In spontaneous combustion, the substance suddenly bursts into flames and starts burning (even without being heated). Spontaneous combustion takes place at room temperature. The heat required for spontaneous combustion is produced inside the substance by its slow oxidation. Spontaneous combustion is usually undergone by those substances which have quite low ignition temperatures. White phosphorus is a substance which undergoes spontaneous combustion. In other words, white phosphorus burns in air at room temperature. So, if we keep a piece of white phosphorus in a china dish, we will see that it catches fire by itself and starts burning (without being heated). **The burning of white phosphorus on its own at room temperature is an example of spontaneous combustion**. The heat required to start this spontaneous combustion is produced internally by the slow oxidation of phosphorus in air.

The spontaneous combustion reactions which take place in nature are very dangerous. For example, the spontaneous combustion of coal dust has resulted in many disastrous fires in coal mines (leading to the death of many persons working in deep coal mines). Forest fires can also be started by spontaneous combustion reactions. Sometimes, due to the heat of the sun (or due to the spark of lightning from the sky), spontaneous combustion of straw and forest wood takes place leading to forest fires. Most of the forest fires are, however, caused due to the carelessness of human beings. When human beings throw away lighted cigarettes in the forest or leave behind a burning campfire after having a picnic in the forest, then forest fires are started. It is, therefore, important that burning cigarette butts, etc., are not thrown in the forest, and campfires are completely extinguished before leaving the forest after a visit.

3. Explosive Combustion (or Explosion)

A very fast combustion reaction in which a large amount of heat, light and sound are produced, is called explosive combustion (or explosion). A large amount of gases is released quickly in an explosive combustion. It is the rapid expansion of these gases which causes a loud sound (or explosion). **The fireworks (crackers, etc.) which we explode during festivals work on the explosive combustion of substances**. When a cracker is ignited with a burning matchstick, the chemicals present in it undergo a sudden (very rapid) combustion producing heat, light and a large volume of gases. The gases produced are heated by the heat evolved in the reaction. The hot gases expand rapidly and cause an explosion (producing a loud sound). Explosive combustion (or explosion) can also take place if pressure is applied on the cracker by hitting it hard. We will now discuss the types of fuels.

FUELS

A material which is burnt to produce heat is called a fuel. Some of the common fuels are : **Wood, Coal, LPG (Cooking gas), Kerosene, Petrol, Diesel, Natural gas and Biogas**. A fuel is a very good source of heat energy. The heat energy produced by burning a fuel can be used directly to cook food and for running motor vehicles and factory machines or it can be converted into electrical energy at thermal power stations. And this electrical energy (or electricity) is then used for various purposes. The fuels which we use for various purposes can be in the form of a solid, a liquid or a gas. Thus, **there are three types of fuels : Solid fuels, Liquid fuels and Gaseous fuels.**

Figure 9. Coal burns to produce heat, so it is a fuel.

(*i*) Some of the examples of *solid fuels* are : Wood, Charcoal, Coal, Coke, Agricultural wastes and Cow-dung cakes (*Uple*).

(*ii*) The examples of *liquid fuels* are : Kerosene, Petrol, Diesel, and Alcohol (Ethanol).

(*iii*) The examples of *gaseous fuels* are : Natural gas, Petroleum gas, Biogas and Coal gas.

We use different types of fuels for various purposes at home, in industry and in transport for running automobiles. Some of the fuels used in homes are wood, charcoal, coal, LPG and kerosene. Some of the fuels used in industry are coal and natural gas. The fuels used in running automobiles (or vehicles) are petrol, diesel and CNG.

Fuel Efficiency : Calorific Value of Fuels

All the fuels produce heat on burning. All the fuels, however, do not produce the same amount of heat. **Different fuels produce different amounts of heat on burning.** Some fuels produce more heat whereas other fuels produce less heat. The efficiency of a fuel is expressed in terms of its calorific value. **The amount of heat produced by the complete burning (or complete combustion) of 1 kilogram of a fuel is called its calorific value.** The calorific value of a fuel is expressed in the unit of **'kilojoules per kilogram'** (which is written in short form as kJ/kg). We know that kerosene is a fuel. Now, when 1 kilogram of kerosene is burned completely, then 45000 kilojoules of heat energy is produced. So, the calorific value of kerosene is

45000 kilojoules per kilogram (or 45000 kJ/kg). The 'calorific value' of a fuel is also known as 'heat value' of the fuel. The calorific values of some of the common fuels are given below.

Calorific Values of Some Common Fuels

Fuel	Calorific value
1. Cow-dung cakes (*Uple*)	6000 to 8000 kJ/kg
2. Wood	17000 to 22000 kJ/kg
3. Coal	25000 to 33000 kJ/kg
4. Biogas	35000 to 40000 kJ/kg
5. Petrol	45000 kJ/kg
6. Kerosene	45000 kJ/kg
7. Diesel	45000 kJ/kg
8. Methane	50000 kJ/kg
9. CNG (Compressed Natural Gas)	50000 kJ/kg
10. LPG (Liquefied Petroleum Gas)	55000 kJ/kg
11. Hydrogen gas	150000 kJ/kg

From the above table we can see that the calorific value of LPG is 55000 kJ/kg. **By saying that the calorific value of LPG is 55000 kJ/kg, we mean that if 1 kilogram of LPG is burnt completely, then it will produce 55000 kilojoules of heat energy.** The calorific value of LPG is much higher than the calorific value of cow-dung cakes and coal. This means that the burning of a given amount of LPG will produce much more heat than burning the same amount of cow-dung cakes or coal. So, if we are asked to heat some water to boil it, then we should prefer to use LPG as fuel rather than cow-dung cakes or coal (because LPG produces much more heat on burning than an equal amount of cow-dung cakes or coal). We will now solve some problems based on calorific value.

Sample Problem 1. In an experiment, 4.5 kg of a fuel was completely burnt. The heat produced was measured to be 180,000 kJ. Calculate the calorific value of the fuel.

(NCERT Book Question)

Answer. Calorific value of a fuel is the heat produced by burning 1 kg of fuel. Now,

Heat produced by burning 4.5 kg fuel = 180000 kJ

So, Heat produced by burning 1 kg fuel $= \frac{180000 \times 1}{4.5}$

$= 40000$ kJ

Thus, the calorific value of the given fuel is 40000 kJ/kg.

Sample Problem 2. Arrange the following fuels in the increasing order of their calorific values (keeping the fuel with the lowest calorific value first) :

Kerosene, Hydrogen gas, LPG, Coal, Wood

Answer. Wood, Coal, Kerosene, LPG, Hydrogen gas
(Lowest) (Highest)

Characteristics of an Ideal Fuel (or Good Fuel)

An ideal fuel (or a good fuel) has the following characteristics :

(*i*) **It has a high calorific value.** That is, it produces a large amount of heat (per unit mass).

(*ii*) **It burns easily in air at a moderate rate.** That is, it burns neither too fast nor too slow.

(*iii*) **It has a proper ignition temperature** (which is neither very low nor very high).

(*iv*) **It does not produce any harmful gases or leaves any residue after burning** (which may pollute the environment).

(*v*) **It is cheap, readily available, and easy to transport.**

Please note that there is perhaps no fuel which can be considered to be an ideal fuel. We should choose a fuel which fulfils most of the requirements for a particular use.

FLAME

If we heat one end of a magnesium ribbon over a burner, we find that the magnesium ribbon burns by producing a brilliant white flame. We have also seen a candle flame, a kerosene lamp flame and a Bunsen burner flame (see Figure 10). Actually, flame is the 'blaze' of a fire. It is called *'jwala'* or *'lapat'* in Hindi.

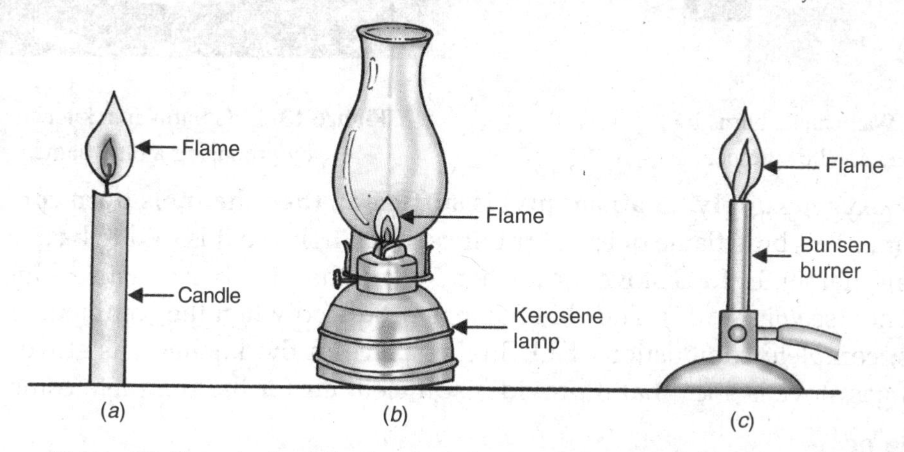

Figure 10. Flames of candle, kerosene lamp and Bunsen burner (Gas burner).

A flame is a region where combustion (or burning) of gaseous substances takes place. All the gases which undergo combustion (or burn) produce flame. But only those solid and liquid fuels which vaporise on being heated, burn with a flame. In other words, only those solid and liquid fuels which form gases on being heated, burn with a flame. LPG and biogas are gases which undergo combustion, so LPG and biogas burn by producing a flame. Wax and camphor are solid substances which vaporise (or form gases) on heating, so wax and camphor burn with a flame. Similarly, kerosene oil and mustard oil are liquids which form vapours (or gases) on being heated, so kerosene oil and mustard oil also burn by producing flames. Thus, **some of the substances which burn by producing flames are : LPG, biogas, wax (in the form of candle), camphor, magnesium, kerosene oil and mustard oil.** Wax candle and kerosene oil lamp have wicks. Molten wax and kerosene oil rise through the wick, get vaporised during burning and form flames. On the other hand, charcoal is a solid fuel which does not vaporise on heating. So, **charcoal does not burn by producing a flame.** Charcoal only glows on combustion. Similarly, coal is a solid fuel which does not vaporise on heating. So, **coal also does not burn by producing a flame.** Coal just glows red on combustion (see Figure 11).

Figure 11. Coal does not burn by producing a flame. It just glows.

When fuels burn, the type of flame produced depends on the proportion of oxygen (of air) which is available for burning the fuel (or for combustion of fuel).

(*i*) When the oxygen supply (or air supply) is insufficient, then the fuels burn incompletely producing mainly a yellow flame. The yellow flame is caused by the glow of hot unburnt carbon particles produced due to incomplete combustion of fuel. This yellow flame produces light, so it is said to be a luminous flame (or light-giving flame). When wax burns in the form of a candle, it burns with a yellow, luminous flame (see Figure 12). Thus, the colour of candle flame is mainly yellow. When kerosene is burned in a lamp, it also burns with a yellow, luminous flame.

Figure 12. Wax candle burns by producing a yellow flame.

Figure 13. LPG burns in a kitchen stove by producing a blue flame.

***(ii)* When the oxygen supply (or air supply) is sufficient, then the fuels burn completely producing mainly a blue flame**. This blue flame does not produce much light, so it is said to be a non-luminous flame (or non light-giving flame). In LPG stove (or kitchen stove), the LPG burns with a blue flame (which is a non-luminous flame) (see Figure 13). The blue flame is produced when the complete combustion of a fuel takes place. Thus, complete combustion of LPG takes place in the kitchen gas stove. The design of the burner of kitchen gas stove is such that it provides sufficient air for the complete combustion of LPG .

Structure of a Flame

A flame consists of three zones (or three parts). These are : innermost zone, middle zone and outer zone. The three zones of a flame have different colours and different temperatures. We will now describe the three zones of a flame in detail by taking the example of candle flame.

***(i)* The innermost zone of a flame is dark (or black)** (see Figure 14). The innermost zone of a flame consists of hot, unburnt vapours of the combustible material (say, wax vapours). The innermost zone is the least hot part of the flame. In other words, we can also say that the innermost zone (or dark zone) is the coldest part of the flame.

***(ii)* The middle zone of a flame is yellow.** It is bright and luminous (light giving) (see Figure 14). The fuel vapours burn partially in the middle zone because there is not enough air for burning in this zone. The partial (or incomplete) burning of fuel in the middle zone produces carbon particles. These carbon particles become white hot and emit light. So, it is the glow of hot carbon particles which makes the middle zone of a flame luminous (or light-giving). These carbon particles then leave the flame as smoke and soot. The middle zone (or luminous zone) of a flame produces a moderate temperature. This zone is the major part of a candle flame.

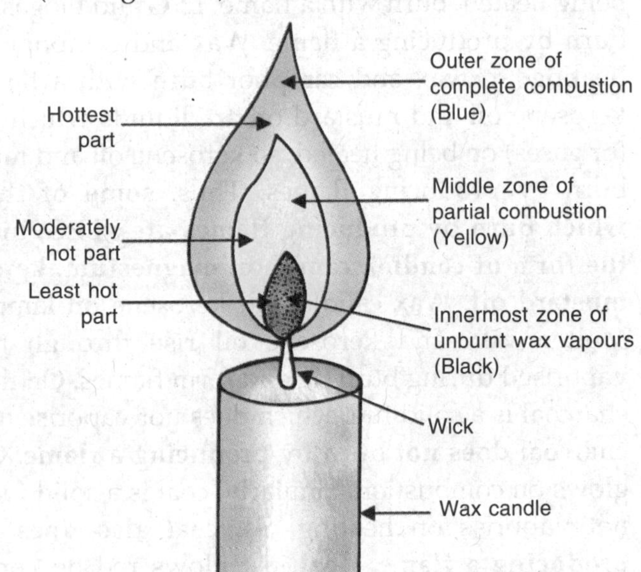

Figure 14. The three zones of a candle flame.

***(iii)* The outer zone of a flame is blue.** It is non-luminous zone (which does not produce much light) (see Figure 14). In the outer zone of a flame, complete combustion of the fuel takes place because there is plenty of air around it. The outermost zone (or non-luminous zone) has the highest temperature in the flame. In other words, the outermost zone (or non-luminous zone) is the hottest part of the flame. The outermost zone of a flame is quite thin as compared to the middle zone.

ACTIVITY TO SHOW THAT THE INNERMOST ZONE OF A CANDLE FLAME CONSISTS OF UNBURNT WAX VAPOURS

Take a wax candle, fix it on a table and light it with a matchstick. Hold a thin glass tube with a pair of tongs and introduce one end of this glass tube in the innermost zone (dark zone or black zone) of the candle flame (see Figure 15). Now bring a lighted matchstick near the other end of the glass tube. We will see a flame at this end of the glass tube (see Figure 15). This can be explained as follows : The innermost zone (or dark zone) of candle flame near the heated wick consists of unburnt wax vapours. Some of these wax vapours enter the glass tube and come out from its other end. When we bring a lighted matchstick near this end of glass tube, the wax vapours coming out of it start burning, producing a flame. This activity shows that the innermost zone (dark zone or black zone) of a candle flame consists of unburnt wax vapours.

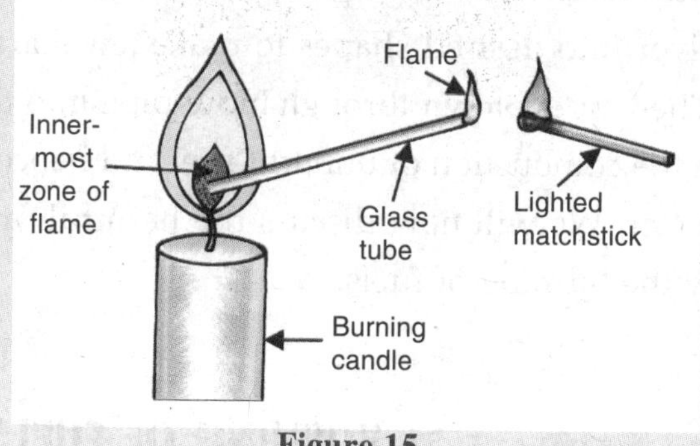

Figure 15.

ACTIVITY TO SHOW THAT THE MIDDLE ZONE OF CANDLE FLAME CONSISTS OF UNBURNT CARBON PARTICLES

Light a candle. Hold a clean glass plate with the help of a pair of tongs and introduce it in the middle zone (or luminous zone) of the candle flame (see Figure 16). Hold the glass plate in this position for about 10 seconds. Then remove the glass plate from candle flame and observe it carefully. We will find that a blackish ring is formed on the glass plate (see Figure 16). This blackish ring is produced due to the deposition of unburnt carbon particles present in the luminous zone of the candle flame. This activity shows that the partial combustion of wax vapours in the middle zone produces unburnt carbon particles.

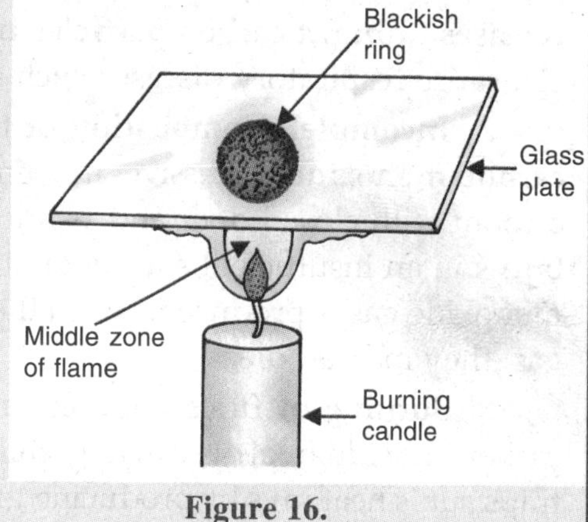

Figure 16.

ACTIVITY TO SHOW THAT THE OUTERMOST ZONE (NON-LUMINOUS ZONE) OF A FLAME IS THE HOTTEST

Take a long copper wire and hold its one end with a pair of tongs. Introduce the other end of copper wire just inside a burning candle flame so that it is in the outermost zone (non-luminous zone) of flame (as shown in Figure 17). Keep the copper wire in this position for about 30 seconds. We will see that the part of copper wire which is in the outermost zone of the flame becomes red hot. This tells us that the outermost, non-luminous zone of a flame has a high temperature. In other words, the non-luminous zone (or outermost zone) of a flame is the hottest part of a flame.

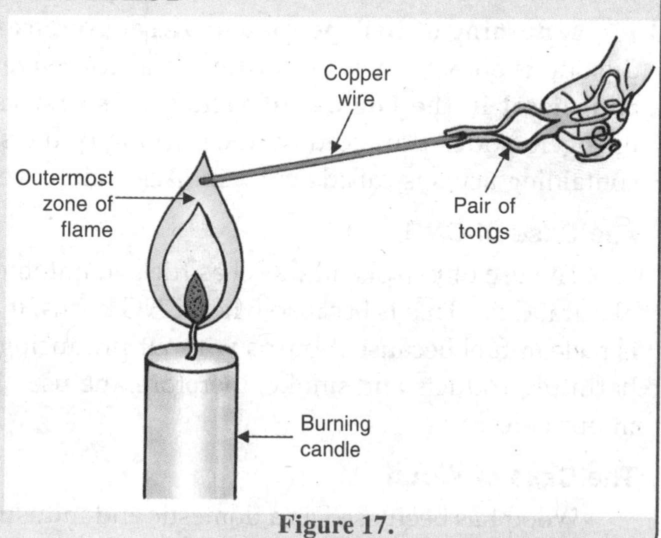

Figure 17.

A kerosene oil lamp produces a flame exactly similar to the candle flame, consisting of the same three zones. Goldsmiths blow air with a blow-pipe to intensify a kerosene lamp flame for melting and moulding the pieces of gold and silver into desired shapes to make jewellery (see Figure 18). When air is blown through blow-pipe into the flame, it helps in the combustion of unburnt fuel and hence makes the flame hotter. We will now discuss the harmful products produced by the burning of fuels.

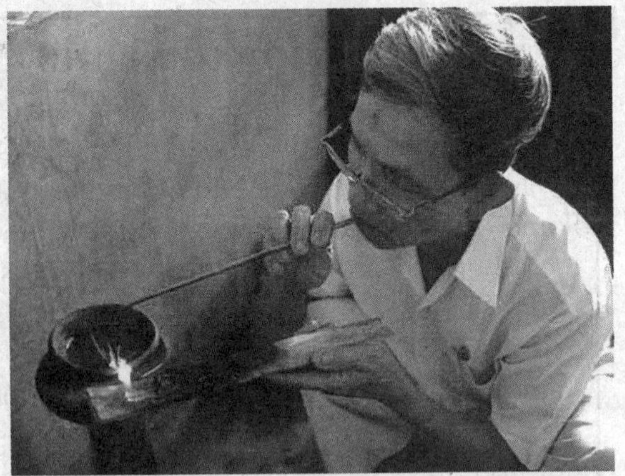

Figure 18. Goldsmith blowing air into a flame through a metallic blow-pipe.

BURNING OF FUELS LEADS TO HARMFUL PRODUCTS

The burning of fuels produces harmful products which pollute the air around us. So, the increasing use of fuels has harmful effects on the environment. The important harmful effects produced by the burning of fuels are as follows :

1. The burning of fuels like wood, coal and petroleum products (kerosene, petrol, diesel, etc.) releases unburnt carbon particles in the air. These fine carbon particles are dangerous pollutants which can cause respiratory diseases such as asthma.

2. Incomplete combustion of fuels (due to insufficient air) produces a very poisonous gas called carbon monoxide. Excessive inhaling of carbon monoxide gas can kill a person. We should never sleep in a room with closed door and windows, and having a coal fire burning inside. This is because when coal burns in an insufficient supply of air in the room (due to closed door and windows), then a lot of carbon monoxide gas is produced. When the persons sleeping in this room breathe in poisonous carbon monoxide gas, they may all die.

3. Burning of fuels releases carbon dioxide into air in the environment. Increased percentage of carbon dioxide in air is causing **global warming**. This happens as follows : Carbon dioxide gas in the air traps sun's heat rays by producing greenhouse effect. **Global warming is the rise in temperature of earth's atmosphere caused by the excessive amounts of carbon dioxide in the air**. Due to rise in the temperature of atmosphere, the ice in polar regions will melt very fast, producing a lot of water. This water may cause a rise in the sea-level leading to floods in coastal areas. The low-lying coastal areas may be completely submerged under water leading to the loss of life and property.

4. Burning of coal, petrol and diesel produces sulphur dioxide gas which goes into the air. Sulphur dioxide is an extremely suffocating and corrosive gas. It may damage our lungs. **The burning of petrol and diesel in the engines of vehicles also releases nitrogen oxides into the air.** Sulphur dioxide and nitrogen oxides produced by the burning of fuels dissolve in rain water and form acids. The rain water containing acids is called acid rain. Acid rain is very harmful for forests, aquatic animals and buildings.

The Case of CNG

The use of petrol and diesel as fuels in automobiles (vehicles) is being replaced by CNG (Compressed Natural Gas). This is because when CNG burns, it produces very small amount of harmful products. CNG is a clean fuel because it burns without producing smoke. Since the burning of CNG produces much less harmful products and smoke, therefore, the use of CNG as fuel in automobiles has reduced air pollution in our cities.

The Case of Wood

Wood has been used as a domestic and industrial fuel for centuries. In many rural parts of our country, people still use wood as a fuel due to its low cost and easy availability. The burning of wood as a fuel has, however, many disadvantages. **Some of the disadvantages of burning wood as fuel are as follows :**

(*i*) The burning of wood produces a lot of smoke which is very harmful for human beings. Smoke causes respiratory diseases.

(*ii*) Trees provide us with many useful substances. When trees are cut down to obtain wood for use as fuel, then all the useful substances which can be obtained from trees are lost.

(*iii*) The cutting down of trees to obtain fuel wood leads to deforestation which is very harmful for the environment.

Wood has now been replaced by coal and other fuels such as LPG. **LPG is a better domestic fuel than wood due to the following reasons :**

(*i*) LPG has a much higher calorific value than wood, so it produces much more heat on burning than an equal mass of wood.

(*ii*) LPG burns without producing any smoke but burning of wood produces a lot of smoke.

(*iii*) LPG burns completely without leaving behind any solid residue but wood leaves behind a lot of ash on burning.

We are now in a position to **answer the following questions :**

Very Short Answer Type Questions

1. Define ignition temperature of a substance.
2. Which of the two has a lower ignition temperature : petrol or kerosene ?
3. Name the most common fire extinguisher.
4. Which is the best fire extinguisher for fires involving electrical equipment and inflammable materials like petrol ?
5. Name one substance which undergoes spontaneous combustion (or burns in air at room temperature).
6. Name the unit in which the calorific value of a fuel is expressed.
7. Which of the following fuels has the lowest calorific value ?
   Diesel, Methane, CNG, Coal, Petrol
8. Which of the following fuels has the highest calorific value ?
   Kerosene, Wood, Hydrogen, Cow-dung cakes, LPG
9. Name the term which is used to express the efficiency of a fuel.
10. Name one solid, one liquid and one gas which burn by producing a flame.
11. Which of the following does not produce a flame on burning ?
    Camphor, Charcoal, Kerosene
12. Name one fuel which burns without producing a flame.
13. How many zones are there in a flame ?
14. Which zone of a candle flame is the hottest ?
15. In a candle flame, what is the colour of : (*a*) innermost zone (*b*) middle zone, and (*c*) outer zone ?
16. Name any harmful product released by the burning of fuels.
17. Name the very poisonous gas produced by the incomplete combustion of fuels.
18. Name the fuel which is gradually replacing petrol and diesel in automobiles.
19. Name two substances having low ignition temperatures and two having high ignition temperatures.
20. Fill in the following blanks with suitable words :
    (*a*) A fuel must be heated to its............before it starts burning.
    (*b*) The most common supporter of combustion around us is............
    (*c*) Fire produced by burning oil cannot be controlled by............
    (*d*) A liquid fuel used in homes is ....
    (*e*) The amount of heat evolved when 1 kg of a fuel is burnt completely is called its............
    (*f*) The substances which vaporise during burning, give ....
    (*g*) Burning of wood and coal causes............of air.

Short Answer Type Questions

21. (*a*) What are fuels ? Name any two common fuels.
(*b*) State any four characteristics of an ideal fuel (or good fuel).

22. (*a*) Define the calorific value of a fuel.
(*b*) "The calorific value of LPG is 55000 kJ/kg". What does it mean ?

23. Can you burn a piece of wood by bringing a lighted matchstick near it ? Explain.

24. Why do you have to use paper or kerosene oil to start fire in wood or coal ?

25. What is meant by rapid combustion ? Give one example of rapid combustion.

26. What is meant by spontaneous combustion ? Give one example of spontaneous combustion.

27. What is meant by explosive combustion (or explosion) ? Give one example of explosive combustion (or explosion).

28. How will you show that air is necessary for combustion ?

29. Can the process of rusting be called combustion ? Give reason for your answer.

30. Why are fires produced by burning oil not extinguished by pouring water ?

31. Explain why, fire caused by electricity should not be extinguished by pouring water.

32. How is the fire produced by burning oil (or petrol) extinguished ?

33. How is the fire caused by electricity extinguished ?

34. A drum full of kerosene catches fire. What is the simplest way to put off this fire ?

35. What is the first thing you should do if a fire breaks out in your home or neighbourhood ?

36. (*a*) What does a Fire Brigade do when it arrives at a place where a building is on fire ?
(*b*) Describe one method of putting out a fire caused by burning wood or paper.

37. Explain why, we are advised not to sleep in a room having closed doors and windows, with a coal fire burning inside.

38. (*a*) What is a flame ? What type of substances, on burning, give a flame ?
(*b*) What is the difference between the burning of a candle and the burning of a fuel like coal ?

39. How does pouring water extinguish a fire ?

40. Explain how, carbon dioxide is able to control fires.

41. If you see a person whose clothes are on fire, how will you extinguish the fire ? Give reason for your answer.

42. Give two examples each of : (*a*) solid fuels (*b*) liquid fuels, and (*c*) gaseous fuels.

43. Name the various zones of a candle flame. Which zone (or part) of a candle flame is the least hot (or coldest) ?

44. Why does a goldsmith blow air into the kerosene lamp flame with a blow-pipe ?

45. In which zone of a candle flame : (*a*) partial combustion of fuel takes place, and (*b*) complete combustion of fuel takes place ?

46. Explain how, the use of CNG in automobiles has reduced pollution in cities.

47. What are the disadvantages of burning wood as fuel ?

48. Give reasons for the following : LPG is a better domestic fuel than wood.

49. Explain why, when a burning candle is covered with an inverted gas jar, the candle gets extinguished after some time.

50. It is difficult to burn a heap of green leaves but dry leaves catch fire easily. Explain.

Long Answer Type Questions

51. (*a*) What are combustible substances ? Name three combustible substances.
(*b*) What are non-combustible substances ? Name three non-combustible substances.

52. (*a*) What is meant by 'combustion' ? Explain with an example.
(*b*) What are the conditions necessary for combustion to take place ?

53. (*a*) Make a labelled diagram of a candle flame.
(*b*) What makes the middle zone of a candle flame luminous (or light-giving) ?

54. What is global warming ? Name the gas whose increasing percentage in air is leading to global warming. State a harmful effect which can be caused by global warming.

55. Explain how, burning of fuels such as coal, petrol and diesel leads to acid rain. How is acid rain harmful ?

Multiple Choice Questions (MCQs)

56. Which of the following substances has the lowest ignition temperature ?
(*a*) kerosene (*b*) spirit (*c*) diesel (*d*) mustard oil

57. One of the following is not a combustible susbtance. This one is :
(*a*) alcohol (*b*) hydrogen (*c*) asbestos (*d*) chaff

58. Which of the following is not used in making matchsticks these days ?
(*a*) potassium chlorate (*b*) white phosphorus
(*c*) antimony trisulphide (*d*) red phosphorus

59. Which of the following undergoes spontaneous combustion ?
(*a*) yellow sulphur (*b*) red phosphorus (*c*) white phosphorus (*d*) brown sulphur

60. Which of the following statement is not correct about carbon dioxide acting as a fire extinguisher for electrical fires ?
(*a*) it is heavier than air (*b*) it is lighter than air
(*c*) it is not combustible (*d*) it does not support combustion

61. Fires in underground coal mines usually occur due to the :
(*a*) explosive combustion (*b*) deliberate combustion
(*c*) spontaneous combustion (*d*) rapid combustion

62. The calorific value of a fuel is 40000 kJ/kg. This fuel is most likely to be :
(*a*) biogas (*b*) methane (*c*) hydrogen gas (*d*) liquefied petroleum gas

63. Which of the following fuels has the highest calorific value ?
(*a*) natural gas (*b*) liquefied petroleum gas (*c*) coal gas (*d*) hydrogen gas

64. On a cold winter night, the persons sleeping in a room with closed door and windows with a coal fire burning inside may die due to the excessive accumulation of :
(*a*) nitrogen monoxide (*b*) nitrogen dioxide (*c*) carbon dioxide (*d*) carbon monoxide

65. Which of the following burns without producing a flame ?
(*a*) camphor (*b*) coke (*c*) cooking gas (*d*) kerosene

66. Which of the following fuels has the lowest calorific value ?
(*a*) kerosene (*b*) CNG (*c*) biogas (*d*) LPG

67. Which of the following is the main cause of global warming ?
(*a*) nitrogen dioxide (*b*) sulphur dioxide (*c*) carbon dioxide (*d*) ozone

68. Which of the following gas does not contribute to the formation of acid rain ?
(*a*) nitrogen monoxide (*b*) carbon monoxide (*c*) sulphur dioxide (*d*) nitrogen dioxide

69. Which of the following is the most environment friendly fuel to be used in automobiles ?
(*a*) petrol (*b*) diesel (*c*) natural gas (*d*) petroleum gas

70. Which of the following does not involve a combustion reaction ?
(*a*) production of heat and light from kerosene in a lantern
(*b*) production of heat and light from hydrogen in a rocket
(*c*) production of heat and light from hydrogen in the sun
(*d*) production of heat and light from wood in a bonfire

71. A heap of green leaves is lying in one corner of a park. The green leaves in the heap burn with difficulty because :
(*a*) they contain a tough material called cellulose
(*b*) they contain a lot of water
(*c*) they contain a green pigment chlorophyll
(*d*) they do not get sufficient oxygen for burning

72. If the clothes of a person working in the kitchen catch fire, then to extinguish the fire :
(*a*) sand should be thrown over the burning clothes
(*b*) water should be thrown over the burning clothes
(*c*) polyester blanket should be used to cover the burning clothes
(*d*) woollen blanket should be used to cover the burning clothes

73. The outermost zone of a candle flame is the :
(*a*) least hot part (*b*) coldest part (*c*) hottest part (*d*) moderately hot part

74. The flame of a kerosene oil lamp (or lantern) has :
(*a*) single zone (*b*) two zones (*c*) three zones (*d*) four zones

75. A lot of dry powder of one of the following chemicals can be released over a fire to extinguish it. This chemical is :
(*a*) plaster of Paris (*b*) baking soda (*c*) washing soda (*d*) bitumen

Questions Based on High Order Thinking Skills (HOTS)

76. An electric spark is struck between two electrodes placed near each other in a closed tank full of petrol. Will the petrol catch fire ? Explain your answer.

77. Give reason for the following :
Paper by itself catches fire easily whereas a piece of paper wrapped around an aluminium pipe does not.

78. Abida and Ramesh want to heat water taken in separate beakers. Abida kept the beaker near the wick in the yellow part of the candle flame. Ramesh kept the beaker in the outermost part of the flame. Whose water will get heated in a shorter time ? Why ?

79. When a lot of dry powder of a substance X is released over a fire, the fire gets extinguished.
(*a*) Name the substance X.
(*b*) How does this substance extinguish the fire ?
(*c*) Name another susbtance which behaves like X.

80. What type of combustion is represented by :
(*a*) burning of white phosphorus in air at room temperature ?
(*b*) burning of LPG in a gas stove ?
(*c*) ignition of a cracker ?
(*d*) burning of coal dust in a coal mine ?

ANSWERS

2. Petrol **3.** Water **7.** Coal **8.** Hydrogen **9.** Calorific value **11.** Charcoal **17.** Carbon monoxide **20.** (*a*) ignition temperature (*b*) air (*c*) water (*d*) kerosene (*e*) calorific value (*f*) flames (*g*) pollution **38.** (*b*) Candle burns with a flame whereas coal burns without producing a flame. Coal just glows on burning **56.** (*b*) **57.** (*c*) **58.** (*b*) **59.** (*c*) **60.** (*b*) **61.** (*c*) **62.** (*a*) **63.** (*d*) **64.** (*d*) **65.** (*b*) **66.** (*c*) **67.** (*c*) **68.** (*b*) **69.** (*c*) **70.** (*c*) **71.** (*b*) **72.** (*d*) **73.** (*c*) **74.** (*c*) **75.** (*b*) **76.** No, the petrol will not catch fire. This is because in a closed tank full of petrol , there is no supporter of combustion like air which can make the petrol catch fire and burn **77.** When paper is heated alone, its ignition temperature is reached quickly due to which it catches fire easily. When the paper wrapped around an aluminium pipe is heated, then the heat supplied to paper is transferred to aluminium pipe by conduction. Due to the continuous transfer of heat from paper to aluminium pipe, the paper does not get heated to its ignition temperature quickly and hence does not catch fire easily **78.** Ramesh's water will get heated in a shorter time because the outermost part of the candle flame is the hottest part of flame **79.** (*a*) Sodium bicarbonate (Sodium hydrogencarbonate) (*b*) The heat of fire decomposes sodium bicarbonate to produce carbon dioxide gas. This carbon dioxide covers the fire like a blanket and cuts off supply of fresh air to the burning susbtance. Due to this the fire gets extinguished (*c*) Potassium bicarbonate (or Potassium hydrogencarbonate) **80.** (*a*) Spontaneous combustion (*b*) Rapid combustion (*c*) Explosive combustion (or Explosion) (*d*) Spontaneous combustion

CHAPTER

7

Conservation of Plants and Animals

The term 'conservation' means 'the process of keeping and protecting something from damage'. So, the conservation of plants and animals means that plants and animals which occur in the forests should be kept in a way that they remain protected in the natural environment in which they are found. Before we go further and discuss the conservation of plants and animals in detail, we should know the meanings of some new terms such as biosphere, wildlife, biodiversity, and ecosystem which will be used in this chapter.

Biosphere is that part of the earth in which living organisms exist (or which supports life). Biosphere includes land surface of the earth, atmosphere of the earth, as well as water bodies on the earth (like rivers, ponds, lakes and oceans). This is because living organisms are found on land, in the atmosphere as well as in water bodies.

The animals living in the natural environment (like forests, etc.) are called wild animals. The wild animals are not domesticated by man. The plants growing in the natural environment on their own, are called wild plants. The wild plants are not cultivated (or grown) by man. **The term 'wildlife' means all the animals and plants which are found naturally in the forests and other natural habitats**. Though the term 'wildlife' also includes wild plants but it is most commonly used for *wild animals and birds*.

The term 'diversity' means 'a variety'. And 'biodiversity' means 'biological variety'. **Biodiversity refers to the variety of organisms (plants, animals and micro-organisms, etc.) found in a particular area or habitat**. In other words, the presence of a large number of species in a particular area (or habitat) is called its biodiversity. Biodiversity denotes the richness of species in a particular area or habitat.

An ecosystem is a 'system' which includes all the living organisms (plants, animals and micro-organisms) of an area and the physical environment (soil, air and water) in which they live. In an ecosystem, the various living organisms interact among themselves (through food chains), and also with the physical environment in which they live. An ecosystem is a self-sustaining unit of living organisms and non-living environment, needing only the input of sunlight energy for its functioning.

The wild animals and birds are comfortable and flourish in their own, natural habitat (which is usually a forest). **The biggest threat to the existence and survival of wild animals and birds (or wildlife) is deforestation**. Let us discuss deforestation first.

DEFORESTATION AND ITS CAUSES

The clearing of forests (by cutting down forest trees) over a wide area is called deforestation. The forest land cleared of trees is used for other purposes. The various **causes of deforestation** (or cutting down of forest trees) are the following :

(*i*) The forest trees are cut down to obtain wood for using as fuel.

(*ii*) The forest trees are cut down to obtain wood (timber) for making doors, windows and furniture.

(*iii*) The forest trees are cut down to obtain wood for making paper.

(*iv*) The forest trees are cut down to obtain more agricultural land for cultivation of crops for the increasing population.

(*v*) The forest trees are cut down to get land for building houses, factories, roads and dams, etc.

Figure 1. Cutting down of trees of the forests on a large scale is called deforestation.

All the above causes of deforestation are the ***man-made causes*** of deforestation. There are also ***natural causes*** of deforestation (which involve the destruction of forests by natural processes). **Some of the natural causes of deforestation are : Forest fires and Severe droughts**. Forest fires can burn down all the trees and other vegetation of the forest and cause deforestation. And when severe droughts occur, the forest trees die out because of lack of water. A great variety of organisms like plants and animals exist on the earth. These plants and animals are essential for the well-being and survival of mankind. Today, a major threat to the survival of organisms (plants and animals) on the earth is deforestation.

Consequences of Deforestation

The destruction of forests by the excessive cutting down of forest trees can make the forests disappear completely one day. The deforestation (or destruction of forests) will have the following consequences for us, other animals and the environment :

(*i*) Deforestation will lead to a shortage of wood and other forest products

The wood obtained from forest trees is used as a fuel ; for making doors, windows and furniture; and for making paper. Deforestation will lead to a shortage of wood for all these purposes. Forests also give us useful products such as honey, gum, sealing wax (lac) and catechu, etc. We will also face shortage of all these forest products if we continue to cut down forest trees.

(*ii*) Deforestation will cause an increase in temperature of earth's atmosphere leading to global warming

Trees (and other plants) use carbon dioxide gas from the atmosphere for the process of food making called 'photosynthesis'. When a lot of trees are cut down during deforestation, then lesser number of trees will be left. The lesser number of trees will use up less carbon dioxide due to which the amount of carbon dioxide in the atmosphere will increase. In this way, *deforestation increases the level of carbon dioxide in the atmosphere.*

Carbon dioxide gas traps the sun's heat rays reflected by the earth (causing greenhouse effect). Trapping of heat rays by carbon dioxide increases the temperature of earth's atmosphere. This will lead to global warming. **The gradual increase in the over-all temperature of earth's atmosphere due to greenhouse effect caused by the increased level of carbon dioxide in the atmosphere is called global warming.** The global warming can melt polar ice caps rapidly producing a tremendous amount of water leading to a rise in the sea level. The raised sea-water level will flood the low-lying coastal areas causing huge loss of life and property.

(*iii*) Deforestation will cause soil erosion making the soil infertile and lead to desertification

The roots of trees (and other plants) bind the particles of top soil together. Due to this binding of soil particles, the wind and water are not able to carry away the top soil easily. In this way, *trees prevent soil erosion*. The tree cover also softens the effect of heavy rains on the forest soil due to which the top soil does not become loose quickly. This is another way in which trees of the forest help prevent soil erosion.

When the forest trees are cut down during deforestation, there are no roots of trees which can bind the soil particles together and prevent them from being carried away by strong winds or flowing rain water. Moreover, since there is no tree cover on the soil to soften the effect of heavy rains, the bare top soil becomes loose quickly by the force of falling rain water and erodes rapidly. Thus, *fewer trees result in more soil erosion*. The top layer of soil is the most fertile. The removal of top layer of soil during soil erosion, exposes the lower, hard and rocky layer of soil. This *lower layer of soil has less humus and it is less fertile*. Plants do not grow well in this less fertile soil. Gradually, the fertile land gets converted into a desert. **The process by which fertile land becomes desert is called desertification.** Thus, deforestation is the cause of desertification.

(*iv*) Deforestation will cause frequent flooding of rivers leading to loss of life and property

The roots of forest trees (and other forest plants) help in absorbing some of the rain water and make it percolate into the ground. This reduces the amount of rain water which rushes quickly into rivers and flooding does not occur. When the forest trees (and plants) are cut down, the percolation of rain water into soil is reduced. A lot of rain water from deforested soil rushes into the rivers quickly, causing floods. In this way, *deforestation decreases the water holding capacity of soil which leads to floods*. Another reason for the floods is the soil erosion caused by deforestation. The eroded soil is carried by flowing rain water into rivers. The eroded soil keeps on collecting on the river bed and decreases the depth of rivers gradually. Due to decreased depth, the water-carrying capacity of the river is reduced. When heavy rains occur, the river is not able to carry away all the rain water quickly. The excess water overflows from the banks of the river into the adjoining areas causing floods. These floods damage standing crops, houses and even drown people living in nearby areas.

(*v*) Deforestation affects the water cycle leading to decrease in rainfall. The decrease in rainfall lowers the groundwater level and could cause droughts

The forest trees put a lot of groundwater, sucked through their roots, into the atmosphere as water vapour by the process of transpiration (evaporation from the leaves). This water vapour helps in bringing rain in that area. When the forest trees are cut down, then the lesser number of trees put less water vapour into atmosphere through transpiration. Since less water vapour is put into the atmosphere, there is less rainfall in that area. When there is less rainfall in an area, then less water percolates into the ground. Due to this, the groundwater level also gets lowered. The shortage of surface water (in ponds, lakes, etc.) and groundwater due to persistent low rainfall in an area can lead to droughts (A prolonged period of abnormally low rainfall leading to severe shortage of water is called drought). From the above discussion we conclude that *as a consequence of deforestation, there will be increased chances of natural calamities such as floods and droughts*.

(*vi*) Deforestation leads to the extinction of many wild animals and plants

Forests are the natural habitats of many wild animals, birds as well as plants. When forest trees and other forest plants are cut down, the natural habitat of wild animals and birds gets destroyed. These homeless wild animals fall prey to human beings and get killed. Moreover, in the absence of forest trees and plants, the wild animals and birds do not get enough food and starve to death. In this way, many animal and bird species become extinct (or vanish) from that area. When forests are cleared during deforestation, many wild plant varieties growing in the forest also get destroyed.

We have just studied that **deforestation reduces rainfall on the one hand and leads to floods on the other.** These are two contradictory situations which can be explained as follows : Deforestation reduces rainfall because less groundwater is put into the atmosphere (as water vapour) through transpiration. Deforestation leads to floods because : (*i*) percolation of rain water into the soil is reduced, and (*ii*) soil eroded by rain water deposits on the river bed and decreases its water-carrying capacity. We will now discuss the conservation of forests and wildlife.

CONSERVATION OF FORESTS AND WILDLIFE

We have just studied that one of the adverse effects of deforestation (large scale cutting down of forest trees) is the destruction of natural habitats of wild animals (and birds). So, deforestation poses a serious threat to the survival of wild animals and their natural environment. The large scale poaching (killing) of wild animals residing in the forests by man (to obtain their skin, etc.) is also a big threat to the survival of many animal and bird species. The death or killing of wild animals disturbs the food chains in which these animals take part, resulting in undesirable consequences for the whole ecosystem. **We should conserve forests and wildlife to preserve biodiversity (variety of species), to prevent endangered species from becoming extinct, and to maintain ecological balance in nature**. Some of the measures which can be taken for the conservation of forests and wildlife are given below :

(*i*) The unauthorised felling (cutting) of forest trees for timber trade and fire-wood should be stopped immediately. This is because depletion of forests destroys the natural habitats of wild animals and birds, and exposes them to the cruelty of man as well as nature.

(*ii*) In case of Government authorised felling of forest trees, for every acre of forest cut down, an equal area of land should be planted with saplings of trees to make up for the loss in the long run.

(*iii*) The natural habitats of wild animals should be preserved by establishing conservation areas such as Biosphere Reserves, Wildlife Sanctuaries and National Parks where the wild animals can flourish in natural surroundings protected from the outside world.

(*iv*) A total ban should be imposed on the poaching (killing) or capturing of any wild animal or bird.

Biosphere Reserves, Wildlife Sanctuaries and National Parks

There are three types of protected areas which have been established (or earmarked) by the Government for the conservation of forests and wild animals. These are :

(*i*) Biosphere Reserves,
(*ii*) Wildlife Sanctuaries, and
(*iii*) National Parks.

Biosphere Reserves, Wildlife Sanctuaries and National Parks are the areas protected by law which have been established throughout our country for the conservation of plants and animals (flora and fauna) present in the area. The human activities such as cutting of forest trees, cultivation of crops, plantations, grazing of cattle, hunting and poaching of wild animals are prohibited (not allowed) in these protected areas. **The purpose of establishing several Biosphere Reserves, Wildlife Sanctuaries and National Parks in India is to conserve wild animals and their natural surroundings (such as forests) so as to maintain a healthy balance in nature, and to prevent the extinction of endangered wild animals**. We will now discuss Biosphere Reserves, Wildlife Sanctuaries and National Parks in detail, one by one.

BIOSPHERE RESERVE

A Biosphere Reserve is a large, protected area of land meant for the conservation of wildlife, biodiversity, and the traditional lifestyle of the tribal people living in the area. In creating the large areas of conservation called Biosphere Reserves, the need of local people to have access to the resources of this area has been kept in mind. So, **a special feature of the protected areas called Biosphere Reserves is that local people or tribals are an integral part (necessary part) of it.** Thus, Biosphere Reserves jointly manage biodiversity and economic activity. We will now describe the basic design of a Biosphere Reserve.

A Biosphere Reserve is a very large conservation area which is divided into three zones : core zone, buffer zone and transition zone (see Figure 2).

***(i)* The innermost zone of a Biosphere Reserve is known as core zone** (see Figure 2). The core zone of a Biosphere Reserve is devoted to strict protection of wildlife. No human activity (or economic activity) is allowed in the core zone of a Biosphere Reserve.

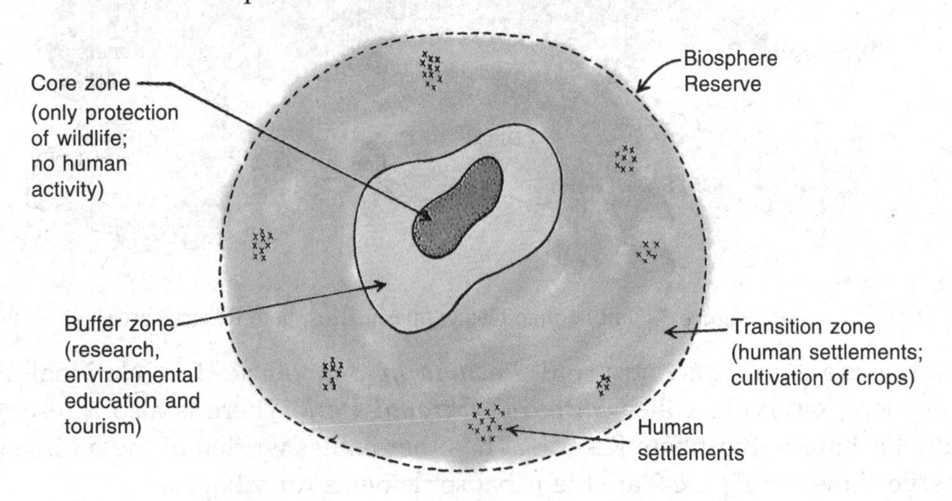

Figure 2. Design of a Biosphere Reserve.

***(ii)* The middle zone of a Biosphere Reserve is called buffer zone** (see Figure 2). Buffer zone surrounds the core zone. In the buffer zone only limited human activity (compatible with conservation) is allowed. For example, research, environmental education and tourism are allowed in the buffer zone of a Biosphere Reserve.

***(iii)* The outermost zone of a Biosphere Reserve is called transition zone** (see Figure 2). In the transition zone, several non-destructive human activities such as settlements (houses) of tribals and cultivation of crops, etc., are allowed which are necessary to sustain the life of tribals.

Some tribal people live in the outermost zone (transition zone) of a Biosphere Reserve who depend on the local natural resources to fulfil the various needs of life. In a Biosphere Reserve, the conservation of wildlife and biodiversity are combined with the sustainable use of natural resources for the benefit of local people (called tribals). No commercial exploitation of natural resources is allowed in a Biosphere Reserve.

A Biosphere Reserve may also contain other protected areas in it. For example, a Biosphere Reserve may contain **Wildlife Sanctuary** and/or **National Park** in it. Biosphere Reserves are open to tourists up to the buffer zone. No tourists are allowed in the core zone of a Biosphere Reserve. There are 14 Biosphere Reserves in India. **The names and locations of some of the Biosphere Reserves of India are given below :**

1. Great Nicobar Biosphere Reserve (Andaman and Nicobar)
2. Kaziranga Biosphere Reserve (Assam)
3. Sunderbans Biosphere Reserve (West Bengal)
4. Kanha Biosphere Reserve (Madhya Pradesh)
5. Pachmarhi Biosphere Reserve (Madhya Pradesh)

We will now describe the **Pachmarhi Biosphere Reserve** in somewhat detail. Pachmarhi Biosphere Reserve is in Madhya Pradesh state of our country (see Figure 3). *The biodiversity found in Pachmarhi Biosphere Reserve is unique. The plants and animals found in Pachmarhi Biosphere Reserve are similar to those of the upper Himalayan Peaks and to those belonging to lower Western Ghats.* Conserving and preserving areas of such biological importance make them a part of our National heritage. The Pachmarhi Biosphere Reserve contains other protected areas in it. In fact, **Pachmarhi Biosphere Reserve contains two Wildlife Sanctuaries and one National Park** (see Figure 3). The two Wildlife Sanctuaries contained in Pachmarhi

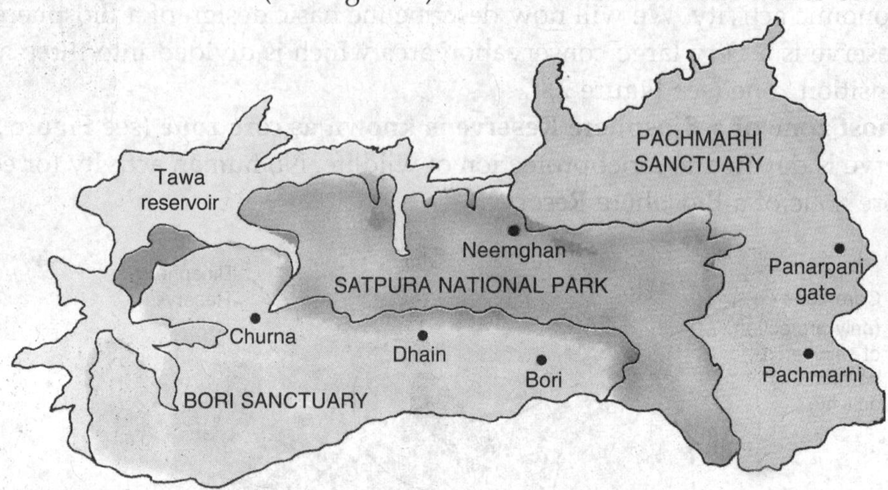

Figure 3. Outline map of Pachmarhi Biosphere Reserve area.

Biosphere Reserve are ***Bori Sanctuary*** and ***Pachmarhi Sanctuary***. The National Park contained in Pachmarhi Biosphere Reserve is called ***Satpura National Park***. There is also a reservoir called 'Tawa Reservoir' in the Pachmarhi Biosphere Reserve. The other names written in the outline map of Pachmarhi Biosphere Reserve shown in Figure 3 are the tribal settlements (or villages).

Role of Biosphere Reserves

The Biosphere Reserves perform the following roles (or functions) :

(*i*) Biosphere Reserves help in the conservation of wildlife (wild animals and plants) of the area.

(*ii*) Biosphere Reserves help to maintain the biodiversity of the area.

(*iii*) Biosphere Reserves preserve the natural ecological conditions (or ecosystems) in the area.

(*iv*) Biosphere Reserves promote the economic development of the area which is compatible with conservation objectives.

(*v*) Biosphere Reserves help to maintain the lifestyle (or culture) of the tribal people living in the area.

(*vi*) Biosphere Reserves prevent the commercial exploitation of the area.

(*vii*) Biosphere Reserves provide opportunities for scientific research, environmental education and tourism.

Flora and Fauna

Some plants and animals typically belong to a particular area. They are not found naturally in other areas. The '***plants***' that grow naturally in a particular area are called '***flora***' of that area. And the '***animals***' which live naturally in a particular area are called '***fauna***' of that area. **The plants and animals of a particular area are called flora and fauna of that area.** Thus, flora and fauna means plants and animals which are found naturally in that area. **Teak, *Jamun*, Fern, (*Sal*) Mango and *Arjun* are the flora of Pachmarhi Biosphere Reserve** (see Figure 4). ***Cheetal*, Wolf, Leopard, *Chinkara*, Blue bull, Barking deer, and Wild dog are the examples of fauna of the Pachmarhi Biosphere Reserve** (see Figure 5).

(*a*) Teak (*b*) *Jamun* (*c*) Fern

Figure 4. Some of the flora (plants) of Pachmarhi Biosphere Reserve.

(*a*) *Cheetal* (*b*) Wolf (*c*) Leopard

Figure 5. Some of the fauna (animals) of Pachmarhi Biosphere Reserve.

Before we go further and give the definition of species, we should know the meaning of the term 'fertile offspring'. A fertile offspring is a baby animal or a baby plant which can reproduce its own kind of organisms in due course of time. We can now define a species as follows : **A species is a group of same kind of organisms which can interbreed to produce fertile offspring.** All the members of a species have common features (or common characteristics). The species can be of animals as well as of plants. Some of the examples of species are : Human (Man), Tiger, Dog, Cat, Cow, Peacock, Cockroach, Mango, *Neem*, Paddy, Pine, Lotus and Sunflower. There are about 6,00,000 species of animals and 4,00,000 species of plants in the world today. The term 'endemic' means something which exists in a 'particular area' (or 'specific area'). Keeping these points in mind, we will now discuss 'endemic species'.

Endemic Species

Endemic species are those species (of plants and animals) which are found exclusively in a 'particular area'. Endemic species are restricted to a certain area. Endemic species are not found naturally anywhere else (in other areas). ***The plants and animals which are found only in a particular area are said to be 'endemic' to that area.*** A particular type of animal or plant species may be endemic to a zone, a state or a country. The plants such as *sal* and wild mango are found naturally only in the Pachmarhi Biosphere Reserve area. **So, *sal* and wild mango are the two examples of the endemic flora (or endemic plants) of the Pachmarhi Biosphere Reserve**. In other words, the plant species (or trees) like *sal* and wild mango are endemic to the Pachmarhi Biosphere Reserve area (see Figure 6). The animals such as Indian giant squirrel (having a big fluffy tail), flying squirrel and bison (a wild ox) are found naturally only in the Pachmarhi Biosphere Reserve area. So, **giant squirrel, flying squirrel and bison are the three examples of endemic fauna (endemic animals) of the Pachmarhi Biosphere Reserve area**. In other words, the animal species like giant squirrel, flying squirrel and bison are endemic to Pachmarhi Biosphere Reserve area (see Figure 7).

Figure 6. Wild mango tree. This is endemic to Pachmarhi Biosphere Reserve area.

Figure 7. Giant squirrel. This is also endemic to Pachmarhi Biosphere Reserve area.

A number of factors are endangering the existence of endemic species. For example, the 'destruction of forests', 'introduction of new species' and 'increasing human population' in the vicinity of forest reserves are affecting the natural habitat of endemic species and endangering their existence. Because of this, some of the endemic species may vanish in the near future.

WILDLIFE SANCTUARY

The term *'sanctuary'* means *'a place of safety'*. So, *'wildlife sanctuary'* means *'the place where wild animals remain safe'*. **A Wildlife Sanctuary is a protected area of land which is created for the protection of wild animals in their natural environment like forests (in which no hunting is permitted).** A Wildlife Sanctuary provides protection and suitable living conditions to the wild animals in their natural habitat. *In a Wildlife Sanctuary killing (poaching) and capturing of wild animals is strictly prohibited*. A Wildlife Sanctuary does not allow any human activity which disturbs the wild animals or their habitat. So, in a Wildlife Sanctuary, animals live in natural environment, protected from people. Some of the threatened wild animals which are protected and preserved in our Wildlife Sanctuaries are : Black buck, White-eyed buck, Elephant, Golden cat, Pink-headed duck, *Gharial*, Marsh crocodile, Python and Rhinoceros.

India has preserved vast tracts of forests and natural habitats for its Wildlife Sanctuaries. The Wildlife Sanctuaries are spread across our country and provide a fascinating diversity of terrain, flora and fauna. Indian Wildlife Sanctuaries have unique landscapes—broad level forests, mountain forests and bush lands in deltas of big rivers. India has more than 500 Wildlife Sanctuaries. Some Wildlife Sanctuaries are specifically named Bird Sanctuaries because their major inhabitants are birds. **The names of some of the Wildlife Sanctuaries of India and their locations are given below :**

1. Sanjay Gandhi Wildlife Sanctuary (Maharashtra)
2. Mudumalai Wildlife Sanctuary (Tamilnadu)
3. Nagarjunsagar Wildlife Sanctuary (Andhra Pradesh)
4. Bharatpur Bird Sanctuary (Rajasthan)
5. Sultanpur Lake Bird Sanctuary (Haryana)
6. Dandeli Wildlife Sanctuary (Karnataka)
7. Thattekad Bird Sanctuary (Kerala)
8. Satkosia Basipalli Wildlife Sanctuary (West Bengal)
9. Lockchao Wildlife Sanctuary (Manipur)
10. Bori Wildlife Sanctuary (Madhya Pradesh)

We have already discussed Biosphere Reserve and now we have discussed Wildlife Sanctuary. Both, Biosphere Reserves and Wildlife Sanctuaries are the legally protected areas for the conservation of wildlife and its environment. **The main differences between a Biosphere Reserve and a Wildlife Sanctuary are given below :**

Biosphere Reserve	*Wildlife Sanctuary*
1. A Biosphere Reserve is spread over a very large area of land.	1. A Wildlife Sanctuary occupies a comparatively smaller area of land.
2. A Biosphere Reserve is for the conservation of biodiversity of the area as well as the economic development of the area.	2. A Wildlife Sanctuary is mainly for the protection of wild animals of the area in their natural habitat.
3. Local people (tribals) form an integral part of a Biosphere Reserve.	3. Local people do not form an integral part of a Wildlife Sanctuary.
4. The human activities such as cultivation of land and building of settlements (houses) are allowed in the outermost zone of a Biosphere Reserve.	4. The human activities such as cultivation of land and building of settlements (houses) are not allowed in a Wildlife Sanctuary.

We have all visited a zoo sometime or the other. The term 'zoo' is the short form of 'zoological garden' or 'zoological park'. **A large garden (or park) where many types of wild animals and birds brought from different parts of the country and the world are kept in cages or enclosures, so that people can see them, is called a zoo.** A 'Wildlife Sanctuary' provides protection to the wild animals and a 'Zoo' also provides protection to the wild animals. **The main differences between a Wildlife Sanctuary and a Zoo are given below :**

Wildlife Sanctuary	*Zoo*
1. In a Wildlife Sanctuary, the wild animals and birds live in their natural habitat in the forest.	1. In a Zoo, the wild animals and birds live in artificial settings such as cages and enclosures.
2. A Wildlife Sanctuary contains wild animals and birds found locally.	2. A Zoo contains wild animals and birds brought from the different parts of the country and from various other countries of the world.
3. A Wildlife Sanctuary is not open to public freely like a zoo. Public can visit a Wildlife Sanctuary only when accompanied by the forest guards.	3. A Zoo is open to public for a fixed time everyday.
4. Wild animals and birds are very comfortable in the natural environment of a Wildlife Sanctuary.	4. Wild animals and birds are not comfortable living in the artificial environment of a Zoo.

We will now discuss National Parks.

NATIONAL PARKS

A National Park is a relatively large area of scenic beauty protected and maintained by the Government to preserve flora and fauna (plants and animals), landscape, historic objects of the area and places of scientific interest. Another purpose of establishing National Parks is to provide **human recreation and enjoyment**. People are allowed to enter National Parks (under certain conditions) to see wild animals roaming about in their natural environment, and historic objects of the area for recreational, enjoyment, inspirational and educative purposes. Basically, a National Park is an area reserved for the conservation of Wildlife within which such public recreational activity is permitted which is compatible with the primary objective of conservation. National Parks are large and diverse enough to protect whole sets of ecosystems.

The exploitation of natural resources in a National Park is strictly prohibited. National Parks are protected from most of the developmental schemes (like construction of dams, etc.) so that their natural attractions are preserved. The human activities such as mining, hunting and fishing, etc., are not allowed inside a National Park. India has more than 80 National Parks. Many of these National Parks were initially

Wildlife Sanctuaries. **Some of the prominent National Parks of India and the States in which they are located are given below :**

1. Corbett National Park (Uttarakhand)
2. Kanha National Park (Madhya Pradesh)
3. Ranthambore National Park (Rajasthan)
4. Gir National Park (Gujarat)
5. Kaziranga National Park (Assam)
6. Sunderbans National Park (West Bengal)
7. Bandipur National Park (Karnataka)
8. Dachigam National Park (Jammu and Kashmir)
9. Sariska National Park (Rajasthan)
10. Satpura National Park (Madhya Pradesh)

We will now discuss the Satpura National Park in detail. **Satpura National Park in Madhya Pradesh is the first Forest Reserve of India.** The finest Indian teak is found in this forest. The objects of historical importance called **rock shelters** are found inside the Satpura National Park. These rock shelters are the evidence of pre-historic human life in these jungles. They give us an idea of the life of the primitive people who lived in this area long ago. A total of 55 rock shelters have been identified in this area. **Rock paintings** are found in these rock shelters. These rock paintings depict the figures of men (and animals) fighting, hunting, dancing and playing musical instruments. ***The rock shelters and rock paintings are one of the major tourist attractions of Satpura National Park.*** Satpura National Park has a large species of wild animals in it. Some of the wild animals found in Satpura National Park are : Tiger, Gaur (Indian bison), Leopard, Indian giant squirrel, *Sambar*, Sloth bear, *Cheetal*, Barking deer, Wild boar, Rhesus monkey, *Langur*, *Nilgai*, Hyena, Wild dog, Porcupine, Marsh crocodile and Four-horned antelope.

Tiger is one of the many wild life species (or wild animals) which are slowly disappearing from our forests (see Figure 8). There is a 'Satpura Tiger Reserve' in the Satpura National Park which is part of Project Tiger. **Project Tiger is a wildlife conservation project which was launched by the Government of India in 1972 to protect the tigers in the country.** The objective of this project was to ensure the survival and maintenance of the tiger population in specially constituted 'Tiger Reserves' throughout India. There are 28 Tiger Reserves in India at present which are governed by 'Project Tiger'. Under 'Project Tiger', a significant increase in the population of tigers has been seen in the Satpura Tiger Reserve of Satpura National Park. Once upon a time, the wild animals such as lions, elephants,wild buffaloes and *barasingha* were also found in the Satpura National Park (see Figure 9). All these wild animals have now vanished from this area due to illegal activities of the people living in the nearby areas. It is a pity that even the protected Forest Reserves are not safe because people living in the neighbourhood encroach upon them and destroy them and the wildlife present in them to make some quick money. This must be stopped immediately. Before we end this discussion on National Parks, we would like to give **the main differences between a Wildlife Sanctuary and a National Park.**

Figure 8. Tigers are found in Satpura National Park.

(*a*) Wild buffalo

(*b*) *Barasingha*

Figure 9. These wild animals are no longer found in Satpura National Park.

Wildlife Sanctuary	*National Park*
1. A Wildlife Sanctuary may or may not be an area of great scenic beauty.	1. A National Park is an area of great scenic beauty.
2. A Wildlife Sanctuary protects and preserves the wild animals in their natural environment.	2. A National Park protects and preserves wild animals and their environment as well as the scenic beauty, historical objects and habitats of scientific interest in the area.
3. A Wildlife Sanctuary is not meant for recreation and enjoyment of the public. It is dedicated to the protection of wild animals only.	3. In a National Park, in addition to protection, wild animals are kept for recreation, enjoyment and educative interests of the public.
4. A Wildlife Sanctuary usually does not allow easy access to the visitors.	4. A National Park allows easy access for the visitors to the land and wildlife inside it.

EXTINCT SPECIES AND ENDANGERED SPECIES

The species which no longer exist anywhere on the earth are called extinct species. Extinct species are those which have died out completely. A species becomes extinct when the last living member of that species dies. The extinct species may be of animals or plants. For example, ***a type of animal which no longer exists on earth is called an extinct animal.*** It means that this animal has no living member on the earth. All the members of this animal species have died out. For example, the animals called 'dinosaurs' which lived on this earth in ancient times have become extinct a long time ago (see Figure 10). **Some of the examples of extinct species of animals are : Dinosaur, Dodo, Cave lion, Caspian tiger, and Irish deer.** All the members of these animals are dead. We can never see these animals again. They have all vanished from this earth for various reasons.

Figure 10. Dinosaur is an extinct animal.

The species which are facing the risk of extinction are called endangered species. Endangered species are animals and plants which are on the verge of vanishing from the earth. They are the animals and plants that exist in small numbers on the earth, and if we do not take quick action to save them, they may be lost for ever (or become extinct). For example, ***the wild animals whose numbers are diminishing to such a low level that they might face extinction soon, are known as endangered animals.*** An animal species becomes endangered either because it is few in numbers or it is being killed by predators or it is being hunted by human beings or its natural habitat is being destroyed by deforestation (leading to lack of shelter, food and water). In fact, the survival of some wild animals is becoming more and more difficult day by day because of the destruction of their natural habitats. **Some examples of endangered animal species are : Tiger, Snow leopard, Great Indian rhinoceros, Asiatic lion, Desert cat, Lion-tailed macaque, Namdapha flying squirrel, and Kashmir stag** (see Figure 11).

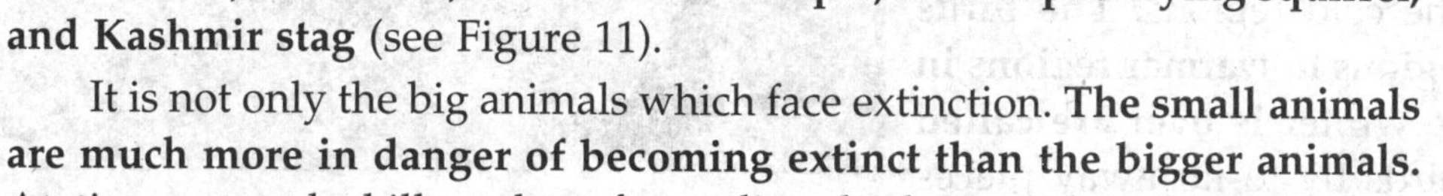

Figure 11. Lion-tailed macaque is an endangered animal.

It is not only the big animals which face extinction. **The small animals are much more in danger of becoming extinct than the bigger animals.** At times, people kill snakes, frogs, lizards, bats and owls without realising their importance in the ecosystem. They might be small in size but their role in the ecosystem is very important which cannot be ignored. The small animals form parts of food chains and food webs which are essential for maintaining a balance in nature. By killing the small animals, we are actually harming ourselves. This point will become more clear from the following example of snake.

Snake is a small, wild animal. The skin of snakes is in great demand for making fancy leather goods, so the snake skin sells at a high price in the market. Now, to make some easy money, some greedy people kill the snakes indiscriminately in large numbers to obtain their skin. This **large scale killing of snakes disrupts the food chains in which snakes occur and creates an imbalance in nature.** For example, snake is a friend of the farmer in the sense that it eats vermins like rats and mice which are pests and damage the crops (see Figure 12). Now, when the snakes are killed in large numbers to obtain their skin, the population of snakes is reduced greatly. Due to lesser number of 'predator' snakes, the population of pests like rats and mice in crop-fields increases. The increased number of rats and mice in the fields damage the standing crops leading to loss in the production of foodgrains. We will now discuss the red data book.

Figure 12. Snake is a friend of the farmer because it eats rats and mice which damage the crops.

RED DATA BOOK

Red Data Book is the 'book' (or publication) which keeps a record of all the endangered animals, plants and other species. Actually, Red Data Book contains a list of species which are in danger of becoming extinct. There are different Red Data Books for plants, animals and other species. Red Data Books are being published in many different countries and provide useful information on the threat status of the various species. There is also a Red Data Book of India. **Some of the endangered species of animals listed in the Red Data Book of India are : Flying squirrel, Indian giant squirrel, *Barasingha*, Black buck, Himalayan musk deer, Great Indian rhinoceros, Snow leopard, and Tiger.** The advantage of maintaining Red Data Book is that we come to know which species of animals, plants, etc., are very small in number and facing the danger of extinction so that timely remedial steps can be taken by the Authorities concerned to prevent their extinction.

MIGRATION

The process of a bird (or other animal) moving from one place to another according to the season, is called migration. In other words, when a bird (or other animal) moves from one place to another in one season and returns in a different season, it is called migration. ***Migration of birds (or other animals) is an adaptation to escape the harsh and cold conditions of their normal habitat in winter so as to survive.*** Let us discuss the migration of birds.

When the winter sets in cold regions of the earth, the climate becomes extremely cold in those regions. The birds, which normally live in these regions, migrate (fly off) to far flung warmer places on earth to escape the extremely cold winter climate and survive (see Figure 13). And when the winter season is over, these birds fly back to their original habitats in the cold regions. **The birds which move from very cold regions to warmer regions in winter, and go back after the winter is over are called migratory birds.** Migratory birds fly to far away places every year during a particular time because of climatic changes. The purpose of migration of birds is to survive

Figure 13. Some birds migrate to warmer regions to escape the extreme cold of their normal habitat during winter.

by escaping the extremely cold conditions of their natural habitat and also for breeding (by laying eggs).

India is one of the destinations of many of the migratory birds coming from the very cold regions of the earth. **One of the most common migratory bird which comes to India every year for a few months is the Siberian crane**. The normal habitat of Siberian crane is Siberia (which is a very cold place). When winter sets in Siberia and it gets extremely cold, the Siberian crane flies thousands of kilometres and comes to warmer places in India such as Bharatpur in Rajasthan, Sultanpur in Haryana, some wet lands of North-East, and some other parts of India. The Siberian cranes stay in the warmer places in India for a few months. The Siberian cranes fly back to Siberia when the winter ends there and climate becomes favourable. We will now discuss the recycling of paper.

RECYCLING OF PAPER

Paper is one of the important products we get from the forests. **Paper is made from wood pulp that is produced from the wood of forest trees.** It has been estimated that 17 full grown trees are needed to make 1 tonne of paper. So, many trees have to be cut down from the forests to make paper. Paper making is another cause of deforestation. So, **we should save paper to save the forest trees**. If each one of us saves just one sheet of paper in a day, we can save many trees in a year. We can save paper by writing fully on both sides of every sheet of paper in our notebook (without leaving any blank sheets). We can also save paper by using chalk and slate for doing rough work (instead of paper notebook). Reuse of paper means that, if possible, we should use the 'used paper' again. For example, the 'used paper envelopes' can be reversed inside out and used again.

The term 'recycling of paper' means to process the waste paper (to make new paper) so that it can be used again. Paper can be recycled from old newspapers, magazines, books, notebooks, and packaging materials after removing ink from them. If all of us keep on collecting old newspapers, magazines, books, notebooks and paper wrappers, etc., and send them to paper mills for recycling through a junk dealer (*kabadiwala*), we will be able to save many forest trees from being cut down. In fact, **paper can be recycled five to seven times for use**. We should 'save paper', 'reuse paper' and 'recycle paper'. By doing this, we will not only save trees but also save energy and water needed for manufacturing paper. The amount of harmful chemicals used in paper making will also be reduced.

From the above discussion we conclude that we should save paper, reuse paper and recycle paper :

(*i*) to save forest trees from being cut down,
(*ii*) to save water used in paper making,
(*iii*) to save energy (electricity) used in making paper, and
(*iv*) to reduce the amount of harmful chemicals used in paper making.

REFORESTATION

The answer to the problem of *deforestation* is *reforestation*. **The planting of trees in an area in which forests were destroyed, is called reforestation** (see Figure 14). The term 'reforestation' means 'to cover again with forest' (by planting new trees). The planted trees should generally be of the same species which were cut down from the forest during deforestation. *We should plant at least as many trees as have been cut down*. In this way, the forests will continue to have sufficient number of trees all the time. Forests are called green wealth of a country. We have already caused tremendous damage to our lush green forests by deforestation. *If we have to retain our 'green wealth' for future generations, then planting of more trees (reforestation) is the only option*. Reforestation can also take place naturally. If the deforested area is left undisturbed for some time, it

Figure 14. New trees being planted in place of cut down trees in a forest. This is called reforestation.

re-establishes itself by the natural growth of trees. This is called natural reforestation. There is no role of human beings in natural reforestation.

Advantages of Reforestation

Reforestation has several advantages. Some of the advantages of reforestation are given below :

(*i*) Reforestation produces a large quantity of raw materials for industry (like paper industry), timber trade, etc.

(*ii*) Reforestation will lead to a decrease in global warming by reducing the amount of carbon dioxide gas in the atmosphere.

(*iii*) Reforestation increases rainfall in an area. This will raise groundwater level and prevent droughts.

(*iv*) Reforestation prevents soil erosion and floods.

(*v*) Reforestation increases the area of earth under forests which is good for the conservation of wildlife.

In order to meet the ever-increasing demand for wood in factories and for shelter (building houses, furniture, etc.), forest trees are being cut continuously. It is quite justified to cut a limited number of trees for genuine developmental purposes. But at the same time, an equal number of new trees should be planted (in place of cut down trees) to maintain a balance in nature and avoid unpleasant consequences of deforestation.

There is a Forest Conservation Act in India. The Forest Conservation Act is aimed at the preservation and conservation of natural forests and at the same time meeting the basic needs of the people living in or near the forests. We are now in a position to **answer the following questions :**

Very Short Answer Type Questions

1. (*a*) Which gas in the atmosphere is utilised by the trees and plants in photosynthesis ?

(*b*) Which gas in the atmosphere traps the heat rays reflected by the earth ?

2. Write one word for the following :

Variety of plants, animals and micro-organisms generally found in an area.

3. What name is given to that part of the earth in which living organisms exist (or which supports life) ?

4. Name the three types of protected areas which have been earmarked for the conservation of forests and wildlife.

5. Name one Wildlife Sanctuary and one National Park which are contained in Pachmarhi Biosphere Reserve.

6. Name any five threatened wild animals which are protected and preserved in our Wildlife Sanctuaries.

7. For what purpose are National Parks in our country established ?

8. Name the first Reserve Forest of India.

9. Name the objects of historical significance found in Satpura National Park.

10. Name two animals which have vanished from Satpura National Park and two animals which are still found there.

11. What name is given to those species :

(*a*) that are on the verge of vanishing from earth ?

(*b*) that have died out completely ?

12. Name the publication which contains record of all the endangered species (plants and animals, etc.).

13. Name any five endangered species of animals listed in the Red Data Book of India.

14. What is the answer to deforestation ?

15. Name the various old paper products which can be recycled.

16. How many full grown trees are needed to make 1 tonne of paper ?

17. State one way in which we can reuse paper.

18. Fill in the following blanks with suitable words :

(*a*) Deforestation increases the level of..........in the atmosphere.

(*b*) Species found only in a particular area is known as .... species.

(*c*) A place where animals are protected in their natural habitat is called..........

(*d*) Satpura National Park is a part of..........Biosphere Reserve.

(*e*) Red Data Book contains a record of..........species.

(*f*) Migratory birds fly to far away places because of .... changes.

(*g*) In reforestation, the planted trees should be generally of the same........ which were cut down in that forest.

(*h*) Paper can be recycled........ to........ times.

Short Answer Type Questions

19. What is meant by deforestation ? What are the causes of deforestation ?
20. What are the consequences of deforestation ?
21. What is desertification ? Name one human activity which may lead to desertification.
22. Explain how, deforestation makes the soil infertile leading to desertification.
23. What is global warming ? Name the gas which is responsible for causing global warming.
24. How does deforestation reduce rainfall on the one hand and lead to floods on the other ?
25. Define the term 'biosphere'.
26. Define (*i*) biodiversity, and (*ii*) ecosystem.
27. What will happen if the natural habitat of a wild animal is destroyed ?
28. What is the purpose of establishing several Biosphere Reserves, Wildlife Sanctuaries and National Parks in India ?
29. State the role of Biosphere Reserves.
30. What is meant by the 'flora' and 'fauna' of an area ? Give two examples of flora and two examples of fauna of Pachmarhi Biosphere Reserve.
31. Which of the following belong to 'fauna' and which belong to 'flora' of the Pachmarhi Biosphere Reserve ? *Sal*, *Arjun*, *Cheetal*, Teak, Leopard, Fern, Blue bull, Barking deer, Mango, Wolf.
32. What is the difference between 'flora' and 'fauna' ?
33. What is meant by 'species' ? Give any five examples of species.
34. What do you understand by 'endemic species' ? Name two plant species and two animal species which are endemic to Pachmarhi Biosphere Reserve area.
35. Name two man-made causes of deforestation and two natural causes of deforestation.
36. What is a Wildlife Sanctuary ? Name any two Wildlife Sanctuaries in India. Where are these Sanctuaries located ?
37. Name any two Bird Sanctuaries in India. Where are these located ?
38. What are the differences between a Biosphere Reserve and a Wildlife Sanctuary ?
39. State the differences between a Wildlife Sanctuary and a Zoo.
40. What is a National Park ? Name any two National Parks of India. Where are these National Parks located ?
41. What are the differences between a Wildlife Sanctuary and a National Park ?
42. What do the rock paintings found in rock shelters of Satpura National Park depict ?
43. What is 'Project Tiger' ? What was the objective of this project ?
44. Why even protected forests are not completely safe for wild animals ?
45. What is meant by extinct species ? Name any two extinct animals.
46. What is meant by endangered species ? Name any two endangered animals.
47. Differentiate between 'endangered species' and 'extinct species'.
48. By taking the example of snakes, explain how by killing small animals, we are actually harming ourselves.
49. What is Red Data Book ?
50. State one advantage of maintaining Red Data Book.
51. What do you understand by the term 'migration' ?
52. What is meant by the migration of birds ? Why do birds migrate ?
53. What are migratory birds ? Name one migratory bird which comes to warmer regions of India every year.
54. Why does Siberian crane come from Siberia to places like Bharatpur in India every year for a few months ?
55. What can be done to retain our 'green wealth' for the future generations ?
56. What is meant by 'reforestation' ? What are the advantages of reforestation ?
57. What are the aims of the 'Forest Conservation Act' in India ?
58. Why should we save, reuse and recycle paper ?
59. Explain how, recycling of paper helps in the conservation of forests.
60. With the help of a labelled diagram, describe the basic design of a Biosphere Reserve.

Long Answer Type Questions

61. (*a*) What is meant by the term 'wildlife' ? Why should forests and wildlife be conserved ?
(*b*) What are the various measures which can be taken to conserve forests and wildlife ?

62. (*a*) What are the various purposes for which the forest trees are cut ?
(*b*) What is the effect of deforestation on wild animals ?

63. (*a*) How does deforestation cause soil erosion ?
(*b*) Explain how, deforestation leads to frequent flooding of rivers.

64. (*a*) Explain how, deforestation leads to reduced rainfall.
(*b*) How does deforestation lead to global warming ?

65. What is a Biosphere Reserve ? Name any two Biosphere Reserves of India. Where are these Biosphere Reserves located ?

Multiple Choice Questions (MCQs)

66. The part of earth in which living organisms exist (or which supports life) is called :
(*a*) lithosphere (*b*) globe (*c*) hydrosphere (*d*) biosphere

67. Deforestation increases the level of one of the following in the atmosphere. This one is :
(*a*) ozone (*b*) carbon dioxide (*c*) oxygen (*d*) water vapour

68. Which of the following is not a part of Pachmarhi Biosphere Reserve ?
(*a*) Bori Wildlife Sanctuary (*b*) Satpura National Park
(*c*) Bandipur National Park (*d*) Pachmarhi Wildlife Sanctuary

69. The most rapidly dwindling natural resource in the world is :
(*a*) water (*b*) soil (*c*) sunlight (*d*) forests

70. Which of the following animal is no longer found in Satpura National Park in Madhya Pradesh ?
(*a*) wild buffalo (*b*) wild boar (*c*) marsh crocodile (*d*) wild dog

71. Those species (of plants and animals) which are found exclusively in a particular area are called :
(*a*) epidemic species (*b*) pelagic species (*c*) pandemic species (*d*) endemic species

72. The migratory bird which comes from Siberia to India for a few months every year is :
(*a*) crow (*b*) koel (*c*) crane (*d*) kingfisher

73. Sanjay Gandhi Wildlife Sanctuary is located in :
(*a*) Rajasthan (*b*) Manipur (*c*) Madhya Pradesh (*d*) Maharashtra

74. Which of the following activities, if not checked in time, may ultimately lead to the rise in sea level causing the flooding of low-lying coastal areas ?
(*a*) desalination (*b*) desertification (*c*) deforestation (*d*) desegregation

75. Snakes are killed in large numbers because :
(*a*) they are very poisonous (*b*) they kill rats
(*c*) their skin is expensive (*d*) they damage the crops

76. The National Park which is located in Rajasthan is :
(*a*) Corbett National Park (*b*) Kanha National Park
(*c*) Satpura National Park (*d*) Sariska National Park

77. Which of the following wild animals is not listed in the Red Data Book of India ?
(*a*) black buck (*b*) flying squirrel (*c*) tiger (*d*) leopard

78. The Siberian crane comes to India every year in winter for a few months :
(*a*) to escape the severe summer (*b*) to escape the severe winter
(*c*) to escape the heavy rains (*d*) to escape from predators

79. Which of the following wild animals is a friend of the farmer ?
(*a*) deer (*b*) rat (*c*) snake (*d*) wild buffalo

80. Which of the following is an extinct species of animals ?
(*a*) tiger (*b*) desert cat (*c*) snow leopard (*d*) dodo

81. Which of the following is not a consequence of deforestation ?
(*a*) increase in soil erosion (*b*) increase in earth's temperature
(*c*) decrease in rainfall (*d*) decrease in floods

82. One of the following is not a part of the fauna of Pachmarhi Biosphere Reserve. This one is :
(*a*) yak (*b*) leopard (*c*) blue bull (*d*) barking deer

83. The indiscriminate killing of which of the following animals can lead to loss in the production of food grains ?
(*a*) birds (*b*) snakes (*c*) grasshoppers (*d*) rats

84. Which of the following activities can help in the conservation of forests ?
(*a*) recycling of cotton shirts (*b*) recycling of silk blankets
(*c*) recycling of jute ropes (*d*) recycling of cardboard cartons

85. Which of the following is an endangered species of animals ?
(*a*) Dinosaur (*b*) Asiatic lion (*c*) Irish deer (*d*) Hyena

Questions Based on High Order Thinking Skills (HOTS)

86. Large scale deforestation leads to the decrease in the amount of gas A in the atmosphere whereas the amount of gas B increases. The increased amount of gas B in the atmosphere causes an effect C which leads to excessive heating of the earth and its atmosphere producing an undesirable phenomenon D. What are A, B, C and D ?

87. A highly poisonous animal X which inhabits crop fields is commonly known as friend of the farmer. It eats up pests like Y and saves the crops from damage. The animal X is killed in large numbers to obtain its Z which sells in the market at a high price. What are X, Y and Z ?

88. The species P and Q of wild animals are found exclusively in a 'particular area'. The species R of wild animals is listed in Red data book of India whereas species S of wild animals no longer exists anywhere on the earth.
(*a*) What name is given to species like P and Q ?
(*b*) What are the species like R known as ?
(*c*) Name one species like R.
(*d*) What is the special name of species like S ?
(*e*) Name one species like S.

89. Consider the following animal species :
Dodo, Yak, Deer, Black buck, Tiger, Crow, Kashmir stag, Dinosaur, Elephant, Lion tailed macaque, Peacock, Snow leopard
Which of these animals are :
(*a*) endemic to mountain habitats ?
(*b*) endangered species ?
(*c*) extinct species ?

90. X is a human activity which will cause an increase in temperature of earth's atmosphere leading to global warming. It can cause soil erosion making the soil infertile and lead to desertification. It can cause decrease in rainfall but increase the flooding of rivers. It can also affect water cycle and lead to the extinction of many wild species. What is X ?

ANSWERS

1. (*a*) Carbon dioxide (*b*) Carbon dioxide **2.** Biodiversity **11.** (*a*) Endangered species (*b*) Extinct species **14.** Reforestation **18.** (*a*) carbon dioxide (*b*) endemic (*c*) wildlife sanctuary (*d*) Pachmarhi (*e*) endangered (*f*) climatic (*g*) species (*h*) five; seven **66.** (*d*) **67.** (*b*) **68.** (*c*) **69.** (*d*) **70.** (*a*) **71.** (*d*) **72.** (*c*) **73.** (*d*) **74.** (*c*) **75.** (*c*) **76.** (*d*) **77.** (*d*) **78.** (*b*) **79.** (*c*) **80.** (*d*) **81.** (*d*) **82.** (*a*) **83.** (*b*) **84.** (*d*) **85.** (*b*) **86.** A: Oxygen ; B : Carbon dioxide ; C : Greenhouse effect ; D : Global warming **87.** X : Snake ; Y: Rats ; Z : Skin **88.** (*a*) Endemic species (*b*) Endangered species (*c*) Tiger (*d*) Extinct species (*e*) Dinosaur **89.** (*a*) Yak ; Kashmir stag ; Snow leopard (*b*) Black buck ; Tiger ; Kashmir stag ; Lion tailed macaque ; Snow leopard (*c*) Dodo ; Dinosaur **90.** Deforestation.

CHAPTER

8

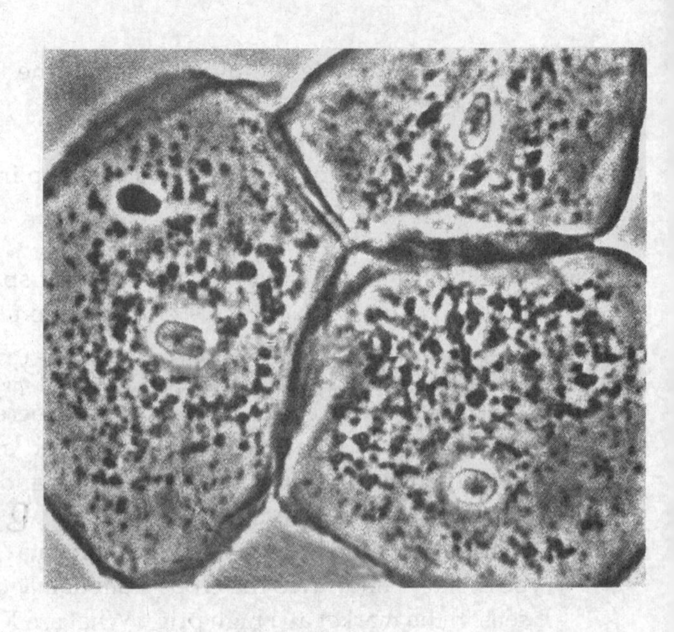

Cell Structure and Functions

All of us have seen the plants like rose, marigold, sunflower, cabbage, mushroom, mango tree, oak tree, coconut tree, banana tree, ferns and many others. We also know the animals like dog, cat, man (human beings), cow, horse, camel, lion, elephant, mosquito, housefly, sparrow, fish, starfish, butterfly, sea anemone, centipede, spider, grasshopper, and many others. In fact, **there is a large variety of organisms (plants and animals) around us**. They have different shapes and sizes. Their food habits and habitats (living places) are also different. **Inspite of great variations in the size, shape, food habits and habitats, etc., all the living organisms have a basic similarity among them : they are made up of tiny units called 'cells'**. Thus, in biology cells are the structural and functional units of life. In order to

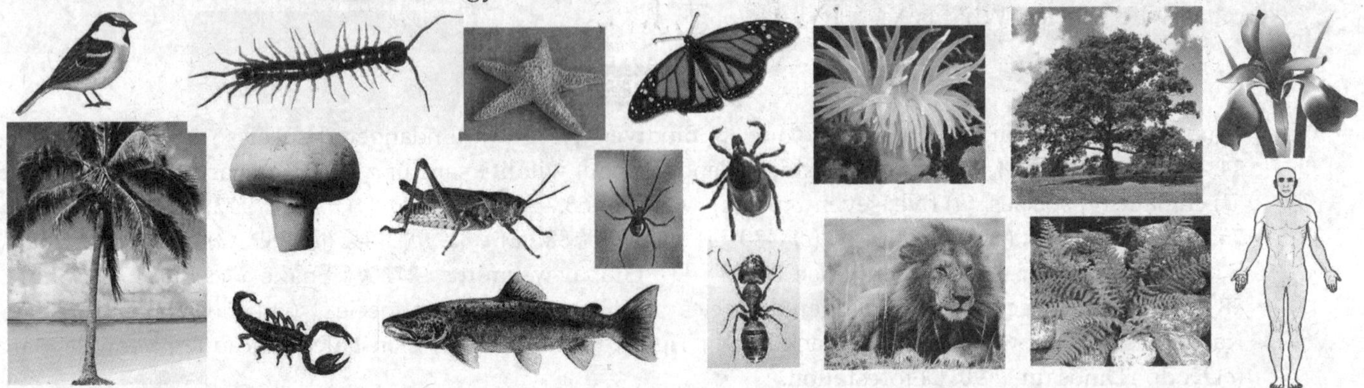

Figure 1. The various organisms (plants and animals) differ in shape, size, food habits and habitats but they have one common feature : All of them are made of tiny units called 'cells'.

understand the structure and functioning of an organism, we should know the structure of cells and functions of the various parts of the cells. In this Chapter, we will study the basic parts of the cells and the role played by these parts in the working of cells. Before we go further and study how cell was discovered, we should know the meaning of the term 'cork' which was used in the discovery of cell. Cork is a part of the bark of a tree. Cork is a dead plant material.

Discovery of Cell

In 1665, an English scientist named Robert Hooke used a microscope to investigate the structure of a thin slice of cork. He observed that cork had tiny compartments in it (see Figure 2). Robert Hooke thought of these compartments as 'small rooms' and called them 'cells' (because cells are small rooms in which prisoners are locked up or in which monks and nuns sleep). From this observation Robert Hooke concluded that the cork is made of tiny cells. The dead plant cells of cork observed by Robert Hooke were found to be empty.

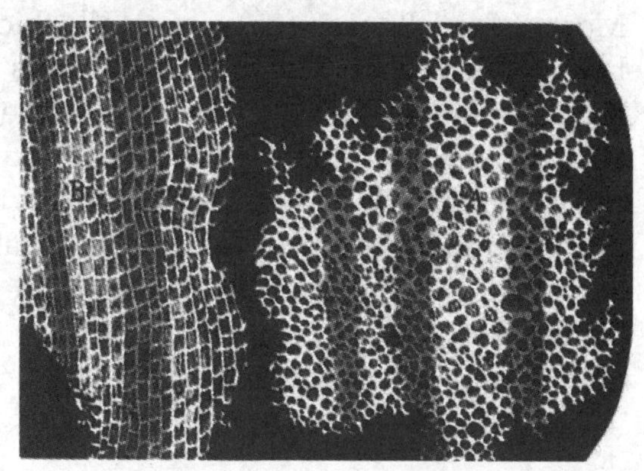

Figure 2. Robert Hooke's drawing of cork cells.

The microscope used by Robert Hooke was primitive. When better microscopes (having greater magnifying power) were made, scientists investigated pieces of living plants and found that like cork, they also had a cell structure. The living plant cells were, however, not empty like cork cells, they were found to contain a number of tiny structures (called organelles). In 1838, a German scientist Schleiden suggested that all plants are made of cells. A year later, another German scientist Schwann suggested that all animals are made of cells. These observations led to a general cell theory of organisms. **The cell theory states that the basic unit of structure and function of all living organisms is the cell**. All cells arise from pre-existing cells by cell division.

The cells have been present in the living organisms (plants and animals) since the origin of life. The cells were, however, not studied or observed for thousands of years because most of them are extremely small and cannot be seen with naked eye. It was only when microscopes having high magnification power were made in the seventeenth century to magnify things greatly that cells could be seen in living organisms. These days, microscopes can enable us to see the objects as small as one-millionth of a metre (10^{-6} m). Such high magnifying power microscopes have helped the scientists to study the minute details of cells. We will now study the cells in detail.

CELLS

Cells are the basic units of life. All the living things (plants and animals) are made from cells. Just as a house is made up of bricks, in the same way, a living organism is made up of cells. Thus, ***cells are the building blocks of plants and animals.*** A baby, a banyan tree and a bacterium look very different from one another, but they are all made up of tiny cells. Our nose, skin, muscles and bones are all made up of cells. The simplest organisms like *Amoeba* consist of only one cell but a complex organism like man (human being) is made up of trillions of cells. We can now define a cell as follows :

(*a*) The wall of a house is made of bricks

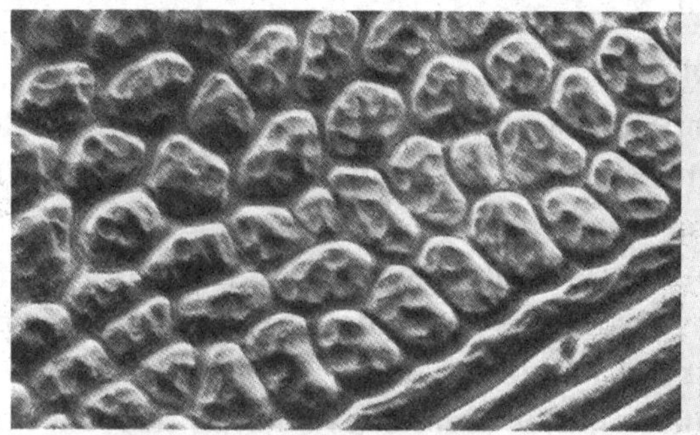

(*b*) The leaf of a plant is made of cells

Figure 3. Just as a house is made of bricks, in the same way, living organisms (plants and animals) are made of cells. Bricks are non-living things whereas cells are living structures.

A cell is the smallest unit of life which has a definite structure and performs a specific function. Most of the cells are very, very small which cannot be seen with naked eye. They can be seen only with the help of microscope. The diagrams of cells which we see in the books are highly magnified (as seen through the microscope). For example, the picture of human cheek cells given on the top of page 126 is magnified 1500 times. It is within the cells of our body that all our life activities occur. **All living cells come only from other living cells.** Since cells are the living units, they require energy. The cells obtain their energy by burning food like glucose in a process called respiration. **The cells are of two main types : Animal cells and Plant cells.** Though many things are common between animal cells and plant cells but they differ in some ways. We will first give the basic parts of the cells and then describe the general structure of an animal cell and a plant cell.

PARTS OF A CELL

Each cell has a number of smaller parts in it. Some of the parts are present in all type of cells (plant cells as well as animal cells). But certain parts are found only in plant cells, they are not present in animal cells.

(*i*) The important basic parts of all the cells are : Cell membrane (or Plasma membrane), Cytoplasm, Nucleus and Mitochondria.

(*ii*) The parts which are present only in plant cells are : Cell wall, Chloroplasts and Large vacuole.

We will now describe all these parts of the cell in detail, one by one. Let us start with the cell membrane.

1. Cell Membrane (or Plasma Membrane)

Every cell is covered by a thin sheet of skin which is called cell membrane (or plasma membrane) (see Figure 6 on page 130). The cytoplasm and nucleus are enclosed within the cell membrane. In other words, cell membrane encloses the living part of the cell called 'protoplasm'. The cell membrane protects the cell and also gives shape to the cell. The cell membrane has tiny pores in it. **The cell membrane controls the movement of substances 'into the cell' and 'out of the cell'.** The dissolved substances such as food (glucose) and oxygen can enter into the cell whereas the waste products such as carbon dioxide can go out from the cell through the pores of the cell membrane. The cell membrane separates the cells from one another and also from the surrounding medium. The cell membrane (or plasma membrane) is a living part of the cell.

2. Cytoplasm

Cytoplasm is a transparent, jelly-like material which fills the cell between nucleus and cell membrane (see Figure 6 on page 130). Cytoplasm is a kind of chemical factory of the cell. Here, new substances are built from materials taken into the cell, and energy is released and stored. In fact, **most of the chemical reactions which keep the cell alive take place in the cytoplasm**. The cytoplasm of a cell has many tiny structures in it. **The various structures present in the cytoplasm of a cell are called 'organelles'.** The most prominent organelle in the cytoplasm is the nucleus. In addition to nucleus, cytoplasm of all the cells contains other organelles such as mitochondria, golgi bodies and ribosomes, etc. The cytoplasm of plant cells also contains chloroplasts (We will study golgi bodies and ribosomes in higher classes). The cytoplasm and the nucleus taken together make up the protoplasm.

3. Nucleus

Nucleus is a large, spherical organelle present in all the cells (see Figure 6 on page 130). In animal cells, nucleus lies in the centre of the cell whereas in plant cells the nucleus may be on the periphery (near the edge) of the cell. Nucleus is the largest organelle in a cell. Nucleus can be stained (with a dye) and seen easily with the help of a microscope. Nucleus is separated from the cytoplasm by a membrane called nuclear membrane. **Nucleus controls all the activities of cell.**The transmission of characteristics (or qualities) from the parents to the offspring is called inheritance. *Nucleus plays a role in inheritance.* This can be explained as follows : **Nucleus contains thread-like structures called chromosomes** (Chromosomes can be seen only when the cell divides). Chromosomes contain genes (which are the units of inheritance). **The function of chromosomes is to transfer the characteristics from the parents to the offsprings through the genes** (which is called inheritance). The nucleus also contains a tiny round structure called **nucleolus** (which can be seen under the high magnifying power of a microscope). We will study the functions of nucleolus in higher classes. **The nucleus containing chromosomes and nucleolus is bound by a membrane called 'nuclear membrane'**. The nuclear membrane has tiny pores for the exchange of materials with cytoplasm. Red blood cells, however, do not have a nucleus.

Before we end the discussion on nucleus, it will be good to say a few words about 'genes' which are held on the chromosomes present in the nucleus of a cell. **Gene is a unit of inheritance in living organisms** (which is transferred from a parent to offspring during reproduction and determines some characteristic feature of the offspring). Thus, genes control the transfer of hereditary characteristics from parents to offsprings. This means that our parents pass on some of their characteristic features to us through their genes. It is the genes inherited (or received) from our parents which are responsible for our various body features. In other words, it is the genes on the chromosomes which decide all sort of things about us, say, the colour of our eyes, the look of our hair, the shape of our nose, our complexion, our appearance, etc. For example, if our father has brown eyes, we may also have brown eyes (due to the gene of brown eye colour coming from him). And if our mother has curly hair, then we may also inherit curly hair (through the gene coming from her). However, usually the different combinations of genes from both the parents (father and mother) result in different characteristics in the offspring. Let us discuss the protoplasm now.

All the living matter in a cell is called protoplasm. Protoplasm is a liquid substance which is present inside the cell membrane. **Protoplasm includes cytoplasm, nucleus and other organelles**. Most of the protoplasm is made up of compounds of only four elements : **carbon, hydrogen, nitrogen and oxygen**. It also contains some other elements. Some of the compounds present in protoplasm are water, carbohydrates, proteins, fats, nucleic acids and minerals salts. A unique combination of elements and compounds provides living nature to protoplasm.

4. Mitochondria

Mitochondria are the tiny rod-shaped or spherical organelles which are found in all the cells (see Figure 6 on page 130). Mitochondria provide energy for all the activities of the cell. This energy is produced by the process of respiration in which food (such as glucose) is broken down by oxygen. Thus, **mitochondria use glucose and oxygen to produce energy**. Mitochondria are found in large numbers in the cytoplasm in all the cells.

5. Cell Wall

The plant cells have a thick cell wall around them (outside the cell membrane) (see Figure 7 on page 131). The cell wall is made of a tough material called cellulose. **Cell wall gives shape and support to the plant cell.** Cell walls also hold the plant cells together and give plants most of their strength. Plants need protection against high wind speed, variations in temperature and atmospheric moisture, etc. Since plants are fixed at a place, so they cannot move to protect themselves from the various changes in their surroundings. The tough cell wall present in plant cells provides protection to plants. Please note that cell wall is present only in plant cells. ***Cell wall is not there in animal cells. The cell wall is a non-living part of the plant cells.***

6. Chloroplasts

Chloroplasts are the green coloured organelles present in the cytoplasm of plant cells (see Figure 7 on page 131). The process of food making by plants known as **photosynthesis takes place in chloroplasts**. The green colour of chloroplasts is due to the presence of a green pigment called **'chlorophyll'** in them. Chlorophyll can absorb sunlight energy. In the chloroplasts, carbon dioxide and water combine in the presence of sunlight energy to produce food such as glucose. And this process of food making is called photosynthesis. Thus, **chloroplasts help in the making of food by green plants**. Please note that chloroplasts are found only in those plant cells which carry out photosynthesis (like the cells in green leaves). So, ***chloroplasts will not be found in the root cells of a plant.***

The organelles containing pigments (coloured matter) present in the cytoplasm of plant cells are called plastids. Plastids can be of many different colours. **The plastids containing green pigment (chlorophyll) are known as chloroplasts.** The green coloured plastids (called chloroplasts) present in the

Figure 4. The green colour of leaf is due to the presence of green coloured plastids (chloroplasts) in its cells.

Figure 5. The red colour of tomatoes is due to the presence of red coloured plastids in their cells.

cells of leaves provide green colour to the leaves (see Figure 4). The plastids can also have pigments of other colours. For example, the red colour of tomatoes is due to the presence of 'plastids having red pigment' in the cytoplasm of its cells (see Figure 5). The different colours of flowers are due to the presence of plastids containing pigments of different colours.

7. Large Vacuole

Vacuole is a space in the cytoplasm of a cell which is enclosed by a membrane and usually contains substances dissolved in water. A vacuole appears as an empty space under the microscope. **All the plant cells have a large vacuole** (see Figure 7 on page 131). **The vacuole is filled with a liquid called 'cell sap', which contains dissolved sugars and salts**. The pressure of liquid (cell sap) pushes on the outer parts of the plant cell keeping the plant cell firm (or turgid). **The function of vacuole in a cell is to store various substances including waste products of the cell**. Most of the animal cells do not have vacuoles. Some animal cells have vacuoles but they are much smaller than those found in plant cells. For example, *Amoeba* is an animal cell which has very small vacuoles. In *Amoeba*, vacuoles contain food particles, so they are known as food vacuoles.

STRUCTURE OF CELL

A cell consists of a jelly-like material enclosed in a thin membrane. The jelly-like material which fills the cell is called 'cytoplasm' and the thin, outer covering of the cell is called 'cell membrane'. A general diagram of an animal cell is shown in Figure 6. There are many bodies of different shapes and sizes inside the cytoplasm. For example, there is a large floating body usually in the centre of a cell which is called the nucleus. **The function of nucleus is to control all the activities of the cell (like cell growth, cell division, etc.).** The nucleus contains a tiny round structure called nucleolus and a fibrous material called chromatin. It also has a nuclear membrane on the outside. Chromatin forms

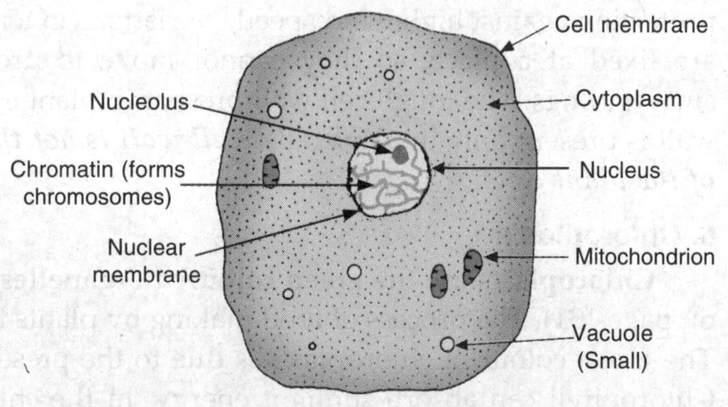

Figure 6. The general diagram of an animal cell.

chromosomes during cell division. These chromosomes transfer characteristics from the parents to the next generation. The cytoplasm and nucleus taken together form the protoplasm of the cell. Protoplasm is the living material of a cell. There are a number of small bodies called 'mitochondria' in a cell (The singular of mitochondria is mitochondrion). **The function of mitochondria is to carry out respiration for releasing energy from food.** Since mitochondria produce energy from food, they are a kind of 'power-house' of the cell.

The tiny air spaces in the cytoplasm of an animal cell are called vacuoles. **The function of a vacuole in an animal cell is to hold air, water and particles of food.** *Most of the animal cells do not have vacuoles.* But some of the animal cells have several small vacuoles. The cell membrane is a thin sheet of skin all around the cell. **The main function of cell membrane is to control the passage of materials which 'go into the cell' or 'go out from the cell'.** The cell membrane also protects the cell and gives shape to the cell. The cytoplasm is a transparent, jelly-like substance. It is a complex solution. **The function of cytoplasm is to carry out all the activities of the life processes (or metabolism).** We will now describe the structure of a plant cell.

We have just studied that the main parts of an animal cell are : cell membrane, cytoplasm, nucleus and mitochondria. All these parts are also present in plant cells. In addition to these, the plant cells have three more parts in them. These are : ***cell wall, chloroplasts* and *large vacuoles*.** These parts make the plant cells different from animal cells. A general diagram of a plant cell is given in Figure 7.

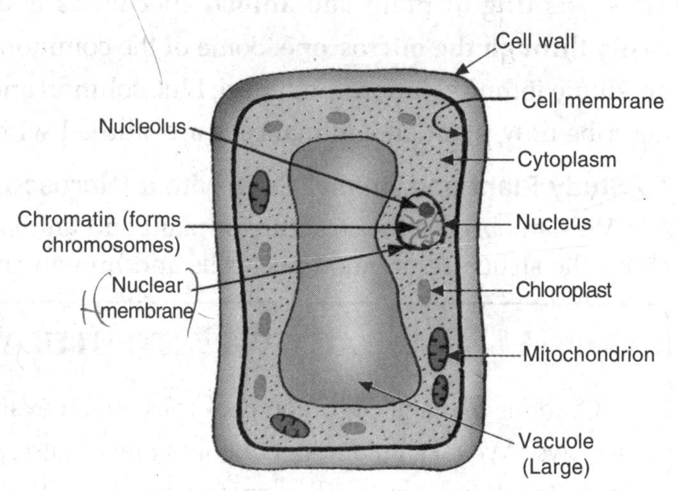

Figure 7. The general diagram of a plant cell.

A plant cell consists of a jelly-like cytoplasm enclosed in a thin cell membrane. There is a large, floating nucleus in the cytoplasm (see Figure 7). The nucleus contains nucleolus and chromatin. It is surrounded by a nuclear membrane. Several mitochondria are also present in the plant cell. The mitochondria present in the plant cell carry out respiration and release energy from food. All the plant cells have a thick cell wall all around the cell membrane (see Figure 7). This cell wall is made of a material called cellulose. **The cell wall protects the plant cell, gives it a fixed shape and makes it rigid (strong).** Even the single-celled organisms like bacteria have cell walls around them. *The cell wall is not present in animal cells.*

The plant cells contain green coloured plastids called chloroplasts. Chloroplasts are the green coloured bodies (or particles) in the cytoplasm of the plant cell. **The chloroplasts make food in green plants by the process of photosynthesis.** *Chloroplasts are not present in animal cells.* Some plant cells may contain plastids of different colours (other than green). The plant cells have very large vacuoles. These vacuoles are filled with cell sap. Cell sap is a solution of sugars and mineral salts. *The animal cells either do not contain vacuoles or contain several small vacuoles.* The animal cells do not contain a large vacuole.

Comparison of Plant Cells and Animal Cells

We have just studied that plant cells and animal cells are similar in many respects but they are also different in some respects. We will now give the important similarities and differences between plant cells and animal cells.

The main similarities between plant cells and animal cells are given below :

1. Plant cells and animal cells have a cell membrane (or plasma membrane) around them.
2. Plant cells and animal cells have cytoplasm.
3. Plant cells and animal cells have a nucleus.

4. Plant cells and animal cells have a nuclear membrane.
5. Plant cells and animal cells have mitochondria.

The main differences between plant cells and animal cells are given below :

Plant cell	*Animal cell*
1. A plant cell has a cell wall around it.	**1.** An animal cell does not have a cell wall around it.
2. A photosynthetic plant cell has chloroplasts in it. Other plant cells have different plastids in them.	**2.** An animal cell does not have chloroplasts or other plastids
3. A plant cell has a large vacuole in it.	**3.** An animal cell has usually no vacuole. Only some animal cells have small vacuoles.

The various parts of a cell are colourless and hence difficult to distinguish. So, in order to see the plant and animal cells with a microscope, they are stained (coloured) with dyes. The dyes stain the various parts of a cell with different colours due to which these parts can be seen more easily under the microscope. Thus, **staining of plant and animal specimens is done to identify the different components of a cell easily through the microscope.** Some of the common dyes (or stains) used in the study of structure of cells are dilute iodine solution, methylene blue solution and safranin. Keeping these points in mind, we will now describe how plant and animal cells are studied with the help of a microscope.

To Study Plant and Animal Cells with a Microscope

We can observe the structure of plant and animal cells in the laboratory by using a microscope. Let us study the structure of onion peel cells and human cheek cells.

ACTIVITY TO STUDY ONION PEEL CELLS

Onion is a plant. So, onion peel cells are an example of plant cells. The onion peel cells can be studied as follows : We take an onion, cut it into pieces and separate the thin inner layer of any fleshy layer with our fingers. This thin layer is called onion peel. A small piece of onion peel is placed on a glass slide. Put a drop of water over the onion peel. Add a little of dilute iodine solution to the slide and cover it with a glass cover slip. When we observe this slide under a microscope, we will see the structure of onion peel cells (as given in Figure 8). The chamber like structures shown in Figure 8 are the cells of onion peel. We can identify the cell membrane, cell wall, nucleus, cytoplasm and vacuole present in the onion peel cell. The boundary of onion peel cell is the cell membrane. The cell membrane is covered by a thick covering which is the cell wall. The dense round body on the periphery (near the edge) of the cell is the nucleus. The jelly like substance between the nucleus and the cell membrane is cytoplasm. And the empty looking space in the centre of onion peel cell is the vacuole (see Figure 8).

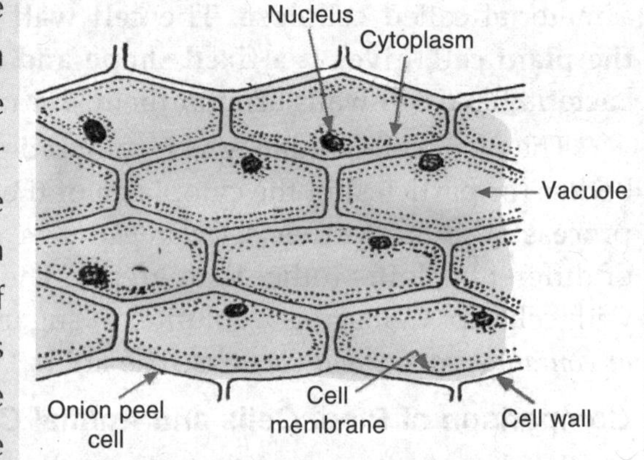

Figure 8. Cells of onion peel (as seen through microscope).

ACTIVITY TO STUDY HUMAN CHEEK CELLS

The human beings are animals. So, the human cheek cells are an example of animal cells. We can study the structure of human cheek cells as follows : We open our mouth and scrape the inner surface of our cheek lightly with a tooth-pick. This scraping is mixed in a drop of water placed on a glass slide. Add a little of methylene blue solution to the slide and cover it with a glass cover slip. If we observe this prepared slide under a microscope, we will see the structure of human cheek cells as shown in Figure 9. We can identify the cell membrane, nucleus and cytoplasm in the human cheek cells. There is no cell wall seen in the human cheek cells (see Figure 9).

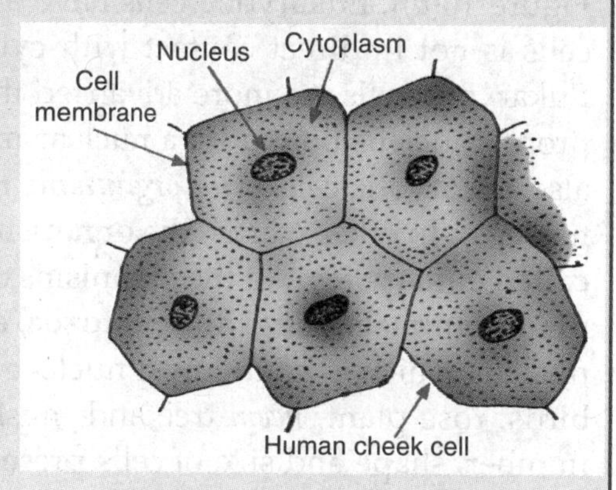

Figure 9. Cells of human cheek (as seen through microscope).

Please note that the cells of plants and animals are extremely small in size due to which we cannot see them with the naked eyes (they can be seen only through a microscope). We can, however, see the plants and animals around us because they are made up of millions and millions of tiny cells joined together. Before we go further and discuss prokaryotic cells and eukaryotic cells, please note that *the material which makes up the nucleus of a cell is called 'nuclear material'*.

PROKARYOTIC CELLS AND EUKARYOTIC CELLS

Prokaryotic cells are more primitive than eukaryotic cells and have a simple structure. The nucleus of prokaryotic cells is not well organised. There is no nuclear membrane around the nuclear material in the prokaryotic cells. **The cells having nuclear material without a nuclear membrane around it, are called prokaryotic cells** [see Figure 10(*a*)]. The nuclear material in a prokaryotic cell is in direct contact with the cytoplasm. In fact, prokaryotic cells have no real nucleus. ***The organisms made of prokaryotic cells are called prokaryotes.*** (pro = primitive ; karyon = nucleus). All the prokaryotes are simple, unicellular organisms. Thus, **the simple unicellular organisms whose cells do not possess a nucleus bound by a nuclear membrane are called prokaryotes.** Prokaryotes have no real nucleus in their cells. Prokaryotes are relatively simple organisms. **Bacteria and Blue-green algae are prokaryotes**. Bacterium is a prokaryote because its cell has no real nucleus [see Figure 10(*a*)]. The nuclear material in a bacterium cell is not surrounded by a nuclear membrane. The nuclear material in a bacterium cell is in direct contact with the cytoplasm. Thus, the nucleus of a bacterium cell is not well organised like the nucleus of cells of multicellular organisms. Prokaryotes were probably the first living things to evolve on the earth.

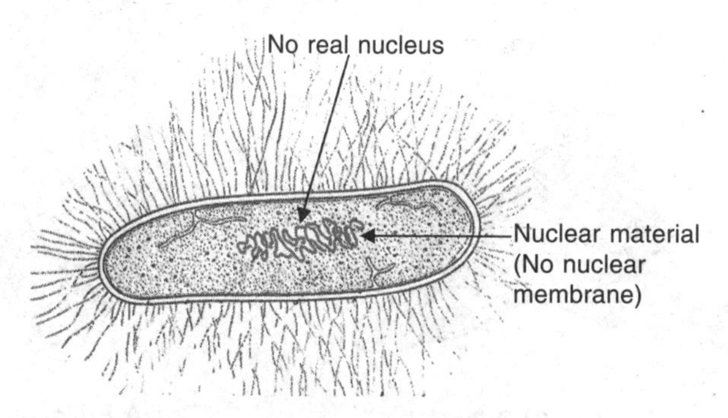

(*a*) Bacterium cell. It has no real nucleus having a nuclear membrane. It is a prokaryotic cell

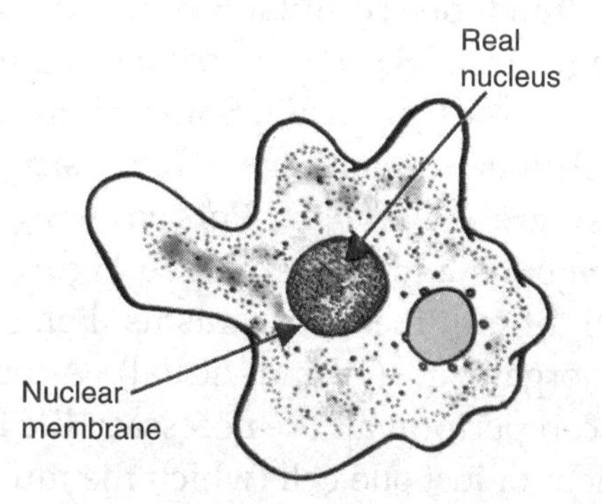

(*b*) *Amoeba* cell. It has a real nucleus having a nuclear membrane around it. It is a eukaryotic cell

Figure 10. Prokaryotic cell and Eukaryotic cell.

The cells having nuclear material enclosed by a nuclear membrane are called eukaryotic cells [see Figure 10(*b*)]. Eukaryotic cells have a proper, well organised nucleus. The nuclear material in eukaryotic cells is not in direct contact with cytoplasm, it is separated from cytoplasm by the nuclear membrane. Eukaryotic cells are more advanced than prokaryotic cells. *Amoeba* cell is a eukaryotic cell because it has a proper nucleus bound by a nuclear membrane [see Figure 10(*b*)]. The onion peel cells and cheek cells are also eukaryotic cells. ***The organisms made of eukaryotic cells are called eukaryotes*** (eu = true ; karyon = nucleus). In other words, **the organisms whose cells possess a nucleus bound by a nuclear membrane are called eukaryotes**. All the organisms other than bacteria and blue-green algae are eukaryotes. For example, **plants, animals, fungi and protozoa, etc., are all eukaryotes**. *Amoeba* is a eukaryote because it has a distinct nucleus which is bound by a nuclear membrane [see Figure 10(*b*)]. A human being, a lion, tiger, dog, cat, birds, rose plant, *neem* tree and mushroom, etc., are all eukaryotes. We will now discuss the variety in number, shape and size of cells present in living organisms.

ORGANISMS SHOW VARIETY IN CELL NUMBER, CELL SHAPE AND CELL SIZE

There are millions of living organisms all of which are made of tiny 'cells'. The various organisms, however, differ in the number of cells which make up their body ; they differ in the shapes of cells in their body ; and they also differ in the size of cells in their body.

(*i*) **Different organisms have different number of cells in their bodies.** This is called variety in the number of cells (or variety in 'cell number').

(*ii*) **The cells in multicellular organisms (multicellular plants and animals) have many different shapes.** This is called variety in shape of cells (or variety in 'cell shape').

(*iii*) **The cells in multicellular organisms can have many different sizes.** This is called variety in size of cells (or variety in 'cell size').

We will now describe the variety in 'number of cells', 'shape of cells' and 'size of cells' in organisms in detail. Before we do that we should keep in mind that : A million is a thousand thousand (which is 1000,000) ; a billion is a thousand million ; and a trillion is a thousand billion.

1. Variety in Number of Cells

All the living organisms are made up of cells. But the number of cells varies from organism to organism. Some organisms (plants and animals) are made up of just 'one cell' (single cell) while other organisms are made up of 'a large number of cells'. This is called variety in the number of cells. ***Depending on the number of cells in the body of an organism, an organism is called unicellular or multicellular.***

The simplest living organisms have only 'one cell' in their body. **The organisms which are made up of only 'one cell' are called unicellular organisms** (uni = one ; cellular = cell). One cell is also called 'single cell'. So, *the unicellular organisms are actually single-celled organisms.* **Some of the examples of unicellular organisms (or single-celled organisms) are : *Amoeba, Paramecium, Euglena, Chlamydomonas* and Bacteria** (see Figure 11). The single cell of all these organisms behaves as a 'complete organism' (or individual). A unicellular organism can perform all the necessary life functions with the help of just one cell (which the multicellular organisms perform with the help of many groups of specialised cells). For example, *Amoeba* is a tiny animal which consists of only one cell [see Figure 11(*a*)]. All the basic functions of life like taking food

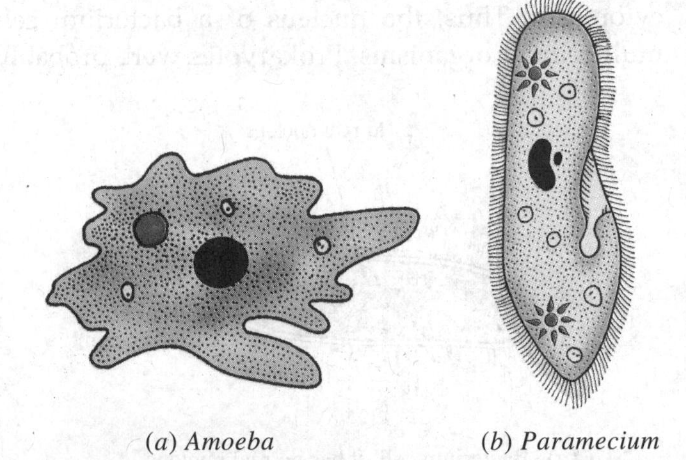

(*a*) *Amoeba* (*b*) *Paramecium*

Figure 11. *Amoeba* and *Paramecium* are unicellular organisms (or single-celled organisms). They are made of only 'one cell' each.

(or capturing food), digestion, respiration, movement, response to environmental changes, removal of waste (excretion), and reproduction, etc., are performed by the single cell of *Amoeba*. Similar functions in multicellular organisms are performed by various organs (or organ systems) which are made up of many, many different types of cells.

The organisms which are made up of many cells are called multicellular organisms (multi = many ; cellular = cell). Depending on its size, a multicellular organism may be made up of millions, billions or trillions of cells joined together. Most of the plants and animals around us are multicellular organisms. For example, **a rose plant, a *neem* tree, a rat, a human being (man) and an elephant are all multicellular organisms** which are made up of many, many cells joined together (see Figure 12). A big and tall tree, a huge animal like an elephant and the body of a human adult has trillions of cells which vary in shapes and sizes. Different groups of cells perform a variety of functions.

(*a*) Tree

(*b*) Human (Man)

Figure 12. A tree and a human being (man) are multicellular organisms. They are made up of many, many cells.

A rat is a small organism whereas an elephant is a big organism. So, a rat has a smaller number of cells in its body than that of an elephant. The smaller number of cells in a rat does not affect the functioning of a rat. In general we can say that : **The number of cells being less in small organisms does not in any way affect the functioning of the small organisms**. In fact, even an organism made of millions, billions or trillions of cells begins its life as a single cell called 'fertilised egg cell' (or 'zygote'). The fertilised egg cell (or zygote) divides and multiplies due to which the number of cells increases as the development of the organism proceeds.

2. Variety in Shape of Cells

There are many types of cells in the bodies of multicellular organisms. These cells differ in shapes. For example, the shape of a nerve cell in animals is very different from the shape of a muscle cell [see Figures 13(*a*) and (*b*)]. **A nerve cell is long and branched (having thread-like projections) whereas a muscle cell is pointed at both ends and has a spindle shape**. Due to this, a nerve cell looks very different from a muscle cell. **The different shapes of cells are related to their functions** which they have to perform in the body of an animal (or plant). For example, **nerve cells receive and transmit messages between brain and other body parts whereas muscle cells bring about movement in body parts**. Nerve cells and muscle cells are actually specialised animal cells because they perform specific functions in the bodies of animals. Some

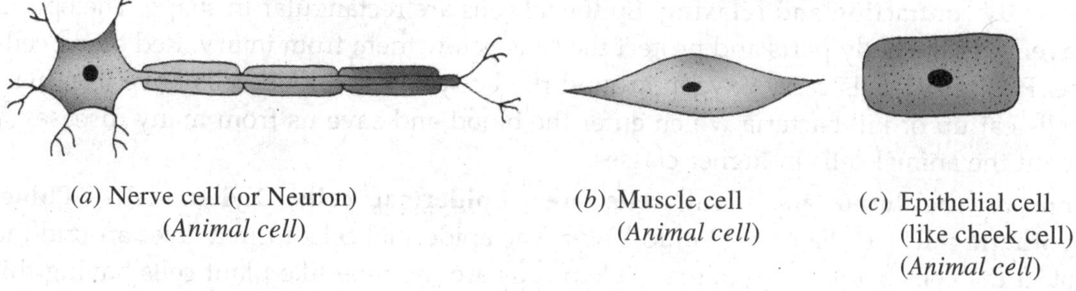

(*a*) Nerve cell (or Neuron) (*Animal cell*)

(*b*) Muscle cell (*Animal cell*)

(*c*) Epithelial cell (like cheek cell) (*Animal cell*)

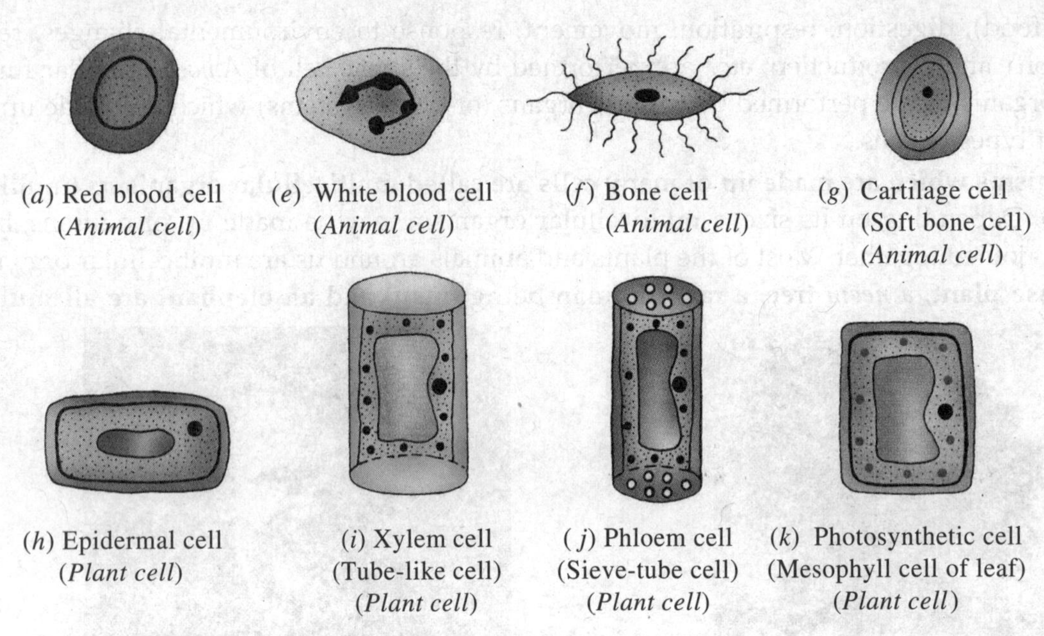

Figure 13. Different types of cells (specialised cells). They all have different shapes.

of the specialised animal cells and plant cells are shown in Figure 13. See how different their shapes are.

Cells are different in shapes and sizes so that they can perform different functions. We say that the cells are specialised (to do different jobs). All the cells shown in Figure 13 are specialised cells. Each one of these cells is adapted to perform some specific function. For example, some cells protect the body parts, some cells help in moving body parts, some cells help in sending messages from brain to body parts (or from body parts to brain), and so on. The human body is made up of about 20 different types of cells and each type of cell performs different functions. **Some of the examples of animal cells (or human body cells) are : Nerve cell (or Neuron), Muscle cell, Epithelial cell, Red blood cell, White blood cell, Bone cell and Cartilage cell** [see Figures 13(*a*) to 13(*g*)].

The shape of a cell helps in its functioning. For example, a nerve cell is very long and has wire-like projections coming out of it [see Figure 13(*a*)]. The large length of nerve cell makes it carry messages over long distances in the body. And the wire-like projections of the nerve cell help it to make many contacts with other nerve cells so that messages from the brain can be sent to all the parts of the body. Thus, **the nerve cells are long and have projections so that they can make contacts with many other nerve cells and carry messages over long distances (between brain and other parts of the body).** And we say that the nerve cells are specially adapted to transmit messages. Since nerve cells receive and transfer messages, they help to control and coordinate the working of different parts of the body.

Let us take the example of muscle cells now. Muscle cells are specially adapted for movement. This happens as follows : The muscle cells have a special property that they can contract (and relax). When the muscle cells contract, they shorten in length. So, the contraction of muscle cells moves the body part (to which they are attached). And when these contracted muscle cells relax, they expand (increase in length), so that the body part comes back to its original position. Thus, **muscle cells bring about the movement of body parts by contracting and relaxing.** Epithelial cells are rectangular in shape. The epithelial cells form a thin layer over the body parts and protect the cells below them from injury. Red blood cells are spherical in shape. Red blood cells carry oxygen around the body. White blood cells have irregular shape. White blood cells eat up or kill bacteria which enter the blood and save us from many diseases. We will study more about the animal cells in higher classes.

Some of the important plant cells are : Epidermal cells, Xylem cells, Phloem cells and Photosynthetic cells [see Figures 13(*h*) to 13(*k*)]. The epidermal cells form a layer around the plant organs and protect the cells below from injury. Xylem cells are the tube-like plant cells having thick and strong walls which carry water and mineral salts from the roots of the plant to its leaves. Phloem cells are also tube-like plant cells having thin walls which carry the food made by leaves to all other parts of the plant. The photosynthetic cells of the plant contain chlorophyll and prepare food by photosynthesis. The mesophyll cells of leaf are the photosynthetic plant cells. These cells in the leaf of a plant are specially adapted for making food by photosynthesis.

All the cells of different shapes which we have described above are those which are present in the bodies of multicellular animals and plants. They are not capable of independent existence. We will now describe the shape of a cell called *Amoeba* which is a complete organism in itself and capable of independent existence.

The shape of an *Amoeba* cell is irregular [see Figure 11(*a*) on page 134]. In fact, **the *Amoeba* cell has no fixed shape. The *Amoeba* cell keeps on changing its shape continuously.** The shape of *Amoeba* changes because *Amoeba* can make its cytoplasm flow in any direction it wants to. The *Amoeba* cell has finger-like projections of varying lengths protruding out of its body which are called pseudopodia (pseudopodia means false feet). *Amoeba* can produce pseudopodia on any side by pushing the cytoplasm in that side. The pseudopodia appear and disappear when *Amoeba* 'moves' or 'feeds'. For example, *Amoeba* moves very slowly with the help of pseudopodia (which keep on appearing and disappearing to make it move). *Amoeba* also uses pseudopodia to catch (or engulf) the food particles from the water in which it lives. Thus, ***Amoeba* derives two advantages by changing its shape : The changing of shape due to the formation of pseudopodia helps *Amoeba* (*i*) in movement, and (*ii*) in capturing food.**

We have just studied that *Amoeba* is a single cell which can change its shape. A white blood cell (WBC) present in human blood is another example of a single cell which can also change its shape. The difference between *Amoeba* cell and white blood cell is that while *Amoeba* cell is a full fledged organism capable of independent existence, white blood cell is merely a cell of human blood which is not a full fledged organism and hence cannot exist independently. It can exist only inside the blood.

Before we describe the variation in the size of cells, we should know the very small unit of measuring length called micrometre (or micron). A micrometre (or micron) is one-millionth of a metre (which is 10^{-6} metre). Let us discuss the sizes of cells now.

3. Variety in Size of Cells

The cells are of many different sizes. The cells in living organisms may be as small as 'a millionth of a metre' (micrometre or micron), or it may be as large as 'a few centimetres'. However, **most of the cells are extremely small in size (or microscopic) and hence cannot be seen with the naked eyes.** Most of the cells found in living organisms have to be enlarged (or magnified) by using a microscope before they can be seen by us. Bacteria cells are extremely small (see Figure 14). The bacteria cells have a length of 0.1 micrometre to 0.5 micrometre. In fact, **bacteria *Mycoplasma* is the smallest cell** having the size (or length) of only 0.1 micrometre. Even the long cells are so thin that they can be seen only with a microscope. For example, the muscle cells in animals are a few centimetres long and the nerve cells are more than a metre long, but they are so thin that they can only be seen with the help of a microscope.

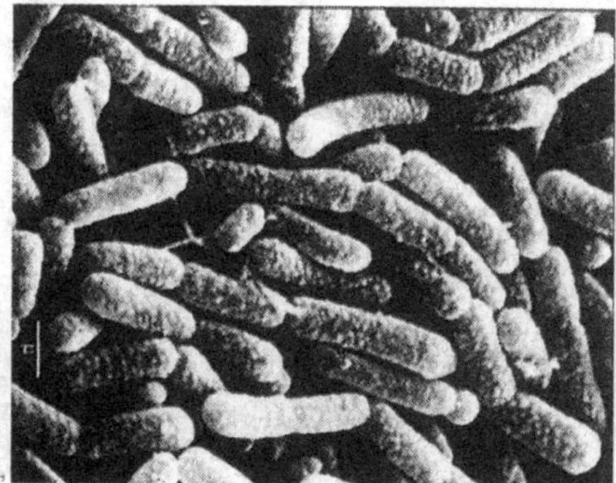

Figure 14. Bacteria cells are extremely small. They can be seen only when magnified through a microscope (The above picture of bacteria cells has been taken with the help of a microscope).

Figure 15. The bird's eggs (like hen's eggs) are very large cells. They can be seen easily with naked eye (without the help of a microscope).

Some of the cells are, however, big and can be seen easily with naked eye. The bird's eggs are very large cells (see Figure 15). Each egg of the bird is a single cell. For example, the hen's egg is a single cell. The hen's egg is quite big. We can see the single cell called hen's egg with naked eyes (without using a microscope). **The biggest cell (which can be seen with the naked eye) is the ostrich egg.** The ostrich egg can be as much as 17 centimetres long (which is 170 millimetres long or 170,000 micrometres long). In the human body, some blood cells are the smallest and the nerve cells are the longest. The sizes (or lengths) of some of the common cells are given below. We are giving the lengths of these cells in millimetres (mm) as well as in micrometres (microns) just for the sake of comparison.

Name of cell	*Size (Length)*		
1. Bacteria *Mycoplasma* (*Smallest cell*)	0.0001 mm	(or	0.1 micrometre)
2. Red blood cell	0.009 mm	(or	9 micrometre)
3. Liver cell	0.02 mm	(or	20 micrometre)
4. Human egg	0.1 mm	(or	100 micrometre)
5. Humming bird egg	13 mm	(or	13,000 micrometre)
6. Hen egg	60 mm	(or	60,000 micrometre)
7. Ostrich egg (*Biggest cell*)	170 mm	(or	170,000 micrometre)

Figure 16. This is the ostrich bird. Its egg is the biggest cell.

We have just said that the biggest cell is the egg of an ostrich bird (see Figure 16). The ostrich egg is actually 170 millimetres long and 130 millimetres wide. So, we can say that the biggest cell is the egg of an ostrich and it measures 170 mm × 130 mm. If we use the unit of centimetres, we can say that an ostrich egg measures 17 cm × 13 cm. **The size of cells has no relation with the size of the body of an animal (or plant).** Now, elephant is a very big animal whereas rat is a quite small animal. It is not necessary that the cells in the elephant be much bigger than those in a rat. **The size of a cell is related to its function.** For example, the nerve cells, both in elephant and rat are long and branched. They perform the same function of transmitting messages (in the form of electrical impulses).

CELLS, TISSUES, ORGANS, ORGAN SYSTEMS AND ORGANISM

1. Cells

A cell is the smallest unit of life which has a definite structure and performs a specific function. All the cells of a multicellular organism are not similar. They are of many different shapes and sizes. Most of the cells are specialised to perform particular functions. They are called 'specialised cells'. For example, in animals, muscle cells are specialised to contract (and relax) so that they can bring about movement in body parts (see Figure 17). In plants, photosynthetic cells are specialised to carry out photosynthesis and make food. There are many types of specialised cells in animals and plants which perform different functions.

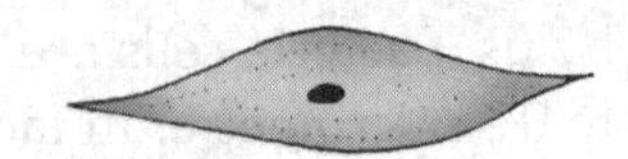

Figure 17. Muscle cell.

2. Tissues

A multicellular organism (animal or plant) is made up of millions of cells. The cells in a multicellular organism do not work as single cells, they work in groups of similar cells. **The group of similar cells which work together to perform a particular function is called a tissue.** For example, in animals, muscle tissue is a group of muscle cells joined together which is specialised to contract (and relax) so as to move body parts (see Figure 18). Thus, muscle tissue brings about movement in the body parts of animals. In plants, photosynthetic tissue is a group of photosynthetic cells joined together which is specialised to do photosynthesis and make food. There are

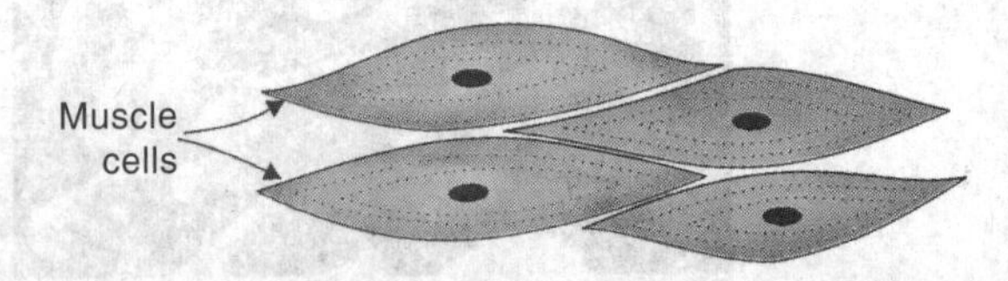

Figure 18. Muscle tissue.

many different types of tissues in both, animals as well as in plants. In fact, there are as many types of tissues as there are cells.

3. Organs

The bodies of animals and plants are made up of different types of tissues. The different tissues in the body of an animal or a plant do not work separately. The different tissues combine together to form 'organs'. These organs perform different tasks for the animal or the plant. We can now say that : **An organ is a collection of different tissues which work together to perform a particular function in the body of an organism**. The multicellular organisms are made up of many different organs which do different jobs for the organism. Some of the organs which are found in the bodies of animals (including human beings) are: Heart, Stomach, Brain, Lungs, Kidneys, Liver, Intestine, Mouth, Eyes and Hands, etc. Some of the

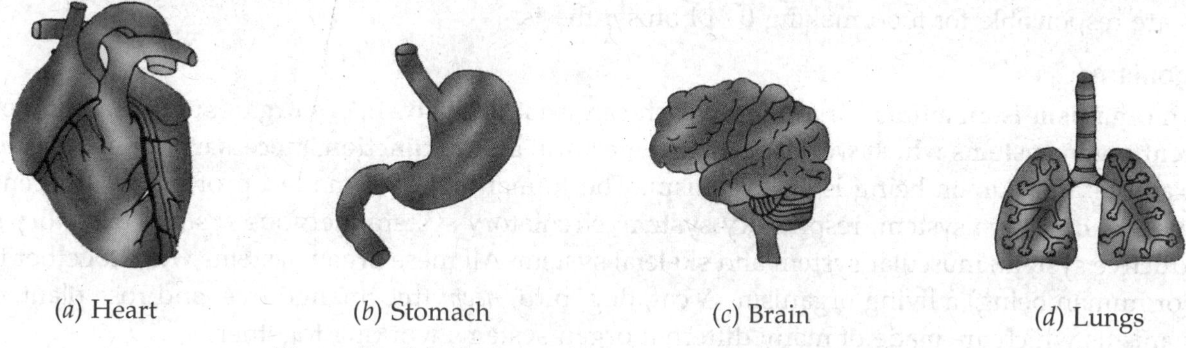

(*a*) Heart (*b*) Stomach (*c*) Brain (*d*) Lungs

Figure 19. Some organs of human body (animal organs).

human organs are shown in Figure 19. The function of heart is to pump blood around the body. The function of stomach is to digest food. The function of brain is to control all the parts of the body. Lungs are the organs of breathing. The function of lungs is to take in oxygen and give out carbon dioxide.

The body of a plant also contains many organs. The various organs found in the body of a plant are: Roots, Stem, Leaves, Flowers and Fruit. Some of the plant organs are shown in Figure 20. Each of these organs performs a particular job for the plant. For example, roots absorb water and dissolved mineral salts from the soil. Stem carries water and minerals from the roots to the leaves, and the prepared food

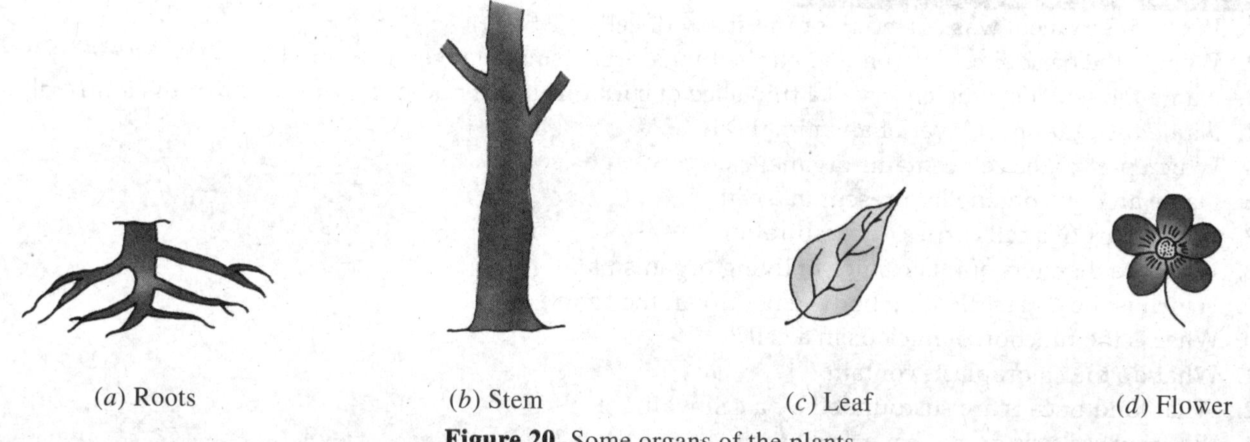

(*a*) Roots (*b*) Stem (*c*) Leaf (*d*) Flower

Figure 20. Some organs of the plants.

from the leaves to other parts of the plant. It also holds branches and leaves. The leaves prepare food for the plant by the process of photosynthesis. The flowers are reproductive organs which lead to the formation of fruits and seeds. The fruit protects the seeds.

4. Organ Systems

The various organs in an organism (animal or plant) do not work separately. The organs work in groups. **A group of interconnected organs which works together to do a big job for the organism, is called an organ system** (or just a 'system'). All the multicellular animals and plants have many organ systems in their bodies to carry out various life processes. For example, the various organ systems of animals (including human beings) are : Digestive system, Respiratory system, Circulatory system, Nervous system, Excretory system (or Urinary system), Reproductive system, Muscular system and Skeletal system. Each organ system performs a specific function for the body. For example, the function of digestive system is to break down the food into simple substances which can be absorbed by the body. The main organs of the digestive system are : Mouth, Oesophagus (Food pipe), Stomach, Small intestine, Large intestine, Rectum and Anus. Each organ of the digestive system performs a different function such as ingestion, digestion, absorption, assimilation and egestion, etc. The plants have two main organ systems : Root system and Shoot system. The root system consists of various types of roots whereas shoot system is made up of organs like stem, leaves, flowers and fruits, etc. Different organs of the 'organ systems' of plants perform specific functions for the plants. For example, roots help in the absorption of water and minerals. And leaves are responsible for food making by photosynthesis.

5. Organism

An organism is an animal or a plant which can exist on its own. An organism is made up of many different organ systems which work together to perform all the functions necessary for maintaining life. For example, the human being is an organism. The human being is made up of many different organ systems like digestive system, respiratory system, circulatory system, nervous system, excretory system, reproductive system, muscular system and skeletal system. All these organ systems work together to make man (or human being) a living organism. A cat, dog, bird, *neem* tree, mango tree, and rose plant, etc., are all organisms which are made of many different organ systems working together.

From all the above discussion, we come to the conclusion that **multicellular organisms (animals and plants) are built like this : cells make up tissues, tissues make up organs; organs make up organ systems and finally organ systems make up an organism.** The organisation within the living body can now be depicted as follows :

Cells → Tissues → Organs → Organ systems → Organism

We are now in a position to **answer the following questions :**

Very Short Answer Type Questions

1. Which instrument was essential for the study of cells ?
2. What is the basic similarity among all the living organisms (plants and animals) ?
3. Name the scientist who observed a thin slice of cork through a microscope and coined the term 'cell'.
4. Name the outermost layer of an animal cell.
5. Which part of the cell contains organelles ?
6. Name any two organelles present in a cell.
7. Which part of a cell carries out respiration ?
8. What are the units of inheritance in living organisms ?
9. Which is the largest floating body generally in the centre of a cell ?
10. What is the function of nucleus in a cell ?
11. What do the chloroplasts contain ?
12. What kind of cells are surrounded by a cell wall ?
13. What sort of cells do not have a cell wall around them ?
14. Name the layer which is outside the cell membrane of a plant cell.
15. How is the outside of a plant cell different from that of an animal cell ?
16. Name the organelle of a plant cell where photosynthesis takes place.
17. Which part of a plant cell protects outside of the cell ?
18. Which part of a plant cell releases energy from food ?
19. What causes the red colour in tomatoes ?
20. Which of the two has a large vacuole : a plant cell or an animal cell ?
21. Name any two parts which are present in a plant cell but not in an animal cell.

22. Which cells transmit messages between the brain and other parts of the body ?
23. Which cells in the human body can contract (and relax) ?
24. Name the spindle-shaped cells present in the human body.
25. State an important characteristic of muscle cells.
26. Name a single cell (other than *Amoeba* cell) which can change its shape.
27. Name one 'single cell' which behaves like a complete organism.
28. Name one cell which can be seen easily with naked eye and one cell which can be seen only through a microscope.
29. State whether the following statements are true or false :
    (*a*) Unicellular organisms have one-celled body.
    (*b*) Muscle cells are branched.
    (*c*) *Amoeba* has irregular shape.
    (*d*) All the cells in our body are alike.
    (*e*) A hen's egg is a group of cells.
    (*f*) The basic living unit of an organism is organ.
30. Name the smallest unit of life.
31. What is a 'cell' in biology ?
32. (*a*) What are the basic parts of all the cells ?
    (*b*) What parts are present only in plant cells ?
33. What controls the flow of substances in and out of a cell ?
34. Which part of a cell controls all the activities of the cell ?
35. Name the animal cell which is long and has thread-like branches.
36. What is the function of chloroplasts in a plant cell ?
37. Name an organism which has no definite shape, and it keeps on changing its shape.
38. Name one cell in human body which is spherical in shape.
39. Which organism has the smallest cell ?
40. Name the biggest cell.
41. Name two animal organs and two plant organs.
42. Name the plant organ which is responsible for :
    (*a*) making of food. (*b*) absorption of water and minerals.
43. Which of the two does not have a true nucleus : prokaryotic cell or eukaryotic cell ?
44. Name one prokaryotic cell and one eukaryotic cell.
45. Fill in the following blanks with suitable words :
    (*a*) Cells were first observed in cork by..............in 1665
    (*b*) What is brick to a house is...... to an organism.
    (*c*) The cytoplasm and nucleus make up the ....
    (*d*) The shape and size of a cell is related to its ....
    (*e*) The bacteria cells are ....to............micrometre in length.
    (*f*) The smallest unit of life is a ....
    (*g*) Tissues make up..............
    (*h*) Organ systems make up an..............
    (*i*) Cells make up..............
    (*j*) Organs make up ....

Short Answer Type Questions

46. Why are plant and animal specimens usually stained with dyes before observing them through a microscope ? Name one stain (or dye) used for this purpose.
47. What is a tissue ? Give two examples of tissues.
48. What is an organ ? Give two examples of organs.
49. What is an organ system ?
    (*a*) Give two examples of organ systems in animals.
    (*b*) Name the two main organ systems in plants.
50. Which of the following are plant organs and which are the animal organs ?
    Brain, Leaf, Lungs, Roots, Stem, Kidneys, Flower, Heart

51. What are the functions of the following organs ?
    (*a*) Heart (*b*) Brain (*c*) Roots (*d*) Leaves
52. What is the shape of red blood cells in human blood ? What function do red blood cells perform ?
53. (*a*) State the difference between prokaryotes and eukaryotes.
    (*b*) Name two prokaryotes and two eukaryotes.
54. (*a*) Why are nerve cells long and have branches ?
    (*b*) What is the other name of a nerve cell ?
55. (*a*) Why could cells not be observed and studied for thousands of years ?
    (*b*) State the cell theory of organisms.
56. Explain the function of mitochondria in a cell.
57. Make a sketch of human nerve cell. What function do nerve cells perform ?
58. Make a sketch of the human muscle cell. What function do muscle cells perform ?
59. What are pseudopodia in *Amoeba* ? What are the functions of pseudopodia ?
60. Where are chromosomes found in a cell ? State their function.
61. What are genes ? Where are genes located ?
62. What is a plastid ? What is the name of green plastids present in plant cells ?
63. What is the size of an ostrich egg ? Is it a single cell or a group of cells ?
64. What is the function of cell wall in a plant cell ?
65. Name two cells which are found in animals and two which are found in plants.

Long Answer Type Questions

66. (*a*) What is cytoplasm ? What is its function ?
    (*b*) What is protoplasm ? Name the four elements which make up major part of protoplasm.
67. (*a*) What are unicellular organisms ? Name two unicellular organisms.
    (*b*) What are multicellular organisms ? Name two multicellular organisms.
68. (*a*) Draw the general diagram of an animal cell and label it.
    (*b*) Draw the general diagram of a plant cell and label it.
    (*c*) Explain why, chloroplasts are found only in plant cells.
69. (*a*) An *Amoeba* cell can change its shape and a white blood cell in human blood can also change its shape. What is the difference between an *Amoeba* cell and a white blood cell ?
    (*b*) Out of *Amoeba* cell and white blood cell, which one is (*a*) eukaryotic cell, and (*b*) prokaryotic cell ?
70. (*a*) State three differences between a plant cell and an animal cell.
    (*b*) Which of the following are prokaryotic cells and which are eukaryotic cells ?
    Amoeba cell, Bacterium cell, Human cheek cell, Blue-green algae cell, Onion peel cell.

Multiple Choice Questions (MCQs)

71. The organelles which provide energy for all the activities of a cell are :
    (*a*) chloroplasts (*b*) mitochondria (*c*) golgi bodies (*d*) ribosomes
72. In a living cell, chromatin is present in :
    (*a*) cytoplasm (*b*) chloroplasts (*c*) nucleus (*d*) vacuole
73. The cell wall in onion peel cell is made of :
    (*a*) starch (*b*) gelatin (*c*) cellulose (*d*) cell sap
74. The group of similar cells which work together to perform a particular function is called :
    (*a*) organ (*b*) organelle (*c*) organism (*d*) tissue
75. A long and branched animal cell is :
    (*a*) muscle cell (*b*) epithelial cell (*c*) nerve cell (*d*) cartilage cell
76. Which of the following is not a unicellular organism ?
    (*a*) Amoeba (*b*) Yak (*c*) Yeast (*d*) Rhizobium
77. Which of the following organelle is not found in the root cells of a plant ?
    (*a*) nucleus (*b*) vacuole (*c*) mitochondria (*d*) chloroplasts
78. The part of a cell which plays a role in inheritance is :
    (*a*) nucleus (*b*) cytoplasm (*c*) plasma membrane (*d*) mitochondria
79. The basic similarity among all the living organisms is that they are made up of :
    (*a*) tissues (*b*) organs (*c*) cells (*d*) organ systems
80. The structural and functional unit of life called cell was discovered by :
    (*a*) Robert Boyle (*b*) Charles Darwin (*c*) Robert Koch (*d*) Robert Hooke

81. Which of the following cell does not have a nucleus ?
(*a*) white blood cell (*b*) red blood cell (*c*) nerve cell (*d*) muscle cell

82. The parts which are not present in an animal cell are :
A. Cell membrane B. Chloroplast C. Cell wall D. Mitochondria
(*a*) A and B (*b*) B and C (*c*) A and D (*d*) B and D

83. All the living matter in a cell is called :
(*a*) endoplasm (*b*) protoplasm (*c*) cytoplasm (*d*) cell sap

84. Which of the following is a plant cell ?
(*a*) cartilage cell (*b*) neuron (*c*) epidermal cell (*d*) epithelial cell

85. The egg cell measuring about 17 cm × 13 cm is most likely that of :
(*a*) hummingbird (*b*) hen (*c*) elephant (*d*) ostrich

86. Which of the following have cell walls ?
A. Epidermal cell B. Epithelial cell C. Mesophyll cell D. Liver cell
(*a*) A and B (*b*) B and C (*c*) A and C (*d*) B and D

87. Which of the following are prokaryotes ?
A. Protozoa B. Blue-green algae C. Fungi D. Bacteria
(*a*) A and B (*b*) B and C (*c*) A and D (*d*) B and D

88. Which of the following human system includes oesophagus ?
(*a*) respiratory system (*b*) circulatory system (*c*) digestive system (*d*) reproductive system

89. Which of the following cells can change their shape ?
A. White blood cell B. Amoeba cell C. Red blood cell D. Euglena cell
(*a*) A and B (*b*) B and C (*c*) A and D (*d*) B and D

90. The non-living part of a tomato cell is its :
(*a*) cell membrane (*b*) nucleus (*c*) chloroplasts (*d*) cell wall

Questions Based on High Order Thinking Skills (HOTS)

91. The parts P and Q are present only in plant cells, they are not present in animal cells. The part P contains a green pigment called R whereas part Q is made of a tough material S. The part P takes part in the food making process whereas part Q gives shape and support to the plant cell.
(*a*) What is (*i*) P, and (*ii*) Q ? (*b*) What is (*i*) R, and (*ii*) S ?

92. A, B, C and D are the basic parts of all the cells. The part A contains thread-like structures called E which transfer the characteristics from parents to their offsprings. The part B uses glucose and oxygen to produce energy whereas part C controls the movement of substances 'into the cell' and 'out of the cell'. The part D is a transparent, jelly-like material. What could A, B, C, D and E be ?

93. X and Y are the two types of cells. The cells X have a well organised nucleus which is separated from the cytoplasm by the nuclear membrane. On the other hand, cells Y do not have a real nucleus, their nuclear material is in direct contact with the cytoplasm.
(*a*) What type of cells are (*i*) X, and (*ii*) Y ?
(*b*) Give one example each of cells like (*i*) X, and (*ii*) Y.

94. The cytoplasm of the cells of a tomato plant contains organelles X having different pigments which impart different colours to the leaves of tomato plant and its fruits.
(*a*) What is the general name of the organelles X ?
(*b*) What is the (*i*) name (*ii*) colour, and (*iii*) function, of organelles X present in the leaves of tomato plant ?
(*c*) What is the colour of organelles X which are present in the ripe fruits of tomato plant ?

95. Cells make up *A* ; *A* make up *B* ; *B* make up *C*, and finally *C* make up an organism. What are *A*, *B* and *C* ?

ANSWERS

12. Plant cells **13.** Animal cells **20.** A plant cell **26.** White blood cell **29.** (*a*) True (*b*) False (*c*) True (*d*) False (*e*) False (*f*) False **43.** Prokaryotic cell **45.** (*a*) Robert Hooke (*b*) cell (*c*) protoplasm (*d*) function (*e*) 0.1 ; 0.5 (*f*) cell (*g*) organs (*h*) organism (*i*) tissues (*j*) organ systems **71.** (*b*) **72.** (*c*) **73.** (*c*) **74.** (*d*) **75.** (*c*) **76.** (*b*) **77.** (*d*) **78.** (*a*) **79.** (*c*) **80.** (*d*) **81.** (*b*) **82.** (*b*) **83.** (*b*) **84.** (*c*) **85.** (*d*) **86.** (*c*) **87.** (*d*) **88.** (*c*) **89.** (*a*) **90.** (*d*) **91.** (*a*) (*i*) Chloroplast (*ii*) Cell wall (*b*) (*i*) Chlorophyll (*ii*) Cellulose **92.** A : Nucleus ; B : Mitochondria ; C : Cell membrane ; D : Cytoplasm ; E : Chromosomes **93.** (*a*) (*i*) Eukaryotic cell (*ii*) Prokaryotic cell (*b*) (*i*) *Amoeba* cell (*ii*) Bacterium cell **94.** (*a*) Plastids (*b*) (*i*) Chloroplasts (*ii*) Green (*iii*) Help in the making of food by leaves of the plant by photosynthesis (*c*) Red colour **95.** A : Tissues ; B : Organs ; C : Organ systems

Reproduction in Animals

All the living organisms grow old with time and ultimately die. In fact, every living organism remains alive on this earth for a limited period of time and then dies. So, new organisms have to be produced in place of those who die. **The production of new organisms from the existing organisms of the same species is known as reproduction.** In most simple words we can say that *reproduction is the creation of new living things (from the existing living things)*. Actually, one of the most important characteristics of living organisms is their ability to reproduce more members of their species. **Reproduction is essential for the survival of a species on this earth.** So, living organisms produce more organisms of their kind to maintain the life of their species on this earth.

The process of reproduction ensures continuity of life on earth. For example, human beings reproduce by giving birth to babies (sons and daughters). These babies grow and ultimately become adults. So, when the old parents die, their sons and daughters keep living on this earth. These sons and daughters also reproduce by giving birth to more babies, and this process goes on and on. Thus, **reproduction by human beings ensures that the human species will continue to exist on this earth for all the time to come.** Similarly, cats reproduce by giving birth to kittens so that their species may live for ever. And dogs reproduce by giving birth to puppies so that their species may continue to live on this earth.

It is clear from the above discussion that for a species of an animal to continue living on this earth, it must reproduce itself. **Reproduction gives rise to more organisms with the same basic characteristics as their parents.** For example, human beings always produce human babies; cats always produce kittens; and hens always produce chicks. If, however, some species of the living organisms cannot reproduce due to certain reasons, then the organisms of this species will gradually die out and disappear from this earth one day. In this chapter we will discuss the various methods of reproduction in animals.

Before we go further, we should know the meaning of the term 'young one' of an animal. **The newly born animal (or newly hatched animal) is called young one**. The young ones of different animals have different names. For example, the young one of humans is called 'baby' (see picture on page 144), the young one of cat is kitten, the young one of dog is puppy, the young one of cow is calf, the young one of horse is colt, the young one of lion is cub, the young one of hen is chick, the young one of frog is tadpole, and the young one of a butterfly is caterpillar. The young one of an animal is also called its offspring.

METHODS OF REPRODUCTION

There are many different ways in which new organisms are produced from their parents. Some organisms like *Amoeba* just split into two parts to produce new *Amoebae*; some organisms like *Hydra* grow out of the parent's body in the form of a bud; some organisms like birds and snakes hatch out of the eggs laid by their parents; whereas some organisms like human babies, kittens and puppies are born from their mother. This means that each species of organisms reproduces in a different way. All the different ways of reproduction can be divided into two main groups : asexual reproduction and sexual reproduction. Thus, **there are two main methods of reproduction in living organisms :**

(*i*) asexual reproduction, and

(*ii*) sexual reproduction.

This means that new living organisms (or animals) can be made either by the method of 'asexual reproduction' or by the method of 'sexual reproduction'. We will now discuss the meaning of asexual reproduction and sexual reproduction. In order to understand this please keep in mind that certain organisms contain **'reproductive cells'** (called **'sex cells'** or **'gametes'**) in their bodies whereas some other organisms do not contain 'reproductive cells' ('sex cells' or 'gametes') in their bodies.

1. Asexual Reproduction

The production of a new organism from a single parent without the involvement of sex cells (or gametes) is called asexual reproduction. It is called asexual reproduction because it does not use special cells called 'sex cells' (or gametes) for producing a new organism. In asexual reproduction, a part of the parent organism separates off and grows into a new organism. Thus, ***in asexual reproduction, only one parent is needed to produce a new organism.*** But no sex cells are involved in asexual reproduction. **Some of the examples of asexual reproduction are : binary fission in *Amoeba*; and budding in *Hydra*.** Please note that asexual reproduction is the simplest method of reproduction. It takes place mainly in those animals whose bodies have a simple structure. In asexual reproduction, the young one (or offspring) produced is an exact copy of the parent.

2. Sexual Reproduction

The production of a new organism from two parents by making use of their sex cells (or gametes) is called sexual reproduction. In sexual reproduction, the sex cell of one parent fuses with the sex cell of the other parent to form a new cell called 'zygote'. This zygote then grows and develops to form a new organism. Thus, ***in sexual reproduction, two parents are needed to produce a new organism.*** The two parents which are involved in sexual reproduction are called male and female. The male and female parents have special organs in them which produce male sex cells and female sex cells respectively (which are required in sexual reproduction). **The humans, fish, frogs, cats and dogs, all reproduce by the method of sexual reproduction.** In sexual reproduction, the young one (or offspring) produced is not an exact copy of the parents.

The basic difference between asexual and sexual reproduction is that only one parent is needed in asexual reproduction whereas two parents are needed in sexual reproduction. Another difference is that no sex cells (or gametes) are involved in asexual reproduction but sex cells (or gametes) take part in sexual reproduction.

In order to understand the sexual reproduction in animals, we should know the meanings of some important terms like male sex, female sex, gametes, sperms, eggs (or ova), fertilisation, and zygote which are involved in sexual reproduction. These are discussed below.

Male and Female

Our father is a male and our mother is a female. We can also say that our father has male sex and our mother has female sex. Now, our father is a man and our mother is a woman. This means that **a man is male** whereas **a woman is female.** Thus, a man is said to have male sex and a woman is said to have female sex. Just like us human beings, other animals also have male and female sexes. As we will learn after a while, being male or female depends on the type of sex cells present in one's body. **An animal having male sex cells called 'sperms' in its body is called male.** On the other hand, **an animal having female sex cells called 'eggs' (or 'ova') in its body is called female.** We will now discuss gametes.

Gametes

Sexual reproduction takes place by the combination of special reproductive cells called 'sex cells'. These sex cells are also known by another name which is 'gametes'. We can now say that : **The cells involved in sexual reproduction are called gametes.** Gametes are of two types : male gametes, and female gametes. **The male gamete in animals is called 'sperm'** and **the female gamete in animals is called 'egg' (or 'ovum').** We will now describe sperms and eggs (or ova) in somewhat detail.

(*i*) SPERMS. The sperms are extremely small cells and we need a microscope to see them. A sperm is about 0.05 mm long. A highly enlarged sketch of sperm is shown in Figure 1. A sperm has a head, a middle piece and a tail. **A sperm is a single cell** with all the usual cell components like nucleus, cytoplasm and cell membrane. The nucleus of sperm cell is tightly packed in its head. The sperm cell has very little cytoplasm. The purpose of tail of sperm is to make it move. The sperm moves by waving its tail from side to side. The term 'sperm' is the short form of the word 'spermatozoon' (which means male sex cell or male gamete).

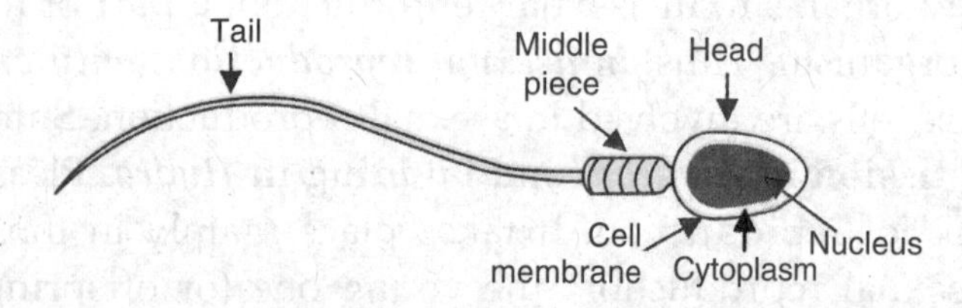

Figure 1. A human sperm (male gamete) (Highly enlarged).

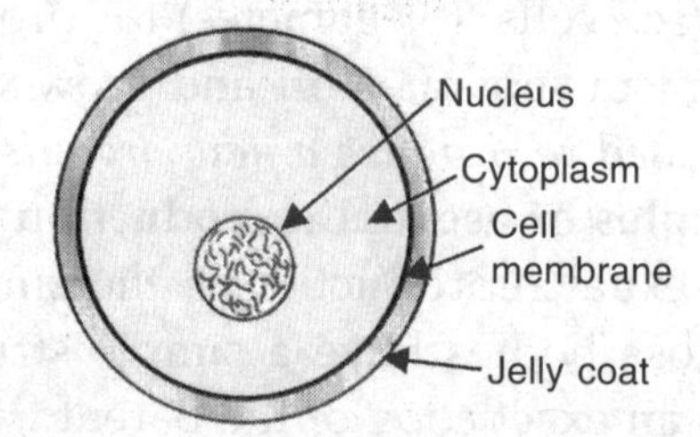

Figure 2. A human egg (or ovum) (female gamete) (Highly enlarged).

(*ii*) EGGS (OR OVA). The eggs (or ova) are also very small and we need a microscope to see them. Eggs (or ova) are, however, much larger than the sperms. The human egg (or ovum) is round and about 0.15 mm in diameter. A highly enlarged sketch of an egg (or ovum) is shown in Figure 2. **The egg (or ovum) is also a single cell** having a nucleus, cytoplasm and a cell membrane. Outside the cell membrane, an egg (or ovum) has a thin layer of jelly called 'jelly coat' which allows only one sperm to enter into it during fertilisation. Please note that a female gamete (or female sex cell) is known by two names : 'egg' and 'ovum'. So, whether we use the name 'egg' or 'ovum', it will mean the same thing. Another point to be noted is that the plural of ovum is ova. Some of the animals like birds (hen, ostrich, etc.) and reptiles (snakes, crocodiles, etc.) have large eggs.

The nuclei of sperm and egg contain chromosomes which carry genes and transmit the characteristics of parents to the offsprings. The fusion of a male gamete (sperm) with a female gamete (egg) gives rise to a new cell called zygote. Thus, **the new cell which is formed by the fusion of a male gamete and a female gamete is called zygote**. **Zygote is a single cell** which contains one nucleus. The nucleus of zygote is formed by the combination of nuclei of 'sperm' and 'egg'. The process of fusion of gametes is called fertilisation. This is discussed on the next page.

Fertilisation

For sexual reproduction to occur, a male gamete must combine (or fuse) with a female gamete. **The fusion of a male gamete with a female gamete to form a zygote during sexual reproduction, is called fertilisation.** Since the male gamete of an animal is called sperm and the female gamete of an animal is called egg (or ovum), therefore, we can also say that : **The fusion of a sperm with an egg (or ovum) to form a zygote during sexual reproduction, is called fertilisation**. Thus, the process of fertilisation produces a new cell called zygote. **The zygote is actually a fertilised egg** (or fertilised ovum).

In human beings, the process of fertilisation is the meeting of a sperm cell from the father with an egg cell from the mother to form a fertilised egg cell called zygote. The process of fertilisation in humans is shown in Figure 3. Figure 3(*a*) shows a sperm and an egg. Figure 3(*b*) shows the head of a sperm entering

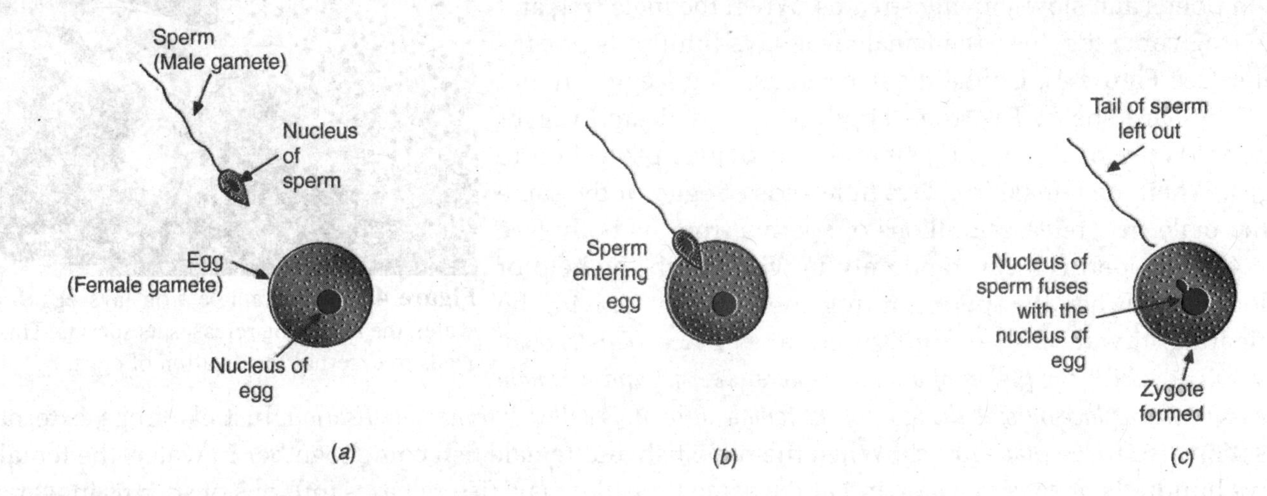

Figure 3. Fertilisation of an egg (or ovum) by a sperm to form a zygote.

the egg. Please note that though many sperms try to enter the egg, only one sperm is successful in entering the egg (We have shown only one sperm in Figure 3 to keep the diagram simple). When the head of sperm enters the egg, then the nucleus of sperm cell fuses (or joins) with the nucleus of egg cell to form a new nucleus. Figure 3(*c*) shows the fusion of nucleus of sperm with the nucleus of egg to form a new cell called 'fertilised egg' or 'zygote' (The tail of successful sperm entering the egg remains outside). **All the multicellular animals (including humans) start their life from a single cell called zygote (through sexual reproduction)**.

Internal Fertilisation and External Fertilisation

We have just studied that the fusion of a sperm with an egg is called fertilisation. Now, **the egg (or ovum) is made in the body of a female animal.** So, the fertilisation of an egg by a sperm can take place either *inside* the body of the female animal or *outside* the body of female animal. This leads to two modes of fertilisation in animals : internal fertilisation, and external fertilisation.

The fertilisation which takes place inside the female body is called internal fertilisation. In internal fertilisation, the female animal's eggs are fertilised by sperms inside her body. In internal fertilisation, the male animal puts his sperms into the female animal's body. And these sperms then fertilise the eggs inside her body. For example, a man puts his sperms inside a woman's body. These sperms then fertilise the egg inside the woman's body. *This type of fertilisation in which the fusion of a male gamete (sperm) and a female gamete (egg) occurs inside the body of the female animal, is called internal fertilisation*. Internal fertilisation takes place in a very large number of animals such as humans, cows, dogs, cats, tigers, lions, rabbits, deer, horse, birds (such as hen, sparrow, crow, pigeon, etc.), reptiles (such as lizard, snake and crocodile, etc.), and insects (such as silk moth, housefly and butterfly, etc.). In all these animals, fertilisation occurs inside the body of the female animal.

The fertilisation which takes place outside the female body is called external fertilisation. In external fertilisation, the female animal's eggs are fertilised by sperms outside her body. In frogs and most fishes, the fertilisation of eggs occurs outside the female animal's body. In external fertilisation, the male and female animals release their sperms and eggs in water where fertilisation takes place by collisions between sperms and eggs. For example, the males and females of frogs and fishes release their sperms and eggs in water in which they live. The sperms then collide with the eggs and fertilise them outside the body of female frog or fish. External fertilisation is very common in aquatic animals such as frog, fish and star fish, etc. In these animals, fertilisation takes place in water. We will now describe the external fertilisation in frogs in somewhat detail.

During spring season (or rainy season), frogs move to the water in ponds and slow-moving streams. When the male frog and female frog come together, the female frog lays hundreds of eggs in water (see Figure 4). Unlike the hen's eggs, frog's eggs are not covered by hard shells. The frog's eggs are very delicate. A layer of jelly holds the frog's eggs together and provides protection to the eggs. When the female frog lays hundreds of eggs, at the same time the male frog releases millions of sperms from its body (see Figure 4). The sperms swim randomly in water with the help of their long tails. When the sperms of frog come in contact with the eggs floating in water, then fertilisation takes place. *This type of fertilisation in which the fusion of a male gamete (sperm) and a female gamete (egg) takes place outside the body of the female animal, is called external fertilisation.* Just like frogs, external fertilisation also takes place in fish. When the male fish and female fish come together in water, the female fish lays hundreds of eggs in water and at the same time, the male fish releases millions of sperms in water. When the sperms come in contact with eggs, the eggs get fertilised. In this way, a female fish's eggs are fertilised by sperms outside its body. This is also an example of external fertilisation.

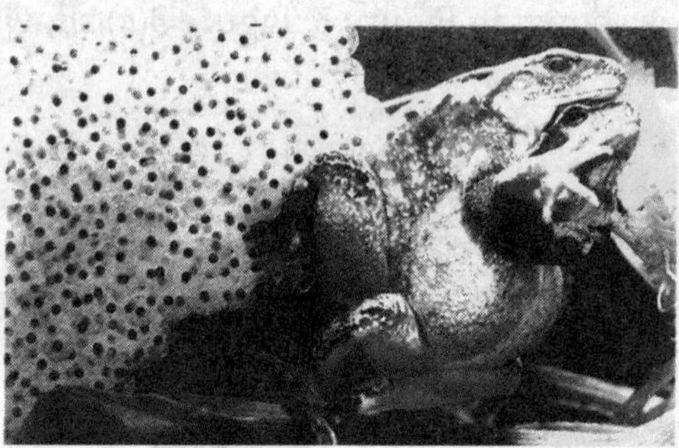

Figure 4. As the female frog lays eggs in water, the male frog releases its sperms. This leads to external fertilisation of eggs.

We will now explain **why the animals like frog and fish lay hundreds of eggs at a time** (whereas a hen lays only one egg at a time). Though the animals like frog (and fish) which undergo external fertilisation, lay hundreds of eggs, all the eggs do not get fertilised and hence do not develop into new frogs (or fishes). This is because of the following reasons :

(*i*) Frog (and fish) lay eggs and release sperms in water. Many of these eggs and sperms are carried away by the movement of water, wind or rainfall, and hence all the eggs do not get fertilised.

(*ii*) Many of the eggs of frogs (and fish) are eaten up by other animals which live in water.

(*iii*) Many of the larvae (young ones) of frog and fish which are hatched from the fertilised eggs are eaten up by other animals in water and fail to develop into adult frogs (or fish).

So, out of hundreds of eggs laid by the female frog (or female fish), only a few will survive to become adult frogs (or adult fish). Since many of the eggs of frogs and fish are carried away by moving water, wind or rain, and many are eaten up by other animals, therefore, the production of a large number of eggs is necessary to ensure the fertilisation of at least a few of them. Out of hundreds of eggs laid by female frog (or female fish) only a few get fertilised and their larvae survive to become adult frogs (or adult fish).

We know that the new cell which is formed by fertilisation is called 'zygote'. And this zygote then grows and develops into a full organism (or baby animal). **The method in which a zygote grows and develops into a full organism also varies in different animals.** For example, in human beings the zygote grows and develops into a baby inside the female body (mother's body). And then the mother gives birth to the baby. Just like humans, the animals like cats and dogs also give birth to their young ones. But the process is entirely different in those animals (or birds) which lay eggs. For example, a hen sits on its fertilised eggs for a considerable time to give them warmth. During this period, the zygote grows and develops to form a complete chick. This chick then comes out of the egg by breaking its shell. It is clear from this discussion that *all the organisms do not give birth to individuals like humans do.*

SEXUAL REPRODUCTION IN ANIMALS

Sexual reproduction is the most common method of reproduction in animals. In sexual reproduction in animals, there are two parents : a male and a female. The two parents have sex organs which produce sex cells or gametes. The male parent has sex organs called 'testes' which produce male sex cells (or male gametes) called sperms. The female parent has female sex organs called 'ovaries' which produce female sex cells (or female gametes) called eggs (or ova). **The whole process of sexual reproduction in animals involves the formation of sperms and eggs, joining together of sperm and egg to form a fertilised egg called 'zygote', and growth and development of zygote to form a baby animal.** The sexual reproduction in animals takes place in the following steps :

(*i*) The male parent produces male gametes called sperms.
(*ii*) The female parent produces female gametes called eggs (or ova).
(*iii*) The sperm enters into the egg. The nucleus of sperm fuses with the nucleus of egg cell to form a new cell called zygote (The zygote is a fertilised egg).
(*iv*) The zygote divides repeatedly to form a hollow ball of hundreds of cells which is called embryo.
(*v*) Embryo grows and becomes a foetus (in which all the main body features of the baby animal have formed).
(*vi*) Foetus grows and develops to form a new baby animal.

SEXUAL REPRODUCTION IN HUMANS

In human beings, there are special reproductive organs to make sperms and eggs ; to bring together sperms and eggs for fertilisation and make a zygote ; and for the growth and development of a zygote into a baby. We will first describe the reproductive systems of human male (man) and human female (woman), and then discuss how reproduction takes place in human beings.

The Male Reproductive System

The human male reproductive system consists of the following organs : Testes, Scrotal sacs, Epididymis, Sperm ducts, Seminal vesicles and Penis. The human male reproductive system is shown in Figure 5. Since the human male is called man, so we can also say that it is the reproductive system of man.

Testes are the oval shaped organs which lie outside the abdominal cavity of a man (see Figure 5). A man has two testes (singular of testes is testis). **Testes are the real reproductive organs in man.** The male sex cells (or male gametes) called 'sperms' are made in testes. Thus, **testes make male gametes called sperms.** Millions of sperms are produced in the testes. The testes are enclosed in two small bags of skin called scrotal sacs (see Figure 5). The testes work more efficiently at a temperature slightly below the body temperature, so they are held outside the body in the scrotal sacs.

The sperms formed in testes come out and go into a coiled tube called epididymis (see Figure 5). The sperms get stored temporarily in epididymis. From epididymis, the sperms are carried by a long tube called sperm duct into organs called seminal vesicles (see Figure 5). The sperms get stored in seminal vesicles. The seminal vesicles

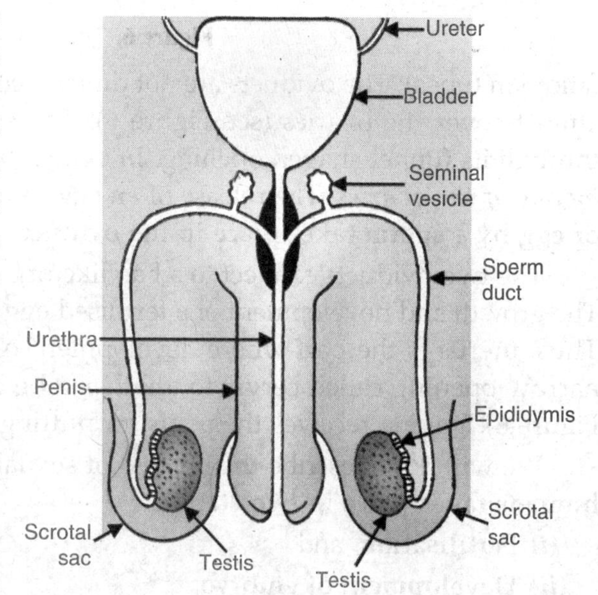

Figure 5. The male reproductive system in humans.

join to another tube called urethra coming from the bladder (see Figure 5). Urethra carries the sperms to an organ called penis which opens outside the body (These sperms are carried in a liquid called semen). The penis is an organ for introducing sperms into the woman. The penis passes the sperms from the man's body to the woman's body for the purpose of reproduction. The male reproductive system is designed to manufacture sperms and to deliver them to the place where one of them will be able to fuse with an egg cell (female gamete) and bring about fertilisation.

The Female Reproductive System

The human female reproductive system consists of the following organs : Ovaries, Oviducts, Uterus, and Vagina. The human female reproductive system is shown in Figure 6. Since the human female is called woman, so we can also say that it is the reproductive system of woman.

Ovaries are the oval shaped organs which are inside the abdominal cavity of a woman (see Figure 6). A woman has two ovaries. **Ovaries are the real reproductive organs in a woman.** The female sex cells (or female gametes) called 'eggs' or 'ova' are made in the ovaries. Thus, **ovaries make the female gametes called eggs (or ova).** Just above the ovaries are the tubes called 'oviducts' (which are also known as

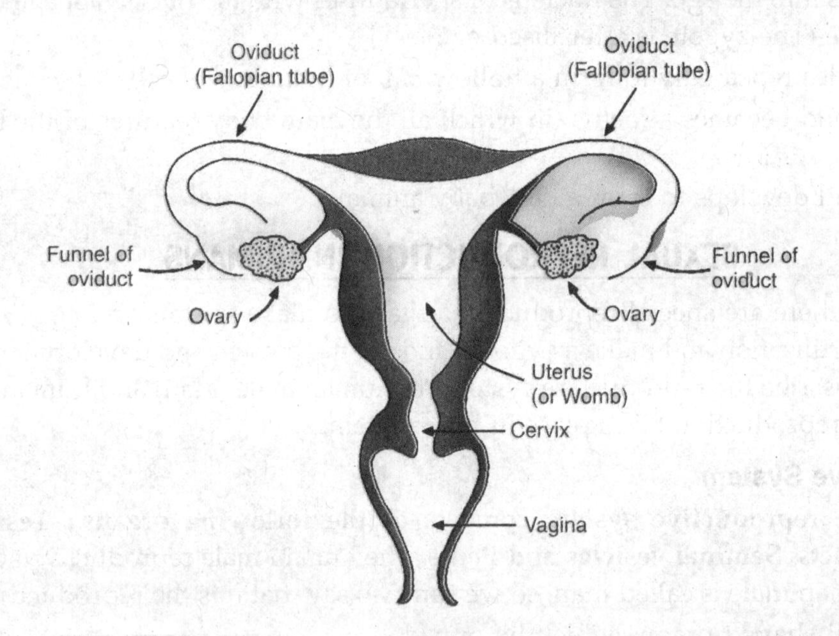

Figure 6. The female reproductive system in humans.

fallopian tubes). The oviducts are not directly connected to ovaries but have funnel shaped openings which almost cover the ovaries (see Figure 6). The egg (or ovum) released by an ovary goes into the oviduct through its funnel-shaped opening. ***In human beings, one mature egg is released into oviduct every month by one of the ovaries.*** The release of an egg (or ovum) from the ovary is called ovulation. **The fertilisation of egg by a sperm takes place in the oviduct.**

The two oviducts connect to a bag like organ called uterus (or womb) at their other end (see Figure 6). The growth and development of a fertilised egg (or zygote) into a baby takes place in the uterus (or womb). Thus, uterus is the part where development of the baby takes place. The uterus is connected through a narrow opening called cervix to another tube called vagina which opens to the outside of the body (see Figure 6). Vagina receives the penis for putting sperms into the woman's body.

We will now describe the process of sexual reproduction in humans. **The process of reproduction in humans takes place in two steps :**

***(i)* Fertilisation,** and

***(ii)* Development of embryo.**

We will discuss both these processes, one by one.

Fertilisation

The sperms made in the testes of man are introduced by penis into the woman's body through vagina. The sperms enter into vagina, pass through the uterus and then go into the oviducts (see Figure 7). The tail of sperms helps them in moving and reach the oviducts. If at the same time, the ovary of woman releases

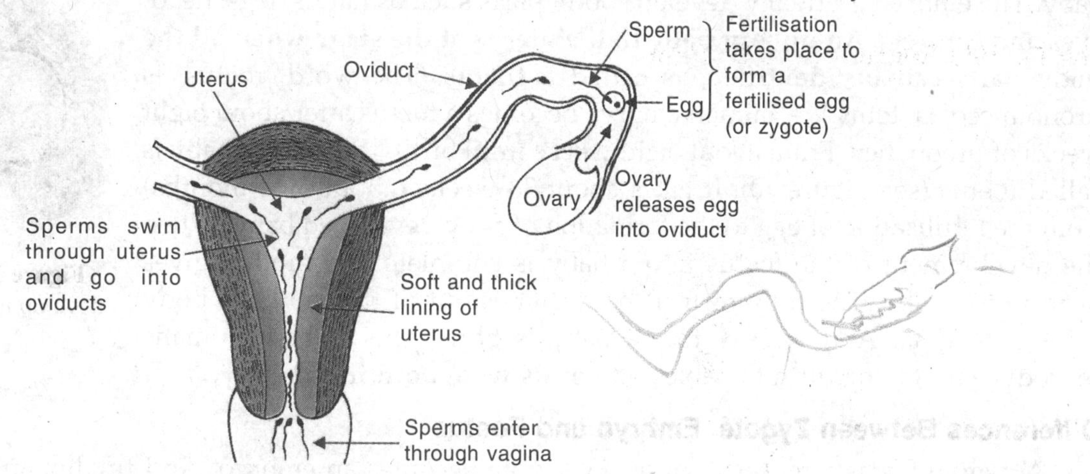

Figure 7. Fertilisation in humans to form a zygote (fertilised egg).

an egg (or ovum), then this egg also goes into the oviduct. One of the sperms enters the egg (see Figure 7). The fusion of sperm with egg is called fertilisation. During fertilisation, the nucleus of sperm fuses with the nucleus of egg cell to form a single nucleus. This results in the formation of a fertilised egg called zygote. Thus, **a sperm combines with the egg in the oviduct and fertilises it to form a zygote**. The zygote (or fertilised egg) is the beginning of the formation of a new baby (or a new individual).

Development of Embryo

The zygote (or fertilised egg) divides repeatedly to make a ball of hundreds of cells. This is called an embryo. The embryo moves down the oviduct into the uterus. The embryo gets embedded in the soft and thick lining of the uterus (see Figure 8). This is called implantation. When the embryo settles down in the uterus, the woman is said to have become pregnant (or said to have conceived). The embryo starts growing

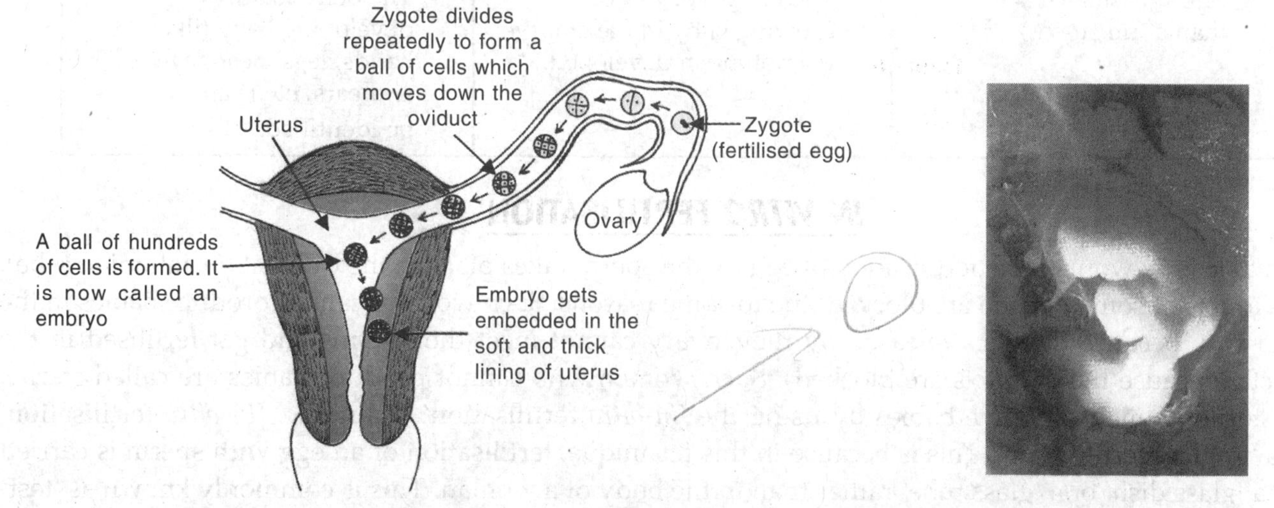

Figure 8. Implantation of embryo in the uterus. **Figure 9.** Human embryo (at 4 to 5 weeks)

into a baby. The embryo gets food and oxygen from the blood vessels in the lining of the uterus. The cells of embryo begin to form specialised groups that develop into different tissues and organs of the baby. Embryo is an early stage in a developing baby. Thus, **an unborn baby at an early stage of development in the uterus is called an embryo**. The body features of the unborn baby are not much developed at the embryo stage. The unborn baby remains an embryo in the first eight weeks of pregnancy. A human embryo at 4 to 5 week's development is shown in Figure 9. We cannot identify any body features (hands, legs, head, eyes and ears, etc.) of the developing baby in this embryo.

The embryo continues to grow and develop in the uterus to form a baby. The embryo gradually develops body parts such as hands, legs, head, eyes and ears, etc. **An unborn baby in the uterus at the stage when all the body parts can be identified, is called a foetus** (The word 'foetus' is pronounced as 'fetus'). A human embryo becomes a foetus after about eight weeks of pregnancy. From about eight weeks until birth, the unborn baby is called foetus (see Figure 10). It takes about 38 weeks (about nine months) from the fertilisation of egg to the formation of fully developed baby. When the development of the foetus into a baby is complete, the mother gives birth to the baby. The fully formed baby comes out of the mother's body through vagina. And we say that a baby is born. This is how humans reproduce by giving birth to babies. **All of us were born in this way.**

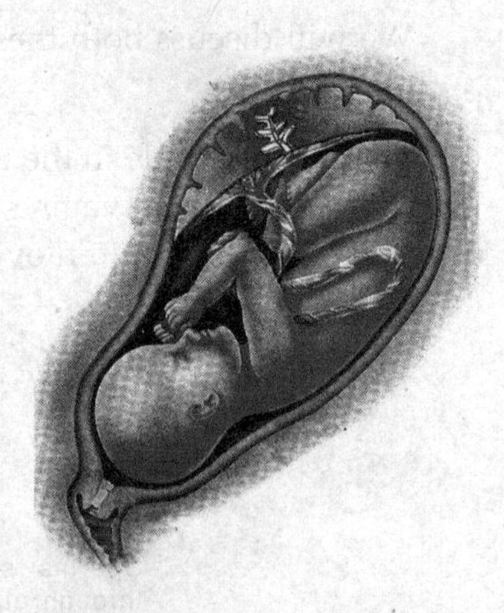

Figure 10. The side view of developing foetus inside uterus.

Differences Between Zygote, Embryo and Foetus

A zygote forms after fertilisation. A zygote becomes an embryo. And finally, an embryo becomes a foetus. The main differences between zygote, embryo and foetus are given below.

Zygote	*Embryo*	*Foetus*
1. A zygote is formed by the fusion of male and female gametes (sperm and egg)	1. An embryo is formed by the repeated cell division of a zygote.	1. A foetus is formed by the growth and development of an embryo.
2. A zygote is the beginning of the formation of a baby.	2. An embryo is an unborn baby in the uterus in the early stages of development (up to 8 weeks)	2. A foetus is an unborn baby in the uterus in the later stages of development (after 8 weeks till birth).
3. A zygote is a single cell. It is smaller than a full stop (.)	3. An embryo is multicellular. The body features of growing baby in the embryo are not much developed.	3. A foetus is also multicellular. The body features of developing baby (like hands, legs, head, eyes and ears, etc.) can be identified.

IN VITRO FERTILISATION

In a healthy woman, the fertilisation of egg by the sperm takes place in the oviduct (or fallopian tube). The oviducts of some women are blocked due to some reasons. Such women cannot produce babies in the normal way because the eggs released by their ovary cannot meet the sperms and get fertilised in the oviducts (because the oviducts are blocked). Such women who cannot produce babies are called sterile. Even sterile women can have babies by using the *'in-vitro* fertilisation' technique. '*In vitro* fertilisation' means 'in glass fertilisation'. This is because in this technique, fertilisation of an egg with sperm is carried out in a 'glass dish' or a 'glass tube' rather than in the body of a woman. This is commonly known as 'test-tube baby' technique. *In Vitro* Fertilisation is known as IVF in short.

The *in vitro* fertilisation technique helps the woman with blocked oviducts in having babies as follows :

1. The eggs are removed from the ovary of woman by laproscopy. In this operation, a small cut is

made in the side of woman's abdomen and an optical fibre tube is inserted into her body so that the doctor can see the ovary and take out the tiny eggs.

2. The woman's husband provides the sperms (in the form of semen).
3. The sperms are mixed with eggs in a glass dish (or glass test-tube) to carry out fertilisation. The fertilised eggs (or zygotes) develop into embryos.
4. After about a week, one or more embryos are placed in the woman's uterus (or womb). If the embryo gets implanted in the uterus successfully, then normal pregnancy occurs and a baby is born after about nine months.

In vitro fertilisation (IVF) technique is used to help those couples (husbands and wives) in having babies who can produce sperms and eggs but fertilisation cannot take place inside the woman's body due to blocked oviducts (or blocked fallopian tubes).

The '*in vitro* fertilisation' technique was initially developed for carrying out the reproduction in farm animals (like cows and buffaloes) in 1950. This technique was used successfully for reproduction in humans in 1978. A large number of babies are now born in our country by using this IVF technique. The success rate of this technique is only about 30 to 40 per cent. **The babies born through *in vitro* fertilisation technique are called test-tube babies because the fertilisation takes place in a glass dish or a glass test-tube.** This is, however, a misnomer (wrong name) because except for a very brief period of about one week of fertilisation and very early development (which is spent in a test-tube), almost the entire growth and development of the baby takes place inside the uterus of the woman as in normal pregnancy.

VIVIPAROUS ANIMALS AND OVIPAROUS ANIMALS

Some animals give birth to young ones. **Those animals which give birth to young ones (or baby animals) are called viviparous animals** (vivi = alive ; parous = bearing). In viviparous animals, the young one develops in the uterus inside the body of the mother (female parent). When the young one is fully developed, then the mother gives birth due to which the alive young one (or baby animal) comes out from the body of the mother. The viviparous animals do not lay eggs outside their body (like the birds and reptiles do). In the viviparous animals, the mother (female parent) gives birth to a fully developed baby animal (called young one). For example, in humans beings, the mother gives birth to a baby, so humans are viviparous animals (see Figure 11). Some other examples of viviparous animals are : Cow, Dog, Cat, Lion, Tiger, Horse, Rabbit, Rat, Elephant and Camel. In fact, all the mammals are viviparous animals.

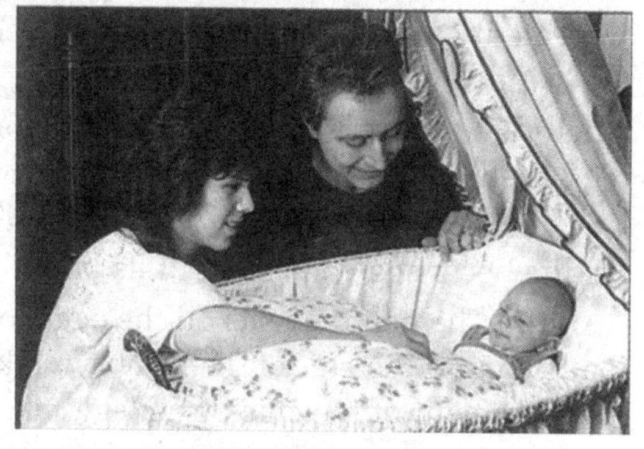

Figure 11. The humans give birth to young ones (babies), so they are viviparous animals.

Some animals lay eggs which later on develop into young ones. **Those animals which lay eggs from which young ones (baby animals) are hatched later on, are called oviparous animals** (ovi = relating to eggs ; parous = bearing). In oviparous animals, the mother (female parent) lays eggs outside its body. The young one of the animal develops inside the egg. When the development of the young one inside the egg is complete, the egg shell breaks open and an alive young one (baby animal) comes out of it. This is called hatching. For example, the hen lays eggs [see Figure 12(*a*)]. The young one of hen (or chick) develops inside the egg. When the development of the chick inside the hen's egg is complete, the egg shell breaks open and an alive chick comes out of it [see Figure 12(*b*)]. Since a hen lays eggs from which its young ones (chicks) are hatched, therefore, the hen is an oviparous animal. Some other examples of oviparous animals are : Sparrow, Crow, Butterfly, Housefly, Frog, Fish, Lizard, Snake, Ostrich, and Crocodile. In fact, all the insects, birds and reptiles are oviparous animals. They are the egg laying animals.

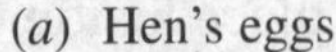

(*a*) Hen's eggs (*b*) A newly hatched chick

Figure 12. The hen lays eggs from which young ones (chicks) are hatched later on, so the hen is an oviparous animal.

It is easy to collect the eggs of oviparous animals because they lay the eggs outside their bodies. So, if we can collect the eggs of an animal easily, then it will be an oviparous animal. We can, however, not collect the eggs from viviparous animals (like human beings, cow, cat and dog, etc.) because they do not lay eggs outside their body. Their eggs are extremely small (or microscopic) which remain inside their bodies. We will now discuss the reproduction in hen.

THE CASE OF HEN

The hen is a bird. **Internal fertilisation takes place in hen also.** But **a hen does not give birth to chicks** (like human beings give birth to babies). We will now describe how chicks are born.

(*i*) After fertilisation takes place inside the body of the hen, the fertilised egg (or zygote) divides repeatedly to form embryo which travels down the oviduct.

(*ii*) As it travels down the oviduct, many protective layers are formed around the embryo. The hard shell that we see in a hen's egg is the outermost protective layer (other protective layers being inside the egg shell).

(*iii*) After the hard egg shell is formed around the developing embryo, the hen finally lays the egg. That is, the hen's egg comes out of its body.

(*iv*) The hen then sits on the eggs to provide sufficient warmth to the eggs for the development of the embryo into the chicks. The embryo takes about 3 weeks (21 days) to develop into a complete chick. The development of the chick from egg takes place inside the hen's egg during this period (see Figure 13).

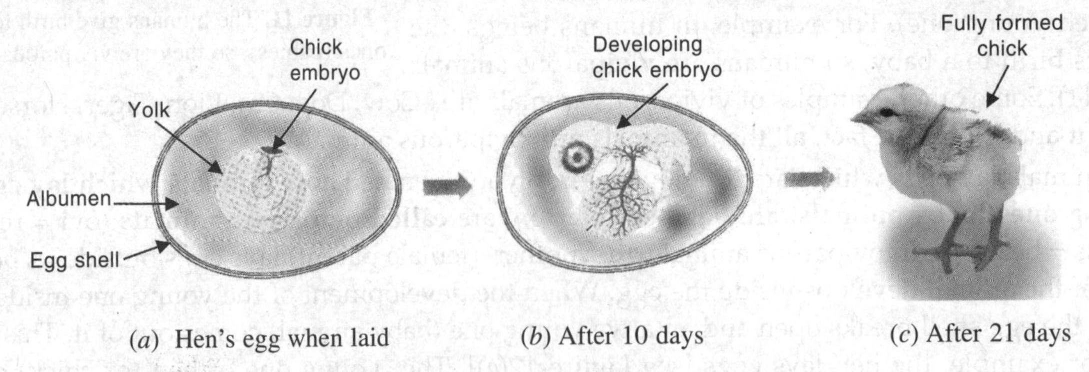

(*a*) Hen's egg when laid (*b*) After 10 days (c) After 21 days

Figure 13. Development of chick in the hen's egg.

(*v*) After the chick is completely developed, the egg shell breaks open automatically and the chick comes out of it. This is how a chick is born.

During its development into chick, the embryo gets all its food from the egg yolk. The albumen present in egg helps to protect the embryo from damage. The embryo obtains oxygen by the diffusion of air through the egg shell and other membranes. The embryos of all the birds and reptiles, etc., develop in the same way (like that of hen) inside their shelled eggs.

In animals (such as frogs, fish, etc.) which undergo external fertilisation, development of the embryo takes place in the eggs, outside the female body. In such animals, the embryos continue to grow within the egg coverings. After the embryos develop into complete young ones, the eggs hatch and the alive young ones come out. From this discussion we conclude that :

(*i*) Some embryos (like those of humans, cats, dogs, etc.) grow inside the mother's uterus till they have completely developed into baby animals, but

(*ii*) Some embryos (like that of birds, reptiles and insects, etc.) grow inside the eggs which the mother has laid, till they have completely developed into baby animals.

Before we go further and discuss 'metamorphosis', we should know the meaning of the terms larva, caterpillar and tadpole. **Larva is an immature form of an animal (like frog or silk moth) formed by the hatching of its eggs.** The embryos present in the eggs of some amphibians (like frogs) and insects (like silk moths) first grow into larvae. The larvae change into adults later on. *The larva of an animal is very different in appearance from that of the adult animal.* **The larva of silk moth is called caterpillar. The larva of frog is called tadpole.** Tadpole is the tailed, aquatic larva of frog breathing through gills and lacking legs, until the later stages of its development.

METAMORPHOSIS

We know that a number of animals (such as birds, frogs and insects, etc.) are hatched from eggs. For example, chicks are hatched from the eggs of hens. Similarly, the frogs and silk moths are also hatched from their eggs. Now, when a chick comes out of the egg of a hen, its over-all appearance is almost the same as that of an adult hen except that it is much smaller than the hen. So, when a small chick grows and develops to become an adult hen (or cock), then there is not much change in its body appearance. That is, the young one of hen (chick) hatched from the egg and the adult hen look alike. But this is not so in the case of animals such as frogs and silk moths, etc. For example, the young ones (larvae or tadpole) in frogs hatched from the eggs look very different from the adult frogs. **The process of transformation from an immature form of an animal like 'larva' to its 'adult form' in two or more distinct stages, is called metamorphosis.** In most simple words, *the change of a larva to an adult animal is called metamorphosis.* Metamorphosis occurs in amphibians (like frogs) and insects (such as silk moth, butterfly, housefly and mosquito, etc.).

Metamorphosis in Frog

The hatching of a fertilised egg of frog produces a very immature young one called tadpole (which is a larva) [see Figures 14(*a*) and (*b*)]. The tadpole (or larva of frog) develops gradually and undergoes many drastic changes in appearance before it forms an adult frog [see Figure 14(*c*)]. The tadpole looks very different from the adult frog. We say that during its life cycle, a frog undergoes metamorphosis. *The transformation of 'larva' into an 'adult' through drastic changes in appearance is called metamorphosis.* **The change from tadpole to frog is an example of metamorphosis.** This can be represented as :

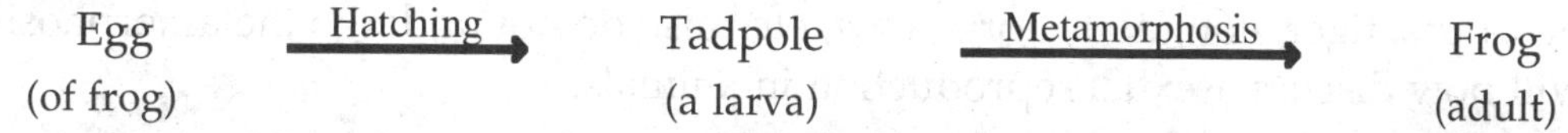

The changes during the formation of an adult frog from its egg can be shown by means of diagrams as follows :

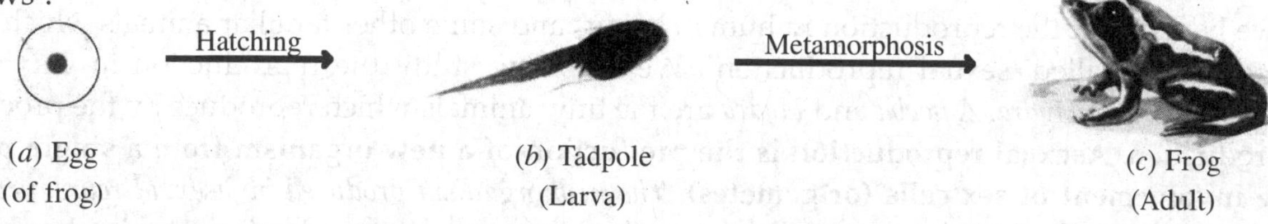

Figure 14. Metamorphosis changes tadpole into frog.

We can see from Figure 14 that a tadpole and the adult frog look very different from each other. The body features which are present in an adult frog are not present in its immature form called tadpole.

Tadpole is adapted to live life in water only. Tadpole has a long tail which helps it to swim easily in water. Tadpole breathes through the gills (like fish). It eats only tiny water plants and small aquatic animals. The tadpole gradually transforms into an adult frog by losing many old features of its body and developing new body features in many stages (which have not been shown in Figure 14). An adult frog is adapted to live life in water as well as on land. The adult frog develops webbed feet which help it to swim in water and also to hop on land (it has no tail). The adult frog has lungs for breathing in air on land (it has no gills). The adult frog can also breathe through its thin, moist skin. The frog has a long and forked tongue which can be flicked out to catch insects as food. The eyes in frog move to the top of the head so that the frog can see above the surface of water. In this way, by shedding old body features and developing new body features, a tadpole changes into a frog. This change is called metamorphosis. Metamorphosis changes the aquatic tadpole (water living tadpole) into an amphibian frog (which can live in water as well as on land).

Metamorphosis in Silk Moth

Silk moth is an insect (like butterfly) which undergoes metamorphosis during its life cycle. The silk moth passes through the larva and pupa stages during its development after hatching from the egg and forms an adult silk moth. The various stages in the development of a silk moth can be represented as follows :

Egg (of silk moth) —Hatching→ Caterpillar (Larva) —Metamorphosis→ Pupa (Chrysalis) —Metamorphosis→ Silk moth (Adult)

The changes during the formation of an adult silk moth from its egg can be shown by diagrams as follows :

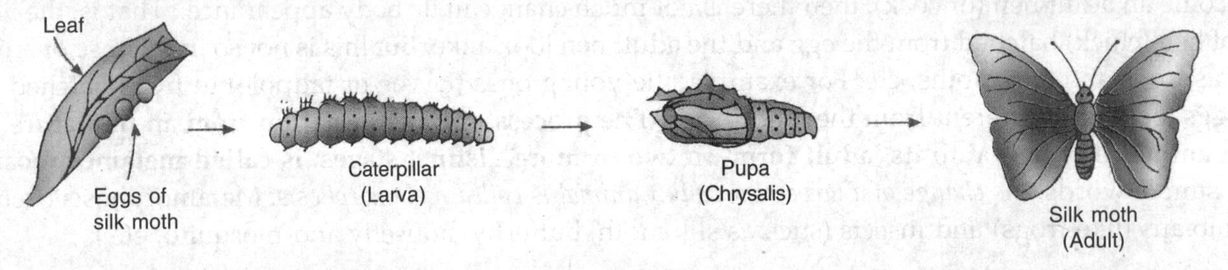

Figure 15. Metamorphosis transforms the caterpillar (larva of silk moth) into an adult silk moth.

Please note that the caterpillar (larva) and pupa stages in the development look very different from the adult silk moth. So, there have been drastic changes in appearance during the transformation of caterpillar into the adult silk moth. Thus, **the change from caterpillar to a silk moth is an example of metamorphosis**.

We (human beings) also observe changes in our body appearance as we grow but these changes are not drastic changes. In human beings, the body parts in adults are similar to those which are present in babies from the time of birth. So, **human beings do not undergo metamorphosis**. The common animals such as cats, dogs, hens, tiger, lion, deer, horse, cow, etc., also do not undergo metamorphosis during their life cycle. We will now discuss asexual reproduction in animals.

ASEXUAL REPRODUCTION IN ANIMALS

So far we have learnt the reproduction in human beings and some other familiar animals which takes place by the process called 'sexual reproduction'. We will now study the reproduction in very small animals like *Amoeba* and *Hydra*. *Amoeba* and *Hydra* are the tiny animals which reproduce by the process of asexual reproduction. **Asexual reproduction is the production of a new organism from a single parent without the involvement of sex cells (or gametes).** *The new organism produced by asexual reproduction is exactly identical to the parent*. The two most common methods of asexual reproduction in animals are :

(i) **Binary fission,** and

(ii) **Budding.**

We will now describe these two methods of asexual reproduction in animals, one by one. Before we discuss the method called binary fission, we should know the meaning of the terms 'fission' and 'binary fission' as used in biology. In biology, fission is the process of asexual reproduction in unicellular organisms (or single-celled organisms). **In the process of fission, the parent organism splits (or divides) to form two (or more) new organisms.** Fission begins with the division of the nucleus followed by the division of cytoplasm to form new organisms. Fission is of two types : binary fission and multiple fission, depending on whether the parent organism splits to form two new organisms or more than two organisms. In this Class, we will study only the binary fission in detail. Now, the word 'binary' means 'two' and the word 'fission' means 'splitting'. So, the term 'binary fission' means 'splitting into two'. Let us now discuss binary fission as the method of asexual reproduction in a microscopic animal called *Amoeba*.

(*i*) Binary Fission

Binary fission is an asexual method of reproduction in organisms. **In binary fission, the parent organism splits (or divides) to form two new organisms.** When this happens, the parent organism ceases to exist and two new organisms come into existence. The unicellular organism (or unicellular animal) called *Amoeba* reproduces by the method of binary fission. This is described below.

Amoeba reproduces by binary fission by dividing its body into two parts. This happens as follows : When the *Amoeba* cell has reached its maximum size of growth, then first the nucleus of *Amoeba* lengthens and divides into two parts. After that the cytoplasm of *Amoeba* divides into two parts, one part around each nucleus. In this way, one parent *Amoeba* divides to form two smaller *Amoebae* (called daughter *Amoebae*). And we say that one *Amoeba* produces two *Amoebae*. The reproduction in *Amoeba* by binary fission is shown in Figure 16.

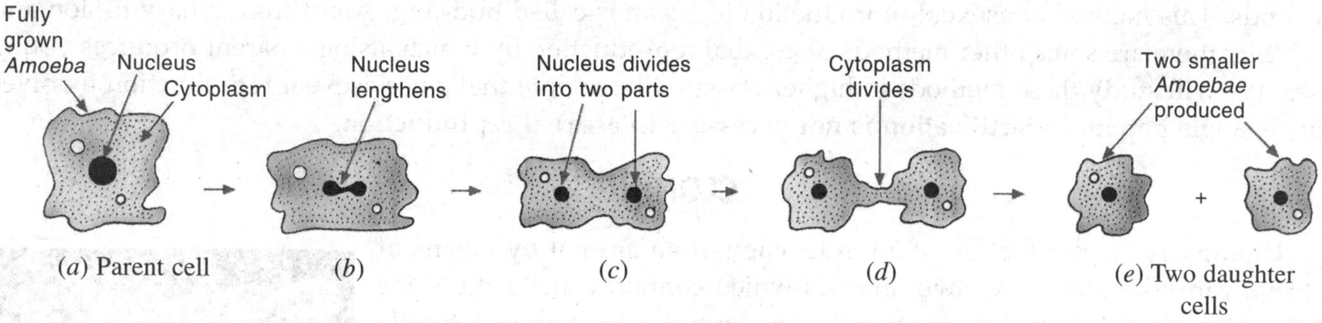

Figure 16. *Amoeba* reproducing by binary fission.

Amoeba takes about an hour to divide into two daughter *Amoebae*. The two daughter *Amoebae* produced here grow to their full size by eating food and then divide again to produce four *Amoebae*, and so on. The daughter *Amoebae* produced by the process of binary fission are identical to the parent *Amoeba*. From the above discussion we conclude that *Amoeba* reproduces by dividing itself into two. This type of asexual reproduction is called binary fission. Another tiny, unicellular animal called *Paramecium* also reproduces by the asexual method of binary fission. In multiple fission, the parent organism splits (or divides) to form many new organisms at the same time. We will discuss this is detail in higher classes.

Before we discuss the next asexual method of reproduction called 'budding', we should know the meaning of the term 'bud'. The 'bud' here means a 'small outgrowth' from the body of an organism (say, an animal). When a bud is formed on the body of an organism, then the nucleus divides into two, and one of the nuclei passes into the bud. Let us discuss the method of 'budding' now.

(*ii*) Budding

Budding is an asexual method of reproduction. **In budding, a small part of the body of the parent organism grows out as a 'bud' which then detaches and becomes a new organism.** The asexual reproduction by budding is observed in animals like *Hydra*, sea-anemones, sponges and corals. We will now describe the asexual reproduction in *Hydra*.

Hydra is a simple multicellular animal [see Figure 17(*a*)]. ***Hydra* reproduces by the process of budding.** This happens as follows : In *Hydra,* first a small outgrowth called 'bud' is formed on the side of its body by the repeated divisions of its cells [see Figure 17(*b*)]. This bud then grows gradually to form a small *Hydra*

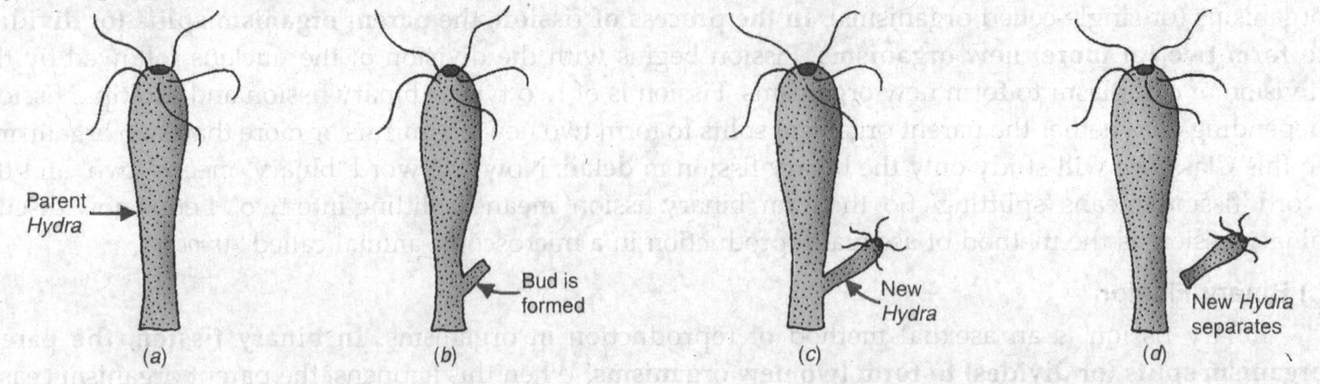

Figure 17. *Hydra* reproducing by the method of budding.

by developing a mouth and tentacles [see Figure 17(*c*)]. And finally the tiny new *Hydra* detaches itself from the body of parent *Hydra* and lives as a separate organism [see Figure 17(*d*)]. In this way, the parent *Hydra* has produced (or created) a new *Hydra*. Thus, *Hydra* reproduces asexually by growing buds from its body. This is called budding. Please note that the bud formed in a *Hydra* is not a single cell. It is a group of cells. If we collect some pond water and observe it through a hand magnifying glass, we will see a number of *Hydrae* with small buds attached to their body.

From the above discussion we conclude that in the tiny animal called *Hydra*, new *Hydrae* develop from the buds. This method of asexual reproduction in *Hydra* is called budding. Apart from binary fission and budding, there are some other methods of asexual reproduction by which a single parent produces young ones. We will study these methods in higher classes. Please note that since asexual reproduction involves only a single parent, so **fertilisation is not necessary in asexual reproduction**.

CLONING

Cloning is the production of an exact copy of an animal by means of asexual reproduction. Any two animals which contain exactly the same genes are called 'genetically identical'. ***An animal which is genetically identical to its parent is called a clone.*** A clone is an exact copy of its parent. The cloning of a large animal was successfully done for the first time by Ian Wilmut and his colleagues at the Roslin Institute in Edinburgh (Scotland). They cloned a sheep named 'Dolly' from its parent sheep called Finn Dorest sheep (a female sheep) (see Figure 18). Dolly (the cloned sheep) was born on 5th July, 1996. Dolly sheep was the first mammal to be cloned.

Figure 18. *Dolly – the cloned sheep.*

The cloning of animals is a special kind of asexual reproduction method. **The cloning in animals is done by the transfer of 'nucleus' of a cell.** This happens as follows : The nucleus of a normal body cell of the animal (which we wish to clone) is transferred into an empty egg cell (whose nucleus has been removed). The newly formed egg cell is allowed to develop normally. An exact copy of the animal (or a clone) is produced.

Before we describe the cloning of Dolly sheep in detail, we should know the meaning of the term 'ewe'. **Ewe is a female sheep.** The two sheep which were involved in the cloning of Dolly sheep were Finn Dorset ewe and Scottish Blackface ewe. We can also call them female Finn Dorset sheep and female Scottish Blackface sheep. As the name suggests, the Scottish Blackface sheep has a black face. Keeping these points in mind, we will now describe the cloning of 'Dolly' sheep from Finn Dorset ewe.

Cloning of 'Dolly' Sheep

The Dolly sheep was cloned in the following way :

(*i*) A normal body cell was removed from the mammary gland of a female Finn Dorset sheep (which was to be cloned) (see Figure 19).

(*ii*) An unfertilised egg cell was taken from a female Scottish Blackface sheep and its nucleus (containing chromosomes) was removed, leaving the egg cell empty (having no nucleus) (see Figure 19).

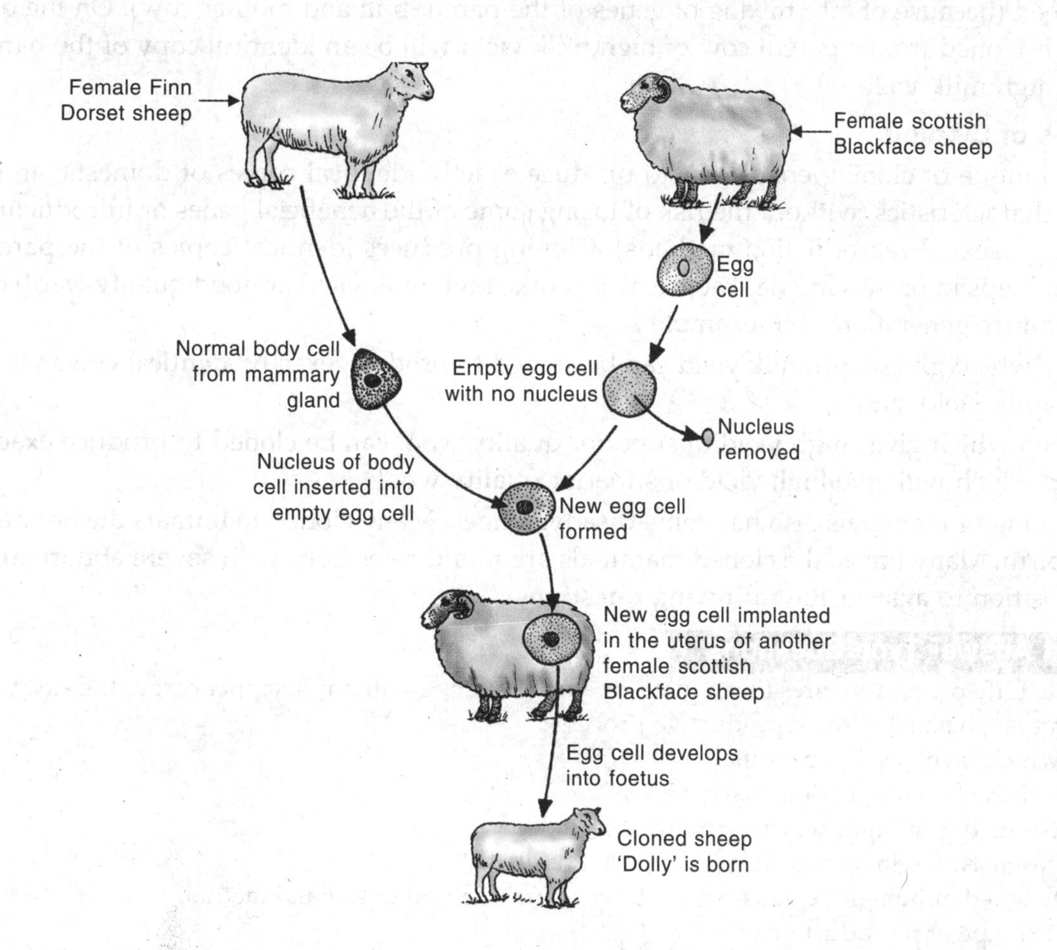

Figure 19. The technique of cloning 'Dolly' sheep from Finn Dorset sheep.

(*iii*) The nucleus of normal body cell of Finn Dorset sheep was inserted into the empty egg cell of Scottish Blackface sheep. In this way a new egg cell was obtained (which had the nucleus of Finn Dorset sheep body cell).

(*iv*) The new egg cell was implanted in the uterus of another female Scottish Blackface sheep making it pregnant. After 148 days, this pregnant Scottish Blackface sheep gave birth to Dolly sheep (see Figure 19).

Though Dolly sheep was given birth by a Scottish Blackface sheep, it was found to be exactly identical to the original Finn Dorset sheep from whose cell 'nucleus' was taken. So, **Dolly was a clone of Finn Dorset sheep** whose 'cell nucleus' was used in developing it. All the body cells of Dolly sheep contained the same set of chromosomes having exactly the same genes as the Finn Dorset sheep cells. So, though Dolly was given birth by a Scottish Blackface sheep but its real mother was Finn Dorset sheep. Since the nucleus from the egg cell of original Scottish Blackface sheep was removed, Dolly did not show any characteristic features of the Scottish Blackface sheep. For example, the Scottish Blackface sheep had a blackface but Dolly sheep did not have a blackface.

Dolly was a clone of Finn Dorset sheep and produced several offsprings of her own through sexual means in due course of time. Dolly had exactly the same genetic material as her real mother Finn Dorset sheep. Dolly died on 14th February 2003 due to a lung disease.

Since Dolly, the cloned sheep, who was born in 1996, cows (and pigs) have been cloned. The cloning of superior cows which give high milk yield has become possible now. If a high milk yielding cow was to be mated with a bull for reproduction, then the resulting offspring cow may or may not be of high milk yielding breed (because of intermixing of genes of the parent bull and mother cow). On the other hand, a cow which is cloned from a parent cow of high milk yield will be an identical copy of the parent cow and hence give high milk yield.

Advantages of Cloning

The technique of cloning enables us to produce exactly identical copies of domestic animals having favourable characteristics (without the risk of losing some of the beneficial genes or introducing unwanted genes through sexual reproduction methods). Cloning produces identical copies of the parent domestic animals and helps in preserving desirable features (like high milk yield or good quality wool) of the parent animal for future generations. For example :

(*i*) a cow which gives high milk yield can be cloned to produce exactly identical cows which will give high milk yield, and

(*ii*) a sheep which gives high yield of superior quality wool can be cloned to produce exactly identical sheep which will give high yield of superior quality wool.

The cloning of mammals also has some disadvantages. Many cloned mammals die before birth or die soon after birth. Many times, the cloned mammals are found to be born with severe abnormalities. We are now in a position to **answer the following questions :**

Very Short Answer Type Questions

1. Which life process ensures that a plant or animal species will not disappear from the earth ?
2. What is the name of the reproductive process :
   (*a*) which involves two parents ?
   (*b*) which involves only one parent ?
3. (*a*) Name two animals which reproduce sexually.
   (*b*) Name two animals which reproduce asexually.
4. State whether human beings reproduce by sexual method or asexual method.
5. Which type of reproduction :
   (*a*) involves gametes ?
   (*b*) does not involve gametes ?
6. Give another term for a fertilised egg.
7. Name the process of the fusion of gametes.
8. Do all animals give birth to individuals like humans ?
9. What is the other name of sex cells ?
10. What are the organs in humans which produce the gametes ?
11. (*a*) What are the male gametes in humans called ?
    (*b*) Name the organs which produce male gametes
12. (*a*) What are the female gametes in humans called ?
    (*b*) Name the organs which produce female gametes.
13. Name the organs which produce sperms in humans.
14. Name the organs which produce eggs (or ova) in humans.
15. What do the testes in a man produce ?
16. What do the ovaries in a woman produce ?
17. Which organ of the human body passes sperms from a man to a woman ?
18. In which female reproductive organ does the embryo get embedded ?
19. Which stage comes earlier in the development of a human baby from zygote : foetus or embryo ?

20. Name the technique which is used to help a woman with blocked oviducts to have a baby.
21. Write the full name of IVF.
22. What is the success rate of IVF technique of reproduction in humans ?
23. What type of fertilisation takes place in a hen ?
24. What term is used for the following ?
The change from tadpole to frog.
25. Name two animals which produce embryos that grow into larvae before transforming into adults.
26. What term is used for 'bulges' observed on the sides of the body of *Hydra* ?
27. What type of fission takes place in *Amoeba* ?
28. Name one animal each which reproduces : (*a*) by binary fission, and (*b*) by budding.
29. Name the asexual method of reproduction : (*a*) in *Hydra*, and (*b*) in *Amoeba*.
30. Name the technique which was used in producing 'Dolly' the sheep.
31. Name the parent sheep of which Dolly was a clone.
32. What name is given to the following ?
An animal which is an exact copy of its parents.
33. What are the two general methods of reproduction in organisms ?
34. State whether the following statements are true or false :
(*a*) Each sperm is a single cell.
(*b*) A new human individual develops from a cell called gamete.
(*c*) Egg laid after fertilisation is made up of a single cell.
(*d*) A zygote is formed as a result of fertilisation.
(*e*) External fertilisation takes place in frog.
(*f*) An embryo is made up of a single cell.
(*g*) Oviparous animals give birth to young ones.
(*h*) Internal fertilisation takes place in hens.
(*i*) The hens give birth to chicks like human beings give birth to babies.
(*j*) *Amoeba* reproduces by budding.
(*k*) Binary fission is a method of asexual reproduction.
(*l*) Fertilisation is necessary even in asexual reproduction.
(*m*) Cloning is a sexual reproduction method in mammals.
35. Fill in the following blanks with suitable words :
(*a*) The process of .... ensures continuity of life on earth.
(*b*) The cells involved in sexual reproduction are called ....
(*c*) Fusion of gametes gives rise to a single cell called ....
(*d*) The process of fusion of gametes is called ....
(*e*) The other name of egg cell is ....
(*f*) A sperm is much .... than an egg cell.
(*g*) In humans, one mature egg (or ovum) is released into oviduct every ....by one of the ovaries.
(*h*) The egg laying animals are called..............animals.
(*i*) The cow is a ....animal whereas ostrich is an ....animal.
(*j*) The change of caterpillar into an adult silk moth is called ....
(*k*) The larva of frog is called..............
(*l*) The two common methods of asexual reproduction in animals are..............and..............
(*m*) Dolly, the sheep, was produced by the technique called..............

Short Answer Type Questions

36. (*a*) What is the basic difference between asexual and sexual reproduction ?
(*b*) Which of the following organisms reproduce by sexual method and which by asexual method ?
Amoeba, Cats, Humans, *Hydra*, Birds
37. What is meant by the terms 'internal fertilisation' and 'external fertilisation' ? Explain with examples.
38. Draw a labelled diagram of the human female reproductive system.
(*a*) Where in the human body does an egg (or ovum) get fertilised ?

(*b*) Where does a fertilised egg (or zygote) develop into a baby in the human body ?

39. What type of fertilisation takes place in the following ?
(*a*) Cow (*b*) Frog (*c*) Humans (*d*) Fish (*e*) Hen

40. Why do female frogs (or female fish) lay hundreds of eggs ?

41. What is meant by an 'embryo' ? Can we identify the body features in an embryo ?

42. Give two differences between a zygote, an embryo and a foetus.

43. Describe the various steps involved in the sexual reproduction in animals.

44. Draw a labelled diagram of the human male reproductive system.

45. Define foetus. After how many weeks of development, a human embryo is said to become a foetus ?

46. What is metamorphosis ? Give two examples of metamorphosis.

47. What is the difference between viviparous animals and oviparous animals ?

48. Which of the following are oviparous animals and which are viviparous animals ?
Frog, Human being, Sparrow, Lizard, Cow, Dog, Hen, Fish, Butterfly, Cat.

49. Give five examples each of animals which develop :
(*a*) inside the mother.
(*b*) inside eggs which the mother lays.

50. Explain how, chicks are born. How much time does the embryo present in hen's egg take to develop into a chick (when provided sufficient warmth) ?

51. In which of the following animals the embryos develop fully inside the mother's body and in which they develop fully in the eggs laid by mother ?
Cow, Ostrich, Frog, Lizard, Deer, Cat, Snake, Tiger, Dog, Hen

52. Name two animals which undergo metamorphosis and two which do not.

53. Which of the following animals undergo metamorphosis and which do not ?
Cow, Butterfly, Silk moth, Humans, Frog, Housefly, Sparrow, Hen, Mosquito, Monkey

54. (*a*) What are gametes ?
(*b*) In which sort of reproduction are gametes involved ?
(*c*) What is formed when two gametes fuse ?
(*d*) What is this act of fusion called ?

55. Match the terms given in column *A* with those given in column *B* :

Column A	*Column B*
(*i*) Sperm	(*a*) Female organ
(*ii*) Ovary	(*b*) Egg tube
(*iii*) Oviduct	(*c*) Male organ
(*iv*) Testes	(*d*) Male gamete

56. Differentiate between internal fertilisation and external fertilisation. What type of fertilisation takes place in (*a*) frog, and (*b*) fox ?

57. What is meant by 'cloning' ? State whether gametes are involved in cloning or not. Name two animals which have been produced by cloning.

58. What is meant by 'reproduction' ? Why is it essential ?

59. Define asexual reproduction. Name two methods of asexual reproduction in animals. Name two animals which reproduce by these asexual reproduction methods.

60. What is a clone ? Name one famous clone.

Long Answer Type Questions

61. (*a*) Explain the term 'fertilisation'. Describe the process of fertilisation in human beings.
(*b*) Draw a labelled diagram to show the fertilisation of a human egg by a sperm to form a zygote.

62. (*a*) What type of couples are helped to have babies by the *in vitro* fertilisation technique ?
(*b*) Describe the '*in vitro* fertilisation' technique of reproduction in humans.

63. (*a*) What are viviparous animals ? Give two examples of viviparous animals.
(*b*) What are oviparous animals ? Give two examples of oviparous animals.

64. How does an *Amoeba* reproduce ? Describe with the help of labelled diagrams.

65. How does a *Hydra* reproduce ? Explain with the help of labelled diagrams.

Multiple Choice Questions (MCQs)

66. Internal fertilisation occurs :
(*a*) in female body (*b*) outside female body (*c*) in male body (*d*) outside male body

67. The number of nuclei present in a zygote is :
(*a*) none (*b*) one (*c*) two (*d*) four

68. A tadpole develops into an adult frog by the process of :
(*a*) fertilisation (*b*) metamorphosis (*c*) embedding (*d*) budding

69. Fertilisation results immediately in the formation of :
(*a*) a zygote (*b*) an embryo (*c*) a placenta (*d*) a foetus

70. Which of the following is not a part of the human male reproductive system ?
(*a*) testes (*b*) oviducts (*c*) seminal vesicles (*d*) epididymis

71. Which of the following is not a viviparous animal ?
(*a*) rat (*b*) lizard (*c*) rabbit (*d*) cat

72. The multicellular organism which reproduces by budding is :
(*a*) *Amoeba* (*b*) Yeast (*c*) *Paramecium* (*d*) *Hydra*

73. In asexual reproduction, two offsprings having the same genetic material and the same body features are called :
(*a*) callus (*b*) twins (*c*) chromosomes (*d*) clones

74. Which of the following animal does not show metamorphosis ?
(*a*) fish (*b*) frog (*c*) silk moth (*d*) mosquito

75. Asexual reproduction is :
(*a*) a fusion of specialised cells
(*b*) a method by which all types of organisms reproduce
(*c*) a method producing genetically identical offsprings
(*d*) a method in which more than one parent are involved

76. Which of the following organisms reproduces by binary fission ?
(*a*) *Hydra* (*b*) Yeast (*c*) *Amoeba* (*d*) *Sea anemone*

77. Which of the following is not an oviparous animal ?
(*a*) snake (*b*) fish (*c*) rat (*d*) frog

78. Tadpole is the larva of :
(*a*) fish (*b*) frog (*c*) mosquito (*d*) butterfly

79. The production of an exact copy of an animal by asexual reproduction is known as :
(*a*) budding (*b*) mating (*c*) cloning (*d*) hatching

80. One of the following is not a part of the human female reproductive system. This one is :
(*a*) ovary (*b*) uterus (*c*) scrotal sacs (*d*) oviducts

81. Reproduction is essential for living organisms in order to :
(*a*) keep the individual organism alive (*b*) fulfil their energy requirements
(*c*) maintain growth (*d*) continue the species for ever

82. One of the following occurs in the reproductive system of flowering plants as well as that of humans. This is :
(*a*) sperm ducts (*b*) anther (*c*) ovary (*d*) style

83. In human males, the testes lie in the scrotal sacs outside the body because it helps in the :
(*a*) process of mating (*b*) formation of sperms
(*c*) easy transfer of sperms (*d*) all the above

84. Characterstics that are transmitted from parents to offsprings during sexual reproduction show :
(*a*) only similarities with parents (*b*) only variations with parents
(*c*) both similarities and variations with parents (*d*) neither similarities nor variations with parents

85. The offspring formed as a result of sexual reproduction exhibits more variations because :
(*a*) sexual reproduction is lengthy process
(*b*) genetic material comes from two parents of different species
(*c*) genetic material comes from two parents of same species
(*d*) genetic material comes from many parents

Questions Based on High Order Thinking Skills (HOTS)

86. Two very small organisms X and Y both reproduce by the method of budding. Organism X is industrially very important because it is used in making alcohol from sugar. It is also used in making bread. Organism Y is a tiny animal having tentacles which lives in water.
(*a*) What is organism X ?
(*b*) Name the process in which X converts sugar into alcohol.
(*c*) To which class of organisms does X belong ?
(*d*) What is organism Y ?
(*e*) Out of X and Y, which organism is multicellular and which one is unicellular ?

87. A unicellular organism P lives in pond water. The organism P has no fixed shape, its shape keeps on changing. It moves and catches its prey with the help of organs Q which keep on appearing and disappearing. The organism P reproduces by a process R. Another organism S also reproduces by this process. Name P, Q, R and S.

88. The animal A which is classified as an amphibian lays eggs in pond water. The hatching of its eggs produces a tailed-form B which looks very different from the animal A. The form B then undergoes a change C and gets converted into animal A.
(*a*) Name (*i*) animal A, and (*ii*) form B.
(*b*) What is the change C known as ?
(*c*) Name the breathing organs of A.
(*d*) What are the breathing organs of B ?

89. X and Y are the two types of animals. The animals like X undergo external fertilisation whereas animals like Y undergo internal fertilisation. The animals like X lay eggs from which baby animals are hatched. On the other hand, in animals like Y, the young one develops inside the uterus of mother which then gives birth to the baby.
(*a*) What is the general name of animals like X ?
(*b*) Give two examples of animals like X.
(*c*) What is the general name of animals like Y ?
(*d*) Write the names of two animals like Y.

90. A is an insect which breeds in ponds of stagnant water. The egg of this insect produces a worm like form B which is entirely different in appearance from the adult insect. The form B undergoes a change C and gets converted into insect A. The female of insect A is a carrier of protozoan D which spreads a disease in humans.
(*a*) What are A, B, C, and D ?
(*b*) Name another insect which also undergoes change C.

ANSWERS

1. Reproduction **6.** Zygote **19.** Embryo **34.** (*a*) True (*b*) False (*c*) True (*d*) True (*e*) True (*f*) False (*g*) False (*h*) True (*i*) False (*j*) False (*k*) True (*l*) False (*m*) False **35.** (*a*) reproduction (*b*) gametes (*c*) zygote (*d*) fertilisation (*e*) ovum (*f*) smaller (*g*) month (*h*) oviparous (*i*) viviparous ; oviparous (*j*) metamorphosis (*k*) tadpole (*l*) binary fission ; budding (*m*) cloning **55.** (*i*) *d* (*ii*) *a* (*iii*) *b* (*iv*) *c* **66.** (*a*) **67.** (*b*) **68.** (*b*) **69.** (*a*) **70.** (*b*) **71.** (*b*) **72.** (*d*) **73.** (*d*) **74.** (*a*) **75.** (*c*) **76.** (*c*) **77.** (*c*) **78.** (*b*) **79.** (*c*) **80.** (*c*) **81.** (*d*) **82.** (*c*) **83.** (*b*) **84.** (*c*) **85.** (*c*) **86.** (*a*) Yeast (*b*) Fermentation (*c*) Fungi (*d*) Hydra (*e*) Multicellular : Y (Hydra) ; Unicellular : X (Yeast) **87.** P : *Amoeba* ; Q : Pseudopodia ; R : Binary fission ; S : *Paramecium* **88.** (*a*) (*i*) Frog (*ii*) Tadpole (*b*) Metamorphosis (*c*) Lungs and Skin (*d*) Gills **89.** (*a*) Oviparous animals (*b*) Frog ; Fish (*c*) Viviparous animals (*d*) Humans ; Dogs **90.** (*a*) A : Mosquito ; B : Larva ; C : Metamorphosis ; D : *Plasmodium* (*b*) Butterfly

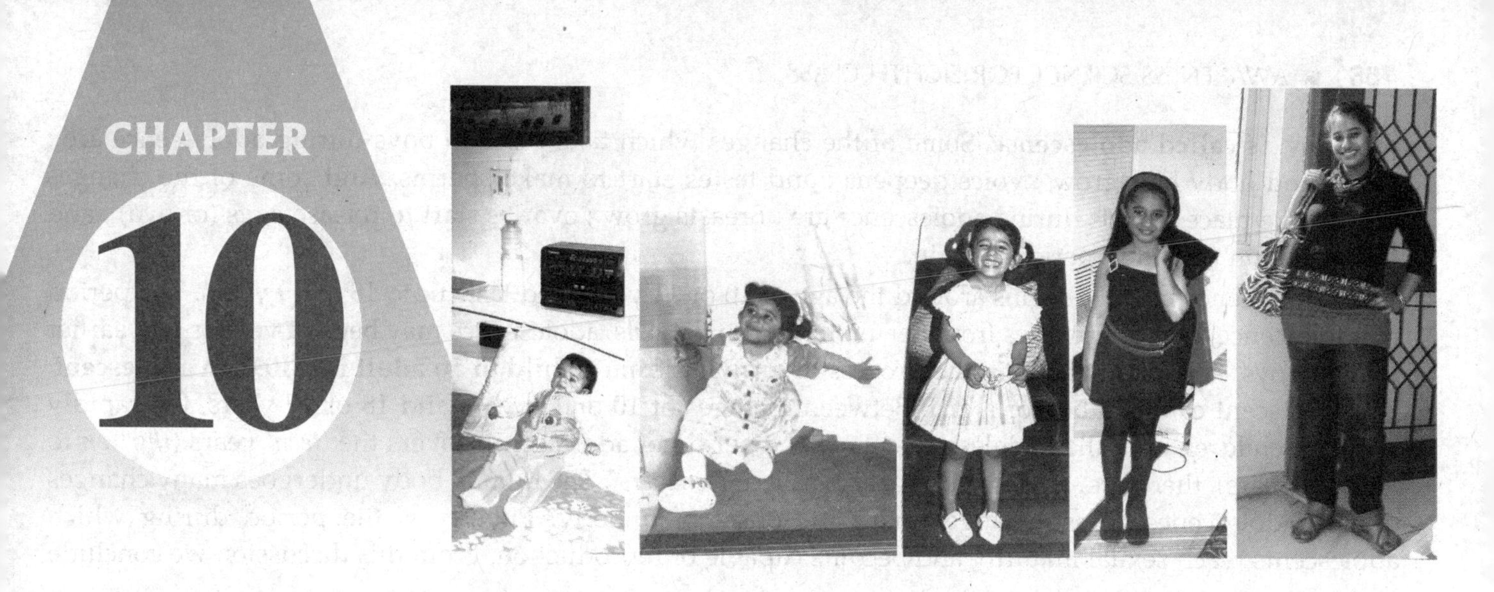

CHAPTER 10

Reaching the Age of Adolescence

In the previous Chapter we have learnt the human male and female reproductive systems. We have also learnt how human beings reproduce by giving birth to babies. Babies grow and become children. Children grow further and become adolescents. And finally, adolescents grow and become adults. It is only after growing up to a 'certain age' that humans are able to reproduce. **The human beings can reproduce (by giving birth to babies) only after a 'certain age' because their reproductive systems start working only after a 'certain age'**. In this Chapter we will learn about the changes which take place in a human 'male body' and a 'female body' after which a person becomes capable of reproduction. In other words, we will learn about the changes which take place in a boy's body and in a girl's body after which they become capable of reproduction. We will also discuss the role played by hormones in initiating the working of a male (boy's) and a female (girl's) reproductive systems, and make them fit for the process of reproduction. So, **it is the hormones which bring about changes that make a child grow into an adult.**

ADOLESCENCE

Growing up is a natural process. Growth begins from the day a baby is born. A baby grows and becomes a child. **A young human being below the age of full physical development is called a child**. When a child crosses the age of 10 years or 11 years, there is a sudden spurt in his (or her) body growth which becomes noticeable. The rapid changes which take place in the body from this age onwards are the part of 'growing up'. These changes indicate that the person is no longer a child but he (or she) is on the way to becoming an adult. **A mature human being who is fully grown and developed is called an adult.**

There is a period of life in human beings when a person is neither a child nor an adult. ***The transitional period of physical and mental development which occurs between childhood and adulthood is called adolescence.*** In most simple words, the period of life between childhood and adulthood is called adolescence. Adolescence is the time when a lot of changes take place in the bodies of boys and girls which make their reproductive systems 'mature'. We can now write another definition of adolescence as follows : **The period of life of a person when the body undergoes a lot of changes leading to reproductive**

maturity, is called adolescence. Some of the changes which take place in boys during adolescence are : Facial and body hair grow ; voice deepens ; and testes start to make sperms. And some of the changes which take place in girls during adolescence are : Breasts grow ; ovaries start to release eggs (or ova) ; and menstruation (monthly periods) begin.

Adolescence usually begins around the age of 10 or 11 years and lasts upto 18 or 19 years. The period of adolescence, however, varies from person to person. In girls, adolescence may begin a year or two earlier than in boys. **A person who is in the process of growing from a child to an adult is called an adolescent**. An adolescent can be a boy or a girl. Between the ages of 10 or 11 years and 18 or 19 years, the rapidly growing children are called adolescents. Since the period of adolescence covers the '*teen*' years (*thirteen* to *nineteen* years), therefore, adolescents are also called '*teenagers*'. The human body undergoes many changes during adolescence. These changes mark the onset of puberty. Puberty is the period during which adolescents reach sexual maturity and become capable of reproduction. From this discussion we conclude that :

(*i*) **Adolescence is the time between childhood and adulthood.**

(*ii*) **Puberty is the time when adolescents become sexually mature.**

We will now discuss puberty in detail.

PUBERTY

Although a baby is born with a full set of reproductive organs (male or female organs), these organs do not function during the first 10 or 12 years of life. Under the influence of hormones produced in the body, the reproductive organs become active at the time known as puberty. In other words, the reproductive systems of humans (boys and girls) start functioning at the time called puberty. At puberty, sex hormones (or gametes) begin to be produced due to which the boys and girls become sexually mature. **The period during which adolescent boys and girls reach sexual maturity and become capable of reproduction is called puberty**. In other words, after attaining puberty, an adolescent (boy or girl) becomes capable of having a baby. Puberty is a period of several years in which rapid physical growth (or rapid body growth) occurs leading to sexual maturity. In fact, puberty is the time when a child's body starts changing into an adult's body.

Puberty tends to start earlier in girls (females) than in boys (males). Generally, girls attain puberty at a lower age of 10 to 13 years while boys reach puberty at a comparatively higher age of 12 to 14 years. **At puberty, many changes occur in the bodies of boys and girls.** In both, boys as well as girls, there is a rapid increase in the rate of growth (height, etc.) during puberty. Puberty is marked by the development of secondary sexual characteristics in boys and girls (such as growth of facial hair and deeper voice in boys; and development of breasts and start of menstruation in girls). The most significant sign of puberty in girls is the beginning of menstruation (or monthly periods).

The various changes which occur in boys during puberty are as follows :

(*i*) Hair grow on the face of boys (in the form of moustache and beard), and on chest.

(*ii*) Voice deepens in boys. It becomes low pitched voice.

(*iii*) Testes start to make sperms.

(*iv*) Testes and penis become larger.

(*v*) Chest and shoulders of boys broaden (become wider).

(*vi*) Body becomes muscular (due to development of muscles).

(*vii*) Hair grow in armpits and in pubic regions (genital area) between the thighs.

(*viii*) Rapid increase in height occurs.

(*ix*) Feelings and sexual drives associated with adulthood begin to develop.

The various changes which occur in girls during puberty are as follows :

(*i*) Breasts develop and enlarge in girls.
(*ii*) Ovaries start to release eggs (or ova).
(*iii*) Menstruation (monthly periods) begin.
(*iv*) Ovaries, oviducts, uterus and vagina enlarge.
(*v*) Hips of girls broaden (become wider). Extra fat is deposited on hips and thighs.
(*vi*) Hair grow in armpits and in pubic regions (genital area) between the thighs (*This change is the same as in boys*).
(*vii*) Rapid increase in height occurs (*This change is the same as in boys*).
(*viii*) Feelings and sexual drives associated with adulthood begin to develop (*This change is also the same as in boys*).

All the changes which occur in boys and girls at puberty are brought about by various hormones. When a boy or girl reaches puberty, he (or she) becomes an adolescent. Adolescence continues until the age of about 18 (or 19) years when the growth stops and the person becomes an adult. We will now describe the changes taking place during puberty in detail.

CHANGES AT PUBERTY

Puberty marks the beginning of the reproductive period when one becomes capable of reproduction. We will now discuss some of the changes which take place in adolescent boys and girls at puberty, in detail.

1. Increase in Height

The most conspicuous change (clearly visible change) during puberty is the sudden increase in the height of boys and girls. At the time of puberty, the long bones (the bones of arms and legs) elongate or lengthen and make a person tall. Thus, children gain a lot of height during puberty. **Initially, girls grow faster than boys** but by about 18 years of age, both boys and girls reach their maximum height. Generally, the maximum height of girls is slightly less than that of boys.

The rate of growth in height varies in different persons. Some boys and girls may grow suddenly at puberty and then slow down, while others may grow gradually. All the parts of the body do not grow at the same rate. Sometimes the arms and legs, or hands and feet of adolescent boys and girls look oversized and out of proportion with the body. But soon the other parts catch up with them in growth. This results in a more proportionate body. The average rate of growth in height of boys and girls with age is given in the following table.

Age	*Percentage of full height in boys*	*Percentage of full height in girls*
8 years	72%	77%
9 years	75%	81%
10 years	78%	84%
11 years	81%	88%
12 years	84%	91%
13 years	88%	95%
14 years	92%	98%
15 years	95%	99%
16 years	98%	99.5%
17 years	99%	100%
18 years	100%	100%

The figures given in second and third column of the above table tell us the percentage of full height which a person (boy or girl) has reached at the age given in the first column. For example, at the age of 10 years, a boy reaches only 78 per cent of his full height whereas a girl reaches 84 per cent of her full height. Please note that the figures on the rate of growth in height given in the above table are only indicative. There may be variations from person to person.

By using the data given in the above table, we can draw a graph to show the variation in the 'percentage of full height' in boys and girls with 'age'. We take 'age' on x-axis and 'percentage of full height' on y-axis, and plot the various readings given in the above table on the graph paper. The graph thus obtained is shown in Figure 1. We can see from the graph shown in Figure 1 that initially the height of girls

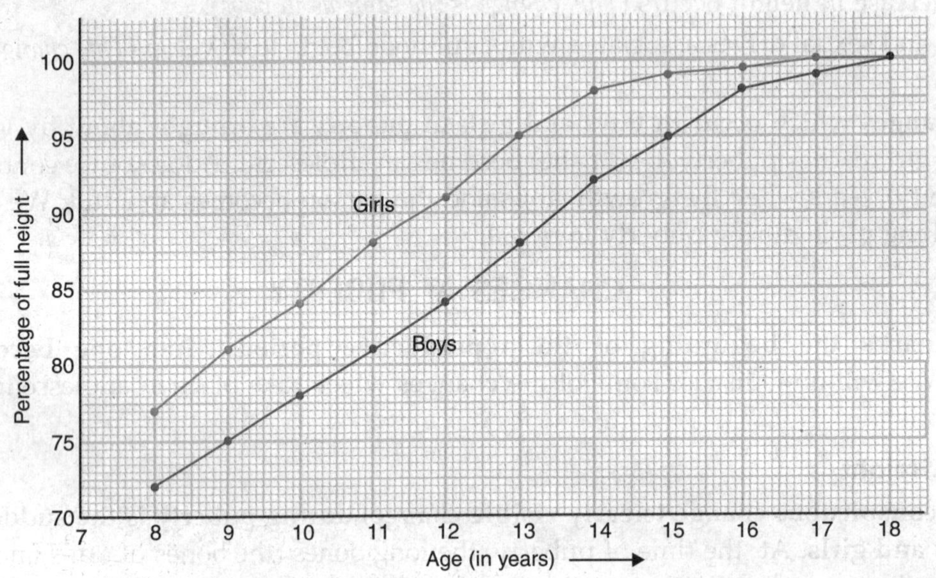

Figure 1. Graph showing the variation of percentage of full height with age in boys and girls.

increases at a faster rate than that of boys but by about 18 years of age, both boys and girls reach their maximum height. By using the graph shown in Figure 1, we can find out the percentage of full height of a boy or girl at any age, say 12.5 years. We can also calculate the full height which a boy or girl might eventually reach. This will become more clear from the following discussion.

If we know the present height of a person (boy or girl) and the percentage of his (or her) full height at present age, we can calculate the full height of the person which he (or she) will eventually reach. This can be done by using the formula :

$$\text{Full height of a person} = \frac{\text{Present height of person}}{\text{Percentage of full height at present age}} \times 100$$

We can use the unit of centimetre or metre for height. The calculation of full height of a person (boy or girl) will become clear from the following example.

Sample Problem. A ten year old boy is 125 cm tall. If the present height of the boy is 78% of his full height, calculate the full height which the boy will eventually reach at the end of growth period.

Solution. Here, Present height of boy = 125 cm

Percentage of full height = 78

So,

$$\text{Full height of boy} = \frac{\text{Present height of boy}}{\text{Percentage of full height}} \times 100$$

$$= \frac{125}{78} \times 100 \text{ cm}$$

$$= 160.25 \text{ cm}$$

Thus, the full height of boy will be 160.25 centimetres.

Please note that **the height of a person depends on the genes inherited from the parents**. For example, if both the parents (or one of the parents) are very tall, the son or daughter is likely to be very tall. The height of a person is more or less similar to that of some family member.

2. Change in Body Shape and Appearance

When a child is small, sometimes it becomes difficult to tell from appearance whether it is a boy or a girl. This is because small boys and girls have the same body shape (see Figure 2). So, it is usually the type of clothes worn by a small child which help us in telling whether it is a boy or a girl (and not its body shape). **When puberty sets in, a time of rapid changes in body shape and appearance starts in boys and girls which make the boys and girls look different from one another** (see Figure 3). Actually, testes in boys and ovaries in girls make different hormones which make the bodies of boys and girls to develop in different ways. So, the changes in body shape occurring in adolescent boys and girls at puberty are different. Some of the **changes in body shape and appearance in boys and girls brought about by the onset of puberty** are as follows :

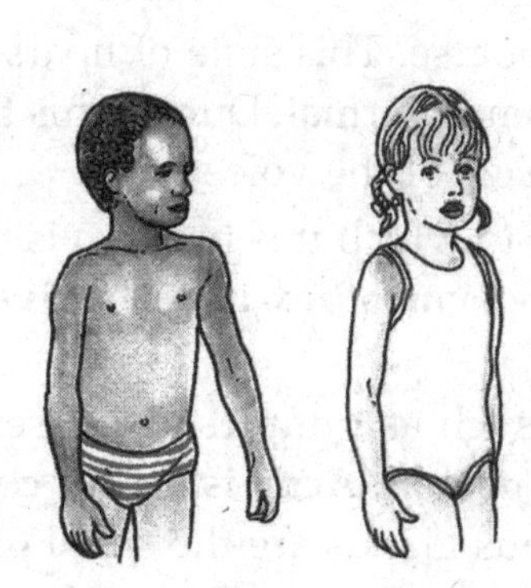

Figure 2. Small boys and girls have the same body shape. So, they look alike.

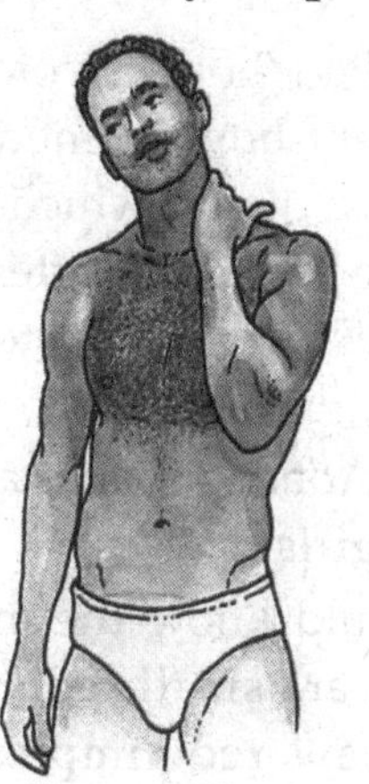

Figure 3. Grown up boys and girls have different body shapes. So, they look different.

(*i*) Boys develop broader shoulders and wider chests than girls (see Figure 3).

(*ii*) Girls develop broader hips than boys. Due to this, the region below the waist becomes wider in girls (see Figure 3).

(*iii*) Boys develop more muscle than girls. So, the body of boys looks more muscular than that of girls.

(*iv*) Girls develop breasts. This also changes the body shape of grown up girls and makes them look different from boys.

(*v*) Boys develop Adam's Apple (a bulge in front of throat) which makes them look different from girls.

(*vi*) Boys develop facial hair (moustache and beard) but the girls do not have facial hair. So, growth of facial hair makes boys look different from girls.

3. Change in Voice

When we are talking, our voice (or sound) is produced by the voice box. The voice box is in our throat. Voice box is also called larynx. At puberty, the voice box (or larynx) begins to grow in boys as well as in girls. **The growth of voice box in boys is much more than the growth of voice box in girls.** Due to this, the voice box in boys becomes much bigger than the voice box in girls. This means that *the grown up boys have a bigger voice box in their throat whereas grown up girls have a smaller voice box in their throats.*

The bigger voice box in boys gives deeper voice (or low pitched voice) to the boys. Thus, the voice of boys changes at puberty and becomes deeper because their voice boxes enlarge too much and become bigger in size. The bigger (or larger) voice box in a grown up boy can be seen as a bulge (or projection) in front of the throat (or neck). **The bulge (or projection) at the front of throat or neck in grown up boys is called Adam's Apple** (see Figure 4). Adam's Apple is formed in grown up boys because of their bigger voice box in the throat. It is called an Adam's Apple after the story of Adam (the first man created by God) and Eve (the first woman created by God) described in Bible in which Adam ate a piece of forbidden fruit (apple) in the Garden of Eden which got stuck in his throat. An Adam's Apple sometimes looks like a small rounded apple just under the skin in front of the throat. Thus, *Adam's Apple is a feature of throat (or neck) of grown up boys.* It is a bulge (or bump) in the throat or neck.

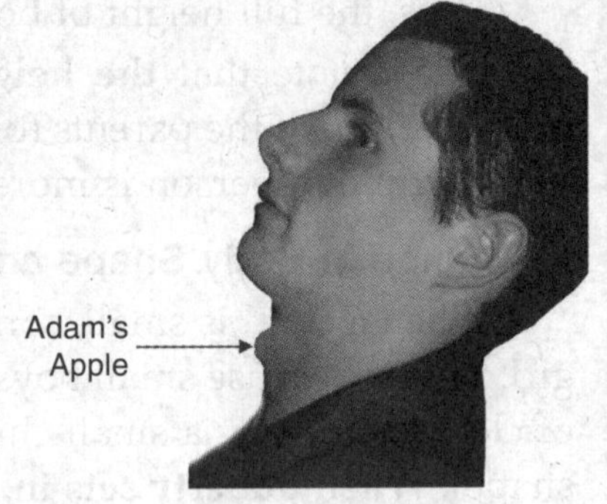

Figure 4. Adam's Apple in a grown up boy.

A rough and harsh voice is called 'hoarse' voice. Hoarse voice is called '*phati awaaz*' in Hindi. In adolescent boys, sometimes the muscles of the growing voice box go out of control due to which their voice becomes hoarse. This state of hoarseness of voice may remain for a few days or few weeks after which the voice becomes normal. During this time, the vocal cords of the voice box (or larynx) get adjusted to the new, bigger size of the voice box.

In girls, the voice box (or larynx) is comparatively small in size due to which it is hardly visible from outside. So, girls do not develop Adam's Apple at puberty. **The smaller voice box in girls gives shrill voice (high pitched voice) to the girls.**

Before we go further, we should know the meaning of the terms such as pimples, acne, sebaceous glands and sweat glands. **Pimples are small red spots on the face of a person. Acne is a skin condition marked by the eruption of numerous red pimples on the face**. Sebaceous glands are the small glands in the skin which secrete an 'oil' (called sebum) through skin pores to lubricate and protect the surface of skin. 'Sebaceous glands' are commonly known as 'oil glands'. Sweat glands are the tiny glands in the skin which secrete 'sweat' (*pasina*).

4. Development of Pimples and Acne on the Face

Many young boys and girls get pimples and acne on the face during puberty. **The pimples and acne are formed due to the increased activity of sebaceous glands and sweat glands present in the skin.** This happens as follows : The level of hormones in boys and girls rises too much at puberty. These hormones stimulate the sebacious glands and sweat glands present in the skin. The increased activity of sebaceous glands (oil glands) causes them to secrete more 'oil' and the increased activity of sweat glands makes them secrete more sweat. The excess oil and sweat get collected in the tiny pores of the skin. The accumulation of oil, sweat and dead skin cells blocks the tiny pores in the skin of the face. Bacteria grow in the mixture of oil, sweat and dead skin cells in the 'blocked skin pores' causing the swelling and redness of skin which leads to the formation of pimples (see Figure 5). Thus, **pimples are caused by the clogged and infected skin pores**. When outbreaks of too many pimples occur often, it is considered acne. Pimples and acne occur most commonly during adolescence. In most boys and girls, pimples and acne diminish over time and tend to disappear by the time one reaches the age of 18 or 19 years. In some, they continue into adulthood.

Figure 5. Many adolescent boys and girls get pimples on the face during puberty.

5. Development of Sex Organs

The onset of puberty brings about complete growth and development of sex organs in boys and girls due to which their reproductive systems start functioning at this stage. For example :

(*i*) At puberty, the male sex organs like testes and penis enlarge and develop completely. **The testes begin to produce male sex cells called 'sperms' at puberty**. This makes the male (boy's) reproductive system functional at puberty.

(*ii*) At puberty, the female sex organs like ovaries, oviducts and uterus enlarge and develop completely. The eggs (or ova) begin to mature at this stage. **The ovaries start releasing mature eggs (or ova) at puberty.** This makes the female (girl's) reproductive system functional at puberty.

6. Reaching Mental, Intellectual and Emotional Maturity

When a person reaches puberty, he (or she) becomes an adolescent. Adolescence continues up to the age of 18 or 19 years. During adolescence, a boy (or girl) reaches mental, intellectual and emotional maturity (Mental is called *maansik*, intellectual is called *baudhik* and emotional is called *bhavatamak* in Hindi).

(*a*) Adolescence is a period which brings maturity in a person's way of thinking. At this stage, adolescent boys and girls spend considerable time thinking about many things occurring in their minds. During adolescence, mental maturity makes the boys and girls more independent than before.

(*b*) Adolescence is a period which also brings intellectual maturity. Due to intellectual development, the boys and girls get into the habit of reasoning, and understanding things objectively. In fact, **adolescence is the time in one's life when the brain has the greatest capacity for learning.**

(*c*) The changes which occur in the body of boys and girls during adolescence may cause emotional swings. These emotional swings show intense feelings such as joy, anger, boredom, worries or sadness which are not based on reasoning or knowledge. The adolescents also become self-conscious (unduly aware of one's actions leading to nervousness). Sometimes, the adolescents feel insecure (not confident) while trying to adjust to the changes in body and mind. There is, however, no reason to feel insecure. **The changes which take place in body and mind during adolescence are a natural part of the growing up process. Every human being has to pass through this stage. Our parents have passed through this stage, now we are passing through this stage, and one day our children will also pass through this stage in life.**

SECONDARY SEXUAL CHARACTERISTICS IN HUMANS

There are two types of sexual characteristics (or sex characteristics) in human beings : Primary sexual characteristics and Secondary sexual characteristics. **The sexual characteristics which are present at birth are called primary sexual characteristics.** Primary sexual characteristics include internal and external sex organs which are present in babies at the time of their birth. The primary sexual characteristics in males (or boys) are : **Testes, Penis** and **Seminal vesicles**, etc. The primary sexual characteristics in females (or girls) are : **Ovaries, Oviducts, Uterus** and **Vagina**, etc. *The primary sexual characteristics are directly involved in reproduction.*

The sexual characteristics controlled by hormones which distinguish between sexually mature males and females (sexually mature boys and girls) but are not directly involved in reproduction, are called secondary sexual characteristics. In secondary sexual characteristics, the body parts (other than sex organs) develop special features which make it easier to distinguish a boy from a girl. For example, the **growth of facial hair** (like moustache and beard) in boys is a secondary sexual characteristic which helps to distinguish between a mature boy and a girl (because facial hair do not grow in girls). Similarly, the **development of breasts** in girls is a secondary sexual characteristic which helps to distinguish a girl from a boy (because boys do not develop breasts). The secondary sexual characteristics start developing at the time of puberty and continue to develop through the period of adolescence.

The main secondary sexual characteristics in males (or boys) are the following :

(*i*) Hair grow on face (in the form of moustache and beard) in boys.
(*ii*) Shoulders and chest broaden (become wider) in boys.
(*iii*) A deeper voice (or low pitched voice) in boys.
(*iv*) Adam's Apple develops in front of throat (or neck) in boys.

The secondary sexual characteristics in boys are produced by the male sex hormone called 'testosterone' made in testes.

The main secondary sexual characteristics in females (or girls) are the following :

(*i*) Development of breasts in girls.

(*ii*) Hips broaden and become more curved and prominent in girls.

(*iii*) A shrill voice (or high pitched voice) in girls.

The secondary sexual characteristics in girls are produced by the female sex hormone called 'estrogen' made in ovaries. We will now discuss the role of hormones in initiating reproductive functions.

HORMONES

Hormones are the chemical substances which co-ordinate the activities of living organisms (including human beings), and also their growth. Hormones are made and secreted by specialised tissues in the body called 'endocrine glands'. The hormones are poured directly into the blood and carried throughout the body by the blood circulatory system. The hormones act on 'specific tissues' or 'specific organs' in the body called 'target sites'. A hormone is produced by an endocrine gland in one part of human body but causes a particular effect in another part of the body. Hormones are of many different types and perform different functions in the body. **The hormones involved in the development and control of the reproductive organs and secondary sexual characteristics are called sex hormones**. The two common sex hormones are **testosterone** and **estrogen** (Estrogen is also spelled as oestrogen).

Exocrine Glands and Endocrine Glands

A gland is a structure which secretes a specific substance (or substances) in the body. A gland is made up of 'a group of cells' or 'tissue'. There are two types of glands in our body : exocrine glands and endocrine glands. **A gland which secretes its product into a duct (or tube) is called an exocrine gland.** So, exocrine glands are the glands having ducts. For example, the salivary gland secretes its product (saliva) into a duct called 'salivary duct', therefore, **salivary gland** is an exocrine gland. **Sweat glands** and **sebaceous glands (or oil glands)** are also exocrine glands which release their secretions through ducts.

A gland which does not have a duct and secretes its product directly into the blood stream, is called an endocrine gland. Thus, endocrine glands are ductless glands. An endocrine gland makes (and secretes) a chemical substance called 'hormone'. Some of the examples of endocrine glands (or ductless glands) in our body are **pituitary gland, thyroid gland and adrenal glands.** Some of the glands in our body function both, as endocrine glands as well as exocrine glands. For example, pancreas, testes and ovaries function as endocrine glands as well as exocrine glands. Endocrine glands release their hormones directly into the blood stream of a person. These hormones reach the concerned body part (called target site) through the blood and act on it.

ROLE OF HORMONES IN INITIATING REPRODUCTIVE FUNCTIONS

The hormones play an important role in initiating the reproductive functions of adolescents. This is because the changes which occur at adolescence are controlled by hormones. In other words, **the onset of puberty is controlled by hormones**. Puberty begins with a sudden increase in the production of hormones in the body which cause a number of changes in boys and girls. We have already studied that testes and ovaries are the real reproductive organs in humans. The testes and ovaries produce sex cells or gametes (sperms, and eggs or ova). In addition to producing gametes (sperms and eggs), testes and ovaries produce sex hormones which bring about changes in the bodies of boys (males) and girls (females) at puberty, and make their reproductive systems functional.

The testes make male sex hormone called testosterone. The ovaries make female sex hormone called estrogen. The production of these sex hormones is under the control of another hormone (called gonadotropic hormone) secreted by **pituitary gland.** The production of sex hormones 'testosterone' and 'estrogen' increases dramatically at the stage of human growth called puberty. This rise in the level of sex hormones is responsible for the changes in the bodies of adolescent boys and girls at puberty.

(*a*) The hormones from pituitary gland stimulate testes to release male sex hormone 'testosterone' at the onset of puberty in boys. **Testosterone hormone performs the following functions :**

(*i*) Testosterone hormone produces male secondary sexual characteristics in boys at puberty (such as deeper voice; growth of facial hair like moustache and beard ; broad shoulders and chest ; Adam's Apple ; and more muscles).

(*ii*) Testosterone hormone causes the growth and development of male sex organs at puberty.

(*iii*) Testosterone hormone causes 'growth spurt' in boys at puberty.

(*b*) The hormones from pituitary gland stimulate ovaries to release the female sex hormone 'estrogen' at the onset of puberty in girls. **Estrogen hormone performs the following functions :**

(*i*) Estrogen hormone produces female secondary sexual characteristics in girls at puberty (such as shrill voice ; development of breasts ; and broader hips).

(*ii*) Estrogen hormone causes the growth and development of female sex organs at puberty.

(*iii*) Estrogen hormone brings about the monthly preparation of uterus for pregnancy.

The pituitary gland secretes many hormones. One of these hormones stimulates the formation of sperms in testes and maturation of egg cells (or ova) in ovaries. A yet another hormone develops the mammary glands (milk producing glands) inside the breasts.

REPRODUCTIVE PHASE OF LIFE IN HUMANS

Adolescents become capable of reproduction at puberty when their testes and ovaries begin to produce gametes (sperms and eggs). Adolescent boys grow and become men. Adolescent girls grow and become women. In men, the capacity to produce male gametes (or sperms) usually lasts throughout life. But in women, the capacity to produce female gametes (eggs or ova) lasts only up to about 45 to 50 years of age. Due to this, *the reproductive phase of life in men (or males) lasts much longer than in women (or females)*. We will now discuss the reproductive phase of life in human females (or women) in detail. **In females (or women), the reproductive phase of life begins at puberty (10 to 12 years of age) and generally lasts till the age of approximately 45 to 50 years.**

(*a*) With the onset of puberty, the eggs (or ova) begin to mature in the ovaries of a woman. One mature egg (or ovum) is released by one of the ovaries of the woman once in about 28 to 30 days. During this period, the inner lining of uterus grows and becomes thick and spongy, and prepares itself to receive the fertilised egg (see Figure 6). So, in case the fertilisation of egg cell occurs by a sperm, the fertilised egg cell begins to divide to form an embryo. The embryo then gets embedded in the thick uterus lining. This results

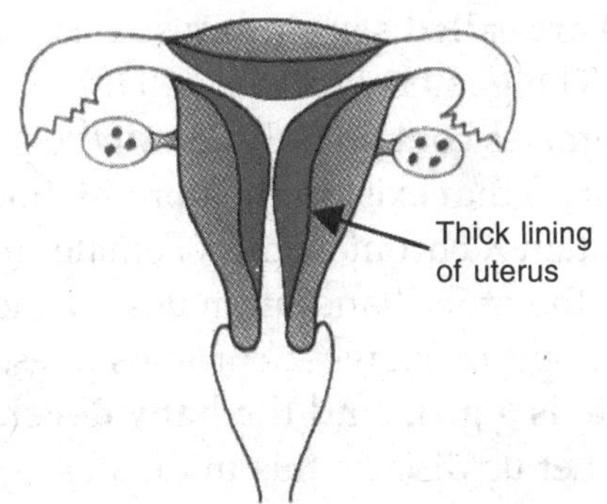

Figure 6. A thick lining grows in the uterus to receive the fertilised egg cell (if any).

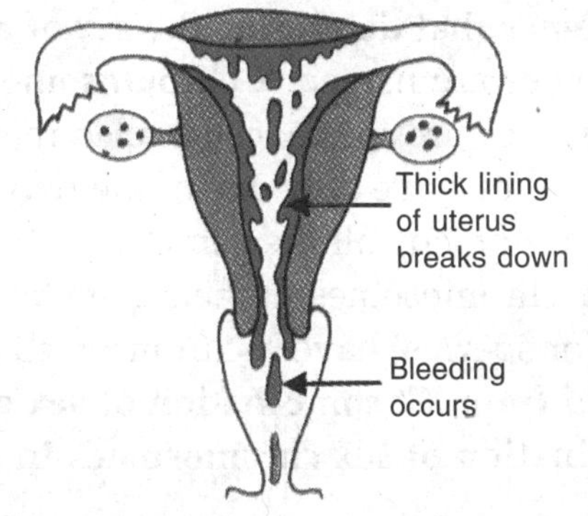

Figure 7. In case the egg cell is not fertilised, the thick uterus lining breaks down leading to bleeding. This is called menstruation (or period).

in pregnancy which ultimately leads to the birth of a baby. ***We will now describe what happens if the fertilisation of egg cell does not take place.***

(*b*) If fertilisation does not occur (due to lack of sperm), then the egg released by the ovary dies within a few days and the thick lining breaks down (because it is no longer required). Since the thick uterus lining contains a lot of blood vessels, therefore, the breaking down (or disintegration) of uterus lining produces blood alongwith other tissues. This blood and other tissues come out of vagina of woman in the form of a 'bleeding'. **The bleeding from the uterus which occurs in a woman (or mature girl) every month (if the egg cell has not been fertilised) is called menstrual flow or menstruation** (see Figure 7). Menstruation occurs once in about 28 to 30 days (which is almost a month). Menstruation occurs every 28 to 30 days because ovulation (release of egg or ovum by the ovary of woman) occurs after every 28 to 30 days. **In everyday language, menstruation is called 'monthly periods' or just 'periods'**. It is also called ***'menses'***. Menstruation (or periods) usually lasts for about 3 to 5 days in a month.

The first menstruation (or menstrual flow) begins at puberty (when the girl or woman is around 10 to 12 years of age). **The first occurrence of menstruation (or periods) at puberty is called menarche**. Menarche is the beginning of the reproductive life of a girl (or woman). In other words, menarche is the time from which a girl (or woman) becomes capable of having a baby. Menstruation stops temporarily when a woman becomes pregnant. Menstruation restarts after the birth of the baby. Menstruation stops permanently when a woman reaches the age of about 45 to 50 years. With the permanent stoppage of menstruation, a woman loses her ability to bear children. **The permanent stoppage of menstruation (or periods) in a woman is called menopause.** Menopause occurs in women at the age of about 45 to 50 years. A woman stops ovulating at menopause and can no longer become pregnant. Menopause is the end of the reproductive life of a woman. We can now say that ***the reproductive life of a woman starts at menarche and ends at menopause.***

Menstrual Cycle

The process of ovulation and menstruation in women is called menstrual cycle (because it occurs again and again after a fixed time period). The menstrual cycle is a period of about 28 to 30 days during which an egg cell (or ovum) matures, the mature egg cell is released by the ovary, thickening of uterus lining takes place, and finally the uterus lining breaks down causing bleeding in women (if the egg cell has not been fertilised). Initially, the menstrual cycle in girls may be irregular (it may not be of 28 to 30 days). It becomes regular after some time. **Menstrual cycle in women is controlled by hormones.** We will now discuss the determination of sex in humans which involves the role of sex chromosomes.

Sex Chromosomes

Chromosomes are present inside the nucleus of every cell of the human body. All the human beings have 23 *pairs* of chromosomes in the nuclei of their normal body cells. Out of these 23 pairs of chromosomes, one pair of chromosomes (or two chromosomes) are called sex chromosomes. **The two chromosomes that determine the sex of an offspring (or baby) are called sex chromosomes**. The two sex chromosomes are named **X chromosome** and **Y chromosome.** The gametes (or sex cells) are not normal body cells, they are special cells called reproductive cells. The gametes contain only 23 *single* chromosomes and hence they have only one of the two sex chromosomes (X or Y) that exist in the normal body cells. In females (or women), all the gametes or all the egg cells (ova) contain X chromosomes. Females (or women) have no Y chromosomes in their gametes (egg cells or ova). On the other hand, in males (or men), half of gametes (or sperms) have X chromsomes and half of gametes (or sperms) have Y chromosomes. **The baby developed from XX combination of sex chromosomes in zygote is a girl. And the baby developed from XY combination of sex chromosomes in zygote will be a boy**. Let us discuss this in detail.

HOW IS THE SEX OF A BABY DETERMINED

When a woman gives birth to a baby it can be a boy (male) or a girl (female). *The sex of a baby to be born is decided at the moment the egg cell of woman (mother) gets fertilised by the sperm of man (father) and pregnancy occurs.* This is because the instructions for determining the sex of baby (whether it will be a boy or girl) are present in the sex chromosomes which are in the nucleus of the fertilised egg called zygote. **The sex of baby is determined by the type of sex chromosomes present in the fertilised egg (or zygote) from which the baby develops.** This will become more clear from the following discussion.

There are two types of sex chromosomes : X chromosomes and Y chromosomes.

(*i*) A female (woman or mother) has only X chromosomes in all her gametes called eggs (or ova). This means that all the female gametes called eggs (or ova) have only X chromosomes.

(*ii*) A male (man or father) has X chromosomes as well as Y chromosomes in male gametes called sperms. Actually, half the male gametes or sperms have X chromosomes and the other half of male gametes or sperms have Y chromosomes.

The sex of a child depends on what happens at fertilisation :

(*a*) If a sperm carrying X chromosome fertilises an egg cell (or ovum) which carries X chromosome, then the zygote formed will have XX combination of sex chromosomes due to which the child born will be a girl (or female) (see Figure 8).

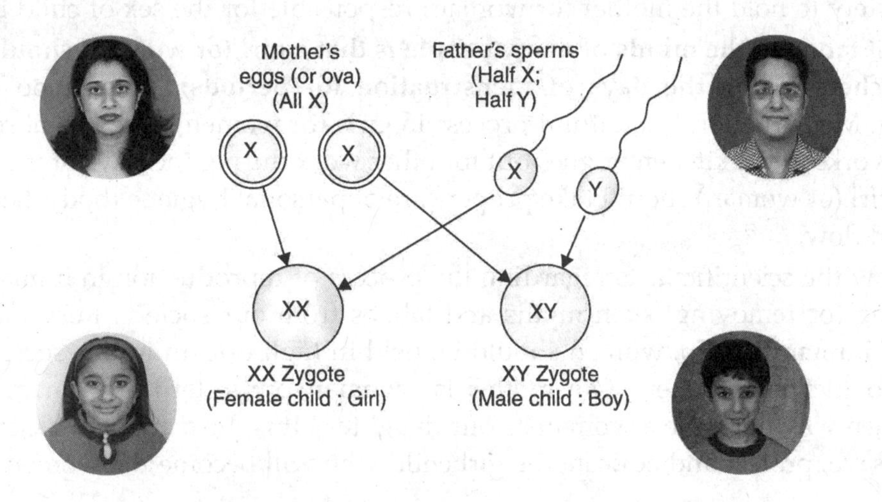

Figure 8. Determination of sex in humans.

(*b*) If a sperm carrying Y chromosome fertilises an egg cell (or ovum) which carries X chromosome, then the zygote formed will have XY combination of sex chromosomes due to which the child born will be a boy (or male) (see Figure 8).

Thus, the sex of unborn child (or unborn baby) depends on whether the zygote formed at the time of fertilisation has XX combination of sex chromosomes or XY combination of sex chromosomes. XX combination of sex chromosomes makes the child a girl whereas XY combination of sex chromosomes makes the child a boy. Please note that **it is the sperm of man which determines the sex of the child**. This is because half of the sperms have X chromosomes and the other half have Y chromosomes. Thus, there is a 50 per cent chance of a boy and a 50 per cent chance of a girl being born to the parents. This is why the human population is roughly half males and half females. **The egg cells (or ova) of woman cannot decide the sex of the child because all the egg cells (or ova) contain the same sex chromosome, X chromosome.**

From the above discussion we conclude that if the father (man or husband) contributes X sex chromosome at fertilisation through his sperm, the baby born will be a girl. On the other hand, if the father (man or husband) contributes a Y sex chromosome at fertilisation through his sperm, then the baby born will be a boy. This means that it is the sex chromosome contributed by father (man or husband) which decides the sex of the baby which the mother (woman or wife) will give birth to. Thus, **father (man or husband) is responsible for the sex of the baby (boy or girl) which is born**. The belief that mother (woman or wife) is responsible for the sex of her baby is absolutely wrong. In many ignorant Indian families, the mother (woman or wife) is held responsible for the birth of a girl child and unnecessarily harassed. Such people should understand that it is the husband who is responsible for the birth of a girl child (and not his wife). Moreover, ***a girl is no less than a boy.***

MYTHS AND TABOOS REGARDING REPRODUCTION

A widely held but false belief is known as myth. And something prohibited by social customs is called taboo. We will now describe some myths and taboos related to the process of reproduction in humans. In the previous Chapter of this book and in this Chapter, we have learnt the scientific facts related to the process of reproduction in humans. There are, however, many wrong notions or beliefs in the minds of many people connected with the process of reproduction (especially in relation to girls or women) which are based on ignorance of facts.

An important myth (wrong belief) in the minds of many people is that mother is responsible for the sex of her child (male or female). This is absolutely wrong. The scientific facts tell us that it is the father (or man) who is responsible for the sex of the child born (whether it is a boy or girl). It is great foolishness on the part of society to hold the mother (or woman) responsible for the sex of child born.

An important taboo in the minds of many people is that a girl (or woman) should not be allowed to work in the kitchen during the days of menstruation (or periods). This taboo (or custom) is also absolutely wrong. Menstruation is a natural process in girls (or women). There is absolutely no harm if a girl (or woman) works in the kitchen or goes out for other work during the days of menstruation. The only thing is that the girl (or woman) should take proper care of personal hygiene (body cleanliness) during the days of menstrual flow.

Since we know the scientific facts regarding the process of reproduction in humans, it is our duty to help in eradicating (or removing) such myths and taboos from our society. Remember, **God has made women to create human life.** So, women should be held in high esteem in our society. Why look down upon women who give birth to men ! Our mother is a woman, our sister is a woman, and a man's wife is also a woman. Then why not have a woman as our daughter ! It is the duty of our Government as well as of our society to save, protect and educate the girl child who will become the woman of tomorrow.

Adolescent Pregnancy

The formal union of a man and a woman by which they become husband and wife is called marriage. In our country, the legal age (as per law) for marriage is 18 years for girls (and 21 years for boys). One of the functions of married life in our society is to have a baby. With the birth of baby, a girl becomes a mother (or she enters motherhood). The legal age for the marriage of girls has been fixed at 18 years because early motherhood (at an age lower than 18 years) causes a lot of problems for the young mother as well as for the baby. **Some of the problems brought about by early marriage in girls leading to early motherhood are given below :**

(*i*) The girls younger than 18 years of age are not prepared physically and mentally for motherhood.
(*ii*) Early marriage and motherhood cause health problems in the mother and the child.
(*iii*) Early marriage and motherhood cause agony (extreme suffering) to the girl as she is not prepared to fulfil the responsibilities of motherhood involved in bringing up the baby.
(*iv*) Early marriage and motherhood curtails (reduces) the chances of higher education for girls.
(*v*) Early marriage and motherhood curtails the employment opportunities for the young girls.

Unfortunately, in some of the areas of our country there is a tradition of child marriages. This tradition is harmful for the growth and development of not only the girls but also for boys. Though our Government is making great efforts to stop child marriages but these efforts can succeed fully only when the people are made to co-operate. So, as responsible citizens, all of us should do our best to stop child marriages in our country. We will now discuss the reproductive health of adolescents.

REPRODUCTIVE HEALTH

According to the World Health Organisation (WHO), **reproductive health is defined as a state of physical, mental and social well-being of a person in all matters relating to the reproductive system at all stages of life**. Though reproductive health is required at all stages of life, but it is more essential during the period of adolescence when the body is growing rapidly and many changes are taking place in it. Some of the important **conditions to maintain good reproductive health during adolescence** are given below :

(*i*) It is necessary to eat balanced diet during adolescence.

(*ii*) It is necessary to maintain personal hygiene during adolescence.

(*iii*) It is necessary to take adequate physical exercise during adolescence.

(*iv*) It is necessary to avoid taking any drugs during adolescence.

We will now describe all these conditions for maintaining good reproductive health in detail, one by one.

1. Nutritional Needs of Adolescents

The food which we eat each day makes up our 'diet'. A diet consists of many food items made from cereal grains (like wheat and rice), pulses, fruits, vegetables, meat, fish, eggs, and milk, etc. **The diet which contains the correct amount of each constituent (such as carbohydrates, fats, proteins, vitamins and minerals) sufficient for the normal growth and development of the body, and keep a person healthy, is called a balanced diet.** Every human being, at any age, needs to have a balanced diet to keep the body healthy. It is all the more important to eat balanced diet (containing the right kinds of foods) during adolescence because a rapid growth and development of the body takes place during this period. A balanced diet helps the bones, muscles and other parts of the body to get adequate nourishment required for rapid growth. The various types of food items which should be included in the diet of adolescents so as to meet their nutritional needs are discussed below.

(*i*) The diet of adolescents should include food items made from cereals (like wheat and rice) which provide carbohydrates for energy. The food items like *chapati* (*roti*), bread, and *poori* (made from wheat flour), and cooked rice and its preparations like *dosa, idli* and *biryani*, etc., provide us carbohydrates for energy. Sugar and jaggery (*gud*) also contain carbohydrates that give us energy.

(*ii*) The diet of adolescents should include fats which also give us energy. The common food items which can be included in the diet for providing fat to our body are butter, *ghee*, cooking oils (like groundnut oil, mustard oil, sunflower oil and coconut oil), *vanaspati ghee*, groundnuts and fatty meat.

Figure 9. Examples of types of food in a balanced diet. These include foods made from cereal grains (wheat and rice), pulses, milk, butter, eggs, meat, fish, fruits and vegetables.

(*iii*) The diet of adolescents should include food items containing proteins which are required for the growth of their body. The common food items which contain a lot of proteins are pulses (*dal*), peas, beans, cheese (*paneer*), eggs, lean meat (meat without fat), fish, milk and groundnuts.

(*iv*) The diet of adolescents should include fruits and vegetables which provide many vitamins and minerals necessary for keeping good health. Fruits and vegetables are called protective foods because they protect our body from many ailments by supplying various vitamins and minerals, and help us stay healthy. For example, citrus fruits like orange, lime and lemon contain vitamin C which builds up body resistance and helps fight infection. Carrots contain vitamin A which is necessary for keeping healthy eyesight. **Iron is a mineral which is necessary for making blood in the body**. The iron-rich foods such as leafy vegetables, Indian gooseberry (*amla*), meat and jaggery (*gud*) are good for adolescents as they help in making blood. Milk contains a lot of calcium mineral which is necessary for making healthy bones of the growing adolescents.

Our Indian meal of *roti* (or rice), pulses (*dal*) and vegetables is a balanced meal. Milk is a balanced food in itself. For infants (small babies), mother's milk provides all the necessary nourishment which they need. Chips and packed (or tinned) snacks, though very tasty, should never be taken in place of regular meals because chips and other such snacks do not have adequate nutritional value. Similarly, other junk foods such as burgers, noodles, vegetable cutlets and soft drinks do not form part of a balanced diet.

2. Personal Hygiene for Adolescents

Keeping our body clean is called personal hygiene. **The maintenance of personal hygiene (or cleanliness) is necessary for adolescents for preventing diseases and keeping good health**. The adolescents can maintain personal hygiene by adopting the following practices :

(*i*) Adolescent boys and girls should take bath regularly (at least once everyday). Though having bath is good for everyone, it is more necessary for adolescents (or teenagers) because the increased activity of sweat glands and oil glands sometimes makes the body smelly. Bathing removes the sweat, oil and dirt, etc., and cleans the body. During bathing, all parts of the body should be washed and cleaned everyday.

(*ii*) Adolescent girls should take special care of cleanliness of the body during the time of menstrual flow (or periods). The girls should keep a track of their menstrual cycle and be prepared for the onset of menstruation.

If personal hygiene (or cleanliness of body) is not maintained by adolescent boys and girls, there are chances of catching bacterial infections. These infections can make a person ill and spoil good health.

3. Physical Exercise for Adolescents

Physical exercise is an activity requiring 'physical effort' which is carried out for the sake of health and fitness. **All the adolescent boys and girls should do physical exercise such as brisk walking, jogging (running), swimming, cycling, dancing, playing outdoor games (like hockey, football, badminton, basketball, etc.) or any other type of exercise, regularly.** Regular physical exercise has the following beneficial effects :

(*i*) Regular physical exercise in fresh air keeps the body fit.

(*ii*) Regular physical exercise also improves the mental health.

(*iii*) Physical exercise protects a person from heart disease, high blood pressure, diabetes and obesity.

(*iv*) Physical exercise builds and maintains healthy muscles, bones and joints.

(*v*) Physical exercise increases our efficiency in studies, sports and work.

(*vi*) Physical exercise improves the general sense of well-being and makes a person feel happier.

4. No Drugs for Adolescents

Drugs are chemical substances which when taken into the body change the functions of the body, influence the mind and sometimes even change the behaviour of the person. Some of the examples of drugs are heroin, cocaine, alcohol, paracetamol and penicillin. Drugs can be swallowed, inhaled or injected into the body.

Adolescence is a period of much activity in the body and mind of young boys and girls. Sometimes an adolescent boy or girl undergoes emotional swings such as feeling confused, insecure, bored, worried, stressed out, tense or angry, etc. We should not worry about such things because these emotional swings are a normal part of the growing up process, and disappear automatically with time. If somebody suggests that you will get relief from confusion, tension, boredom and worries, etc., and feel better by taking some drugs, just say 'No' (unless the drugs have been prescribed by a doctor). Drugs are very powerful chemical substances and should be used only under a doctor's supervision. Adolescent boys and girls should avoid drugs to maintain physical, mental and social well-being which are necessary to live a purposeful, fruitful and satisfying life in this world. Some of the **harmful effects of taking drugs** are as follows :

(*i*) Drugs are addictive. If you take drugs once, you feel like taking them again and again. Soon the person feels he (or she) cannot live without drugs and becomes a drug addict.

(*ii*) Drug addicts become irritable and lose interest in their studies or jobs. They may drop out of school or college, or lose job.

(*iii*) Drugs do physical harm to the body. Drugs can damage brain, liver, lungs and kidney. In this way, drugs ruin health and happiness.

(*iv*) People taking drugs run the risk of accidents because they get confused.

(*v*) The sharing of syringes for injecting drugs spreads AIDS disease among the drug addicts.

AIDS stands for Acquired Immune Deficiency Syndrome. **AIDS is a dangerous disease which is caused by a virus called HIV** (HIV stands for Human Immunodeficiency Virus). This virus is present in the blood of AIDS patients. The AIDS disease-causing virus (HIV) can be easily passed on from an infected person to a healthy person in a number of ways given below :

(*i*) AIDS virus (HIV) can pass from an infected person to a healthy person by the sharing of syringes used for injecting drugs.

If a drug addict has AIDS virus in his blood and he uses a syringe for injecting drugs into his body, then the needle of this syringe will get contaminated with AIDS virus (see Figure 10). When this syringe with contaminated needle is then used by a healthy person for injecting drugs, the AIDS virus present on the injection needle enters into his blood. In this way, the healthy person injecting drugs into his body will also get AIDS disease.

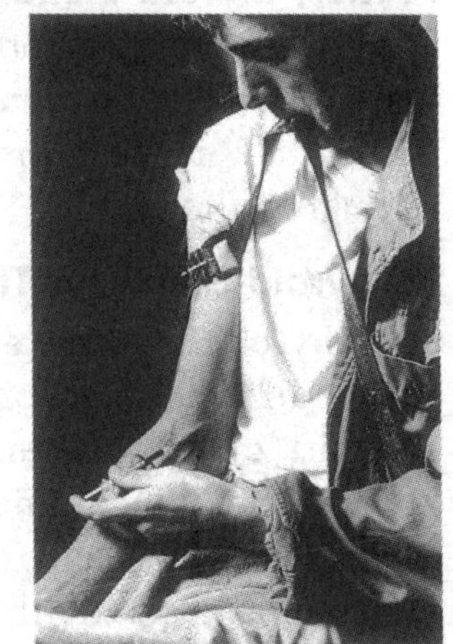

Figure 10. A drug addict injecting himself with a drug. This is the beginning of a life of misery and probably an early death.

(*ii*) AIDS virus (HIV) can be transmitted to a healthy person through sexual contact with a person infected with HIV.

(*iii*) AIDS virus (HIV) can be transmitted to an infant (small baby) from the infected mother through her milk.

AIDS is mainly a sexually transmitted disease. **There is no cure for AIDS disease**. AIDS patients die easily even from simple diseases because AIDS virus weakens the immunity of their body due to which their body cannot fight disease-causing germs.

HORMONES OTHER THAN SEX HORMONES

We have just discussed two sex hormones called testosterone and estrogen. Testosterone hormone is secreted by endocrine glands called 'testes' (which are found only in males) whereas estrogen hormone is secreted by endocrine glands called 'ovaries' (which are found only in females). There are many hormones other than sex hormones. **Some of the important hormones (other than sex hormones) are : Growth hormones, Thyroxine, Insulin and Adrenaline**. Growth hormone is secreted by pituitary gland ; thyroxine hormone is secreted by thyroid gland ; insulin hormone is secreted by pancreas; whereas adrenaline hormone is secreted by adrenal glands. Pituitary gland, thyroid gland, pancreas and adrenal glands are found in both, males as well as females. The positions of major endocrine glands in the human body are shown in Figure 11. We will now describe the functions of the hormones (other than sex hormones) secreted by various endocrine glands.

1. Pituitary Gland (or Pituitary)

Pituitary is an endocrine gland. **Pituitary gland is attached to the base of the brain** (see Figure 11). The pituitary gland secretes a number of hormones. **One of the hormones secreted by pituitary gland is the growth hormone** (or human growth hormone). The growth hormone controls the growth of the human body (like the growth of bones and muscles). Proper amount of growth hormone is necessary for the normal growth of the body. **A person having the deficiency of growth hormone in childhood remains very short and becomes a dwarf.** On the other hand, a **person having too much growth hormone becomes very tall (or a giant).**

Pituitary gland is the most important endocrine gland in the body. **Pituitary gland is called 'master gland'** because many of the hormones which it secretes control the functioning of other endocrine glands in the body. For example, **pituitary gland secretes hormones that make other endocrine glands such as testes, ovaries, thyroid gland, and adrenal glands to secrete their hormones**. In other words, the testes, ovaries, thyroid gland and adrenal glands secrete their hormones when they receive orders from the pituitary gland through its hormones.

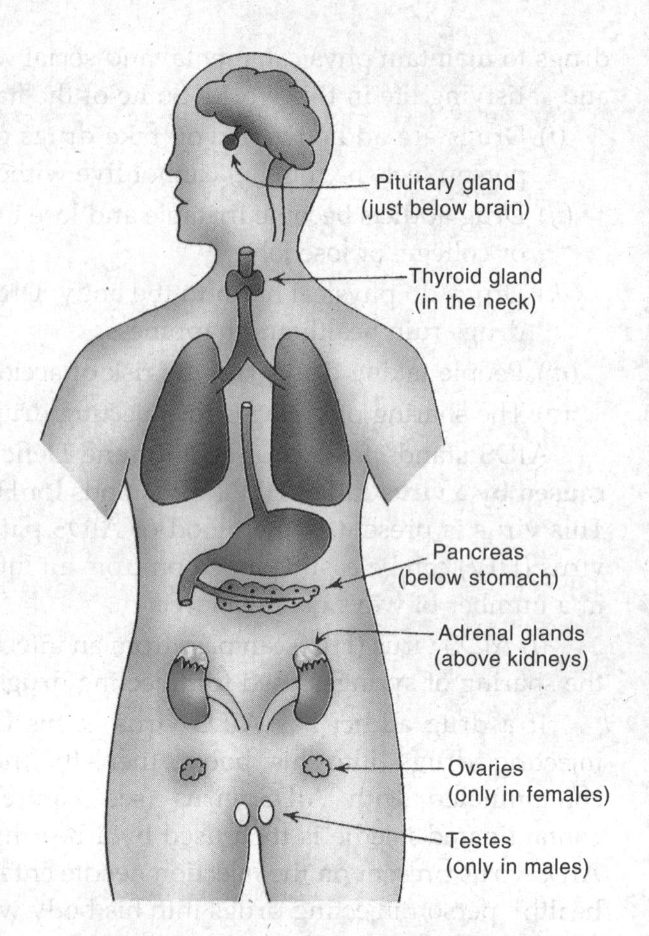

Figure 11. The positions of major endocrine glands in the human body.

2. Thyroid Gland (or Thyroid)

Thyroid is a large endocrine gland in the neck. It is attached to the wind pipe in our body (see Figure 11). **Thyroid gland makes a hormone called thyroxine** (which contains iodine). Thyroid gland secretes its hormone (thyroxine) when it receives instructions from the pituitary gland through its hormones. Thyroxine hormone controls the rate of body's metabolism (the chemical processes which occur continuously in human body to maintain life). *Thyroid gland needs a constant supply of iodine in order to produce thyroxine hormone.* This iodine normally comes from the diet we eat. Since iodine is necessary for the making of thyroxine hormone, therefore, *a deficiency of iodine in the diet can cause a deficiency of thyroxine hormone in the body.*

The deficiency of thyroxine hormone causes a disease known as goitre. Since the deficiency of thyroxine hormone is caused by a deficiency of iodine in the diet, so we can also say that *the deficiency of iodine in the body causes a disease known as goitre.* **The main symptom of goitre disease is that the neck of the person suffering from goitre appears to be swollen** (see Figure 12). The person appears to have a big and bulging neck. Goitre is a disease of the thyroid gland. If people do not have enough iodine in their diet, they will get goitre disease caused by the lack of thyroxine hormone in the body. **People are advised to use iodised salt for cooking food so as to prevent goitre disease.** Iodised salt contains appropriate amount of iodine compounds (such as potassium iodide). Iodised salt can provide all the iodine needed by thyroid gland to make sufficient thyroxine hormone for our body. Since there will be no deficiency of thyroxine hormone in the body, goitre cannot develop. So, *the importance of consuming iodised salt is that it prevents goitre disease.*

Figure 12. A woman suffering from thyroid disease called 'goitre'. Please note the swollen neck of the woman.

3. Pancreas

Pancreas acts as an endocrine gland as well as an exocrine gland. **Pancreas is just below the stomach in our body** (see Figure 11). Pancreas secretes the hormone called insulin. **The function of insulin hormone is to lower the blood sugar level (or blood glucose level).** Deficiency of insulin hormone in the body causes a disease known as diabetes. Diabetes disease is characterised by large quantities of sugar in the blood (and even urine). The insulin hormone controls the metabolism of sugar. If, due to some reason, pancreas does not produce and secrete sufficient amount of insulin into blood, then the sugar level in the blood rises. The high sugar level in the blood can cause many harmful effects to the body of a person. The person having high sugar level in blood (or diabetes) is called a diabetic. Diabetic persons are advised by doctors to take less sugar in their diet. Common diabetes can be controlled by controlling diet, reducing weight, doing regular physical exercise and taking medicines. The persons having severe diabetes are treated by giving injections of insulin.

4. Adrenal Glands (or Adrenals)

Adrenals are endocrine glands. There are two adrenal glands in our body which are located on the top of two kidneys (see Figure 11). **The adrenal glands produce adrenaline hormone** (Adrenaline is also written as adrenalin). **The adrenaline hormone prepares our body to function at maximum efficiency during emergency situations (like danger, fear, shock, surprise, anger or excitement, etc.)** This happens as follows : When we are faced with a dangerous situation (say, like being chased by a ferocious dog), then the adrenal glands secrete more adrenaline hormone into our blood. This adrenaline hormone speeds up the heartbeat, increases breathing rate, raises blood pressure, and causes liver to release more stored glucose into the blood. ***All these actions of adrenaline hormone produce a lot of energy in our body very, very quickly.*** And this energy helps us to run away very fast from the dog to save ourselves. In this way, the adrenaline hormone prepares our body to function at maximum efficiency during the emergency situation of being chased by a ferocious dog. It is also the adrenaline hormone which prepares our body to fight an enemy (say, a burglar in our house) by providing a lot of energy in a very short time. A lot of adrenaline hormone is also secreted by adrenal glands when we are angry or excited. The rapid output of energy thus caused helps us to cope with these extreme emotional situations.

The adrenal glands also produce and secrete another hormone called 'aldosterone'. **The aldosterone hormone secreted by adrenal glands maintains the correct salt balance in the blood**. The adrenals (or adrenal glands) secrete their hormones when they receive instructions from the pituitary gland through its hormones. We will now discuss the role of hormones in completing the life history of frogs and insects.

ROLE OF HORMONES IN COMPLETING THE LIFE HISTORY OF FROGS AND INSECTS

We have already studied the life history of frog. In a frog, the tadpole (or larva) hatched from the eggs passes through certain stages to become a frog. The change from tadpole (or larva) to the adult frog is called metamorphosis. **In a frog, metamorphosis is brought about by thyroxine hormone** (produced by the thyroid gland). The production of thyroxine hormone requires the presence of iodine in water. So, if the pond water in which the tadpoles are growing does not contain sufficient iodine to make enough thyroxine hormone, there will be deficiency of thyroxine hormone due to which tadpoles cannot undergo metamorphosis and hence cannot become adult frogs. It is not only the frogs which require thyroxine hormone to complete their life history, in fact, **all the amphibians need thyroxine hormone to undergo metamorphosis and change from larvae into adults**.

We have also studied the life history of a silk moth. The caterpillar (or larva) has to pass through various stages and undergo metamorphosis to become an adult silk moth. Silk moth is a kind of insect. There are a large number of other insects too. All the insects undergo metamorphosis to change from larvae to adult forms. **The process of metamorphosis in insects (such as silk moth) is controlled by insect**

hormones. In other words, the change from larva to adult insects during metamorphosis is brought about by insect hormones (which are made in their endocrine glands). We are now in a position **to answer the following questions :**

Very Short Answer Type Questions

1. What is the special name of the period of life between childhood and adulthood ?
2. Name another term for 'adolescents'.
3. State the age at which adolescence usually begins and the age up to which it lasts.
4. Which is the most conspicuous change (clearly visible change) in boys and girls during puberty ?
5. How does the height of boys and girls change during puberty ?
6. Name one factor on which the height of a person depends.
7. A nine year old boy is 120 cm tall. If the present height of the boy is 75% of his full height, calculate the full height which the boy will eventually reach at the end of growth period.
8. How do shoulders and chest change in boys during puberty ?
9. How do hips of girls change during puberty ?
10. What is the other name of 'voice box' ?
11. Which of the two have a smaller voice box : grown up boys or grown up girls ?
12. What is the common name of the bulge (or projection) at the front of throat in grown up boys ?
13. What substance is secreted by sebacious glands ?
14. Name the time period in one's life when the brain has the greatest capacity for learning.
15. Name two glands present in the skin whose increased activity causes pimples and acne.
16. What name is given to those sexual characteristics in humans :
    (*a*) which are present in babies at birth ?
    (*b*) which develop in mature boys and girls ?
17. Name any two glands which function as endocrine glands as well as exocrine glands.
18. What is the other name of ductless glands ?
19. What term is used for the secretions of endocrine glands ?
20. Name : (*a*) female sex hormone, and (*b*) male sex hormone.
21. Name the hormone which develops secondary sexual characteristics :
    (*a*) in females (girls).
    (*b*) in males (boys).
22. State the function of male sex hormone 'testosterone'.
23. Name the endocrine gland which controls the production of sex hormones 'testosterone' and 'estrogen'.
24. What is the name of the process in which the thickened uterus lining alongwith its blood vessels is removed from the body of a woman through vaginal bleeding ?
25. Which comes earlier in the life of a woman : menarche or menopause ?
26. (*a*) Around which age menarche occurs in girls ?
    (*b*) Around which age menopause occurs in women ?
27. State one situation (other than menopause) when ovulation and menstruation stop in a woman.
28. Which of the two has a shorter reproductive phase of life : men or women ?
29. Which of the following combinations of sex chromosomes produces a male child (or boy) ?
    XX or XY
30. Which of the two, sperm or egg cell (ovum), decides the sex of the child ?
31. What is the legal age for the marriage of (*a*) boys, and (*b*) girls, in our country ?
32. Name a natural balanced food for infants.
33. During which time the adolescent girls should take special care of personal hygiene ?
34. Write the full name of (*i*) AIDS, and (*ii*) HIV.
35. Write the name of the virus which causes AIDS disease.
36. Name the endocrine gland attached to the base of brain.
37. Name the endocrine gland which secretes the growth hormone.
38. Name the substance which is needed continuously by thyroid gland to make thyroxine hormone.
39. State the main symptom of goitre.

40. Name the hormone whose deficiency in body causes goitre.
41. Name the gland which secretes insulin hormone.
42. Name the disease caused by the insufficient production of insulin hormone by pancreas.
43. Name the hormone which prepares our body for action to face emergency situations like danger, fear, anger or excitement.
44. Name the endocrine gland which secretes hormone that maintains correct salt balance in the blood.
45. Name a hormone (other than adrenalin) secreted by adrenals.
46. Name the hormone whose deficiency causes diabetes.
47. Name the hormone which is required for the metamorphosis of larvae (tadpoles) into adult frogs.
48. Name the hormone which brings about metamorphosis in silk moth and changes caterpillar (larva) into adult silk moth.
49. State whether the following statements are true or false :
(*a*) The estrogen hormone develops deeper voice in males at puberty.
(*b*) Menstruation stops in men permanently around the age of 45 to 50 years.
(*c*) Man (father) is responsible for the sex of the baby.
(*d*) The sex chromosomes in all the gametes (sperms) of a man are not identical.
(*e*) Half the female gametes contain X chromosomes and the other half contain Y chromosomes.
(*f*) It is the sperm which determines the sex of a child.
(*g*) Each normal cell in our body contains 23 chromosomes.
(*h*) A girl should not be allowed to work in the kitchen during menstruation.
(*i*) A woman (mother) is responsible for the sex of her child.
(*j*) Metamorphosis in insects (like silk moth) is brought about by thyroxine hormone.
50. Fill in the following blanks with suitable words :
(*a*) Sexual maturity is reached at ....
(*b*) The changes which occur in boys and girls at puberty are brought about by various.............
(*c*) Initially, girls grow...........than boys.
(*d*) The changes occurring in body and mind during adolescence are a natural part of .... up.
(*e*) The features which help us to distinguish between a mature boy and a girl are called............sexual characteristics.
(*f*) The onset of puberty is controlled by..............
(*g*) Menstrual cycle in women is controlled by..............
(*h*) Each normal body cell in humans contains..........pairs of chromosomes out of which ....pair is of sex chromosomes.
(*i*) X and Y chromosomes are called..............chromosomes.
(*j*) Early marriage and motherhood cause health problems in the ....and the...........
(*k*) The protective foods are.............and.............
(*l*) Thyroid and adrenals secrete their hormones when they receive instructions from...........gland through its hormones.
(*m*) Metamorphosis in amphibians is brought about by .... hormone.

Short Answer Type Questions

51. Define adolescence. State some of the changes which take place in boys and girls during adolescence.
52. What is puberty ? Who attains puberty at an earlier age in human beings : male (boy) or female (girl) ?
53. Write two changes that happen during puberty to : (*a*) both boys and girls (*b*) boys only (*c*) girls only.
54. (*a*) State two ways in which the body shape of boys changes during puberty.
(*b*) State two ways in which the body shape of girls changes during puberty.
55. State one change in the body shape of grown up boys and one change in the body shape of grown up girls which makes them look different.
56. (*a*) What is the effect of a larger voice box in grown up boys ?
(*b*) What is the effect of a smaller voice box in grown up girls ?
57. What is Adam's Apple ? Which of the two usually have an Adam's Apple : grown up boys or grown up girls ?

58. What change in the voice of boys takes place when they reach puberty ? What is the cause of this change ?
59. (*a*) Name two male sex organs and two female sex organs which develop completely at puberty.
(*b*) Name two things which may develop on the face of adolescent boys and girls during puberty due to the increased activity of sebacious glands and sweat glands in the skin.
60. (*a*) State two secondary sexual characteristics in mature boys and two in mature girls.
(*b*) Which of the following are secondary sexual characteristics ?
Ovaries, Moustache, Penis, Broad hips, Vagina, Beard, Breasts, Adam's Apple, Wide shoulders, Uterus, Testes, Deeper voice.
61. Which of the following are endocrine glands ?
Salivary gland, Thyroid gland, Adrenal gland, Sweat gland, Pituitary gland, Sebaceous gland (Oil gland)
62. (*a*) What name is given to the onset of menstruation in human females (or girls) ? At what age does this occur ?
(*b*) What name is given to the permanent stoppage of menstruation in women ? At what age does this occur ?
63. Who is responsible for the sex of the unborn child : father or mother ? Why ?
64. What will be the sex of the child born :
(*a*) if X chromosome carrying sperm fuses with an egg cell carrying X chromosome ?
(*b*) if Y chromosome carrying sperm fertilises an ovum containing X chromosome ?
65. What are hormones ? Where are hormones made in the human body ?
66. What are sex hormones ? Name two sex hormones.
67. Define (*i*) menarche, and (*ii*) menopause.
68. What is menstruation ? Explain.
69. What are sex chromosomes ? Name the two types of sex chromosomes.
70. Explain how sex is determined in the unborn baby.
71. What are the various ways in which AIDS virus (HIV) can be transmitted ?
72. Explain how, the use of drugs helps in spreading AIDS disease.
73. State the harmful effects of taking drugs.
74. Name one hormone secreted by pituitary gland. State the function of this hormone.
75. Explain why, people are advised to use iodised salt in cooking food.
76. What is goitre ? What causes goitre ?
77. Which hormone lowers the blood sugar in humans ? Name the gland which secretes this hormone.
78. Name the hormone secreted by pancreas. What is the function of this hormone ?
79. What is the function of adrenaline hormone in the body ?
80. Name six endocrine glands in the human body. Also name the hormones secreted by each one of these glands.
81. Where are the following glands located in the human body ?
(*i*) Pituitary (*ii*) Thyroid (*iii*) Pancreas (*iv*) Adrenals
82. What will happen if the water in which tadpoles are growing does not contain sufficient iodine ?
83. What is acne ?
84. What are pimples ? How are pimples formed ?
85. Name two food items each which provide mainly : (*a*) carbohydrates for energy (*b*) fats for energy (*c*) proteins for growth (*d*) vitamins and minerals for good health
86. Why is iron mineral needed by our body ? Name some of the iron-rich foods.
87. State any two practices which can be adopted by adolescents to maintain personal hygiene.
88. Why is it more necessary for adolescents to take bath regularly (at least once everyday) ?
89. What will happen if personal hygiene (cleanliness of body) is not maintained by adolescents ?
90. State the various ways in which early marriage and motherhood is harmful to the girls.

Long Answer Type Questions

91. (*a*) What changes take place in boys during puberty ?
(*b*) What changes take place in girls during puberty ?
92. (*a*) What is meant by the 'primary sexual characteristics' in humans ? State two primary sexual characteristics in boys and two in girls.

(*b*) What is meant by 'secondary sexual characteristics' in humans ? Explain with the help of two examples.

93. Describe menstrual cycle. What is the duration of menstrual cycle in women ?

94. (*a*) What is an exocrine gland ? Name two exocrine glands.
(*b*) What is an endocrine gland ? Name two endocrine glands.

95. (*a*) What is meant by 'reproductive health' ?
(*b*) State the conditions necessary to maintain good reproductive health during adolescence.

Multiple Choice Questions (MCQs)

96. Reproductive age in women starts when their :
(*a*) menstruation starts (*b*) breasts start developing
(*c*) body weight increases (*d*) height increases

97. The sex of a child is determined by :
(*a*) the presence of an X chromosome in egg (or ovum).
(*b*) the presence of a Y chromosome in sperm.
(*c*) the age of father and mother.
(*d*) the length of the mother's pregnancy.

98. The legal age for the marriage of boys in our country is :
(*a*) 16 years (*b*) 18 years (*c*) 21 years (*d*) 24 years

99. The legal age for the marriage of girls in our country is :
(*a*) 16 years (*b*) 18 years (*c*) 21 years (*d*) 24 years

100. The right meal for adolescents consists of :
(*a*) chips, noodles, coke (*b*) *chapati*, *dal*, vegetables
(*c*) rice, noodles and burger (*d*) vegetable cutlets, chips, and lemon drink.

101. Adolescents should be careful about what they eat because :
(*a*) proper diet develops their brains.
(*b*) proper diet is needed for the rapid growth taking place in their body.
(*c*) adolescents feel hungry all the time.
(*d*) taste buds are well developed in teenagers.

102. AIDS disease is caused by :
(*a*) bacteria (*b*) virus (*c*) worms (*d*) protozoa

103. Which of the following human disease can be prevented by the same hormone which brings about metamorphosis in frogs ?
(*a*) diabetes (*b*) anaemia (*c*) goitre (*d*) rickets

104. Which of the following is a mis-matched pair ?
(*a*) adrenaline : pituitary gland (*b*) estrogen : ovary
(*c*) pancreas : insulin (*d*) testosterone : testis

105. The hormone which is associated with male puberty is called :
(*a*) oestrogen (*b*) adrenaline (*c*) testosterone (*d*) progesterone

106. The dramatic changes in body features associated with puberty are mainly because of the secretions of :
A. Thyroxine B. Estrogen C. Adrenaline D. Testosterone
(*a*) A and B (*b*) B and C (*c*) A and C (*d*) B and D

107. Which of the following endocrine gland does not occur as a pair in the human body ?
(*a*) adrenal (*b*) pituitary (*c*) testis (*d*) ovary

108. Iodine is necessary for the synthesis of one of the following hormones. This hormone is :
(*a*) adrenaline (*b*) auxin (*c*) insulin (*d*) thyroxine

109. Which of the following is not a primary sexual characteristic in females ?
(*a*) fallopian tubes (*b*) ovaries (*c*) seminal vesicles (*d*) vagina

110. Which of the following is the secondary sexual characteristic in females ?
(*a*) low pitched voice (*b*) broadening of chest
(*c*) broadening of hip (*d*) developing Adam's apple

111. Which of the following can lead to menstruation in a 21 year old woman during ovulation ?
A. Sperms available for fertilisation B. Oviducts blocked

C. Sperms not available for fertilisation D. Oviducts not blocked
(*a*) A and B (*b*) B and C (*c*) A and C (*d*) B and D

112. The faulty functioning of an endocrine gland can make a person very short or very tall. This gland is :
(*a*) thyroid (*b*) pituitary (*c*) adrenal (*d*) pancreas

113. Which of the following hormone prepares our body for action in emergency situations ?
(*a*) testosterone (*b*) insulin (*c*) thyroxine (*d*) adrenaline

114. A doctor advised a person to take injection of insulin because :
(*a*) his blood pressure was high (*b*) his heart beat was high
(*c*) his blood sugar was high (*d*) his thyroxine level in blood was high

115. Pimples and acne are fomed due to the increased activity of :
A. Adrenal glands B. Sebaceous glands C. Thyroid gland D. Sweat glands
(*a*) A and B (*b*) B and C (*c*) A and C (*d*) B and D

Questions Based on High Order Thinking Skills (HOTS)

116. When a human female reaches a certain age, then vaginal bleeding occurs for a few days after regular intervals of time :
(*a*) What is this process known as (*i*) in scientific terms, and (*ii*) in everyday language ?
(*b*) What is the human female said to have attained at this stage ?
(*c*) What does the onset of this process in human female signify ?
(*d*) At what particular event in the life of a human female this process stops temporarily but starts again ?
(*e*) What name is given to the event when this process stops permanently ?

117. A woman is in her reproductive phase of life. Even when the sperms are available at the time of release of a mature ovum by her ovary, fertilisation does not take place. Due to this, the woman is unable to bear a baby.
(*a*) What is the most likely defect in the reproductive system of this woman ?
(*b*) Name the technique by which this woman can also have a baby.

118. Raj Kumar and his wife Sunita have been blessed with a baby girl. On the other hand, Kripa Shanker and his wife Vimla have been blessed with a baby boy. What type of sex chromosome has been contributed :
(*a*) by Raj Kumar (*b*) by Sunita (*c*) by Kirpa Shanker (*d*) by Vimla

119. A gland W is located just below the stomach in the human body. The gland W secrets a hormone X. The deficiency of hormone X in the body causes a disease Y in which the blood sugar level of a person rises too much. The person having high blood sugar is called Z. What are W, X, Y and Z ?

120. It was observed that the tadpoles are not growing into adult frogs in a village pond. A scientist who was asked to find the reason for this concluded that the pond water did not contain sufficient amount of a mineral P which could make a hormone Q. The hormone Q can cause a phenomenon R in tadpoles so that they become adult frogs.
(*a*) What are (*i*) P (*ii*) Q, and (*iii*) R ?
(*b*) Name one human disease which can also be caused due to the deficiency of P in the food and water.

ANSWERS

2. Teenagers **7.** 160 cm **25.** Menarche **27.** Pregnancy **28.** Women **29.** XY **30.** Sperm **32.** Mother's milk **49.** (*a*) False (*b*) False (*c*) True (*d*) True (*e*) False (*f*) True (*g*) False (*h*) False (*i*) False (*j*) False **50.** (*a*) puberty (*b*) hormones (*c*) faster (*d*) growing (*e*) secondary (*f*) hormones (*g*) hormones (*h*) 23 ; 1 (*i*) sex (*j*) mother ; child (*k*) fruits ; vegetables (*l*) pituitary (*m*) thyroxine **64.** (*a*) Female (Girl) (*b*) Male (Boy) **96.** (*a*) **97.** (*b*) **98.** (*c*) **99.** (*b*) **100.** (*b*) **101.** (*b*) **102.** (*b*) **103.** (*c*) **104.** (*a*) **105.** (*c*) **106.** (*d*) **107.** (*b*) **108.** (*d*) **109.** (*c*) **110.** (*c*) **111.** (*b*) **112.** (*b*) **113.** (*d*) **114.** (*c*) **115.** (*d*) **116.** (*a*) (*i*) Menstruation (*ii*) Periods (*b*) Puberty (*c*) That the reproductive system of human female has started working (*d*) Beginning of pregnancy (*e*) Menopause **117.** (*a*) The oviducts of woman are blocked (due to which the ovum released by her ovary cannot meet the sperm and get fertilised) (*b*) *In Vitro* Fertilisation (IVF) **118.** (*a*) X (*b*) X (*c*) Y (*d*) X **119.** W : Pancreas ; X : Insulin ; Y : Diabetes ; Z : Diabetic **120.** (*a*) (*i*) Iodine (*ii*) Thyroxine (*iii*) Metamorphosis (*b*) Goitre

CHAPTER 11

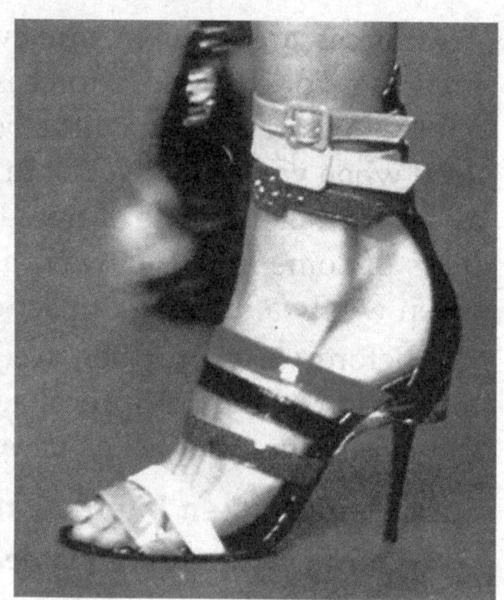

Force and Pressure

When an athlete wants to throw a very heavy ball during shot put contest so that it may go as far as possible, he has to *push* the ball hard. We say that the athlete applies a large *force* to the ball to make it move through a large distance. When a person wants to draw water from a well, he has to *pull* the rope attached to the bucket full of water. We again say that the person applies a *force* to the rope attached to the bucket of water to make it move and come out of the well. And if, by chance, the sharp heel of a woman's shoe falls on our foot, it hurts our foot too much and produces a lot of pain. We say that the force of woman's weight produces a large *pressure* on our foot because all her weight falls on a small area of our foot due to sharp heel. It is this large pressure on our foot which hurts the foot. On the other hand, if the broad heel of a man's shoe falls on our foot, it hurts our foot much less. This is because due to broad heel, the force of man's weight falls on a large area of our foot due to which the *pressure* produced on our foot is much less. In this Chapter we will discuss the various types of forces and the effects produced by these forces. We will also study the pressure exerted by solids, liquids and gases, including atmospheric pressure. Let us start by discussing force.

FORCE – A PUSH OR A PULL

When we want to open a door, we have to push the door handle with our hand. And when we want to close the door, we have to pull the door handle with our hand. This means that to move an object, it has either to be pushed or pulled. **A push or pull on an object is called force** (see Figure 1). The direction in which the object is pushed or pulled is called the direction of force. We open or close a door by applying force. When we push the door to open it, then we apply a force on the door in a direction away from us. And when we pull the door to close it, then we exert a force on the door in a direction towards us.

Figure 1. A push or pull on an object is called force.

Forces are used in our everyday actions like pushing, pulling, lifting, stretching, twisting and pressing. For example, a force is used when we push (kick) a football; a force is used when we pull a door; a force is used when we lift a box from the floor; a force is used when we stretch a rubber band; a force is used when we twist a wet cloth to squeeze out water, and a force is used when we press the brake pedal of a car. The fallen leaves of trees fly away with wind because the force of wind pushes them away. Even the roofs of some huts fly away during a storm because the force of strong winds pushes them away. And when we fly a kite, we can actually feel the force (or push) of the wind on it.

Before we go further, we should know the meaning of the term 'interaction' as used in physics. Interaction means 'reciprocal action'. Interaction involves two objects. *In interaction, each object acts in such a way as to have an effect on the other object*. In most simple words, interaction means to act on each other.

Force is Due to an Interaction

An interaction of one object with another object results in a force between the two objects. In other words, a force arises due to the interaction between two objects. At least two objects must interact with each other for a force to come into play (and show its effect). If there is no interaction between two objects, no force can show its effect. This will become more clear from the following examples.

Suppose a man is standing behind a stationary car [see Figure 2(*a*)]. Since there is no interaction between the man and the car, no force acts on the car and hence the car does not move. Now, suppose the

(*a*) A man standing behind a stationary car. No interaction : no force acts on the car

(*b*) The man pushing the car. Interaction occurs : a force acts on the car and makes it move

Figure 2.

man pushes the car with his hands due to which the car starts moving [see Figure 2(*b*)]. When the man pushes the car, there is an interaction between the man and the car. During this interaction, a force arises which acts on the car and makes it move in the direction of applied force. Please note that the man has to push the car to make it move. Here the two objects (or things) which are interacting for the force to come into play and show its effect are the 'man' and the 'car'. In the above example of a stationary car and man only man is capable of applying force to the stationary car.

If both the objects are capable to applying force on each other, then the interaction between them can be of 'pushing' or 'pulling'. For example, the two boys shown in Figure 3 are interacting and applying force on each other by pushing each other. On the other hand, the man and cow shown in Figure 4 are also

Figure 3. Two boys are interacting and applying force on each other by pushing each other.

Figure 4. The man and cow are interacting and applying force on each other by pulling each other.

interacting but they are applying force on each other by pulling each other. In this case, the man is pulling cow, and the cow is also pulling the man.

Force Has Magnitude as Well as Direction

An adult man can apply large force on an object whereas a child can apply only a small force on an object. This means that a force can be larger or smaller than the other force. **The strength of a force is expressed by its magnitude.** The magnitude of a force is expressed in the SI unit of force called **'newton'** (whose abbreviation is N). **1 newton is the force which can make an object of 1 kilogram mass to move at a speed of 1 metre per second.** Alongwith the magnitude of force, the direction in which a force acts is also to be specified (or taken into account). This is because if the magnitude or direction of the applied force changes, the effect produced by force also changes. When two forces act on an object, then two cases arise : either the forces act in the *same direction* or the forces act in *opposite directions*.

***(i)* If the two forces applied to an object act in the same direction, then the resultant force acting on the object is equal to the sum of the two forces.** In other words, when two forces act in the same direction, their effective magnitude increases. This will become clear from the following example. Suppose there is a heavy box which one man can move only by pushing it very hard. Now, if two men push this heavy box in the same direction, it becomes much easier to move the heavy box (see Figure 5). This is because when the two men apply their forces of push together in the same direction, the two forces add up to provide a much bigger force. And this bigger force can move the heavy box very easily.

Figure 5. Two men pushing a heavy box in the same direction and making it move easily.

Figure 6. Two men pushing the heavy box in opposite directions. The box will move in that direction in which stronger push is applied.

***(ii)* If the two forces applied to an object act in the opposite directions, then the net force acting on the object is equal to the difference between the two forces.** In other words, when the two forces act in opposite directions, their effective magnitude decreases. This will become more clear from the following example. Suppose there is a heavy box lying on the ground. Let the two men push this box from opposite directions (one from the left side and the other from the right side) (see Figure 6). Assuming that one of the men is stronger of the two and applies a larger pushing force than the other man, we can say that the box will move in that direction in which a larger force is applied by the stronger man. The box will, however, move very slowly in this case because the net force acting on the box in this case is equal to the difference in the magnitudes of the two forces applied by the two men. And this net force is small.

***(iii)* If the two forces applied to an object are equal in magnitude and act in opposite directions, then the net force acting on the object is zero (or nil).** Since the net force acting on the object is zero (or nil) in this case, the object does not move at all, it remains in the same position. For example, if the two men push a heavy box from opposite directions by applying exactly equal forces, then the heavy box will not move at all. It will

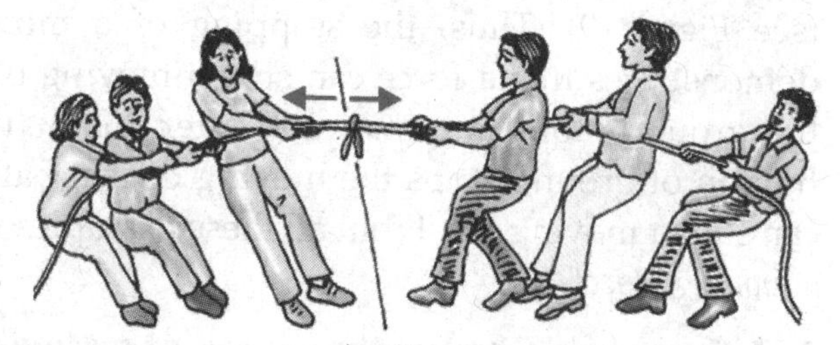

Figure 7. The rope does not move to either side when both the teams pull at it with equal and opposite forces.

remain where it was. A familiar example of this case of equal forces acting in opposite directions is provided by the game called 'tug of war'. In this game, two teams of persons pull at a rope in opposite directions (see Figure 7). Members of both the teams try to pull rope in their direction. When the two teams pull the rope equally hard (by applying equal forces), the rope does not move in any direction (see Figure 7). At this point of time, the two teams are applying equal but opposite forces to the rope due to which the net force acting on the rope is zero and hence the rope does not move. Actually, the equal and opposite forces applied on the rope cancel each other's effect. Ultimately, the team which pulls the rope harder (by applying greater force) wins the game.

EFFECTS OF FORCE

A force cannot be seen. A force can be judged only by the effects it can produce in various objects around us. **A force can produce the following effects :**

(i) A force can move a stationary object.
(ii) A force can stop a moving object.
(iii) A force can change the speed of a moving object.
(iv) A force can change the direction of a moving object.
(v) A force can change the shape (and size) of an object.

We will now give examples of all these effects produced by a force when it acts on objects.

1. A Force can Move a Stationary Object

Take a rubber ball and place it on a table top. Now gently push the ball along the surface of table. We will observe that the ball begins to move. Thus, a ball at rest (or stationary ball) begins to move when a force (of push) is applied to it. While taking penalty kick in a football match, the player applies a force on the stationary football. Before being hit, the football is at rest, and its speed is zero. The force applied by the player makes the football move towards the goal. Thus, if we kick a stationary football kept on the ground, then the football starts moving (see Figure 8). In this case, the force of our foot makes the stationary football move. Similarly, the force of engine makes a stationary car to move. From these examples we conclude that *a force can move a stationary object*.

Figure 8. The force of kick makes a stationary football move.

2. A Force can Stop a Moving Object

Take a rubber ball and place it on a table. Push the rubber ball gently so that it starts moving. Now, place your palm in front of the moving ball. We will observe that the moving ball comes to a stop. Actually, the palm held in the path of moving rubber ball applies a force to the moving ball. This force stops the moving ball. Thus, a moving ball stops when a force is applied to it. In a football match, when the goalkeeper dives or jumps up to save the goal, he applies a force to the moving football with his hands. This force applied by goalkeeper helps in stopping the moving football and saves a goal being scored (see Figure 9). Thus, the stopping of a moving football by a goalkeeper demonstrates that a force can stop a moving object. A cricket ball moving on the ground stops automatically after some time. In this case, the force of friction of ground stops the moving cricket ball. Similarly, the force of brakes can stop a moving car. From all these examples we conclude that *a force can stop a moving object*.

Figure 9. The force applied by goalkeeper makes the moving football stop.

3. A Force can Change the Speed of a Moving Object

Suppose we are moving on a bicycle at a certain speed. Now, if someone pushes the moving bicycle from behind, then the speed of bicycle increases and it will move faster. On the other hand, if someone pulls the moving bicycle from behind, then the speed of bicycle decreases and it will move slower. Thus, a push or pull can change the speed of a moving bicycle. But a push or pull is called force. So, we can say that **a force can change the speed of a moving bicycle (or any other moving object).** *If the force is applied in the direction of motion of the object, its speed increases.* On the other hand, *if the force is applied in the direction opposite to the direction of motion of an object, then its speed decreases.*

When a ball is dropped from a height, its speed goes on increasing. The speed of a falling object (like a ball) increases because the earth applies a pulling force on it which is called the force of gravity. It is the force of gravity of the earth which pulls a falling object towards its centre and increases its speed. On the other hand, when a ball is thrown upwards, then its speed goes on decreasing. This is because the earth applies a pulling force of gravity on the ball in the downward direction (opposite to the motion of the ball). When a hockey player hits a moving ball, the speed of ball increases (see Figure 10). When we pedal the bicycle faster, then the speed of bicycle increases. And when we apply brakes to the moving bicycle, then the speed of bicycle decreases. From all these examples we conclude that *a force can change the speed of a moving object.*

Figure 10. The force applied by a hockey player on a moving ball changes the speed of ball.

4. A Force can Change the Direction of a Moving Object

In a cricket match, when a moving cricket ball is hit by a bat, then the direction of cricket ball changes and it goes in another direction (see Figure 11). In this case, the force exerted by bat changes the direction of a moving cricket ball. In the game of carrom, when we take a rebound, then the direction of striker changes. This is because the edge of the carrom board exerts a force on the striker. If we blow air from our mouth on the smoke rising up from a burning *agarbatti*, then the direction of motion of smoke changes. In this case, the force exerted by the blowing air changes the direction of moving smoke. From these examples we conclude that *a force can change the direction of motion of a moving object.*

Figure 11. The force applied by a batsman changes the direction of moving cricket ball.

Before we go further, we should understand the meaning of the term 'state of motion'. The state of motion of an object is described by its speed and the direction of motion. A change in either the speed of an object, or the direction of its motion, or both, is called a change in its state of motion. **A force can change the state of motion of an object.**

5. A Force can Change the Shape and Size of an Object

If we take a light spring and pull it at both the ends with our hands, then the shape and size of the spring changes (see Figure 12). In this case, the force of our hands changes the shape and size of the spring (The turns of the spring become farther apart and its length increases). Here are some more examples in which a force changes the shape (or size) of an object. The shape of dough (kneaded wet flour) changes on pressing with a rolling pin (*belan*) to make *chapatis*. When we press the dough with a rolling pin, we apply force. So, we can say that the shape of dough changes on applying force. The shape of kneaded wet

(*a*) Original shape and size of spring

(*b*) Force (of pulling) changes the shape and size of the spring

Figure 12.

clay changes when a potter converts it into pots of different shapes and sizes. This happens because the potter applies force on the kneaded wet clay. The shape of a toothpaste tube (or an ointment tube) changes when we squeeze it because we apply force while squeezing it. When we hammer a piece of aluminium metal, its shape changes and an aluminium sheet is formed. This change in shape occurs because we apply force while hammering. When we sit on a sofa with springs, then the springs of the sofa get compressed and their shape and size changes. This happens because our weight applies a force on the springs and compresses them. Similarly, the shape of a sponge, tomato, balloon or tennis ball changes on pressing. And the shape and size of a rubber band changes on stretching. The shape and size of a balloon changes when it is filled with air (or water) because the weight of air (or water) exerts force on the walls of the balloon from inside. From these examples, we conclude that *a force can change the shape and size of an object.* We will now discuss the various types of forces.

TYPES OF FORCES

There are five common types of forces which we notice in our everyday life. These are :

(*i*) **Muscular force,**

(*ii*) **Frictional force (or Friction),**

(*iii*) **Magnetic force,**

(*iv*) **Electrostatic force,** and

(*v*) **Gravitational force.**

All type of forces can be divided into two main groups : Contact forces and Non-contact forces. Please note that the state of one object 'physically touching' another object is known as 'contact'. So, when one object physically touches another object, they are said to be 'in contact'. A force can act on an object with or without being in contact with it. We will now describe contact forces and non-contact forces in detail, one by one.

CONTACT FORCES

A force which can be exerted by an object on another object only through 'physical touching' is called a contact force. The examples of contact forces are :

(*i*) **Muscular force,** and

(*ii*) **Frictional force (or Friction).**

Let us now discuss muscular force and frictional force in detail, one by one.

1. Muscular Force

Suppose a book is lying on a table. To lift this book from the table, some force is required. When we lift this book from the table by hand, the force is exerted by the muscles of our arm. Similarly, when we kick a football, the force is exerted by the muscles of our leg. **The force exerted by the muscles of the body is called muscular force.** Both, human beings and animals exert muscular force to do work. The human beings exert muscular force for performing the various day to day activities like walking, running, jumping, climbing, lifting, pushing, pulling, kicking, stretching, squeezing, twisting and pressing the objects. A boy pushing a cart on a road, a porter carrying a load on a wheel-barrow, a child riding a bicycle and a person drawing water from a well are all examples of the use of muscular force. It is the muscular force which enables us to perform all the activities involving the movement or bending of our body parts. The muscular force also acts inside our body to perform various functions required for our survival. For example, in the process of digestion, it is the muscular force which pushes the food through the alimentary canal. The expansion and contraction of

Figure 13. This weightlifter is using muscular force to lift the weights.

lungs when we inhale and exhale air during breathing, also involves muscular force. And the beating of heart also takes place by the muscular force produced by cardiac muscles.

The animals like an ox, camel and horse provide the force of their muscles and help the man in doing harder jobs. For example, ox pulls the plough by using its muscular force. Similarly, a horse and camel exert their muscular force in pulling the carts. An elephant pulls heavy logs of wood by using the muscular force. And in arctic regions (snow-bound regions), reindeers pull the sledges (wheel-less vehicles) by using their muscular force.

Figure 14. The muscular force of horse pulls the cart.

Since muscular force can be applied to an object only when our body (or body of an animal) is in contact with the object, therefore, muscular force is a contact force. For example, when we lift a book from a table by applying muscular force, then our hand is in contact with the book. We cannot lift the book from table without touching it. Our contact (or that of an animal) with an object for applying muscular force can also be through a stick or a piece of rope, etc. For example, when we hit a hockey ball, we apply muscular force to ball through the hockey stick. And when we draw a bucket of water from a well, then we apply the muscular force through a rope (tied to the bucket).

2. Frictional Force (or Friction)

A ball moving on the ground slows down gradually and stops after covering some distance. We know that a force is required to stop a moving body. This means that a force is exerted by the ground on the moving ball which opposes its motion and brings it to a stop. This force which opposes the motion of ball on the ground is known as frictional force. We can now define the frictional force as follows : **The force which always opposes the motion of one body over another body is called frictional force (or friction).** The frictional force acts between the two surfaces which are in contact with each other. A ball moving on the ground slows down and then stops due to frictional force between the ball and the ground. In this case the two surfaces in contact are the surface of the ground and the surface of the ball. Frictional force is also known as 'force of friction' or just 'friction'. We will now give some more examples from everyday life where frictional force is involved.

(*i*) If we stop pedalling a running bicycle, it slows down gradually and stops after covering some distance. The bicycle moving on the road slows down and finally comes to a stop due to the frictional force between the tyres of the bicycle and the road. This frictional force opposes the motion of bicycle and brings it to a stop. In this case, the two surfaces in contact are the surface of the road and the surface of tyres of bicycle. Similarly, a car or scooter moving on the road comes to rest (or stops) due to frictional force when its engine is switched off.

(*ii*) When we stop rowing a boat which is moving in water, it slows down and then stops. The boat moving in water slows down and finally stops due to the frictional force between the boat and water. The frictional force exerted by water opposes the motion of boat and makes it stop. In this case the two surfaces in contact are the surface of boat and the surface of water.

The frictional force (or force of friction) always acts on all the moving objects, and its direction is always opposite to the direction of motion. **Since frictional force arises only when the surfaces of two objects are in touch with each other, therefore, frictional force is an example of a contact force.**

Please note that it is not necessary that the agent applying a force on an object should always be in contact with it. There are some forces which act on an object even from a distance. We will now discuss such non-contact forces.

NON-CONTACT FORCES

A force which can be exerted by an object on another object even from a distance (without touching each other) is called a non-contact force. There is no physical contact between the object which exerts the force and the object on which force is exerted. The examples of non-contact forces are :

(*i*) Magnetic force,
(*ii*) Electrostatic force, and
(*iii*) Gravitational force.

We will now describe the magnetic force, electrostatic force and gravitational force, one by one. Let us start with the magnetic force.

1. Magnetic Force

A magnet attracts things made of iron (or steel). So, if we bring a magnet near iron nails or pins, the magnet pulls them towards it (see Figure 15). Since the iron nails and pins move towards the magnet, it means that the magnet exerts a force on them. **The force exerted by a magnet is called magnetic force.** The magnetic force acts even from a distance. The magnet exerts a magnetic force on objects made of iron, steel, nickel and cobalt. It does not exert a force on objects made of copper, wood or glass. When we bring a magnet slowly near some iron pins, then at a certain small distance, the iron pins suddenly jump towards the magnet and stick to it (even when the magnet has not touched them). **Since a magnet can exert its magnetic force on iron objects from a distance (even without touching them), therefore, magnetic force is a non-contact force.** The magnetic force between a magnet and an iron (or steel) object is always that of *attraction*.

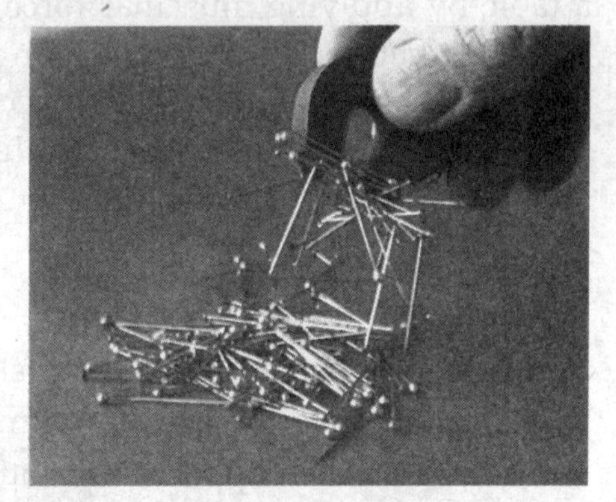

Figure 15. A magnet attracting pins made of iron.

A magnet also exerts a magnetic force on another magnet placed near it. The magnetic force between two magnets placed near one another can be that of *'attraction'* or *'repulsion'* depending upon which poles of the two magnets are facing one another. We will now describe the magnetic force between two magnets kept near each other. Before we do that please note that a north pole of magnet and another north pole are like poles. Similarly, a south pole and another south pole are like poles. On the other hand, a north pole and a south pole are unlike poles.

We have already studied in Class VI that the like poles of two magnets repel each other whereas unlike poles of magnets attract each other. This means that **there is a magnetic force of repulsion between the like poles of two magnets**. On the other hand, **there is a magnetic force of attraction between the unlike poles of two magnets**. We do not have to bring the two magnets in contact with each other for observing the magnetic force between them. A magnet can exert a force on another magnet even from a distance, without touching the other magnet (or without being in contact with the other magnet). So, **the magnetic force exerted by one magnet on the other magnet is an example of a non-contact force.** We will now perform an activity to show the magnetic force of attraction as well as repulsion between two bar magnets.

ACTIVITY

Take two similar bar magnets. Place one bar magnet over three wooden rollers (such as three round pencils) kept on a table as shown in Figure 16. The south pole of bar magnet kept on rollers is on the left side (see Figure 16). We hold the other bar magnet in our hand and bring its north pole gradually near the south pole of bar magnet placed on rollers (making sure that the two magnets do not touch each other).

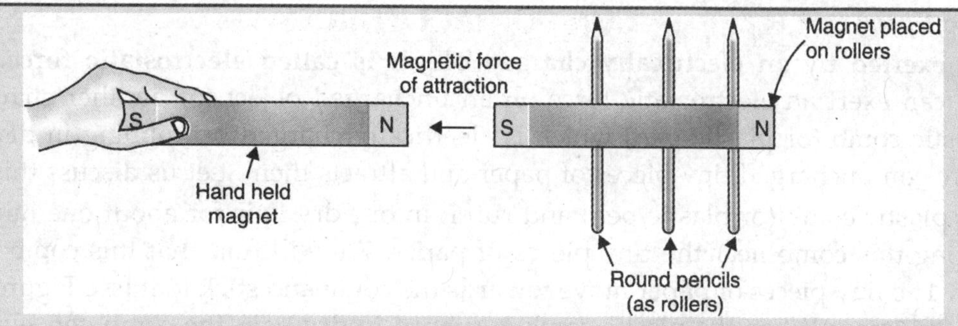

Figure 16. Observing the magnetic force of attraction between unlike poles (N-pole and S-pole) of two bar magnets.

We will observe that when the north pole of hand held magnet is brought near the south pole of other magnet, the magnet placed on rollers begins to move towards the hand held magnet) (see Figure 16). This shows that the north pole of hand held magnet is exerting a magnetic force of attraction on the south pole of magnet kept on rollers which makes it move towards the hand held magnet. Thus, unlike poles (N-pole and S-pole) of magnets exert a magnetic force of attraction.

Now bring the south pole of hand held magnet gradually near the south pole of magnet placed on rollers as shown in Figure 17. We will observe that when the south pole of hand held magnet is brought near

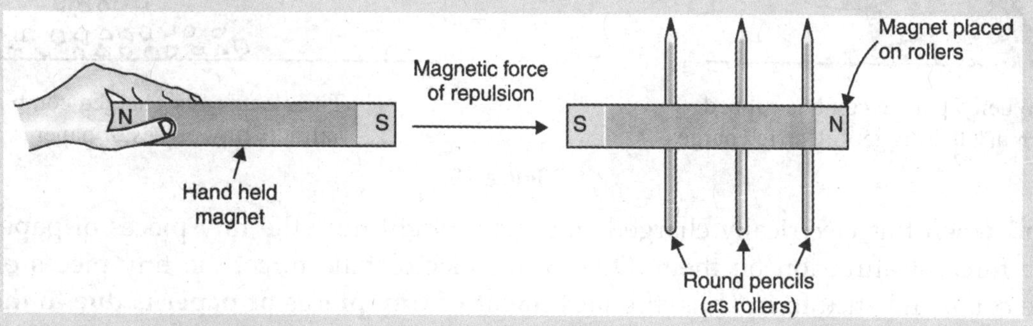

Figure 17. Observing the magnetic force of repulsion between like poles (S-pole and S-pole) of two bar magnets.

the south pole of other magnet, the magnet placed on rollers begins to move away from the hand held magnet (see Figure 17). This shows that the south pole of hand held magnet is exerting a magnetic force of repulsion on the south pole of magnet placed on rollers which makes it move away from the hand held magnet. Thus, like poles (S-pole and S-pole) of magnets exert a magnetic force of repulsion.

The magnetic force is widely used in our everyday life. For example, many toys work due to the magnetic force exerted by magnets fixed inside them. The magnetic strips help to keep the refrigerator door closed tightly. Certain stickers stick to the iron objects like steel almirahs, refrigerators, etc., without any glue. Such stickers contain tiny magnets and hence stick to the iron objects due to magnetic force. The magnetic force exerted by powerful electromagnets is used to separate scrap iron objects from among a heap of waste materials.

Before we go further and study electrostatic force, we should know the meaning of the terms 'electric charge', 'electrostatic' and 'electrically charged object'. The electric charges are of two types : positive electric charges and negative electric charges. Electric charges can be produced by friction when one object is rubbed over another object. The term 'electrostatic' refers to 'stationary electric charges'. An object having stationary electric charges (or static electric charges) on it is called electrically charged object. A comb is said to have acquired electrostatic charge after it has been rubbed with dry hair. Such a comb is an example of a charged object. An electrically charged object exerts an electrostatic force. An electrostatic force is actually an electrical force. It is called electrostatic force because the electric charges involved in it remain static (or stationary). The electric charges remain confined to the charged object and do not move.

2. Electrostatic Force

The force exerted by an electrically charged object is called electrostatic force. An electrically charged object can exert an electrostatic force on an uncharged object (or another charged object). For example, a plastic comb (or plastic pen) which is electrically charged by rubbing in dry hair, exerts an electrostatic force on uncharged tiny pieces of paper and attracts them. Let us discuss this in detail.

We take a plastic comb (or plastic pen) and rub it in our dry hair for about one minute [see Figure 18(*a*)]. Then bring this comb near the tiny pieces of paper. We will find that this comb attracts the tiny pieces of paper. The tiny pieces of paper move towards the comb and stick to it [see Figure 18(*b*)]. This can be explained as follows : When the plastic comb is rubbed in dry hair, the comb gets electric charges by

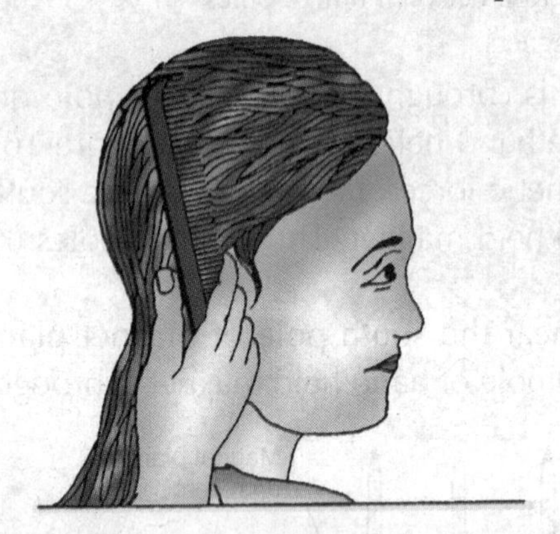

(*a*) When a plastic comb is rubbed in dry hair, it gets electric charges

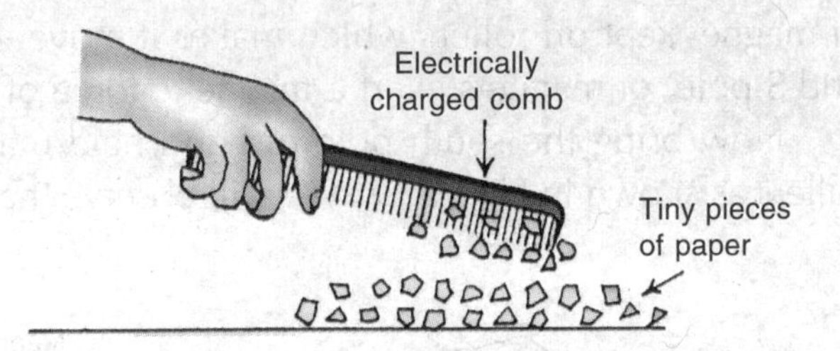

(*b*) The electrically charged comb attracts tiny pieces of paper

Figure 18.

friction. And when this electrically charged comb is brought near the tiny pieces of paper, it exerts an electrostatic force of attraction on them. Due to this electrostatic force, the tiny pieces of paper move towards the comb and stick to it. Thus, **the movement of tiny pieces of paper is due to the electrostatic force exerted by electric charges acquired by the comb on rubbing in hair.** It is the electrostatic force which is responsible for raising our body hair when we try to take off a terylene or polyester shirt in dry weather. If an inflated balloon is rubbed with a piece of synthetic cloth (like polyester or nylon cloth) and then pressed against a wall, the balloon sticks to the wall. It is the electrostatic force which is responsible for the attraction between the balloon and the wall. This happens as follows : When the balloon is rubbed with a piece of synthetic cloth, the balloon gets electrostatic charges and hence gets electrically charged. The electrically charged balloon attracts the uncharged wall and sticks to it.

The electrostatic force can be exerted by a charged object on another object from a distance (even when they are not in touch with each other). So, electrostatic force is an example of a non-contact force. For example, an electrically charged plastic comb exerts electrostatic force of attraction on tiny pieces of paper even from a distance (without touching them), and attracts them. Just like magnetic force, an electrostatic force can be of attraction or repulsion depending on the situation. The electrostatic force between any charged object (positively charged object or negatively charged object) and an uncharged object is always of attraction. The electrostatic force between two charged objects having like charges (positive and positive or negative and negative) is that of repulsion. On the other hand, the electrostatic force between two charged objects having unlike charges (positive and negative) is of attraction. We will study this in detail later on. Electrostatic force is utilised in reducing air pollution by removing dust, soot and fly-ash particles from the smoke coming out of chimneys of factories.

3. Gravitational Force

Newton said that every object in this universe pulls every other object with a certain force. **The pull exerted by objects possessing mass is called gravitational force.** The gravitational force between two objects is a force of attraction and it acts even if the two objects are not connected by any means. **It is the gravitational force between the sun and the earth which holds the earth in its orbit around the sun.** And it is the gravitational force between earth and the moon which holds the moon in its orbit around the earth.

The gravitational force between two ordinary objects having small masses (say, two stones) is very weak and cannot be detected easily. The gravitational pull (or force) becomes strong only if one of the objects has a huge mass, like the earth. Thus, the gravitational pull (or force) between the earth and a stone is very large which can be detected easily. This point will become clear from the following discussion.

If we drop a stone from some height, it falls down towards the earth. This is because the earth exerts a gravitational force of attraction on the stone and pulls it down. In fact, the earth pulls all the objects towards its centre. **The force with which the earth pulls the objects towards it, is called the force of gravity (or just gravity).** Gravity is a natural force of attraction (or pull). We are all pulled to the earth by the force of gravity. For example, when a child falls from a tree, it is the force of gravity which brings the child down to earth. Similarly, when a diver dives into a swimming pool, it is the force of gravity which brings him down. The force of gravity acts on all the objects. **The force of gravity causes all the objects to fall towards the earth.** For example, a leaf falls from the tree due to gravity. A fruit also falls down from a tree due to the force of gravity (see Figure 19). Rain falls down to earth due to the force of gravity. When we open a tap, water begins to flow down towards the ground due to the force of gravity. And water in rivers flows downwards also due to the force of gravity.

Figure 19. The English scientist Isaac Newton had given the idea for gravitational force when he saw an apple fall from a tree.

When a coin slips from our hand, it falls to the ground due to the force of gravity acting on it. Actually, when a coin is held in our hand, it is at rest (not in motion). As soon as the coin is released from our hand, it begins to move downwards and ultimately falls on the ground. This downward movement of coin is brought about by the gravitational force of earth (or force of gravity) which acts in the downward direction, towards the centre of earth. Similarly, when a pen slips from our hand, it falls to the ground because of the force of gravity acting on it (the pen does not go upwards !). And a ball thrown upwards also falls back to the earth due to the force of gravity (or gravitational force of earth). Please note that **the gravitational force of earth (or gravity) acts on objects from a distance (without there being a physical contact), therefore, gravitational force or gravity is an example of non-contact forces.**

PRESSURE

Pressure is produced when a force acts on an object. In most cases, an object exerts force on another object by making contact with it or by touching it. We will now describe how the effect of force applied by one object on another object depends on area of contact between the two objects. This will give us the definition of pressure. This point will become clear from the following example.

If we push hard on a piece of wood with our thumb, the thumb does not go into the wood [see Figure 20(*a*)]. But if we push a drawing pin into the wood with the same force of our thumb, the drawing pin goes into the wood [see Figure 20(*b*)]. These observations can be explained as follows : Our thumb does not go into the wood because the force of thumb is falling on a large area of the wood due to which the 'force per unit area' (or pressure) on the wood is small. The drawing pin goes into the wood because due to the sharp tip of the drawing pin, the force of thumb is falling on a very small area of the wood due to which the 'force per unit area' (or pressure) on the wood becomes very large. It is clear from this example that *pressure is the force acting on a unit area of the object (here wood).* The force of thumb produces small pressure when it acts on a large area of wood but the same force of thumb produces much greater pressure when it acts on

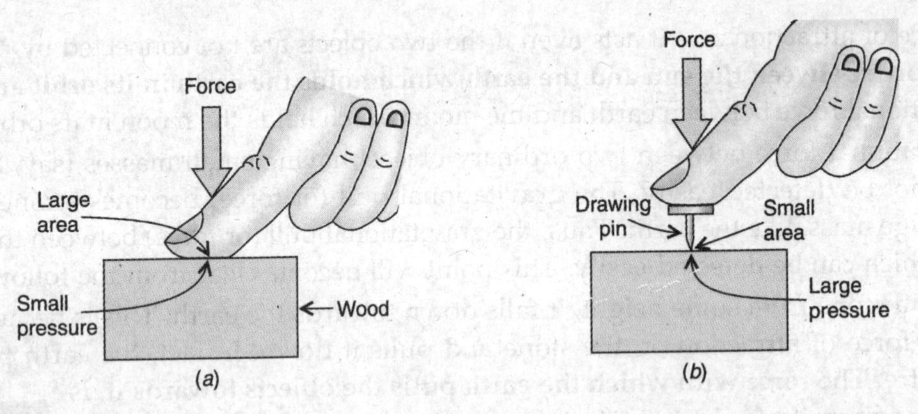

Figure 20.

a very small area of wood through the tip of drawing pin. Thus, **the effect of a force depends on the area of the object on which it acts.** We can now define pressure as follows :

Pressure is the force acting on a unit area of the object. To obtain the value of pressure, we should divide the force acting on an object by the area of the object on which it acts. So, the formula for calculating pressure is :

$$\textbf{Pressure} = \frac{\textbf{Force}}{\textbf{Area}}$$

This formula gives the relation between pressure, force and area. We will now give the units in which pressure is measured. The SI unit of measuring force is newton (N), and the SI unit of measuring area is 'square metre' (m^2), therefore, **the SI unit of measuring pressure is 'newtons per square metre' (N/m^2) which is also called pascal (Pa).**

$$1 \text{ pascal} = 1 \text{ newton per square metre}$$

or

$$1 \text{ Pa} = 1 \text{ N/m}^2$$

Pascal is a very small unit of pressure, so a bigger unit of pressure called 'kilopascal' is also used. The symbol of kilopascal is kPa. **One kilopascal is equal to one thousand pascals.** That is,

$$1 \text{ kilopascal} = 1000 \text{ pascals}$$

or

$$1 \text{ kPa} = 1000 \text{ Pa}$$

If in the above formula for pressure, we put the value of force in newtons (N) and the value of area in square metres (m^2), then we will get the value of pressure in newtons per square metre (N/m^2) or pascals (Pa). Please note that whether we express the pressure in the units of N/m^2 or Pa, it means the same thing. We will now solve a numerical problem based on pressure.

Sample Problem. A force of 100 N is applied to an object of area 2 m^2. Calculate the pressure.

Solution. Here, Force = 100 N

And, Area = 2 m^2

Now, putting these values in the formula for pressure :

$$\text{Pressure} = \frac{\text{Force}}{\text{Area}}$$

We get :

$$\text{Pressure} = \frac{100 \text{ N}}{2 \text{ m}^2}$$

$$= 50 \text{ N/m}^2 \quad \text{(or 50 Pa)}$$

Thus, the pressure is 50 newtons per square metre or 50 pascals.

We have just studied that the formula for calculating pressure is :

$$\text{Pressure} = \frac{\text{Force}}{\text{Area}}$$

Now, since the 'Area' is in the denominator of pressure formula, it means that larger the area of surface, smaller will be the pressure produced by a given force. And smaller the area of surface, larger will be the pressure produced by the same force. The area of the pointed end of drawing pin in Figure 20(*b*) is very small due to which the force of thumb produces a large pressure on the piece of wood. This large pressure pushes the drawing pin into the piece of wood.

From the above discussion we conclude that **the pressure depends on two factors :**

(*i*) **Force applied,** and

(*ii*) **Area over which force acts.**

The same force can produce different pressures depending on the area over which it acts. For example, **when a force acts over a *large* area of an object, it produces a *small* pressure.** But **if the same force acts over a *small* area of the object, it produces a *large* pressure.** Please note that the weight of a body is also a force. And it always acts in the downward direction.

Explanation of Some Everyday Observations on the Basis of Pressure

We have just studied that 'pressure is the force per unit area'. This definition of pressure can be used to explain many observations of our daily life. An important point to be kept in mind in this regard is that **the same force produces less pressure if it acts on a large area but it can produce high pressure if it acts on a small area.**

1. Why School Bags have Wide Straps

A school bag (or shoulder bag) has wide strap made of thick cloth (canvas) so that the weight of bag may fall over a large area of the shoulder of the child producing less pressure on the shoulder. And due to less pressure, it is more comfortable to carry the heavy school bag. On the other hand, if the school bag has a strap made of thin string, then the weight of school bag will fall over a small area of the shoulder. This will produce a large pressure on the shoulder of the child and it will become very painful to carry the heavy school bag. A person employed to carry heavy luggage or other loads is called a porter. We see many porters at a Railway Station. Porters place a thick, round piece of cloth on their head when they have to carry heavy loads (like heavy suitcases, etc.) of passengers. By placing the thick, round piece of cloth, the porters increase the area of contact of the load with their head. Since the load now falls on a larger area of head, the pressure on head is reduced and it becomes easier to carry the heavy load (see Figure 21).

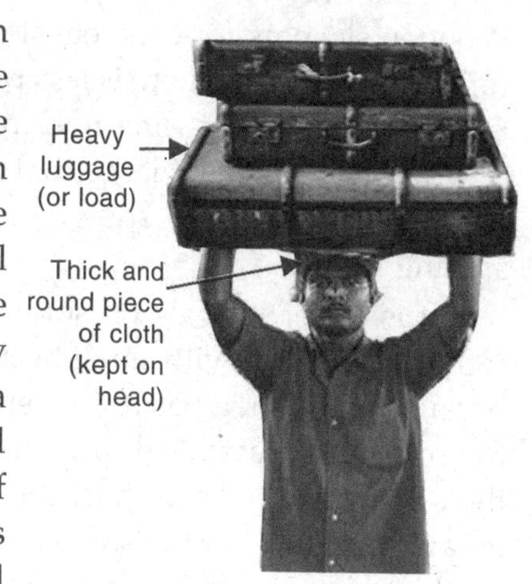

Figure 21. A porter carrying a heavy load on his head.

2. Why a Sharp Knife Cuts Better than a Blunt Knife

A sharp knife has a very thin edge to its blade. A sharp knife cuts objects (like vegetables) better because due to its very thin edge, the force of our hand falls over a very small area of the object producing a large pressure. And this large pressure cuts the object easily. On the other hand, a blunt knife has a thicker edge. A blunt knife does not cut an object easily because due to its thicker edge, the force of our hand falls over a larger area of the object and produces lesser pressure. This lesser pressure cuts the object with difficulty. A razor blade has very sharp edges so that it may produce a large pressure to cut things easily. An axe has also a sharp edge to cut things (like a log of wood) easily.

Figure 22. A sharp knife cuts well because of high pressure under the blade.

3. Why the Tip of a Needle is Sharp

The tip of a sewing needle is sharp so that due to its sharp tip, the needle may put the force on a very small area of the cloth, producing a large pressure sufficient to pierce the cloth being stitched. A knife, a razor blade and an axe are the cutting tools whereas a sewing needle is a piercing tool. The tools meant for cutting and piercing always have sharp edges so that they may create large pressures on the objects on which they act so as to do the required job.

4. Why the Depression is Much More when a Man Stands on the Cushion than when He Lies Down on it

When a man stands on a cushion then only his two feet (having small area) are in contact with the cushion. Due to this the weight of man falls on a small area of the cushion producing a large pressure. This large pressure causes a big depression in the cushion. On the other hand, when the same man is lying on the cushion, then his whole body (having large area) is in contact with the cushion. In this case the weight of man falls on a much larger area of the cushion producing much smaller pressure. And this smaller pressure produces a very little depression in the cushion.

The tractors have broad tyres so that there is less pressure on the ground and the tyres do not sink into comparatively soft ground in the fields. A wide steel belt is provided over the wheels of army tanks so that they exert less pressure on the ground and do not sink into it. Wooden sleepers (or concrete sleepers) are kept below the railway line so that there is less pressure of the train on the ground and railway line may not sink into the ground. The snow shoes have large, flat soles so that there is less pressure on the soft snow and stop the wearer from sinking into it.

Figure 23. A tractor with very wide tyres.

It is easier to walk on soft sand if we have flat shoes rather than shoes with small heels (or pencil heels). This is because a flat shoe has a greater area in contact with the soft sand due to which there is less pressure on the soft sand. Because of this the flat shoes do not sink much in soft sand and it is easy to walk on it. On the other hand, a small heel (or sharp heel) has a small area in contact with the soft sand and so exerts a greater pressure on the soft sand. Due to this greater pressure, the small heels tend to sink deep into soft sand making it difficult for the wearer to walk on soft sand. We will now discuss the pressure exerted by liquids.

PRESSURE EXERTED BY LIQUIDS

The substances such as water, oil, alcohol, petrol and mercury, etc., are liquids. **All the liquids exert pressure** on the base (or bottom) and walls of their containers. Though pressure is exerted by all the liquids but since water is the most common liquid around us, therefore, we will be using water as the liquid in our activities on pressure exerted by liquids.

All the liquids have weight. So, when we pour a liquid into a vessel, then the weight of liquid pushes down on the base of the vessel (or bottom of the vessel) producing a pressure. We say that the liquid exerts pressure on the base of the vessel (in which it is kept). Figure 24 shows a vessel containing a liquid in it. The weight of this liquid pushes on the base of the vessel and exerts a pressure on the base of the vessel (or bottom of the vessel). The pressure exerted by a liquid is given by the formula :

$$\text{Pressure} = \frac{\text{Force}}{\text{Area}}$$

In this case, the 'force' is the 'weight of the liquid' and 'area' is the 'area of the base of the vessel' in which the liquid is placed.

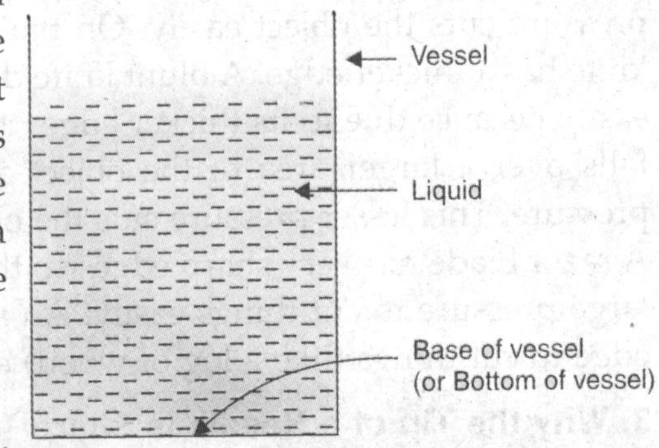

Figure 24.

Please note that the vessel in which a liquid is kept is also known as its container. We will now describe the dependence of pressure exerted by a liquid on depth in the liquid.

Dependence of Pressure on Depth in a Liquid

The pressure in a liquid is not the same at all depths. The pressure exerted by a liquid changes with depth in the liquid. Actually, **the pressure exerted by a liquid increases with increasing depth inside the liquid.** The pressure exerted by a liquid is small just under the surface of the liquid. But as we go deeper in a liquid, the pressure of liquid increases. Actually, *as the depth of liquid increases, the weight of liquid column pushing down from above increases, and hence the pressure also increases.* We will now describe an activity to show that the pressure of a liquid depends on its depth.

ACTIVITY TO SHOW THAT THE PRESSURE OF A LIQUID INCREASES WITH DEPTH

The fact that the pressure of a liquid increases with depth can be demonstrated by using the apparatus shown in Figure 25. A tall vessel has three short and thin tubes *A, B* and *C* fitted at different depths from the top of the vessel. The three tubes are of equal diameters and corks are fitted into them. The vessel is filled with water and then all the corks are removed quickly. On removing the corks, the water from uppermost tube *A* is found to travel the shortest distance from the base of the vessel, the water from middle tube *B* goes a little farther away whereas the water from the lowermost tube *C* shoots out farthest of all (see Figure 25). This can be explained as follows :

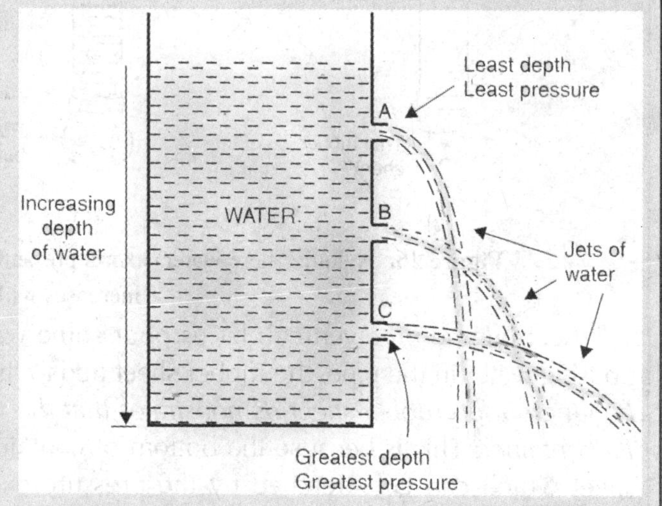

Figure 25. Pressure in a liquid increases with depth.

The depth of water near tube *A* is small so the water comes out from tube *A* with smaller pressure and falls nearer the bottom of the vessel. The depth of water near tube *B* is greater, so the water comes out with greater pressure from tube *B* and falls farther away from the base of vessel. The depth of water near tube *C* is the greatest, so the water comes out of the tube *C* with the greatest pressure and goes farthest from the vessel. This means that as the depth of water increases from *A* to *B* to *C*, the pressure of water gradually increases.

From the above activity we conclude that as the depth of water in the vessel increases, the pressure exerted by water also increases. In general, we can say that **the pressure of a liquid increases with depth.** In other words, **the greater the depth of a point in a liquid, the greater is the pressure.** Thus, the pressure of water will be less at a depth of 10 metres below the surface of sea but it will be much greater at a depth of 20 metres below the surface of the sea.

If we look carefully at Figure 25, we will find that the tubes *A, B* and *C* are fitted in the walls of the vessel and water comes out of these tubes under pressure. So, another conclusion which can be drawn from the above activity is that **liquids also exert pressure on the walls of the vessel in which they are stored. The sideways pressure exerted by liquids also increases with the depth of the liquid.** The sideways pressure of a liquid on the walls of a vessel is almost zero at its surface. As the depth of liquid increases, the sideways pressure on the walls of the vessel gradually increases and it becomes maximum near the bottom of the vessel. The wall of a dam is made thicker at the bottom (than at the top) so as to tolerate very high sideways pressure exerted by deep water stored in the reservoir of dam. We will now describe some activities to demonstrate the pressure exerted by liquids (like water) on the bottom of their container as well as on the walls of their container (or sides of their container).

ACTIVITY TO DEMONSTRATE THAT A LIQUID EXERTS PRESSURE ON THE BOTTOM OF ITS CONTAINER

Take a transparent plastic pipe about 15 cm long having a diameter of about 3 cm. Also take a thin sheet of rubber (like that from a rubber balloon). Stretch the thin rubber sheet and tie it tightly over one end of the plastic pipe [see Figure 26(*a*)]. Now, in this activity, the plastic pipe is the container which has a stretchable bottom made of a thin rubber sheet.

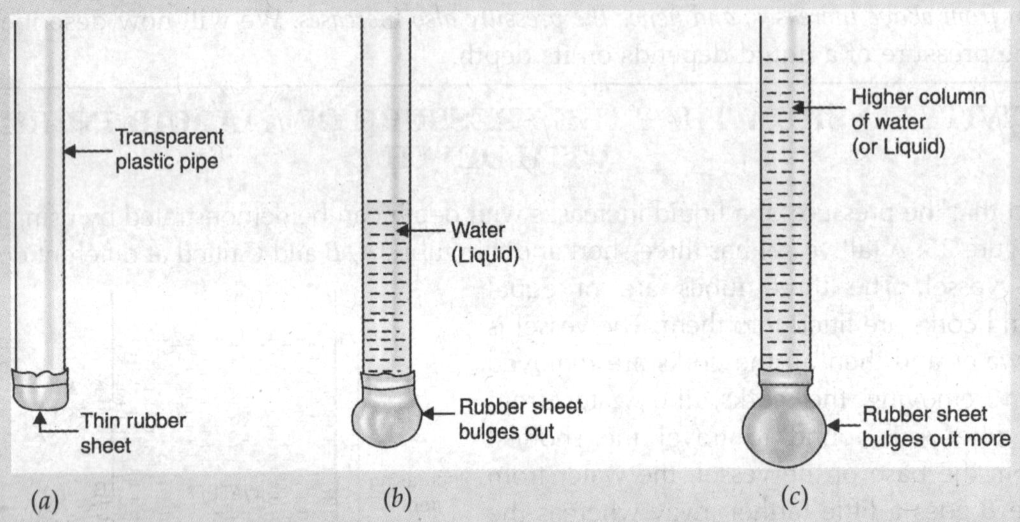

Figure 26. A liquid (here water) exerts pressure on the bottom of its container. This liquid pressure increases with increasing depth.

Keeping the pipe vertical, let us pour some water in the pipe from the top end. We will find that on pouring water in the pipe, the rubber sheet tied to its bottom stretches and bulges out [see Figure 26(*b*)]. *The bulging out of rubber sheet demonstrates that the water poured in pipe exerts a pressure on the bottom of its container.* This is because the bottom of container (pipe) in this case is made of a flexible, thin rubber sheet which can get stretched by the pressure exerted by water to form a bulge (The hard bottoms of containers like metal vessels or glass vessels, however, do not get stretched or bulge by the pressure of water kept in them). Let us now pour some more water in the plastic pipe so that the height of 'water column' in the pipe increases [see Figure 26(*c*)]. We will find that as the height of water column increases, the bulge in the rubber sheet also increases, showing that the pressure of water on the bottom of its container (rubber sheet) has increased. In fact, *greater the height of water column in the pipe, greater will be the bulge in its rubber sheet bottom* (*showing the greater pressure exerted by water*). From this activity we conclude that :

(*i*) A liquid exerts pressure on the bottom of its container, and

(*ii*) The pressure exerted by a liquid depends on the height of the liquid column (above the bottom of the container).

ACTIVITY TO DEMONSTRATE THAT A LIQUID EXERTS PRESSURE ON THE WALLS OF ITS CONTAINER (OR SIDES OF ITS CONTAINER)

Take a plastic bottle (like a water bottle or a soft drink bottle). Fix a small glass tube a little above the bottom of the plastic bottle [see Figure 27(*a*)]. We can do this by heating one end of the glass tube slightly over a burner and then inserting this hot end quickly into the wall of the plastic bottle. We should seal the joint of glass tube with plastic bottle by using molten wax so that water does not leak from the joint. Tie a thin sheet of rubber (like that from a rubber balloon) tightly on the open end of glass tube

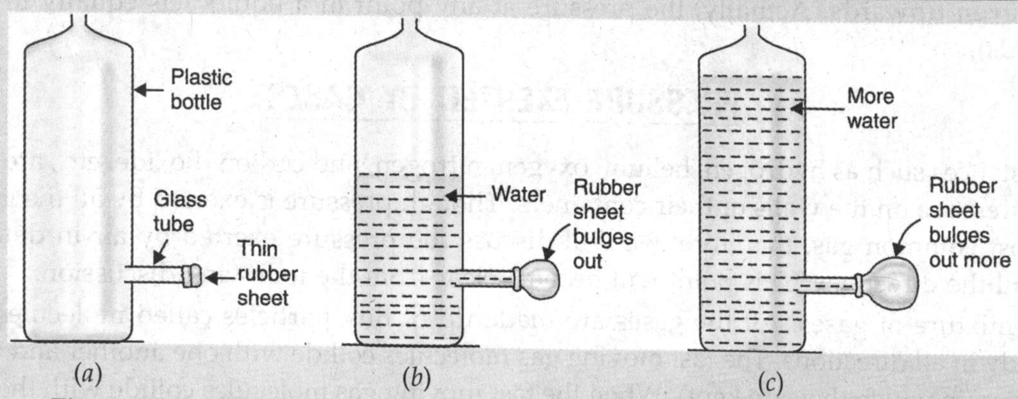

Figure 27. A liquid (here water) exerts pressure on the walls (or sides) of its container. This liquid pressure increases with increasing depth.

[see Figure 27(*a*)]. Now, fill half of plastic bottle with water. We will observe that on filling water, the rubber sheet tied to the mouth of glass tube gets stretched and bulges out [see Figure 27(*b*)]. *The bulging out of rubber sheet tied to the glass tube fixed in the wall of plastic bottle demonstrates that water present in plastic bottle exerts pressure on the wall of the bottle (or side of the bottle).* It is the sideways pressure exerted by water which inflates the thin rubber sheet forming a bulge. If we pour more water in the plastic bottle to increase its depth, we will see that the bulge in the rubber sheet increases [see Figure 27(*c*)]. This indicates that the pressure exerted by water increases with increasing depth. From this activity we conclude that :

(*i*) A liquid exerts pressure on the walls (or sides) of its container, and

(*ii*) The pressure exerted by a liquid on the walls (or sides) of its container increases with increasing depth.

ACTIVITY TO DEMONSTRATE THAT A LIQUID EXERTS EQUAL PRESSURE AT THE SAME DEPTH

Take a plastic bottle. Make two small holes of equal size on the two opposite sides of the plastic bottle some distance above the bottom of the bottle. The holes should be at exactly the same height from the bottom of the plastic bottle. Now fill the bottle with water. We will observe that the two jets of water coming out of the two holes fall at the same distance away from the base of the bottle on its either side (see Figure 28). The two jets of water can fall at equal distance on the two sides of the bottle only if the pressure of water at the depth of two holes in the bottle is equal. From this activity we conclude that :

(*i*) A liquid exerts pressure on the walls (or sides) of its container, and

(*ii*) A liquid exerts equal pressure at the same depth.

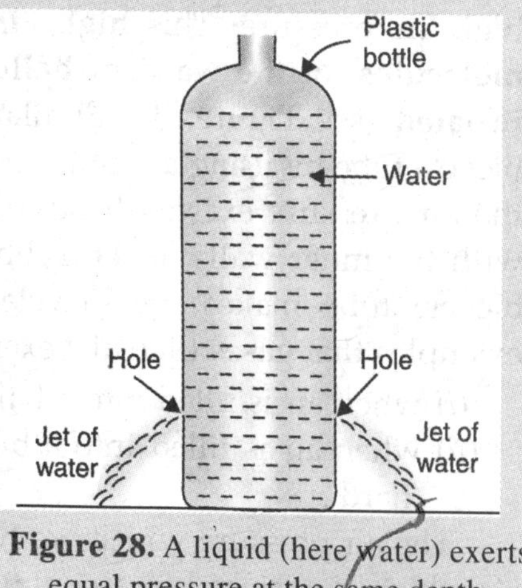

Figure 28. A liquid (here water) exerts equal pressure at the same depth.

Many times we see a fountain of water rushing out of the leaking joints (or holes) in the pipes of main water supply line in our city. It is due to the very high pressure exerted by water on the sides (or walls) of the pipes that such a fountain of water is formed. So, **the formation of fountains of water from the leaking pipes of water supply pipeline tells us that water exerts pressure on the walls of its container (here the walls of water-carrying pipes).**

So far we have studied that a liquid exerts downward pressure on the base (or bottom) of its container and sideways pressure on the walls of the container. Though it may seem awkward but a liquid exerts pressure in the upward direction also. So, in general we can say that : **A liquid exerts pressure in all**

directions—even upwards. Actually, the pressure at any point in a liquid acts equally in all directions (even upwards).

PRESSURE EXERTED BY GASES

The substances such as hydrogen, helium, oxygen, nitrogen, and carbon dioxide, etc., are gases. **All the gases exert pressure** on the walls of their containers. Though pressure is exerted by all the gases but since air is the most common gas, therefore, we will discuss the pressure exerted by air in detail. Air exerts pressure in all the directions. This point will become clear from the following discussion.

Air is a mixture of gases. All the gases are made up of tiny particles called molecules which move around quickly in all directions. The fast moving gas molecules collide with one another and with the walls of the container (in which they are kept). When the fast moving gas molecules collide with the walls of their container, they exert a force on the walls of the container. This force produces gas pressure (or air pressure). We can now say that : **Air pressure arises due to the constant collisions of the tiny molecules of the gases present in air with the walls of the container (or vessel) in which it is enclosed.** Now, if a certain mass of air is compressed into a smaller volume (say, by using a pump), then the number of collisions of air molecules per unit area increases, and hence the pressure exerted by air goes up. And if the container of air has stretchable walls (like a rubber balloon), then the higher pressure exerted by air inflates the container (increases the size of the container and makes it look bigger). This point will become more clear from the following examples.

Take a rubber balloon and fill air into it with mouth (or by using a pump). We will find that on filling in air, the balloon gets inflated (it expands and becomes bigger in size) (see Figure 29). This can be explained as follows : When we put a lot of air in the balloon, then the number of gas molecules in the balloon increases too much. The large number of gas molecules cause too many collisions with the walls of the balloon from inside and create a high air pressure. This **high air pressure produced by the gas molecules on the walls of balloon causes it to expand and get inflated** (see Figure 29). Similarly, when air is pumped into a bicycle tube by using a pump, the bicycle tube gets inflated due to the air pressure exerted by the collisions of gas molecules in air with the inner walls of the rubber tube. This air pressure in the bicycle tube makes the bicycle tyre feel hard. Thus, the two examples that gases (like air) exert pressure are :

Figure 29. Air filled in these balloons exerts pressure and inflates them.

(*i*) when air is filled into a balloon with our mouth, the balloon gets inflated (gets bigger in size), and
(*ii*) when air is filled into a bicycle tube with a pump, the tube gets inflated and makes the tyre feel 'hard'.

The air pressure which we have just described is due to the motion of molecules of gases present in air which is enclosed in a container (like balloon or football). We will now describe another type of air pressure called 'atmospheric pressure' which is due to the weight of air (or its gases) present in the atmosphere above the surface of earth.

ATMOSPHERIC PRESSURE

We live on the earth and there is a lot of air above us. **The layer of air above the earth is called atmosphere.** The air in our atmosphere extends up to about 300 kilometres above the surface of earth. The atmosphere contains a tremendous amount of air. Air has weight, so the atmosphere consisting of tremendous amount of air has enormous weight. The weight of atmosphere exerts a pressure on the surface of the earth and on all the objects on the earth, including ourselves. This pressure is known as atmospheric pressure. We can now say that : Atmospheric pressure is the air pressure which is exerted by the weight of air present in the atmosphere. In other words, **atmospheric pressure is due to the weight of air present in the atmosphere above us.** Just as the pressure in a liquid acts in all directions, in the same way, **atmospheric pressure also acts in all directions (even upwards !).**

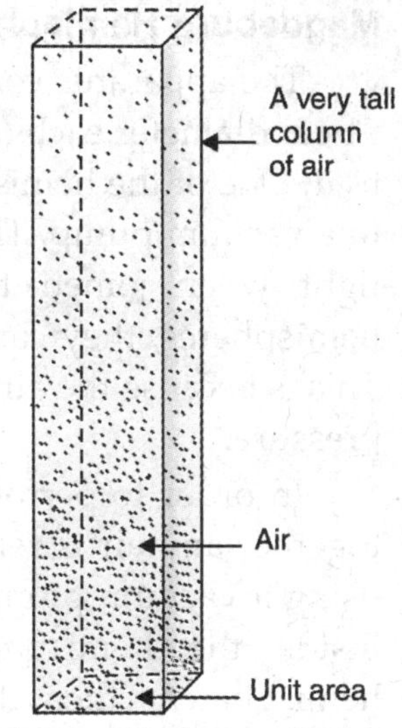

Figure 30. Atmospheric pressure is equal to the weight of air present in a very tall column of unit area (1 square metre area).

Magnitude of Atmospheric Pressure

We know that 'pressure is force per unit area'. Now, if we imagine a unit area of earth's surface and a very tall column 'filled with air' standing on it, then the weight of air in this column will be the atmospheric pressure at that place (see Figure 30). Since the SI unit of area is 'square metre', therefore, we can also say that atmospheric pressure is equal to the weight of air present in a very tall column of air standing on 1 square metre area of the earth.

On the surface of earth, the atmospheric pressure is maximum at the sea-level. This is because the column of air above us is tallest at the sea-level. The atmospheric pressure on the surface of earth (at the sea-level) is 101.3 kilopascals—which is equivalent to the weight of ten elephants pressing on each square metre area ! Thus, the magnitude of atmospheric pressure is very large. **As we go up in the atmosphere from the surface of earth, the atmospheric pressure goes on decreasing**. This is because as we go up in the atmosphere, the weight of air above us goes on decreasing (due to which the pressure also goes on decreasing). So, ***the atmospheric pressure on the top of a high mountain will be much less than at its base.***

Please note that though the SI unit of pressure is pascal (Pa) but atmospheric pressure is usually measured in the unit of 'millimetres of mercury' (mm of mercury) for the sake of convenience in measuring it. Thus, the common unit for expressing and measuring atmospheric pressure is 'millimetres of mercury' (mm of mercury). The atmospheric pressure on the surface of earth (at the sea-level) is 760 mm of mercury.

ACTIVITY TO SHOW THE EXISTENCE OF ATMOSPHERIC PRESSURE

We will now describe an activity to show the existence of atmospheric pressure. In this activity we will use atmospheric pressure to hold water in an inverted glass tumbler. This activity can be performed as follows : A glass tumbler is filled to the brim with water and covered with a piece of thick and smooth card. We press the card hard so that there is no air in the glass tumbler. Keeping the card in position with one hand, we invert the glass tumbler full of water. The hand supporting the card is then withdrawn slowly. We will see that the piece of card does not fall though the tumbler is full of water and exerts pressure on the card in the downward direction (see Figure 31). *The card does not fall because the atmospheric pressure acts on the card in the upward direction and holds the card in place.* The upward atmospheric pressure acting on the card is greater than the downward pressure of water on the card.

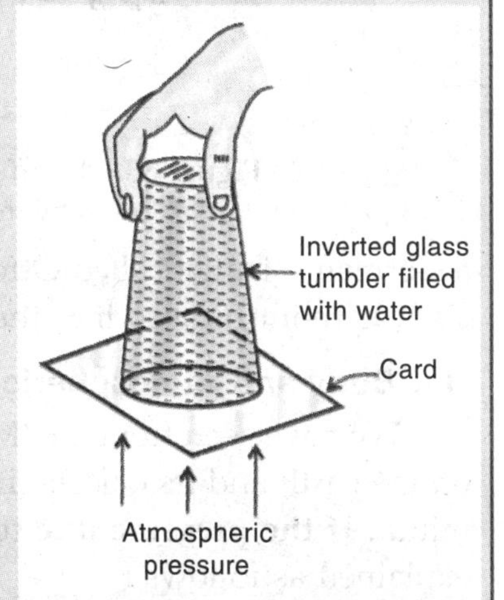

Figure 31. Here the atmospheric pressure acting upwards holds the card in place and prevents water from running out.

Before we describe the next experiment to show the existence of large atmospheric pressure around us, we should know the meaning of the term 'hemispheres'. 'Hemispheres' mean 'half spheres'. Joining together of two hemispheres makes one complete sphere. Let us describe the experiment now.

Magdeburg Hemispheres Experiment to Show Large Atmospheric Pressure

The apparatus consists of two hollow copper hemispheres *A* and *B* of 51 cm diameter each (see Figure 32). Each hemisphere has a hook attached to it. One of the hemispheres has also a side tube which can be connected to a vacuum pump. The two hemispheres are tight fitting and become air-tight when joined together. When air is present inside the joined hemispheres, they can be easily separated by pulling with a small force. This is because the air present inside the joined hemispheres also exerts its pressure.

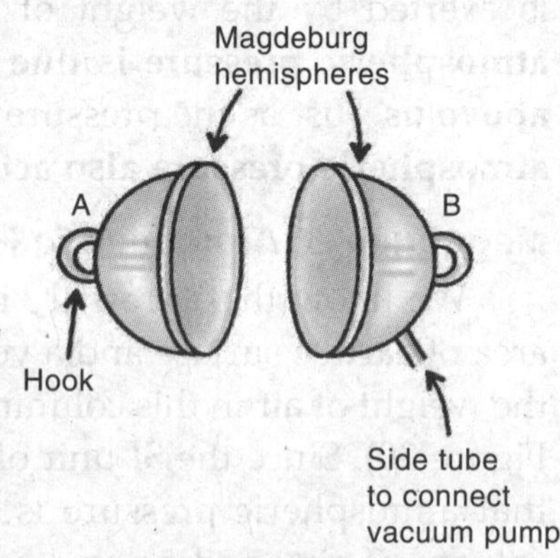

Figure 32. Magdeburg hemispheres.

In order to perform the experiment, the two hemispheres are joined together and air is removed completely from the space between them by using a vacuum pump (see Figure 33). When all the air is removed from inside the hemispheres (or when vacuum is created inside the hemispheres), then the two hemispheres cannot be separated even by pulling with a large force. This is due to the fact that since there is no air inside, the unopposed atmospheric pressure acting over the whole surface of hemispheres from outside presses them very, very hard and does not allow them to be separated. The effect of atmospheric pressure on the evacuated hemispheres was so great that even two teams of eight horses each pulling in opposite directions could not separate the two hemispheres. This is shown in Figure 33. This experiment with hemispheres and horses was conducted by

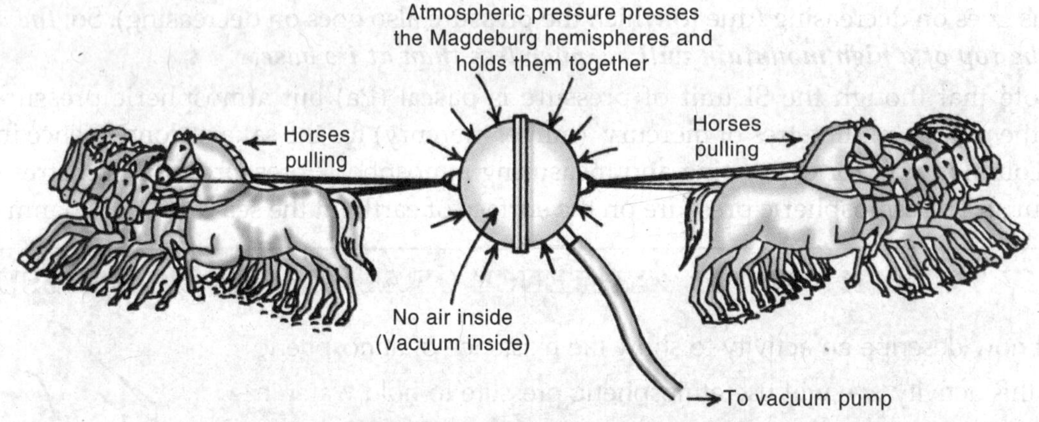

Figure 33. The effect of atmospheric pressure on evacuated hemispheres is not overcome even by the force of eight pairs of horses pulling in opposite directions.

a German scientist called Otto von Guericke in the town of Magdeburg in the year 1640. As soon as some air was re-introduced into the evacuated hemispheres, the hemispheres fell apart.

Our Body and Atmospheric Pressure

We have seen from the 'Magdeburg hemispheres' experiment that the pressure exerted by atmosphere on the earth and its objects (including us) is very, very large. So, an important question now arises in our mind : **If the pressure due to atmosphere is so great, then why are we not crushed by it ?** This can be explained as follows :

Our body has a liquid called 'blood' which flows through blood vessels into each and every cell of our body. Our blood itself exerts a pressure called 'blood pressure' which is slightly greater than the atmospheric pressure. **Since the atmospheric pressure acting on our body from outside is balanced by the blood pressure acting from inside, we do not get crushed.** Actually, the atmospheric pressure is so finely balanced by our blood pressure that we do not feel any discomfort. We even do not feel the existence of atmospheric pressure at all. We will now describe **the effects of low atmospheric pressure on our body.** The atmospheric pressure is maximum on the surface of the earth. When we go to high altitudes (say, a high mountain), then the atmospheric pressure decreases. So, **at high altitudes, the atmospheric pressure**

becomes much less than our blood pressure. Since our blood is at a higher pressure than outside pressure, therefore, some of the blood vessels in our body burst and **nose bleeding takes place at high altitudes.** Thus, nose-bleeding usually occurs in those persons who trek to high mountains (where the atmospheric pressure is much less than our blood pressure).

Applications of Atmospheric Pressure in Everyday Life

We use many simple devices in our everyday life which work on the existence of atmospheric pressure. For example, the devices such as a drinking straw, a syringe, a dropper and a rubber sucker work on the existence of atmospheric pressure (or air pressure) around us. We will now describe all these devices, one by one. Let us start with a drinking straw.

1. DRINKING STRAW. The drinking straw is a very thin pipe which is used to drink soft drinks (like Coca-Cola and Pepsi). **The drinking straw works on the existence of atmospheric pressure.** This can be explained as follows : The lower end of drinking straw is dipped in the soft drink (see Figure 34). When we suck at the upper end of the straw with our mouth, the pressure of air inside the straw and in our mouth is reduced. But the pressure acting on the surface of the soft drink is equal to atmospheric pressure. So, *the greater atmospheric pressure acting on the surface of the soft drink pushes the soft drink up the straw into our mouth* (see Figure 34).

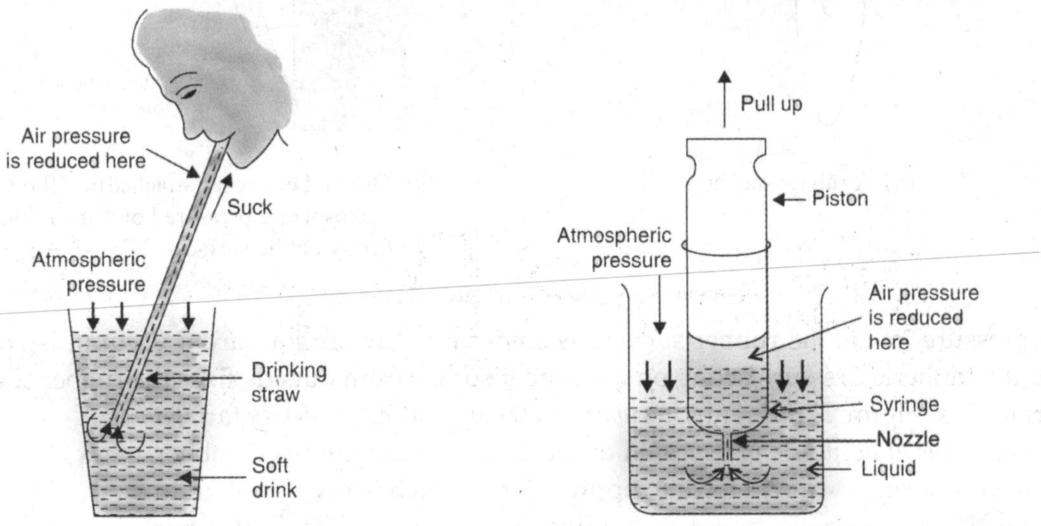

Figure 34. A drinking straw.

Figure 35. A syringe.

2. SYRINGE. A glass tube (or plastic tube) with a nozzle and piston for sucking in and ejecting liquid in a thin stream is called a syringe (A syringe may also be fitted with a hollow needle for giving injections). **The syringe works on the existence of atmospheric pressure.** When the nozzle of a syringe is dipped in a liquid and its piston is withdrawn, the pressure inside the syringe is lowered. *The greater atmospheric pressure acting on the surface of the liquid pushes the liquid up into the syringe* (see Figure 35).

3. DROPPER. The dropper is a short glass tube with a rubber bulb at one end and a nozzle at the other end (see Figure 36). A dropper is used for measuring out drops of a liquid (such as a liquid medicine). **A dropper works on the existence of atmospheric pressure.** When we press the rubber bulb of the dropper by keeping its nozzle dipped in the liquid, air present in the glass tube and bulb is seen to escape in the form of bubbles. Due to this, the air pressure inside the glass tube and rubber bulb of dropper is very much reduced. When we now release the rubber bulb of dropper, the much greater atmospheric pressure acting on the surface of liquid, pushes the liquid up into the dropper tube. Thus, *the rise of liquid (say, water) in a dropper is due to the atmospheric pressure.* If we remove the filled

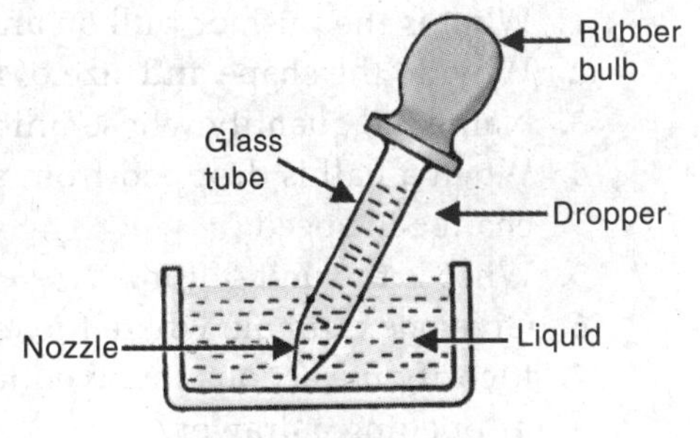

Figure 36. A dropper.

dropper from the container of liquid and press its rubber bulb slowly, then the drops of liquid will come out of the nozzle of dropper tube. Just like a dropper, **the filling of ink in a fountain pen is also based on the existence of atmospheric pressure.**

4. RUBBER SUCKER. A rubber sucker is a device made of rubber (or plastic) that sticks firmly to flat and smooth surfaces on pressing. A rubber sucker looks like a small, concave-shaped rubber cup [see Figure 37(*a*)]. A rubber sucker is also called a 'suction cup' because it sticks to a surface by suction (The production of partial vacuum by the removal of air is called suction). When we press the rubber sucker on a flat, smooth surface, its concave rubber cup gets flattened to a large extent, pushing out most of the air from beneath it [see Figure 37(*b*)]. Since very little air remains inside the flattened rubber sucker, therefore,

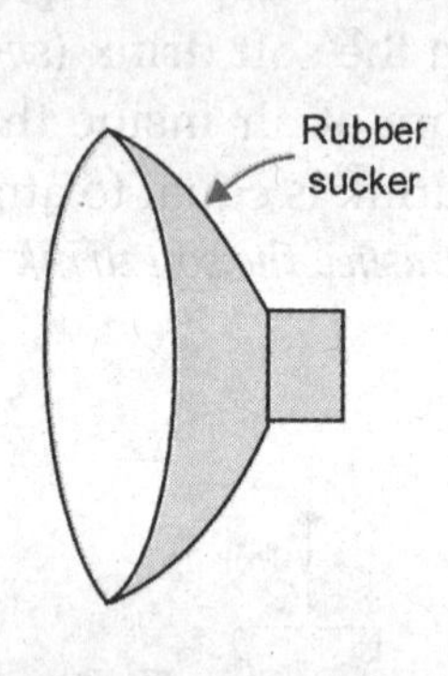

(*a*) A rubber sucker

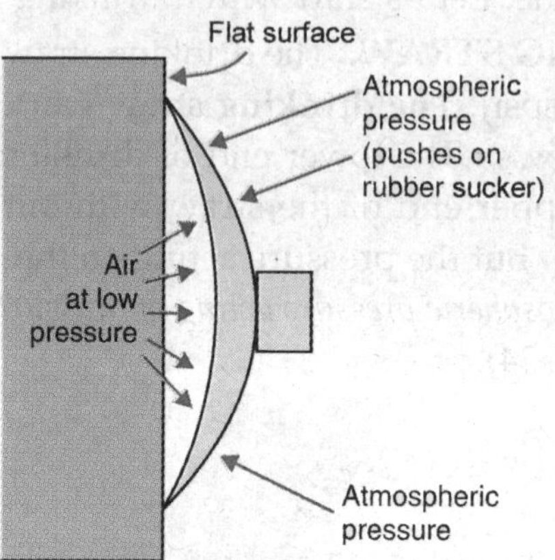

(*b*) The rubber sucker attached to a flat surface. Atmospheric pressure holds the rubber sucker firmly on the surface

Figure 37.

the air pressure inside the rubber sucker becomes very low (and a partial vacuum is created). The much greater atmospheric pressure acting on the rubber sucker from outside fixes the rubber sucker firmly on the flat surface [see Figure 37(*b*)]. Thus, *a rubber sucker stays attached firmly to a flat surface due to the atmospheric pressure.* In order to pull away the fixed rubber sucker from a surface, we will have to apply a force which is large enough to overcome the atmospheric pressure holding it onto the surface. **Rubber suckers are usually used to hold objects together with the help of suction.** For example, rubber suckers are used to hold glass 'table tops' onto the wooden frames of tables. Rubber suckers are also used for making suction hooks which are fixed on walls, doors and almirahs, etc., to hold various things. A suction hook is a rubber (or plastic) sucker having a plastic (or metal) hook attached to it. A suction hook is shown in Figure 38. We are now in a position to **answer the following questions :**

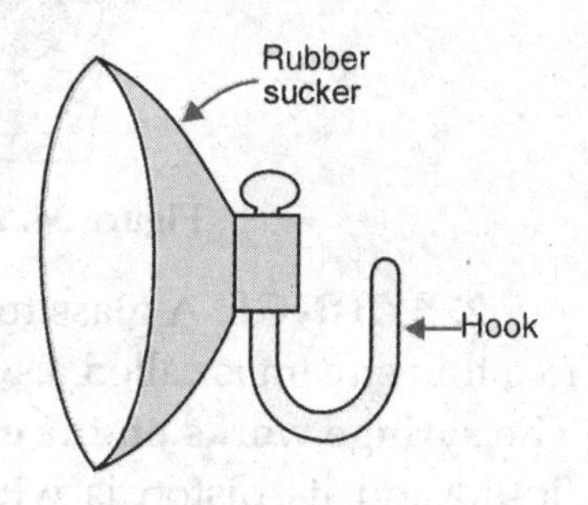

Figure 38. A suction hook.

Very Short Answer Type Questions

1. What is the push or pull on an object known as ?
2. Why do the shape and size of a balloon change when filled with air or water ?
3. Name the quantity whose unit is 'newton' (N).
4. When a ball is dropped from a height, its speed increases gradually. Name the force which causes this change in speed.
5. What is the unit of force ?
6. Give one example where force changes the shape of an object.
7. Identify the actions involved in the following situations as push or pull, or both :
   (*a*) opening a drawer.

(b) a cricket ball hit by a batsman.
(c) drawing a bucket of water from a well.
(d) moving a book placed on a table.
(e) a football player taking a penalty corner.
(f) moving a wheel barrow.

8. (a) Name two contact forces.
(b) Name two non-contact forces.

9. When a plastic pen is rubbed in dry hair, it attracts tiny pieces of paper. Which force is involved in this process ?

10. A small device pulls iron nails from a distance. Which type of force is involved in this process ?

11. Which force can be used to gather iron pins scattered on the floor ?

12. Name the force which always opposes motion.

13. Which force makes a rolling ball stop on its own ?

14. An inflated balloon was pressed against a wall after it has been rubbed with a piece of synthetic cloth. It was found that the balloon sticks to the wall. What force might be responsible for the attraction between the balloon and the wall ?

15. What name is given to the force acting on a unit area of an object ?

16. Name the quantity whose one of the units is pascal (Pa) ?

17. What conclusion do you get from the observation that a fountain of water is created at the leaking joint of pipes of the main water supply line ?

18. What type of pressure is involved in the filling of a liquid in a syringe ?

19. What substance present in our body balances the atmospheric pressure acting on us ?

20. Where will the atmospheric pressure be greater—at ground level or at the top of high mountain ?

21. Name any two devices used in everyday life which work on the existence of atmospheric pressure.

22. If a vacuum is created between two Magdeburg hemispheres joined together, they cannot be separated easily. What presses the hemispheres together ?

23. What makes a balloon get inflated when air is filled in it ?

24. Name the substance whose weight produces atmospheric pressure.

25. Where is the pressure greater, 10 m below the surface of the sea or 20 m below the surface of sea ?

26. What force acting on an area of 0.5 m^2 will produce a pressure of 500 Pa ?

27. Can a liquid exert pressure upwards ?

28. Can a liquid exert pressure sideways ?

29. State whether the following statements are true or false :
(a) The pressure exerted by a liquid depends on the area of base of its container.
(b) A drinking straw works on the pressure exerted by the liquid filled in a soft drink bottle in which it is placed.

30. Fill in the following blanks with suitable words :
(a) To draw water from a well, we have to ....at the rope.
(b) If the two forces applied to an object are equal and act in opposite directions, the net force acting on the object will be ....
(c) Force could be a ....or a ....
(d) Force has magnitude as well as ....
(e) A force arises due to..........between two objects.
(f) A charged body..........an uncharged body towards it.
(g) The north pole of a magnet..........the north pole of another magnet.
(h) Force acting on a unit area is called ....
(i) The pressure exerted by a liquid ....with depth.
(j) A drinking straw works on the existence of ....pressure.
(k) Atmospheric pressure..........with increasing height.

Short Answer Type Questions

31. Define 'state of motion' of an object. Name the 'agent' which can change the state of motion of an object.

32. Give two examples of situations where you push or pull to change the state of motion of objects.
33. What is meant by saying that 'force is due to an interaction' ? Give an example to illustrate your answer.
34. In a tug of war, when the two teams are pulling the rope, a stage comes when the rope does not move to either side at all. What can you say about the magnitudes and directions of the forces being applied to the rope by the two teams at this stage ?
35. What is force ? State the various effects of force.
36. (*a*) Give one example where force moves a stationary object.
    (*b*) State one example where force stops a moving object.
37. (*a*) Give one example where force changes the speed of a moving object.
    (*b*) Give one example where force changes the direction of a moving object.
38. Why does the shape of an ointment tube change when we squeeze it ?
39. What happens to the springs of a sofa when we sit on it ?
40. Name the various types of forces.
41. What is muscular force ? Give one example of muscular force.
42. Which of the following are non-contact forces ?
    Magnetic force, Frictional force, Gravitational force, Muscular force, Electrostatic force.
43. Give two examples from everyday life which show that air exerts pressure.
44. What is a rubber sucker ? How does it work ? State any one use of a rubber sucker.
45. Why do mountaineers usually suffer from nose-bleeding at high altitudes ?
46. Describe one activity to show the existence of atmospheric pressure.
47. Explain why, water comes out more slowly from an upstairs tap than from a similar tap downstairs.
48. What is meant by gravitational force (or force of gravity) ? Give its one example.
49. Calculate the pressure when a force of 200 N is exerted on an area of : (*a*) 10 m^2 (*b*) 5 m^2
50. Which force do the animals apply while moving, chewing and doing other activities ?
51. Which force is responsible for raising our body hair when we try to take off a terylene or polyester shirt in the dry weather ?
52. Name the type of forces involved in the following :
    (*a*) A horse pulling a cart.
    (*b*) A sticker attached to steel almirah without glue.
    (*c*) A coin falling to the ground on slipping from hand.
    (*d*) A plastic comb rubbed in dry hair picking up tiny pieces of paper.
    (*e*) A moving boat coming to rest when rowing is stopped.
53. Why does a sharp knife cut objects more effectively than a blunt knife ?
54. Explain why, wooden (or concrete) sleepers are kept below the railway line.
55. Explain why, a wide steel belt is provided over the wheels of an army tank.
56. Explain why, the tip of a sewing needle is sharp.
57. Explain why, snow shoes stop you from sinking into snow.
58. Explain why, when a person stands on a cushion, the depression is much more than when he lies down on it.
59. Explain why, porters place a thick, round piece of cloth on their heads when they have to carry heavy loads.
60. Give one practical application of magnetic force.

Long Answer Type Questions

61. (*a*) What is meant by a contact force ? Explain with the help of an example.
    (*b*) What is meant by a non-contact force ? Explain with the help of an example.
62. (*a*) Define frictional force (or friction).
    (*b*) Explain why, frictional force is said to be a contact force.
    (*c*) Explain why, magnetic force is said to be a non-contact force.
63. (*a*) Define pressure. What is the relation between pressure, force and area ? State the units in which pressure is measured.
    (*b*) Explain why, school bags are provided with wide straps to carry them.

64. (*a*) What is meant by atmospheric pressure ? What is the cause of atmospheric pressure ?
(*b*) Why are our bodies not crushed by the large pressure exerted by the atmosphere ?
(*c*) Explain why, atmospheric pressure decreases as we go higher up above the earth's surface.

65. (*a*) How does the pressure of a liquid depend on its depth ? Draw a labelled diagram to show that the pressure of a liquid (say, water) depends on its depth.
(*b*) Explain why, the walls of a dam are thicker near the bottom than at the top.

Multiple Choice Questions (MCQs)

66. Which of the following is not an example of muscular force ?
(*a*) a porter carrying a load on a wheel-barrow. (*b*) an apple falling from a tree.
(*c*) a child riding a bicycle. (*d*) a person drawing water from a well.

67. Which of the following is not an example of the force of gravity ?
(*a*) a leaf falling from a tree. (*b*) a boy pushing a cart on a level plane.
(*c*) a diver jumping into a swimming pool. (*d*) a stone falling from the top of a cliff.

68. When we press the bulb of a dropper with its nozzle kept in water, air in the dropper is seen to escape in the form of bubbles. Once we release the pressure on the bulb, water gets filled in the dropper. The rise of water in the dropper is due to :
(*a*) pressure of water (*b*) gravity of the earth
(*c*) shape of rubber bulb (*d*) atmospheric pressure

69. A rectangular wooden block has length, breadth and height of 50 cm, 25 cm and 10 cm, respectively. This wooden block is kept on ground in three different ways, turn by turn. Which of the following is the correct statement about the pressure exerted by this block on the ground ?
(*a*) the maximum pressure is exerted when the length and breadth form the base
(*b*) the maximum pressure is exerted when length and height form the base
(*c*) the maximum pressure is exerted when breadth and height form the base
(*d*) the minimum pressure is exerted when length and height form the base

70. Which of the following are contact forces ?
A. Friction B. Gravitational force C. Magnetic force D. Muscular force
(*a*) A and B (*b*) B and C (*c*) A and D (*d*) B and D

71. If we release a magnet held in our hand, it falls to the ground. The force responsible for this is :
(*a*) muscular force (*b*) magnetic force (*c*) electrostatic force (*d*) gravitational force

72. Which of the following force is utilised in reducing air pollution by removing dust, soot and fly-ash particles from the smoke coming out of chimneys of factories ?
(*a*) magnetic force (*b*) gravitational force (*c*) electrostatic force (*d*) frictional force

73. The same force F acts on four different objects having the areas given below, one by one. In which case the pressure exerted will be the maximum ?
(*a*) 20 m^2 (*b*) 50 m^2 (*c*) 10 m^2 (*d*) 100 m^2

74. Which of the following represent correct values for the normal atmospheric pressure ?
A. 101.3 kilopascals B. 76 mm of mercury C. 101.3 pascals D. 76 cm of mercury
(*a*) A and B (*b*) B and C (*c*) A and D (*d*) B and D

75. Which of the following does not work on the existence of atmospheric pressure ?
(*a*) rise of iodine solution in the glass tube of dropper
(*b*) rise of cold drink in a long plastic straw
(*c*) sticking of suction hook on the wall of a room
(*d*) rise of mercury in glass tube of thermometer

76. The magnitude of force is expressed in the unit of force called :
(*a*) pascal (*b*) kelvin (*c*) newton (*d*) magdeburg

77. Which of the following change appreciably when a batsman hits a moving cricket ball ?
A. Shape B. Direction C. Size D. Speed
(*a*) A and B (*b*) B and C (*c*) A and C (*d*) B and D

78. Which of the following is not an effect of force ?
(*a*) a force can change the speed of a moving object
(*b*) a force can change the direction of a moving object

(*c*) a force can change the composition of a moving object
(*d*) a force can change the shape and size of an object

79. Which of the following is not a non-contact force ?
(*a*) electrostatic force (*b*) gravitational force (*c*) frictional force (*d*) magnetic force

80. Which of the following scientists gave the idea of the existence of gravitational force ?
(*a*) Einstein (*b*) James Watt (*c*) Faraday (*d*) Newton

81. Some mustard oil is kept in a beaker. It will exert pressure :
(*a*) downwards only (*b*) sideways only (*c*) upwards only (*d*) in all directions

82. A pressure of 10 kPa acts on an area of 0.3 m^2. The force acting on the area will be :
(*a*) 3000 N (*b*) 30 N (*c*) 3 N (*d*) 300 N

83. The magnitude of atmospheric pressure is equal to the pressure exerted by a :
(*a*) 76 mm tall column of mercury
(*b*) 760 mm tall column of alcohol
(*c*) 76 cm tall column of mercury
(*d*) 760 cm tall column of mercury

84. The atmospheric pressure is usually measured in the unit of :
(*a*) newtons per square metre (*b*) pascal
(*c*) cm of mercury (*d*) mm of mercury

85. When a force of 5 N acts on a surface, it produces a pressure of 500 Pa. The area of surface then must be :
(*a*) 10 cm^2 (*b*) 50 cm^2 (*c*) 100 cm^2 (*d*) 0.01 cm^2

Questions Based on High Order Thinking Skills (HOTS)

86. Two tiny holes are made in a plastic bucket, one near the middle part and the other just above bottom. When this bucket is filled with water, the water rushes out from the bottom hole much faster than from the upper hole. What conclusion do you get from this observation ?

87. What is common in the working of the devices such as a drinking straw, a syringe, a dropper and a rubber sucker ?

88. A rocket has been fired upwards to launch a satellite in its orbit. Name the two forces acting on the rocket immediately after leaving the launching pad (Ignore the frictional force due to air resistance).

89. One student says that water exerts pressure on the bottom of the bucket but another student says that water exerts pressure on the sides of the bucket. What would you like to say ?

90. Name the forces acting on a plastic bucket containing water held above ground level in your hand. Discuss why the forces acting on the bucket do not bring a change in its state of motion.

ANSWERS

1. Force **7.** (*a*) Pull (*b*) Push (*c*) Pull (*d*) Push ; Pull (*e*) Push (*f*) Push ; Pull **10 .** Magnetic force **14.** Electrostatic force **19.** Blood **20.** At ground level **22.** Atmospheric pressure **25.** 20 m below the surface of sea **26.** 250 N **27.** Yes **28.** Yes **29.** (*a*) False (*b*) False **30.** (*a*) pull (*b*) zero (*c*) push ; pull (*d*) direction (*e*) interaction (*f*) attracts (*g*) repels (*h*) pressure (*i*) increases (*j*) atmospheric (*k*) decreases **49.** (*a*) 20 Pa (*b*) 40 Pa **50.** Muscular force **51.** Electrostatic force **52.** (*a*) Muscular force (*b*) Magnetic force (*c*) Gravitational force (*d*) Electrostatic force (*e*) Frictional force **66.** (*b*) **67.** (*b*) **68.** (*d*) **69.** (*c*) **70.** (*c*) **71.** (*d*) **72.** (*c*) **73.** (*c*) **74.** (*c*) **75.** (*d*) **76.** (*c*) **77.** (*d*) **78.** (*c*) **79.** (*c*) **80.** (*d*) **81.** (*d*) **82.** (*a*) **83.** (*c*) **84.** (*d*) **85.** (*c*) **86.** Pressure exerted by water increases with increasing depth **87.** All these devices work on the existence of atmospheric pressure **88.** Upward force applied by the rocket engine and Downward gravitational force applied by the earth **89.** Water exerts pressure on the bottom of the bucket as well as on the sides of the bucket **90.** Upward muscular force applied by hand and Downward gravitational force applied by earth ; The two forces being equal and opposite balance each other and hence do not bring a change in the state of motion.

CHAPTER

12

Friction

When we push a box lying on the floor with a small amount of force, it does not move at all (see Figure 1). It means that the surface of floor, on which the box is resting, exerts some force on the box which acts in a direction opposite to the force of our push. In other words, some force is acting on the bottom of the stationary box which opposes its motion (due to which the box does not move). This natural force between the floor and the bottom of the box which opposes the motion of box on the floor is friction (see Figure 1). Let us take another example. A ball moving on the ground slows down gradually and stops after covering some distance. We know that a force is required to stop a moving ball. This means that a force is exerted by the ground on the moving ball which opposes its motion and brings it to a stop. This force which opposes the motion of ball on the ground is also friction. We can now define friction as follows : **The force which always opposes the motion of one object over another object in contact with it, is called friction**. Friction occurs between the two surfaces which are in contact with each other. For example, when we push a heavy box kept on floor, the force of friction occurs between the surface of floor and bottom of the box. And in the case of a ball rolling on ground, friction occurs between the surface of ground and the surface of ball. In fact, friction acts on both the surfaces in contact with each other. Friction is a force which occurs when the two objects tend to slide over each other and even when they are actually sliding (moving) over each other.

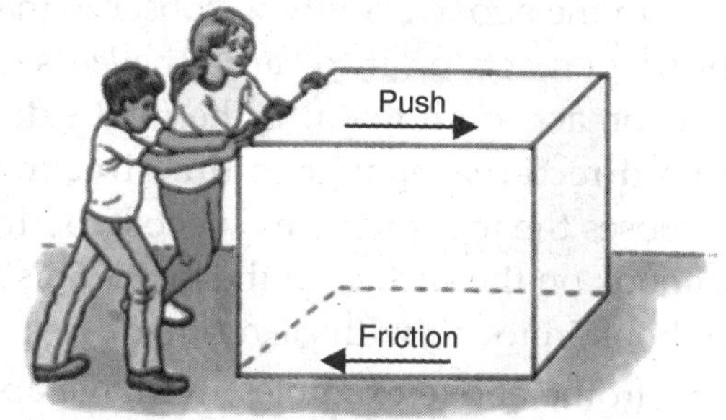

Figure 1. When we push the box with a small force, the friction between floor and bottom of box prevents it from moving.

Direction of Force of Friction

The force of friction always opposes the motion of one object over another object. So, **the force of friction acts in a direction opposite to the direction in which an object moves** (or tends to move). This will become more clear from the following activity.

ACTIVITY

Place a book on the table. Give a push to this book towards the right side [as shown in Figure 2(*a*)]. We will find that the book moves through some distance to the right side and then stops. Since the book moving towards right side stops on its own, this means that the force of friction is acting on it in the opposite direction (towards left) which is opposing its motion and making it stop. Thus, when the force is applied by our push to move the book towards *right* side, then the force of friction acts towards *left* side (in opposite direction to motion of book) [see Figure 2(*a*)].

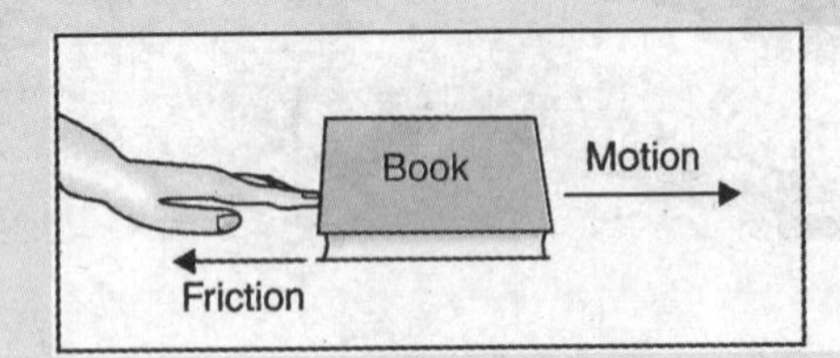

(*a*) When book moves towards *right* side, friction acts towards *left* side

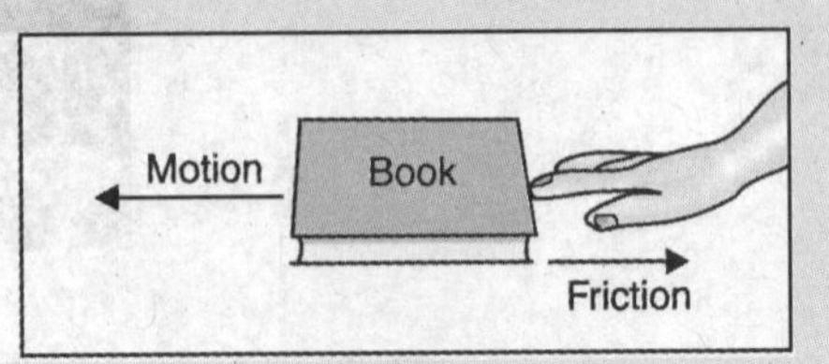

(*b*) When book moves towards *left* side, friction acts towards *right* side

Figure 2. The force of friction always acts in a direction opposite to the direction of motion of object.

Let us now give a push to the book towards the left side [as shown in Figure 2(*b*)]. We will find that the book moves some distance towards the left side and then stops. Since the book moving towards left side stops on its own, this means that the force of friction is acting on it in the opposite direction (towards right) which is opposing its motion and making it stop. Thus, when the force is applied by our push to move the book towards *left* side, then the force of friction acts towards the *right* side (in opposite direction to the motion of book) [see Figure 2(*b*)].

In the above activity we observe that when the motion of the book is towards *right* side, then the force of friction acts on it towards the *left* side. And when the motion of book is towards *left* side, the force of friction acts on it towards the *right* side. So, from this activity we conclude that **the force of friction acts in a direction opposite to the direction of motion of an object.** This is why the force of friction always opposes the motion of an object. In both the cases described above, the force of friction opposes the motion of book on the surface of the table. This force of friction occurs between the surface of table and the surface of book in touch with each other.

In the above examples, we have applied the force of push of our hand to move a book lying on a horizontal table top. The book can also be moved by the force of gravity of earth provided the table is tilted a little. So, if we tilt the table somewhat, then the book kept on it will start sliding down slowly. In this case, the force of gravity is acting in the downward direction (see Figure 3). Since the book is moving in the downward direction, the force of friction must act on it in the opposite direction—upward direction (as shown in Figure 3).

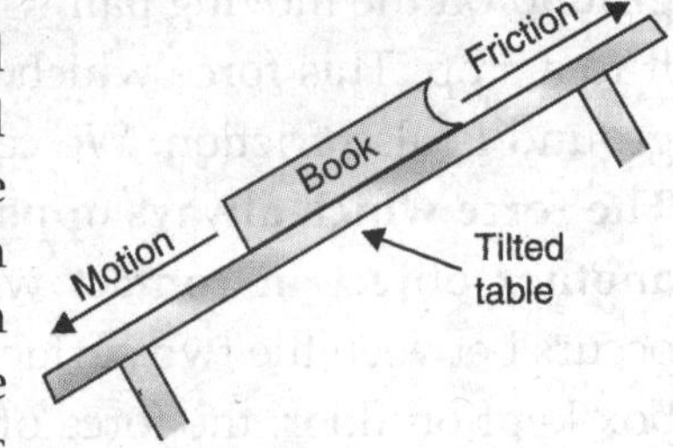

Figure 3.

Cause of Friction

Every object has a rough surface, though the surface may appear to be smooth to the naked eye. When we see through a microscope, it is found that the surfaces of all the objects have rough edges. Some of the particles on the surface of objects are in the form of tiny hills while others form grooves (see Figure 4). The tiny hills and grooves on the surfaces of objects are called 'irregularities of surfaces'. When we try to move one object over another object, the 'irregularities' present on their surfaces get entangled (or locked) with one another (see Figure 4). The interlocking of irregularities of the two surfaces opposes the motion of one object over the other and gives rise to force of friction. Thus, **friction is caused by the interlocking of irregularities in the surfaces of the two objects which are in contact with each other.** When we attempt to move one object over the other, we have to apply a force to overcome interlocking of the irregularities in their surfaces. More the roughness of a surface, larger is the number of irregularities on its surface and hence greater will be the friction. Thus, *the force of friction is greater if very rough surfaces are involved.*

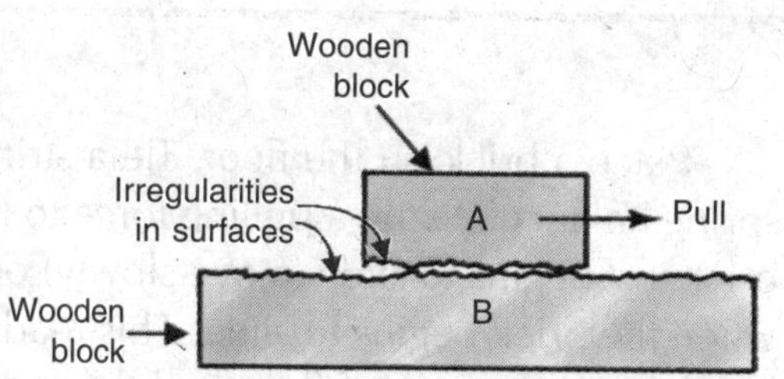

Figure 4. Friction arises due to irregularities (or roughness) in the surfaces of objects.

In Figure 4 we have shown a highly magnified diagram of the surfaces of two wooden blocks *A* and *B* kept one over the other. Please note the rough surfaces of these wooden blocks having lot of irregularities. When we pull the upper wooden block *A* over the lower wooden block *B* by applying a force, then the irregularities of their rough surfaces get entangled (or locked) with one another. This interlocking of surfaces gives rise to force of friction which opposes the motion of upper block *A* over the lower block *B*. We are able to move block *A* over block *B* because we apply sufficient muscular force while pulling to undo the interlocking of surfaces and overcome the opposing frictional force. **The block *A* moves over block *B* only when the pulling force applied by us becomes greater than the force of friction holding them together.** Friction is small for smooth surfaces (like glass and ice). Friction is much greater for rough surfaces (like sand paper and concrete).

Before we go further, we should know what a spring balance is because it will be used to perform some activities based on friction. **The spring balance is a device which is used for measuring force acting on an object** (see Figure 5). The spring balance contains a coiled spring which gets stretched when a force is applied to its free end (having a hook). The extent by which the spring gets stretched is a measure of the force applied. Larger the stretching of spring, greater will be the magnitude of force applied. The stretching of spring or magnitude of force is indicated by a pointer attached to the spring which moves on a graduated scale. The reading on the scale of spring balance (as indicated by the position of pointer) gives us the magnitude of force. When the spring balance is held vertically (as shown in Figure 5), it is said to measure the weight of an object hung from its hook (because weight of an object is also a force). And when a spring balance is held horizontally (attached to an object and pulled), it can be used to measure the force being applied to pull the object on a horizontal surface.

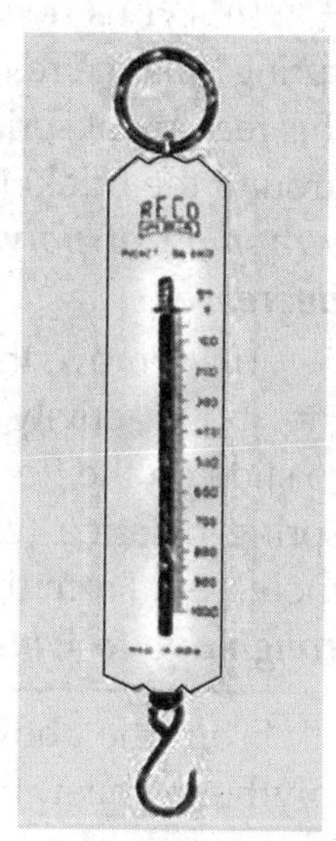

Figure 5. A spring balance.

FACTORS AFFECTING FRICTION

It has been found by experiments that **the friction between two surfaces depends on two factors :**

(*i*) **the nature of the two surfaces (smoothness or roughness of the two surfaces).**

(*ii*) **the force with which two surfaces are pressed together.**

The force of friction, however, does not depend on the 'amount of surface area' of the two objects which is in contact with each other. We will now study how the friction depends on the nature of two surfaces as well as on the force with which the two surfaces are pressed together.

1. Dependence of Friction on the Nature of Two Surfaces

Friction is not the same for all the surfaces. **Friction depends on the smoothness or roughness of the two surfaces which are in contact with each other.** When the two surfaces in contact are smooth, then the friction between them will be small (because the interlocking of smooth surfaces is less). As the degree of roughness of the two surfaces in contact increases, the friction also increases. And when the two surfaces in contact are very rough, then the friction between them will be very large (because the interlocking of very rough surface is too much). We can study the dependence of friction on the nature of surfaces by performing some activities as follows.

ACTIVITY 1

Place a brick on the floor. Tie a string (strong thread) around the brick and connect it to the hook of a spring balance. Apply a pulling force to the brick by pulling the other end of spring balance by hand till the brick just begins to slide (move slowly) on the floor (see Figure 6). Note down the reading on spring balance when the brick begins to slide. This reading of spring balance will give us the magnitude of force of friction between the surface of floor and the surface of brick (which are in contact with each other).

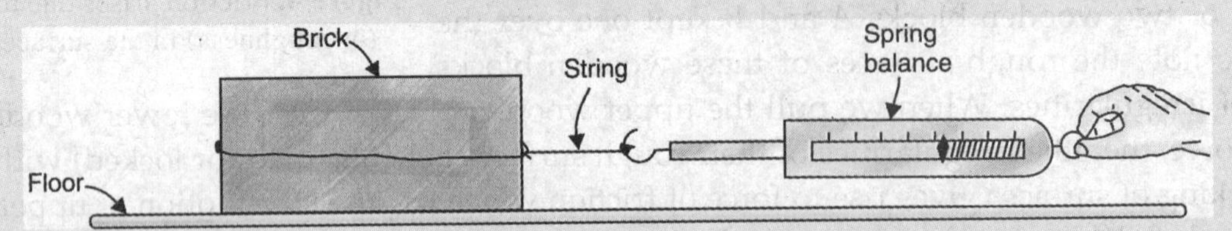

Figure 6. Activity to study the dependence of friction on the nature of surfaces.

(*i*) Let us now wrap a piece of polythene around the brick and repeat the above activity. We note the spring balance reading when the polythene wrapped brick just begins to slide on the floor. We will find that this reading of spring balance is *smaller* than the first reading of spring balance (when there was no polythene around the brick) indicating that the force of friction has decreased. From this observation we conclude that ***wrapping of polythene sheet makes the surface of brick smooth due to which the friction with floor decreases.***

(*ii*) Remove the polythene sheet from the brick. We now wrap a jute cloth around the brick and repeat the above activity once again. We note the spring balance reading when the jute wrapped brick just begins to slide on the floor. In this case we find that the reading of spring balance is *greater* than the first reading of spring balance (when nothing was wrapped around the brick) indicating that the force of friction has increased. From this observation we conclude that ***wrapping of jute cloth makes the surface of brick* more *rough due to which the friction with floor increases.***

From the above activity we learn that when the surface of brick is made more smooth by wrapping polythene sheet, the friction with floor decreases. On the other hand, when the surface of brick is made more rough by wrapping a jute cloth, then the friction with floor increases. Thus, **the friction depends on the nature of two surfaces**. That is, the friction depends on the smoothness or roughness of the two surfaces.

ACTIVITY 2

Make an inclined plane on a smooth marble floor by keeping a wooden board in tilted position with the help of a brick placed behind the wooden board (see Figure 7). Mark a horizontal line *AB* with a ball pen on the upper half of the inclined wooden board (as shown in Figure 7). Hold a pencil cell on the line *AB* marked on the inclined wooden board and then release it. The pencil cell will move down rapidly from the inclined board and travel a certain distance on the marble floor before coming to rest (or stopping) (see Figure 7). The moving pencil cell stops due to friction exerted by the marble floor. We note the distance covered by the pencil cell on the marble floor (from the base of the inclined board). This distance will give us an idea of the friction between marble floor and pencil cell.

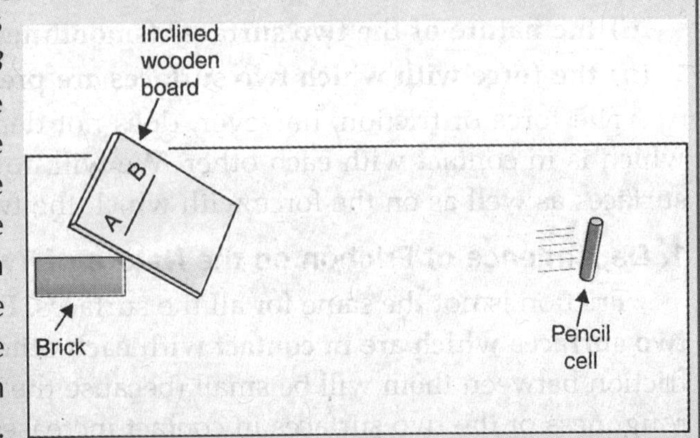

Figure 7. The pencil cell travels different distances on different surfaces (because the friction on different surfaces is different).

(*i*) Let us now put some water on the marble floor to make it wet. We repeat the above activity by releasing pencil cell from the same line *AB* of inclined board.

Again the pencil cell will travel a certain distance on wet marble floor before coming to rest. We note the distance travelled by the pencil cell on wet marble floor. We will find that ***the pencil cell travels a larger distance on the wet marble floor indicating that the friction on wet marble floor is less (than that on dry marble floor).***

(*ii*) We make the wet marble floor dry by wiping it. Let us now spread a sheet of newspaper on the dry marble floor. Repeat the above activity by releasing the pencil cell from the same marked line *AB* on the inclined board and note the distance travelled by the pencil cell before coming to rest. We will find that ***the pencil cell travels less distance on the newspaper sheet (than that on the dry marble floor) indicating that the friction exerted by newspaper is more than that exerted by dry marble floor.***

(*iii*) Let us remove the newspaper and spread a towel on the marble floor. Repeat the activity once again by releasing the pencil cell from the same marked line *AB* on the inclined board and note the distance travelled by the pencil cell before coming to rest. We will find that ***the pencil cell travels the least distance on the towel indicating that the friction exerted by towel is even greater than that of newspaper.***

From the above activity we see that when the surface of marble floor is made more smooth by making it wet with water, the friction with pencil cell decreases. On the other hand, when the surface of marble floor is made more rough by covering it with newspaper or towel, then the friction increases. Thus, **the friction depends on the nature (smoothness or roughness) of the two surfaces.**

2. Dependence of Friction on the Force With Which Two Surfaces are Pressed Together

Friction is caused by the interlocking of irregularities of the two surfaces when one object is placed over another object. If the two surfaces of objects are pressed together harder by a greater force, then the friction will increase (because pressing together two surfaces of objects with a greater force will increase the interlocking in the two surfaces). This will become more clear from the following activity.

ACTIVITY

Suppose we have two boxes of the same size but one box is light and the other box is heavy (see Figure 8). If we push both the boxes on the floor, one by one, we will find that we have to apply only a small force

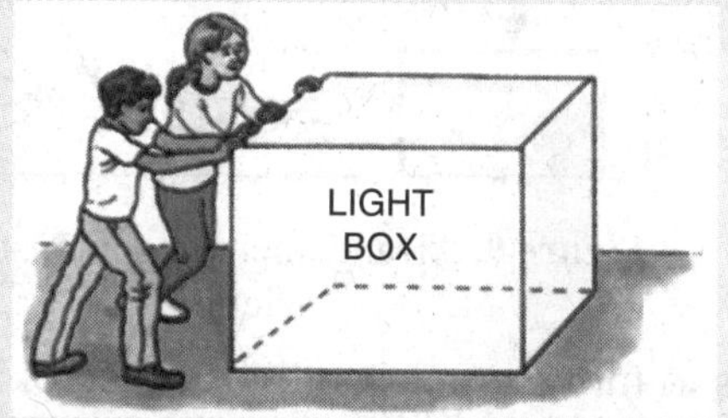

(*a*) It is easier to push a light box (having less weight) because it presses the floor with less force and hence friction is less

(*b*) It is difficult to push a heavy box (having more weight) because it presses the floor harder (with greater force) and friction is much more

Figure 8.

to make the lighter box move on the floor but a much larger force has to be applied to make the heavier box move on the floor. This shows that there is less friction between the light box and floor but much more friction between the heavy box and the floor (see Figure 8). We know that the weight of a box is also a force (which acts in the downward direction). Now, *because of its smaller weight, the light box presses on the floor with less force and hence the friction between lighter box and floor is less* [see Figure 8(*a*)]. This lesser force of friction allows the lighter box to be moved easily by applying a smaller push. On the other hand, *because of its greater weight, the heavy box presses on the floor with a greater force. Since the surfaces of heavy box and floor are pressed together harder (with a greater force) the friction between them increases and becomes much greater* [see Figure 8(*b*)]. This greater friction does not allow the heavy box to be moved on floor by applying a small force. A much larger force of our push has to be applied to make the heavy box move on floor.

From the above activity we conclude that **the friction between two surfaces depends on the force with which the two surfaces are pressed together**. In general, *greater the weight of an object which moves over another surface, greater will be the friction between them.* The force of friction increases when the two surfaces are pressed together harder (because of the greater weight of one of the objects) because then the interlocking of the irregularities of their surfaces increases. We will now discuss static friction, sliding friction and rolling friction.

STATIC FRICTION, SLIDING FRICTION AND ROLLING FRICTION

Friction is of three types :

(i) **Static friction,**

(ii) **Sliding friction,** and

(iii) **Rolling friction.**

We will now describe the three types of friction in detail, one by one. Let us start with the static friction.

1. Static Friction

The maximum frictional force present between any two objects when one object just tends to move or slip over the surface of the other object, is called static friction. Static friction is a kind of starting friction because an object just tends to start moving, it does not actually move. The object remains static (or stationary) in this case. We can demonstrate the static friction between a wooden block and a table top by performing an activity as follows.

ACTIVITY

A wooden block (having a hook attached to its one side) is kept on the horizontal surface of a table. A spring balance is attached to the hook of wooden block (as shown in Figure 9). We pull the spring balance to the right side with a small force of our hand. This will exert a force on the wooden block. The wooden block, however, does not move because its motion is being opposed by the force of friction which acts in the opposite direction (to the left side). As we increase the pulling force applied to the wooden block (through spring balance), the friction also goes on increasing. But ultimately, when the applied force becomes a little more than the maximum frictional force, the wooden block just tends to move or slip on the surface of table. This means that the frictional force acting between the wooden block and table top has a maximum value beyond which it cannot increase. The force which we are exerting in making the wooden block just tend to move or slip is equal (but opposite) to the force of friction. In Figure 9 we can see that when the wooden block just tends to move or slip on the table top, the spring balance shows a reading of 5 N force (5 newtons force). This means that the magnitude of static friction between this wooden block and table top is of 5 newtons.

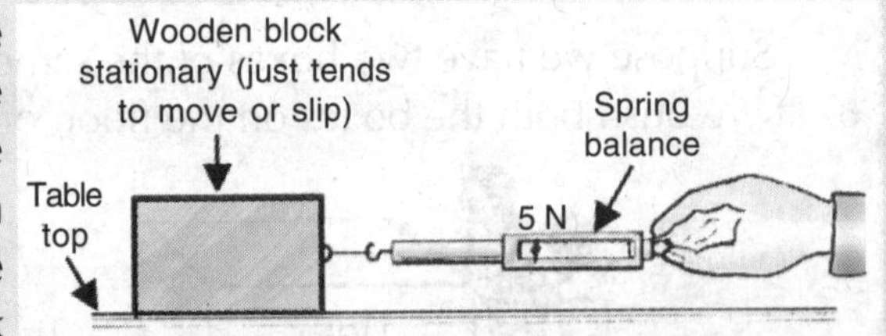

Figure 9. Arrangement to demonstrate static friction.

From the above discussion we conclude that the force required to overcome friction at the instant an object just tends to start moving from rest, is a measure of static friction. Static friction comes into play when we try to move a stationary object (which is at rest). **Please note that in the case of static friction, the object is actually not moving or sliding over the other object, it only tends to move or slide.**

2. Sliding Friction

The frictional force present when one object moves slowly (or slides) over the surface of another object, is known as sliding friction. Thus, sliding friction comes into play when an object is sliding (moving slowly but continuously) over another object.

ACTIVITY

We can demonstrate the sliding friction by extending the above activity further. In the above activity we have seen that when a certain force (equal and opposite to static friction) is applied by us, the wooden block just tends to move or slip on the table.

Let us now increase the force applied to pull the wooden block a little more (by pulling the spring balance more). We will see that the wooden block begins to slide (or move slowly) on the table top (see Figure 10). The force required to keep an object moving slowly (or sliding) with the same speed is a measure of the sliding friction. It has been found that the force required to keep the wooden block sliding (once it has started sliding) is less than the static friction. In other words, when an object starts sliding, then the friction is less. This means that **the sliding friction is smaller than the static friction**. For example, in Figure 10 we can see that when the wooden block starts sliding, then the spring balance shows a reading of 4 N force (4 newtons force). This means that the magnitude of sliding friction is 4 newtons (which is smaller than the static friction of 5 newtons).

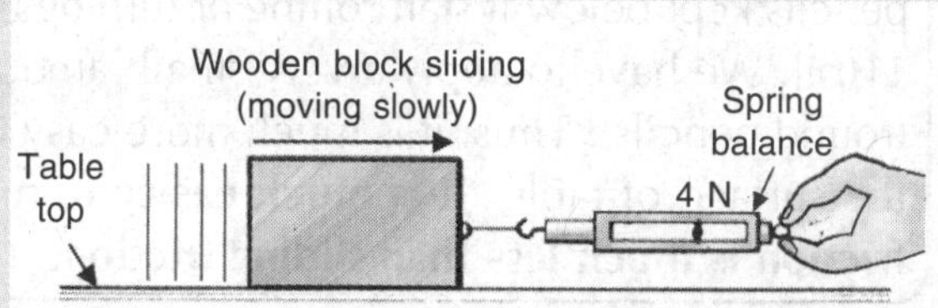

Figure 10. Arrangement to demonstrate sliding friction.

Since the sliding friction is smaller than the static friction, it is easier to keep an object moving which is already in motion than to move the same object from rest (or stationary position). Let us see **why sliding friction is smaller (or less) than the static friction**. This can be explained as follows : When an object has already started moving (or sliding), the irregularities on its surface do not get enough time to lock into the irregularities on the surface of the other object completely. Since the interlocking of the two surfaces is less when an object has already started moving, therefore, the sliding friction is smaller than the static friction.

3. Rolling Friction

When an object (like a wheel) rolls over the surface of another object, the resistance to its motion is called rolling friction. It is always easier to 'roll' than to 'slide' an object over another object. So, **rolling friction is much less than sliding friction**. Thus, ***rolling reduces friction.*** We will now describe an activity to demonstrate that it is much easier to roll an object over another object than to slide it. It will also show that rolling friction is much smaller than sliding friction.

ACTIVITY

Let us keep a thick book on a table. Now push the book with your hand [see Figure 11(*a*)]. We will find that it is not very easy to push the thick book lying on the table and make it move (or slide). We have to apply fairly large amount of force to make the book move (or slide) on the table. This is because when the book lying directly on the surface of table moves, then sliding friction comes into play. The large sliding

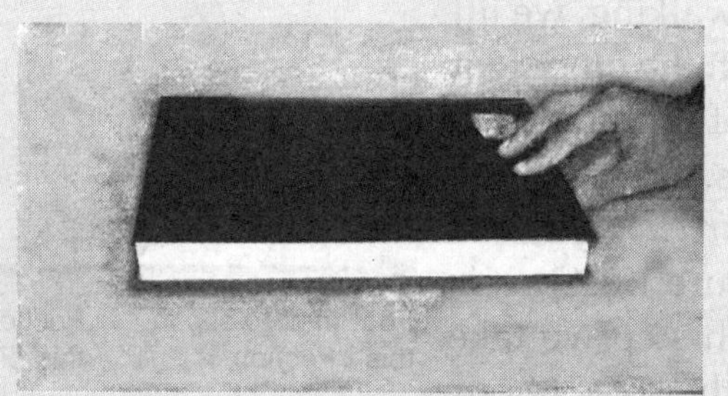

(*a*) Motion of book directly on the surface of table : Sliding friction

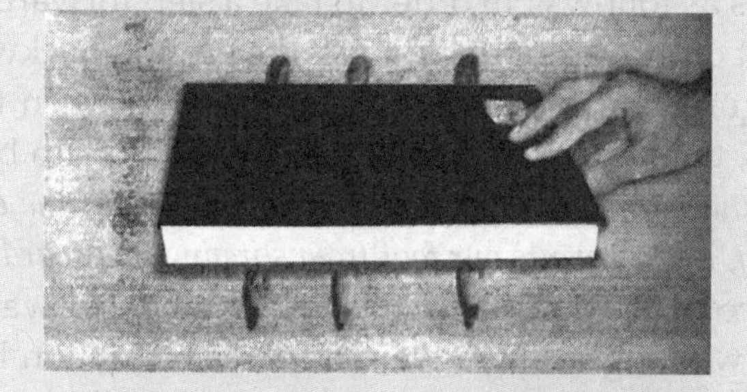

(*b*) Motion of book on round pencils (with pencils rolling on the surface of table): Rolling friction

Figure 11. Rolling friction is much less than sliding friction.

friction between the surface of table and bottom of thick book makes the book comparatively difficult to move. Now, take three round pencils (round pencils are cylindrical in shape and act as rollers). Place the round pencils parallel to one another on the table top. Let us place the same thick book over the round pencils [as shown in Figure 11(*b*)]. Now push the book with your hand. When we push the book, the round pencils kept below it start rolling or turning (like wheels) and make the book move forward easily [see Figure 11(*b*)]. We have to apply a very small amount of force to make the book move when it is placed on rollers (round pencils). Thus, it is much more easy to move the book placed on rollers than to slide it directly over the surface of table. **It is much easier to move an object kept on rollers than to slide it because rolling friction is much less than sliding friction.**

From the above discussion we conclude that when an object rolls over another object, then rolling friction comes into play between them. Rolling friction is much smaller than sliding friction. Since rolling friction is much less (than sliding friction) it is very easy and convenient to pull heavy luggage (like heavy suitcases, etc.) fitted with rollers. **Heavy machines can be easily moved from one place to another by placing round logs of wood under them and then pushing with the force of hands.** The round logs of wood act as rollers (a kind of wheels) and make it much easier to move the heavy machine kept on them. Figure 12 shows a large block of stone being moved by the use of rollers. Please note that the rollers turn around like the wheels and take the block of stone forward. Thus, rollers are a kind of wheels. We can now understand why wheel is said to be one of the greatest inventions of mankind. This is because wheels greatly reduce friction (by rolling) and hence can be used to move even heavy objects (like cars, buses, trucks and trains, etc.) rather easily.

Figure 12. Rollers between the stone block and ground reduce friction. It becomes rolling friction.

FRICTION IS A NECESSARY EVIL

Frictional force plays an important role in our daily life. In some cases, friction is **useful** and we want to keep it but in other cases friction is **harmful** and we wish to reduce it. We will now discuss the advantages and disadvantages of friction in detail, one by one.

ADVANTAGES OF FRICTION

Friction (or frictional force) is necessary because it helps us in performing many of our daily life activities. Friction is useful to us because of the following advantages.

1. Friction Enables Us to Walk Without Slipping

We are able to walk on ground because friction between the sole of our shoes and ground prevents us from slipping over the ground. This happens as follows : In order to take a step forward during walking, we lift one foot off the ground and push the ground backwards with the other foot (see Figure 13). If there were no friction between the sole of our shoe and ground, then our shoe on the ground would slip backwards. *Since we push the ground backwards, the force of friction acts in the opposite direction, forward direction, and prevents our foot from slipping backwards* (see Figure 13). So, it is the force of friction which makes us move forward at each step we take during walking. If there were no force of friction between the soles of our shoes and the ground, it would not be possible to walk (because our shoe would slip everytime we tried to walk).

Figure 13. Friction enables us to walk on ground.

Walking on slippery ground is difficult because the frictional force on slippery ground is much less which may not be sufficient to prevent us from slipping. It is also difficult to walk on a well polished floor (or on ice) because the friction on these smooth surfaces is very small (which cannot prevent us from slipping). If a person throws a bucket of water on a smooth marble floor, it would become even more difficult for us to walk on this wet marble floor. This is because the friction on wet marble floor (having a layer of water on it) becomes very small which cannot prevent us from slipping. So, many times people slip on wet marble floor and fall down (see Figure 14). **When we accidently step on a banana peel thrown on the road, we usually slip and fall down**. This is because the inner side of banana peel being smooth and slippery reduces the friction between the sole of our shoe and the surface of road. And this small frictional force is not enough to prevent our foot from slipping backwards.

Figure 14. This boy slips on wet marble floor and falls down due to very small friction.

2. Friction Enables a Car to Move on Road Without Skidding

The friction between tyres of a car and the road enables a car to move forward on road without skidding. This happens as follows : When the engine of car makes the wheel of a car turn, the tyre pushes the road backwards at its point of contact with the road (see Figure 15). The friction between tyre of car and the surface of road acts in the forward direction and prevents it from skidding (see Figure 15). In this way, friction provides the forward force which drives the car. ***If there were no friction between car tyres and road, then the wheels of car would spin at the same place but the car would not move forward at all. The car would stay where it was.*** It becomes somewhat difficult to drive and control a car on wet road. This is because the presence of water on the surface of a wet road reduces friction and makes it slippery. And because of reduced friction, there are more chances of skidding of car (especially when brakes are applied suddenly). So, driving a car on wet roads needs extra care. Just like a car, **friction also enables a bicycle to move along a road**. In fact, all the vehicles are able to move on road because of the presence of friction between their tyres and the surface of road.

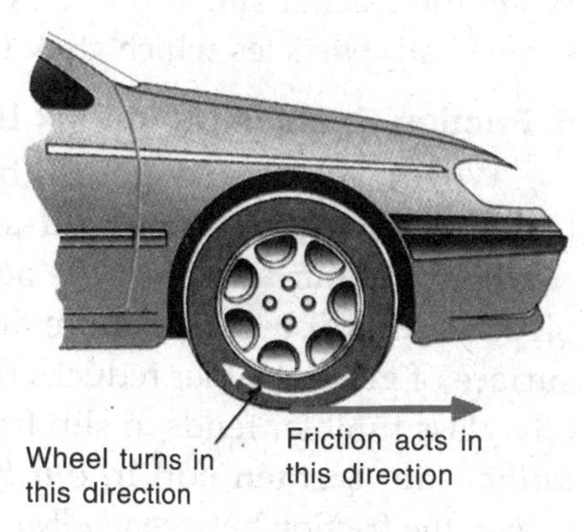

Figure 15. Friction enables a car to move on the road.

3. Friction Enables Us to Apply Brakes and Slow Down or Stop a Moving Car

The brakes of a car work by friction. In a disc brake, a steel disc attached behind each car wheel spins between two small brake pads. When the brakes are applied to the running car by pushing the brake pedal, the brake pads press against the discs of the rotating car wheels. This produces friction between brake pads and the discs, making the wheels to slow down and ultimately stop. **The brakes of a bicycle also work by friction**. When the bicycle is running, the brake pads of bicycle do not touch the wheels (there is a gap between them). But when we press the brake lever to apply brakes to the running bicycle, then the brake pads rub on the rims of the bicycle wheels (see Figure 16). *The friction between brake pad and rim prevents the wheel from moving ahead.* Due to this, the running bicycle slows down and finally stops. *If there were no friction, then once a vehicle started moving, it would never stop (because brakes would not work without friction).* In fact, had there been no friction between the tyres of vehicles and the road, the vehicles could not be started or stopped or turned to change the direction of motion.

Figure 16. Friction between brake pad and rim of wheel stops the running bicycle.

4. Friction Enables Us to Write and Draw on Paper

We are able to write and make drawings on paper because there is friction between the tip of pencil

(or pen) and paper. The pencil has a thin, black core made of carbon (or graphite) which is called 'pencil lead'. When we write with a pencil, friction with paper rubs off carbon particles from the 'pencil lead' which stick to the paper and leave black marks on paper (which we see as our writing on paper) (see Figure 17). Similarly, when we write with a pen, the particles of ink rub off from the pen's refill due to friction with paper, stick on the paper and leave marks of writing on paper. If there were no friction between the pencil (or pen) and paper, writing and drawing on paper would not have been possible. **We cannot write with a pencil on a glass sheet because the glass surface is very smooth due to which the friction between the tip of pencil and glass surface is much less. This friction is not sufficient to rub off black graphite particles from the tip of pencil.** A teacher is able to write on blackboard with a chalk due to friction between the blackboard and the chalk. When the teacher starts to write on the blackboard with a chalk, the rough surface of blackboard rubs off some chalk particles which stick to the blackboard and appear as writing on the blackboard.

Figure 17. Friction between tip of pencil and paper enables us to write on paper.

5. Friction Enables Us to Pick Up and Hold Things in Our Hands

We are able to pick up this book and hold it in our hands due to friction between the book and the hands. Similarly, **we can hold a glass tumbler in our hands because of friction (between the glass tumbler and our hands)**. If the outer surface of a glass tumbler is oily or greasy (having a film of cooking oil on it), then it becomes more difficult to hold it. This is because the presence of a film of oil on the outer surface of glass tumbler reduces the friction between glass tumbler and our hands. Due to less friction, the oily glass tumbler tends to slip from our hand and it becomes more difficult to hold. **It is easier to hold a *kulhar* (an earthen pot) in our hand than a glass tumbler**. This is because due to the rough surface of *kulhar*, the friction between *kulhar* and our hand is much more which makes it easier to hold it. On the other hand, due to the smooth surface of glass tumbler, the friction between glass tumbler and hand is much less which makes it comparatively difficult to hold it. If there were no friction, it would not be possible to hold a book, a glass tumbler, a *kulhar* or any other object in our hands.

6. Nails Can be Fixed in a Wall (or Wood) Due to Friction

We are able to fix nails in a wall due to friction. When we hammer a nail into the wall, it is the friction between the surface of nail and wall which holds the nail tightly in the wall. Without friction, nails could not be fixed in a wall to hold things. **Nails and screws are also held in wood by friction**. If there were no friction, then nails and screws could not be used to hold pieces of wood together and hence we could not make any furniture (like table, chair, etc.). Thus, friction enables nails and screws to hold things together. **Friction enables knots to be tied in strings (ropes, etc.)**. In other words, knots in ropes are held together by friction. **Friction enables a person to climb a tree or pole** without sliding down all the time. An oily or greasy pole has much less friction due to which it becomes difficult to climb up a greasy pole. **Friction enables a ladder to be leaned against a wall** and not slip down to the floor. **Friction helps in the construction of buildings**. Without friction, no building could be constructed. **Friction enables the belts to drive machines in factories**. Without friction, belts would not drive machines.

7. Friction Enables Us to Light a Matchstick

When we rub a matchstick against the rough side of a matchbox, then friction between the head of matchstick and rough side of matchbox produces heat. This heat burns the chemicals present on the head of matchstick due to which the matchstick lights up (see Figure 18). So, the matchstick catches fire and starts burning. *The burning of a matchstick would not be possible without friction*. It is

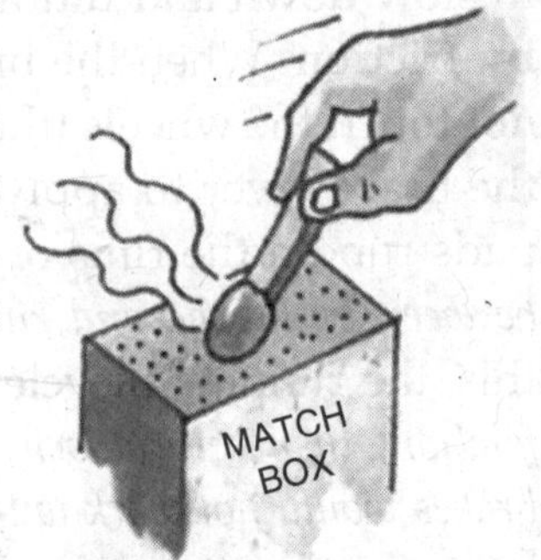

Figure 18. Striking a matchstick head on a rough surface produces fire by friction.

difficult to light a matchstick by striking it on a smooth surface because enough friction is not provided by a smooth surface to produce sufficient heat (for the match stick to catch fire and light up).

8. Friction Enables Us to Cut Wood with a Saw

We are able to cut wood because there is friction between the saw blade and log of wood. If there were no friction between the saw blade and log of wood, then cutting of wood would not be possible.

DISADVANTAGES OF FRICTION

Friction causes objects to wear away. Tiny pieces break off a surface when it rubs against another surface. The wear and tear due to friction can happen quickly or slowly depending on the nature of materials of the two surfaces. Friction (or frictional force) is harmful to us and considered an evil because of the following disadvantages.

1. Friction Wears Away the Soles of Our Shoes

When we walk on the road, there is friction between the soles of our shoes and the surface of road. When the soles of our shoes rub against the rough surface of road, then tiny pieces of the soles keep on breaking off slowly due to which the soles of our shoes wear out gradually (see Figure 19). The wear and tear of the soles of our shoes ultimately causes holes in the soles. The shoes get damaged and become unfit to wear.

Figure 19. The soles of our shoes wear out due to friction with road.

Figure 20. The tyres of vehicles (like cars) wear out due to friction with road.

2. The Tyres of Vehicles Wear Out Gradually Due to Friction

When the vehicles (like cars, buses, trucks, etc.) run on the road, there is friction between the surface of tyres and the surface of road. The rubbing of tyres with road keeps on breaking tiny pieces of rubber from the tyre's surface gradually. Ultimately, all the treads present on the surface of a tyre are worn out and the tyre becomes baldy (see Figure 20). Such baldy tyres have to be replaced by new tyres. Thus, the tyres of vehicles wear out due to friction with the road. **The tyres of bicycles also wear out gradually due to friction with the road.**

3. Friction Wears Out the Rubbing Machine Parts

There are many moving parts in machines (like gears, etc.) which rub against each other constantly. Due to friction, the rubbing parts of a machine wear out gradually (see Figure 21). Friction also wears out moving parts like ball bearings of bicycles (and other machines). The worn out or damaged parts of machines have then to be replaced by new ones.

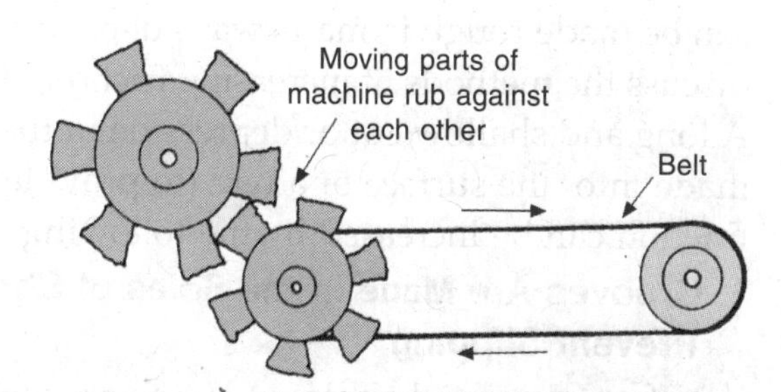

Figure 21. Moving machine parts wear out due to friction.

4. Friction Wears Out the Brake Pads of Vehicles Gradually

When the brakes of a vehicle (like car) are applied, a lot of friction is produced between the brake pads and moving part of the wheel (like disc). This friction wears out the brake pads gradually. Due to this, the brake pads of vehicles have to be replaced quite often (otherwise the brakes will not function properly). The brake pad of a bicycle is made of rubber. The new, soft rubber pad of a bicycle is shown in Figure 22(*a*). In a few months, the bicycle brake pad gets worn out due to friction between brake pad and moving rim of bicycle wheel (on applying brakes). The worn out brake pad of the bicycle is shown in Figure 22(*b*). The brake pads of a bicycle have to be replaced quite often because they wear away due to friction.

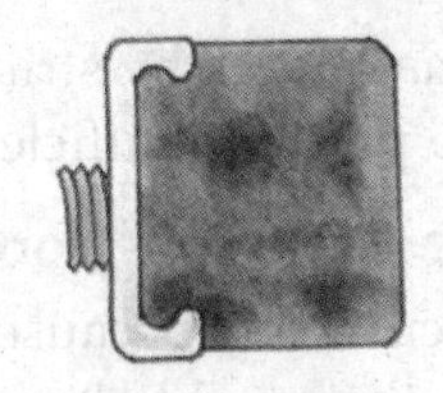

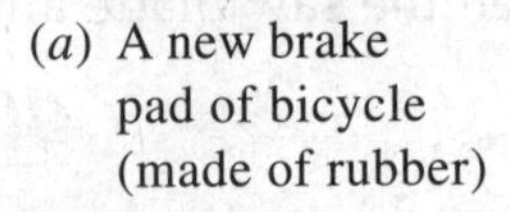

(*a*) A new brake pad of bicycle (made of rubber)

(*b*) Worn out brake pad a few months later

Figure 22.

5. Friction Wears Out Steps of Staircases in Buildings and Foot Over-Bridges

When a lot of people use the staircase in a building every day, the friction between soles of their shoes and the stone steps wears away hard stone steps very, very slowly (see Figure 23). Many times we see worn out steps of foot over-bridges at Railway Stations. The steps of foot over-bridges at Railway Stations (and other crowded places) wear out due to the friction caused by the shoes of extremely large number of people who use these over-bridges all the time.

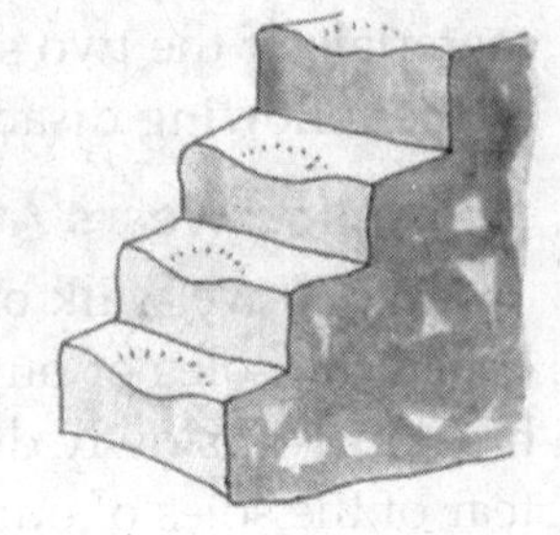

Figure 23. Friction wears away even hard stone steps (though very, very slowly).

6. Friction Produces Heat Which May Damage Machines

If we rub our hands together quickly for a few seconds, they feel warm. This is because friction between the hands produces heat (which makes them feel warm). Similarly, when we operate a mixer and grinder for a few minutes, its jar becomes hot. Here again heat is produced by friction. When the moving parts of a machine rub together, a lot of heat is produced due to friction between them. This heat may damage the machine gradually.

7. Friction Reduces the Efficiency of Machines

Some of the energy supplied to run a machine is wasted in overcoming friction between its moving parts and some of the energy is wasted in heat generated by the machine. This wastage of energy reduces the efficiency of a machine.

8. Friction Slows Down Motion

Friction reduces the motion of moving parts of a machine. In fact, all the moving things (such as cars, buses, aeroplanes, boats and ships, etc.) are slowed down by friction.

METHODS OF INCREASING FRICTION

In some cases friction is useful to us. In such cases we even want to increase friction to make it more useful to us. Friction can be increased by making the surface of an object 'rough'. The surfaces of objects can be made rough in many ways depending on the situation where these objects are to be used. Before we discuss the methods of increasing friction, we should know the meaning of the terms 'grooves' and 'treads'. A long and shallow cut or depression in the surface of a hard material is called groove. A series of patterns made into the surface of a tyre (to provide grip on the road) is called tread.
Friction can be increased by the following methods.

1. Grooves Are Made in the Soles of Shoes to Increase Friction and Prevent Slipping

If we observe the soles of our shoes, we will find that some grooves have been made in the soles by the makers of shoes (see Figure 24). The grooves are made in the soles of shoes to increase friction with the ground (or floor) so that the shoes get a better grip even on a slippery ground and we can walk safely (without the risk of slipping).

Figure 24. Soles of shoes have grooves to increase friction with ground (or floor).

2. Treads are Made in the Tyres of Vehicles to Increase Friction and Prevent Skidding of Vehicles on Wet Roads

When a road is wet (as during rains), there is a layer of water on the roads. The presence of water on the surface of road reduces friction between the tyres of vehicle and the surface of road due to which tyres lose their grip on the road. The reduced friction between tyres and road increases the chances of skidding of vehicle on wet road (especially when brakes are applied suddenly). Due to this it becomes difficult to control the running vehicle on a wet road. The treads on the surface of tyres are designed in such a way that they push away water from under the tyre (as the tyre rolls over a wet road) (see Figure 25). When water is pushed away from the surface of road beneath the tyre, friction between road and tyre increases and improves the tyre's grip on the wet road. The risk of skidding of a fast running vehicle is very much reduced. Thus, the treaded tyres of cars, buses, trucks, motor cycles, scooters and bicycles provide better grip on the road (especially on wet and slippery road). The treads of tyres do not help much in dry weather.

Figure 25. Tyres are designed with treads so that water on the road may squirt out from between the treads and increase friction between tyres and wet road.

3. Spikes are Provided in the Shoes of Players and Athletes to Increase Friction and Prevent Slipping

The players and athletes have to run fast. So, greater friction is required between the soles of their shoes and ground to prevent slipping. To increase friction, spikes are provided in the soles of shoes worn by players and athletes (see Figure 26). Spikes are the pointed nails which get into the ground and increase friction between shoe and the ground. This prevents the slipping of player or athlete on running.

Figure 26. Spikes in the shoes of players and athletes increase friction and prevent slipping.

4. Gymnasts Apply Some Coarse Substance on Their Hands to Increase Friction for Better Grip

Kabaddi players rub their hands with dry soil to increase friction and get a better grip on their opponent players (so that they may not slip out of hands).

5. Machine Belts are Made of Special Materials to Increase Friction and Drive Machine Wheels Properly

In many machines (like flour mills), belts are used to drive wheels for running the machines. Greater friction is required between the belts and machine wheels so that the belts can drive machine wheels properly without slipping off the wheels. To increase friction, the machine belts are made of special materials having rough surfaces.

METHODS OF REDUCING FRICTION

In some cases friction is harmful to us. In such cases, we wish to reduce friction so as to make it less harmful to us. For example, friction between the moving parts of machines causes a lot of wear and tear to the machine parts. It also leads to the production of undesirable heat and loss of energy. So, we should make efforts to reduce the friction (or minimise the friction) so as to prevent much damage. Friction can be reduced by the following methods.

1. Friction Can be Reduced by Making the Surfaces Smooth by Polishing

We know that friction is due to the roughness of surfaces. So, if we make the surfaces smooth by polishing, then friction will be reduced. For example, a slide in the park is polished to make its surface smooth and reduce friction. Due to reduced friction of a polished, smooth slide, children can slide down easily (see Figure 27). However, even highly polished objects look rough when seen through a microscope, so there is always some friction even on polished objects.

Figure 27. Polishing makes a slide smooth and reduces friction.

2. Friction Can be Reduced by Applying Lubricants (like Oil or Grease) to the Rubbing Surfaces

Friction between two surfaces in contact with each other is due to the interlocking of uneven surfaces having irregularities on them [as shown in Figure 28(*a*)]. When oil or grease is applied between the moving

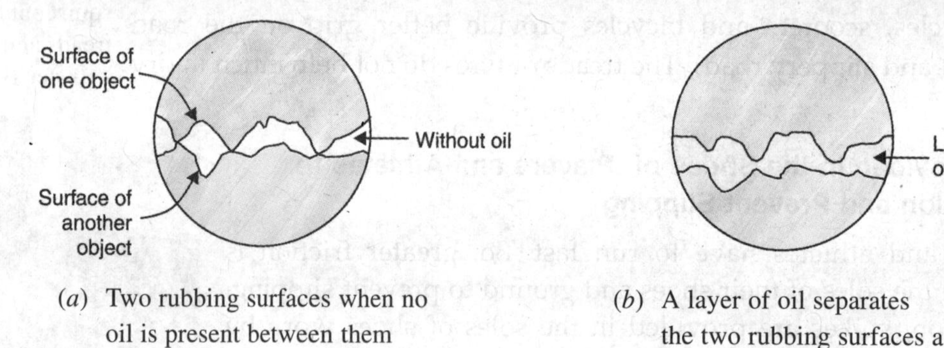

(*a*) Two rubbing surfaces when no oil is present between them

(*b*) A layer of oil separates the two rubbing surfaces a little and reduces friction

Figure 28. Views of rubbing surfaces of two objects as seen through a microscope.

parts of a machine, a thin layer of oil (or grease) is formed between the two rubbing surfaces. This layer of oil separates the two rubbing surfaces a little bit due to which their interlocking is reduced to a large extent [see Figure 28(*b*)]. Since the applying of oil (or grease) helps in avoiding interlocking between the two rubbing surfaces to a large extent, friction is reduced and movement becomes smooth. In fact, when oil or grease is applied to the moving parts of a machine, then their surfaces do not rub directly against each other, they rub through a layer of oil or grease (which is very smooth).

The substances which reduce friction are called lubricants. Oil, grease, graphite and fine powder are lubricants. The applying of lubricants (such as oil, grease, etc.) to a machine is called lubrication. A lubricant (like oil) reduces friction by helping the surfaces slide over each other smoothly. Thus, **friction can be reduced by lubrication.** Machines are lubricated with oil or grease to reduce friction. A well lubricated machine runs more smoothly and lasts longer. A bicycle mechanic and a motor mechanic uses grease between the moving parts of these machines to reduce friction and increase efficiency. Sometimes the hinges of a door make rattling noise when we open or close the door (due to increased friction caused by rusting). **When a few drops of oil are poured on the hinges of a door, the friction is reduced and the door moves smoothly** (without making any noise). We sprinkle fine powder as 'dry lubricant' on a carrom board to reduce friction. Please note that though we can reduce friction but we cannot make it zero (or nil) even if the surfaces are highly polished or large amounts of lubricants are applied. ***Friction can never be entirely eliminated.***

In some machines, it may not be advisable to use oil, etc., as lubricant. In such machines, an air cushion between the moving parts is used to reduce friction. For example, the frictional drag from the sea on a hovercraft is reduced by a cushion of compressed air (A vehicle or craft which travels over land or water on a cushion of air provided by a downward blast, is called a hovercraft).

3. Friction Can be Reduced by Using Wheels to Move Objects

It is quite difficult to move a heavy suitcase by dragging it on the ground because the sliding friction between the heavy suitcase and the ground is very large. Now, if this heavy suitcase is fitted with small wheels (called rollers), then it can be pulled very easily. This is because when we attach wheels, then sliding friction is converted into rolling friction. And rolling friction between the wheels of suitcase and the ground is much less. Thus, **friction can be reduced by attaching wheels (or rollers) to a heavy suitcase (or any other heavy object which is to be moved).** Due to very small rolling friction, even a child can pull a heavy suitcase fitted with small wheels (or rollers) (see Figure 29). From this discussion we conclude that friction can be reduced by attaching wheels to move the objects. So, **wheels are often used to reduce friction**. In fact, all the moving vehicles (like bicycles, cars, buses and trucks, etc.) are fitted with wheels to reduce friction with the road (so that they can move easily).

Figure 29. The use of wheels (or rollers) reduces friction.

Before we go further, we should know the meaning of the device called 'ball bearing'. **Ball bearing is a device which consists of a ring of small metal balls** (see Figure 30). The small metal balls of a ball bearing can roll freely. ***Ball bearings are designed to make the moving parts of a machine to roll over each other rather than slide.*** The ball bearing is introduced between the two surfaces which have to rotate over each other. For example, the axle is fixed on the inner side of a ball bearing and wheel is fixed to the outer side of the ball bearing. The ball bearing reduces friction by making the two surfaces (axle and wheel) to roll over each other. This happens due to the rolling action of small metal balls present inside the ball bearing.

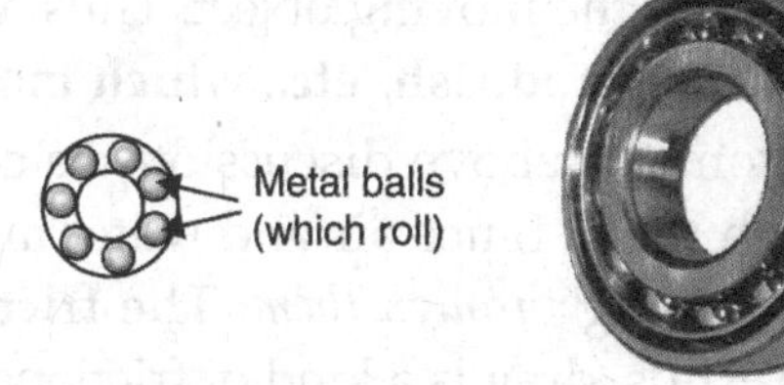

Figure 30. A ball bearing.

4. Friction Can be Reduced by Using Ball Bearings Between the Moving Parts of Machines

We have just studied that friction can be reduced by using wheels. Even a wheel produces some friction where its central hole (or hub) rubs with the axle [see Figure 31(*a*)]. To reduce friction still further, wheels are mounted on 'ball bearings'. The ball bearing is fixed between the hub of wheel and axle.

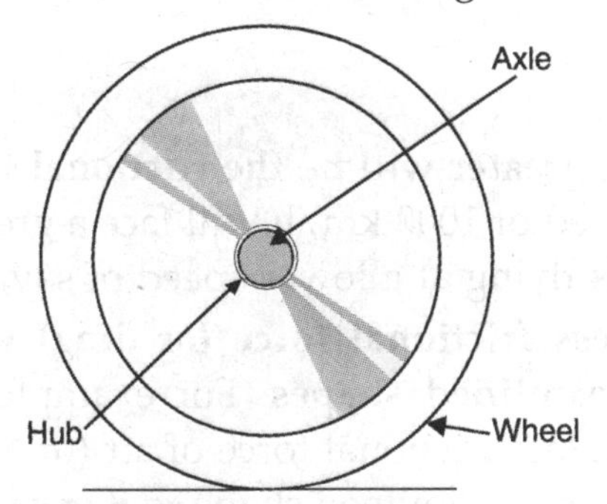

(*a*) Wheel produces some friction where its hub (central hole) rubs against the axle while rotating

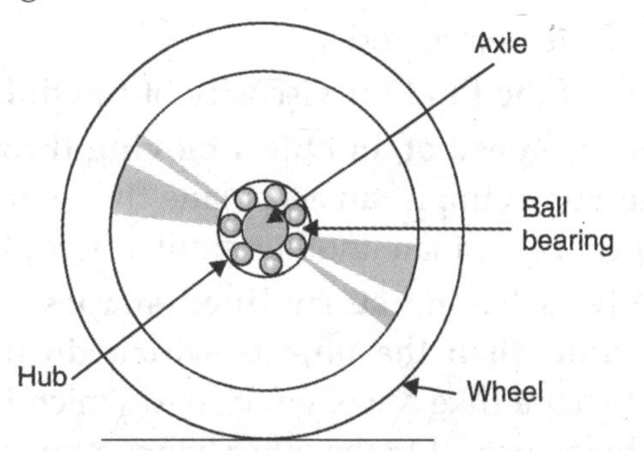

(*b*) When a ball bearing is put between the hub of wheel and axle, friction is reduced

Figure 31. The use of ball bearings in machines reduces friction because then rolling friction comes into play.

When the wheel revolves, the balls of ball bearing roll and reduce friction [see Figure 31(*b*)]. So, ***the use of ball bearing makes the wheel roll smoothly over the axle.*** Since rolling friction is much smaller than the sliding friction, sliding friction is replaced in most of the machines by rolling friction by using ball bearings. Thus, **in most of the machines, friction is reduced by using ball bearings**. For example, the wheels of a bicycle turn on sets of ball bearings. These ball bearings reduce friction because they roll rather than slide.

Ball bearings are used between the hubs and axles of machines such as bicycles, motor cars, and ceiling fans, etc., (to reduce friction).

FLUID FRICTION : FRICTION EXERTED BY LIQUIDS AND GASES

Before we describe fluid friction, we should know the meaning of the term 'fluid'. *Those substances which are able to flow easily are called fluids.* Fluids have no fixed shape. **Liquids and gases are fluids** (because they can flow easily). The most common liquid around us is water, so water is a fluid. The most common gas (or mixture of gases) around us is air, so air is also a fluid. Thus, **water and air are the most common fluids**. *There is friction whenever an object moves through a fluid. It is called fluid friction.*

Air is very light and thin, yet it exerts a frictional force on objects moving through it (which opposes their motion). When an object moves through the air, it pushes the air out of the way and air pushes back on the object. This push of air on the moving object creates friction which tends to slow down the moving object. Thus, **air exerts frictional force on cars, buses, aeroplanes, rockets, and birds, etc., moving through it**. Similarly, water (and other liquids) exert force of friction on objects which move through them and oppose their motion. When an object moves through water, it pushes the water out of the way and the water pushes back on the object. This push of water on the moving object creates friction which tends to slow down the moving object. Thus, **water exerts frictional force on objects like boats, speedboats, ships, submarines and fish, etc., which move through it**.

From the above discussion we conclude that air and water exert force of friction on objects moving through them. Since air and water are fluids, so in general we can say that : *Fluids exert force of friction on objects moving through them.* **The frictional force exerted by a fluid (air or water) is called drag (or drag force)**. Thus, drag is a kind of frictional force exerted by a fluid (like air or water) which opposes the motion of an object through that fluid. *Drag force acts in a direction opposite to the direction of motion of the object.* So, drag slows down the object moving through fluids and makes speeding up harder. **Typical examples of drag forces are the air resistance force experienced by a car or an aeroplane when they move at high speeds, and the water resistance force experienced by a speedboat moving rapidly in the sea**.

The magnitude of frictional force (or drag) exerted by a fluid on an object moving through it depends on four factors :

(*i*) speed of the object,
(*ii*) shape of the object,
(*iii*) size of the object, and
(*iv*) nature of the fluid (or viscosity of the fluid).

Higher the speed of an object moving through a fluid, greater will be the frictional force (or drag) acting on it. For example, an aeroplane flying at a higher speed of 1000 km/h will face a greater frictional force (or drag) of air than another similar aeroplane which is flying at a lower speed of say, 600 km/h.

The objects having streamlined shapes face much less frictional force (or drag) when moving through a fluid than the objects which do not have streamlined shapes. For example, a car has a streamlined shape (like a wedge) due to which it faces much less frictional force of air (or air drag) while running at high speed. On the other hand, a bus does not have a streamlined shape so it encounters a much greater frictional force (or drag) from air while running at the same speed.

Larger the size of an object moving through a fluid, greater will be the frictional force (or drag) acting on it. For example, a big aeroplane flying at a particular speed will face more frictional force of air (or drag) than a small aeroplane flying at the same speed.

Higher the viscosity (or thickness) of fluid, greater will be the frictional force (or drag) acting on an object moving through it. For example, water is much more viscous (or thick) than air, so there will be much more frictional force (or drag) on an object when it moves through water than when it moves through air.

Disadvantages of Fluid Friction

The main disadvantages of fluid friction are as follows :

(*i*) Fluid friction reduces the speed of objects moving through the fluids (by opposing their motion). It makes speeding up harder.

(*ii*) When objects move through fluids (air or water), they lose some of their energy in overcoming the fluid friction. This decreases their efficiency.

For example, when a car is running on the road, then some of the energy (or petrol) of the car is used up or lost in overcoming the friction of air which opposes its motion. Similarly, when a speedboat rushes through water, then some of its energy (or diesel) is used up or lost in overcoming the friction of water. So, in order to improve speed and to reduce the loss of energy (or fuels), efforts are made to reduce or minimise the fluid friction (or drag).

Before we discuss how fluid friction can be reduced we should know the meaning of the term 'streamlined' or 'streamlined shape'. **A 'body shape' which offers very little resistance to the flow of air or water around it, is called streamlined (or streamlined shape).** A streamlined shape is like a thin wedge (or thin triangular object) lying on its base and sloping upwards gradually.

Method of Reducing Fluid Friction

The fluid friction (or drag) can be reduced or minimised by giving special shape called 'streamlined shape' to the objects which move through fluids (like air or water). When an object having a streamlined body shape moves very fast, then the fluid (air or water) can flow past the moving object smoothly, reducing the fluid friction (or drag). For example, **cars are built with streamlined body shape to reduce air resistance (or drag) caused by air** (see Figure 32). A car with streamlined shape moves through the air easily (without facing much air resistance) and hence consumes less petrol than another car of same size running at the same speed that has a shape which gives it more air resistance (or drag). More streamlined the shape of a car, less petrol it will consume.

Figure 32. This car has a streamlined shape. Air can easily flow past this car. So, this car will face much less air friction (air resistance or drag).

An aeroplane has a streamlined shape to reduce air friction (air resistance or drag) that it encounters when flying at high speed through the sky. The shape of an aeroplane is similar to that of a bird in flight (see Figure 33). Both, the aeroplane and the bird have a streamlined body in the middle, two thin wings (one on each side of body), and a tail. The streamlined shape of an aeroplane has been built by scientists

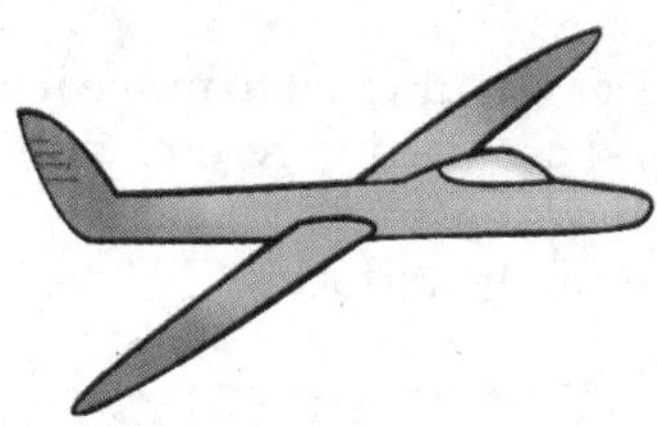

(*a*) An aeroplane

(*b*) A bird in flight

Figure 33. An aeroplane and a bird in flight have similar shapes (streamlined shapes).

and engineers whereas the streamlined shape of a bird has evolved in nature. In fact, the scientists and engineers got the idea for making streamlined shapes of various moving bodies from the living things in nature (such as birds and fish). **The rockets are also built with streamlined shapes so that they encounter the minimum air resistance (or drag) due to air** when they fly off at extremely high speeds. From the above discussion we conclude that cars, aeroplanes and rockets are streamlined to reduce friction with the air (air resistance or drag).

Most of the fish have streamlined body shape which helps them to move through water easily without facing much friction (or drag) from water. For example, **dolphins are streamlined by nature to reduce friction with water** in which they move. The streamlined shape of dolphin helps it to move easily through water. **The objects such as boats, speedboats, ships and submarines which move in water are also built with streamlined body shapes to reduce the frictional force of water (or drag)** and make them move easily through water (see Figure 34). Thus, boats, speedboats, ships and submarines are streamlined to reduce frictional force (or drag) due to water. We are now in a position to **answer the following questions :**

Figure 34. A speedboat has a streamlined shape to move easily through water.

Very Short Answer Type Questions

1. Name the force which always opposes motion.
2. Why does a fast car slow down if its engine is switched off ?
3. Which type of surfaces produce (*a*) least friction, and (*b*) too much friction ?
4. What is the direction of force of friction acting on a moving object ?
5. Name a device which is used to measure force acting on an object.
6. What is a spring balance ?
7. Out of sliding friction, static friction and rolling friction :
   (*a*) which one is the smallest ?
   (*b*) which one is the largest ?
8. Which type of friction comes into play when a book kept on cylindrical pencils is moved by pushing ?
9. Why is it more difficult to walk properly on a well-polished floor ?
10. Why is it difficult to walk on a wet marble floor ?
11. Which force is responsible for the wearing out of car tyres ?
12. What prevents you from slipping every time you take a step forward ?
13. Name the force which helps things to move and stop.
14. What enables us to fix nails in a wall and knots to be tied ?
15. What makes the steps of foot over-bridges at Railway Stations to wear out slowly ?
16. What is done to increase friction between the tyres and road ?
17. Why do gymnasts apply a coarse substance to their hands ?
18. Why do *kabaddi* players rub their hands with dry soil ?
19. Name the device which is used between the hubs and axles of bicycle wheels to reduce friction.
20. What is the purpose of using ball bearings in machines ?
21. Name any two machines in which ball bearings are used.
22. Name the device which is attached to heavy luggage (such as a heavy suitcase) to move it easily by pulling.
23. Name one example from everyday life where wheels (or rollers) are used to reduce friction.
24. Why does oiling the axles of a bicycle make the bicycle move more easily ?
25. State one way in which the friction between wheel and its axle can be reduced.
26. Name two common lubricants.
27. Why do we sprinkle fine powder on carrom board ?
28. Which force gets reduced when the two surfaces in contact are polished to make them smooth ?
29. Why is the surface of a slide polished to make it smooth ?
30. Name the force which increases when the two surfaces in contact are made more rough.
31. What is the special name of frictional force exerted by fluids (like air or water) ?
32. What is the name of 'special shape' which is given to objects moving through air (or water) to reduce drag ?
33. Why are grooves provided in the soles of shoes ?
34. Why are treads made in the surface of tyres ?
35. Fill in the following blanks with suitable words :
   (*a*) Friction always opposes.............between the surfaces in contact with each other.
   (*b*) Sliding friction is............than the static friction.

(*c*) Friction produces................
(*d*) Friction prevents our foot from..............over the ground.
(*e*) Sprinkling of powder on the carrom board.............friction.
(*f*) Ball bearings reduce friction because they...............rather than slide.
(*g*) The friction when something moves through a liquid or gas is called.........
(*h*) Cars and speedboats are...........to reduce drag.
(*i*) Shapes that are designed to reduce air resistance are called............shapes.
(*j*) Objects which can move quickly through the water have a ....shape.
(*k*) The shape of an aeroplane is similar to that of a............in flight.

Short Answer Type Questions

36. When we try to push a very heavy box kept on ground, it does not move at all. Which force is preventing this box to move forward ? Where does this force act ?
37. Suppose your writing table (or desk) is tilted a little. A book kept on the table starts sliding down. Draw a diagram to show the direction of force of friction acting on the book.
38. Which will cause more friction : a rough surface or a smooth surface ? Why ?
39. Explain why, sliding friction is less than static friction.
40. What is meant by 'rolling friction' ?
41. Iqbal has to push a lighter box and Seema has to push a similar heavier box on the same floor. Who will have to apply a larger force and why ?
42. Why does a man slip when he steps on a banana peel thrown on the road ?
43. Car wheels often spin on icy roads. Explain why.
44. Explain why :
(*a*) a pencil will write on paper but not on glass.
(*b*) climbing a greasy pole is very difficult.
45. Why does a matchstick light when we strike it on a rough surface ?
46. Why is it difficult to light a matchstick by striking it on a smooth surface ?
47. Which parts wear away first in shoes ? Give a reason for your answer.
48. Why do brake pads of bicycles have to be replaced quite often ?
49. A pencil eraser loses tiny pieces of rubber each time you use it. Why does this happen ?
50. What happens when you rub your hands vigorously for a few seconds ? Why does this happen ?
51. Explain how, friction enables us to walk without slipping.
52. Which is easier to hold in hand : a *kulhar* (earthen pot) or a glass tumbler ? Why ?
53. How does a bicycle stop when its brakes are applied ?
54. Explain why, the soles of our shoes wear out gradually.
55. Why do tyres of cars wear out gradually ?
56. State two advantages and two disadvantages of friction.
57. Explain why, sportsmen use shoes with spikes.
58. How will you reduce friction between those machine parts which rub against each other ? Give the simplest method.
59. What is meant by lubrication ? Why is it important ?
60. Explain why, wheels are so useful.
61. Why are lubricants (oil or grease) applied to rubbing surfaces of machines ?
62. Explain with the help of diagrams, how the use of oil reduces friction between two surfaces in contact with each other.
63. Why are cars, aeroplanes and rockets streamlined ?
64. Explain why, a speedboat has a streamlined shape.
65. What are fluids ? Name two common fluids.

Long Answer Type Questions

66. (*a*) Define friction. What are the factors affecting friction ? Explain with examples.
(*b*) What is the cause of friction ? Explain with the help of a labelled diagram.
67. (*a*) Give examples to show that friction depends on the nature of two surfaces in contact.
(*b*) Give an example to show that friction depends on the force with which the two surfaces are pressed together.

68. (*a*) What is the difference between static friction and sliding friction ? For a given pair of objects, which of the two is greater ?
(*b*) How can a very heavy machine be moved conveniently from one place to another in a factory ? (No crane is available for this purpose).

69. (*a*) What is drag ? Give two examples of a drag force.
(*b*) How can you reduce the drag on something moving through the air ?

70. (*a*) What is meant by 'streamlined shape' ? Name an object which usually has a streamlined shape.
(*b*) Explain why, objects moving in fluids should have streamlined shape.

Multiple Choice Questions (MCQs)

71. A boy runs his toy car on dry marble floor, wet marble floor, newspaper and towel spread on the floor. The force of friction acting on the car on different surfaces in increasing order will be :
(*a*) Wet marble floor, Dry marble floor, Newspaper, Towel
(*b*) Newspaper, Towel, Dry marble floor, Wet marble floor
(*c*) Towel, Newspaper, Dry marble floor, Wet marble floor
(*d*) Wet marble floor, Dry marble floor, Towel, Newspaper

72. Four children were asked to arrange forces due to rolling, static and sliding frictions in a decreasing order. The correct arrangement is :
(*a*) Rolling, Static, Sliding
(*b*) Rolling, Sliding, Static
(*c*) Static, Sliding, Rolling
(*d*) Sliding, Static, Rolling

73. A big wooden box is being pushed on the ground from east to west direction. The force of friction due to ground will act on this box towards :
(*a*) north direction
(*b*) south direction
(*c*) east direction
(*d*) west direction

74. A spring balance can be used to measure :
A. Mass of an object
B. Force acting on an object
C. Density of an object
D. Weight of an object
(*a*) A and B
(*b*) B and C
(*c*) B and D
(*d*) only D

75. The friction between two surfaces does not depend on one of the following. This one is :
(*a*) amount of surface area of the two objects which is in contact with each other
(*b*) weight of the object which tends to move on the surface of other object
(*c*) degree of smoothness of surfaces of two objects in contact with each other
(*d*) degree of roughness of surfaces of two objects in contact with each other

76. If the sliding friction betwen two surfaces is found to be 8 N, then the static friction between these two surfaces is most likely to be :
(*a*) 5 N
(*b*) 10 N
(*c*) 4 N
(*d*) 2 N

77. Which of the following is not an advantage of friction ?
(*a*) it enables drawing to be made on paper
(*b*) it enables fallen things to be picked up
(*c*) it enables rubber pads to be rubbed off.
(*d*) it enables vehicles to move on ground

78. Which of the following statements is incorrect ?
(*a*) static friction is greater than rolling friction
(*b*) sliding friction is less than rolling friction
(*c*) rolling friction is less than static friction
(*d*) static friction is greater than sliding friction

79. If the static friction between two surfaces X and Y is found to be 20 N, then the rolling friction between these two surfaces should most likely be :
(*a*) 25 N
(*b*) 20 N
(*c*) 5 N
(*d*) 50 N

80. If the static friction between two surfaces P and Q is measured to be 50 N, then the sliding friction between these two surfaces is most likely to be :
(*a*) 75 N
(*b*) 45 N
(*c*) 55 N
(*d*) 65 N

81. Which of the following will produce the maximum friction ?
(*a*) rubbing of sand paper on glazed paper
(*b*) rubbing of sand paper on glass table top
(*c*) rubbing of sand paper on aluminium frame
(*d*) rubbing of sand paper on sand paper

82. Four similar cars having exactly the same mass are running at the same speed on the same road when brakes are applied at the same time. The cars come to a stop after covering distances of 5 m, 5.5 m, 4.8 m and 5.2 m respectively. The friction between the brake pads and discs will be the maximum in the car which travels the distance of :
(*a*) 5 m (*b*) 5.5 m (*c*) 4.8 m (*d*) 5.2 m

83. The weight of an object can be measured by a :
(*a*) beam balance (*b*) analytical balance (*c*) spring balance (*d*) physical balance

84. A book is lying on the horizontal table top. If we tilt the table a little, then the book starts sliding down slowly. This happens because :
(*a*) sliding friction is greater than static friction
(*b*) sliding friction is less than force of gravity
(*c*) static friction is greater than sliding friction
(*d*) force of gravity is less than sliding friction

85. A body shape which offers very little resistance to the flow of air (or water) around it is called :
(*a*) trimlined shape (*b*) steamlined shape (*c*) streaklined shape (*d*) streamlined shape

86. Which of the following should be used to reduce friction on a carrom board ?
(*a*) a lubricating oil (*b*) a dry lubricant (*c*) a layer of grease (*d*) a ball bearing

87. Which of the following does not have a streamlined shape ?
(*a*) aeroplane (*b*) boat (*c*) bird (*d*) bus

88. The frictional force exerted by a fluid is called :
(*a*) brag (*b*) drab (*c*) drag (*d*) tread

89. A person has applied some mustard oil on his hands. Which of the following objects will become most difficult for him to hold in his hand ?
(*a*) Earthen cup (*kulhar*) (*b*) thermocol tumbler (*c*) glass tumbler (*d*) wooden cup

90. Ball bearing is a device which usually converts :
(*a*) rolling friction into sliding friction (*b*) static friction into sliding friction
(*c*) sliding friction into rolling friction (*d*) rolling friction into static friction

Questions Based on High Order Thinking Skills (HOTS)

91. When a pencil cell is released from a certain point on an inclined wooden board, it travels a distance of 35 cm on floor *A* before it comes to rest. When the same pencil cell is released from the same point on the same inclined board, it travels a distance of 20 cm on floor *B* before coming to rest. Which floor, *A* or *B*, offers greater friction ? Give reason for your answer.

92. A car is moving towards North. What will be the direction of force of friction acting on this car due to surface of road ?

93. You spill a bucket of soapy water on a marble floor accidently. Would it make easier or more difficult for you to walk on the floor ? Why ?

94. What kind of friction comes into play :
(*a*) when a block of wood kept on table moves slowly ?
(*b*) when a block of wood kept on table just tends to move (or slip) ?
(*c*) when a block of wood kept on cylindrical iron rods moves ?

95. Explain why, it is easier to drag a mat on floor when nobody is sitting on it but much more difficult to drag the same mat when a person is sitting on it.

ANSWERS

1. Friction **4.** Opposite to the direction in which the object moves **35.** (*a*) motion (*b*) less (*c*) heat (*d*) slipping (*e*) reduces (*f*) roll (*g*) drag (*h*) streamlined (*i*) streamlined (*j*) streamlined (*k*) bird **71.** (*a*) **72.** (*c*) **73.** (*c*) **74.** (*c*) **75.** (*a*) **76.** (*b*) **77.** (*c*) **78.** (*b*) **79.** (*c*) **80.** (*b*) **81.** (*d*) **82.** (*c*) **83.** (*c*) **84.** (*b*) **85.** (*d*) **86.** (*b*) **87.** (*d*) **88.** (*c*) **89.** (*c*) **90.** (*c*) **91.** Floor B offers greater friction (because it makes the moving pencil cell stop at a lesser distance of 20 cm) **92.** Towards South (opposite to direction of motion of car) **93.** More difficult to walk ; Soapy water reduces friction on marble floor too much **94.** (*a*) Sliding friction (*b*) Static friction (*c*) Rolling friction **95.** When a person is sitting on the mat, then due to the weight of the person, the mat and floor are pressed together harder, increasing the friction too much.

CHAPTER 13

Sound

The sensation felt by our ears is called sound. **Sound is a form of energy. Sound is that form of energy which makes us hear.** We hear many sounds around us in everyday life. At home we hear the sounds of our parents talking to us. We also hear the sounds of telephone bell, radio, television, stereo-system, mixer-grinder and washing machine. At school we hear the sounds of our teachers, classmates, and the school bell. We hear the sounds of scooters, motorcycles, cars, buses and trucks on the roads. And the sound of a flying aeroplane is heard from the sky.

At a music concert, we hear the sounds produced by various musical instruments like *sitar, veena,* violin, guitar, *tanpura,* piano, harmonium, flute, *shehnai, tabla* and cymbals, etc. In a garden we hear the sounds of chirping of birds and rustling of leaves of the trees in breeze. In rainy season, we hear the sound of rain drops falling on the roof and the sound of thundering of clouds. In a factory we hear the different types of sounds made by various big and small machines. And at night, when most of the sounds cease (stop), we can still hear the faint sounds of the ticking of a clock and the buzzing of a mosquito.

Each sound is special to the object which produces it. For example, the ticking of a clock, the click of an electric switch and the whirr (sound of rotation) of a ceiling fan are the characteristic sounds produced by a clock, an electric switch and a ceiling fan, respectively. We rarely forget a sound having heard it once and quickly recognise this sound when we hear it again. We can even recognise a person from the sound of his voice even without seeing his face. For example, we can usually tell from the voice which singer is singing a particular song on radio even though we do not see the face of the singer. Thus, every sound has a unique 'quality' which distinguishes it from other sounds. Sound plays an important role in our daily life. It helps us to communicate (or talk) with others. We will now discuss how sound is produced. This is called production of sound.

SOUND IS PRODUCED BY VIBRATING OBJECTS

When an object moves 'backwards and forwards' (to-and-fro) rapidly, we say that the object 'vibrates' or that the object is 'vibrating'. *Sound is produced when an object vibrates (moves to-and-fro rapidly).* In other words, **sound is produced by vibrating objects.** So, whenever we hear a sound, then some object must be vibrating to produce that sound. The energy required to make an object vibrate and produce sound is provided by some outside source (like our hand, wind, etc.). ***In some cases, the vibrations of a sound producing object are quite large which we can see with our eyes. But in other cases, the vibrations of the sound producing object are so small that we cannot see them easily, we have to feel the vibrations of such an object by touching it gently with the fingers of our hand.*** We will now give some examples to show that sound is produced by vibrating objects.

1. Let us ring a bicycle bell and touch it gently with our fingers. We will find that a ringing bicycle bell (which is producing sound) is shaking back and forth continuously. We say that the bicycle bell is vibrating. Now, if we hold the ringing bicycle bell tightly with our hand, it stops vibrating, and the sound also stops coming. From this we conclude that **sound is produced by a vibrating bicycle bell.** When the bicycle bell vibrates, it produces sound; and when the bicycle bell stops vibrating, the sound also stops coming from it. *In general, we can say that a body must vibrate to produce sound.* The sound of a school bell is produced by the vibrations of **iron or brass plate** when it is hit by a hammer.

2. Stretch a rubber band and tie it tightly between two nails fixed on a table [as shown in Figure 1(*a*)]. In this position the rubber band is not vibrating and hence not producing any sound.

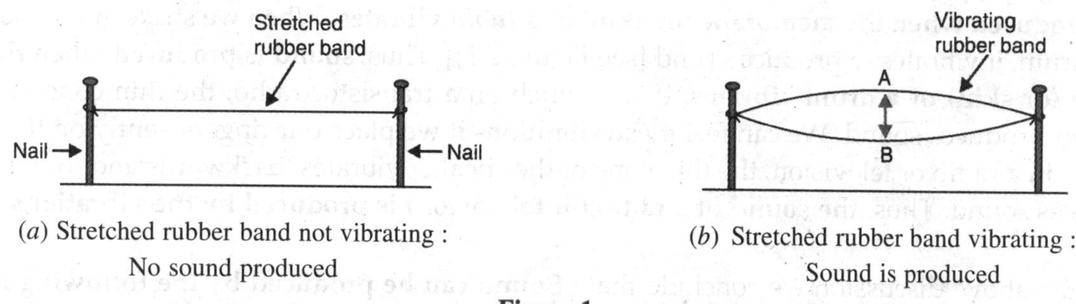

(*a*) Stretched rubber band not vibrating : No sound produced

(*b*) Stretched rubber band vibrating : Sound is produced

Figure 1.

Let us pluck the stretched rubber band in the middle with our finger. The rubber band starts vibrating and produces sound. We can see the vibrations of the stretched rubber band. The rubber band vibrates (moves backwards and forwards rapidly) between the positions *A* and *B* [see Figure 1(*b*)]. Thus, **sound is produced when a stretched rubber band vibrates.** Now, if we hold the vibrating rubber band tightly with our hand, the rubber band stops vibrating and the sound being produced by it also stops.

3. *Sitar* is a musical instrument. It has many stretched strings (stretched wires). If we pluck the string (wire) of a *sitar* in the middle, the *sitar* makes a sound. If we now put our finger gently on the *sitar* string, we can feel the string vibrating. Thus, **sound is produced when a *sitar* string vibrates.** The sound of a *veena* is also produced by the vibrations of **stretched strings.**

4. When we talk, we make sound. This sound is made by the vibrations of two vocal cords present in our voice box fixed in the throat. This can be shown as follows : Let us hold the fingers of our right hand gently on our throat and talk to one of our friends (or sing a song). When we are talking (making sound) our fingers feel that something is moving or vibrating inside the throat. Actually, when we talk, air from the lungs passes up the wind-pipe. This air makes the vocal cords in our voice box to vibrate rapidly. And vibrating vocal cords produce the sound (of our talk). Thus, **sound is produced when our vocal cords vibrate.** Please note that the vibrations of our vocal cords are small, so we can only feel them with our hands. We cannot see these vibrations with our eyes. Mosquitoes and bees make a buzzing sound by vibrating their wings very, very rapidly.

5. If we blow across the mouth of an empty test-tube, then a whistling sound is produced [see Figure 2(*a*)]. This sound is produced by the vibrations of air present in the test-tube. Thus, **sound is produced when the air column enclosed in a tube vibrates.** The sound of a flute (*bansuri*) is produced by the vibrations of air column enclosed in the flute tube. The sound of a bursting balloon is produced by the vibrations of **air** enclosed in the balloon (when it escapes).

(*a*) Sound is produced when the air column inside the test-tube vibrates.

(*b*) Sound is produced when the stretched membrane (or skin) of the drum vibrates.

Figure 2. Sound is produced by vibrations of objects.

6. If we hit the stretched membrane (or skin) of a *tabla*, the membrane starts vibrating and produces a sound. Now, if we put a few small pebbles on the membrane of this sound producing *tabla*, the pebbles will start jumping up and down showing that the *tabla* membrane is vibrating while producing sound. Thus, **sound is produced when the membrane (or skin) of a *tabla* vibrates.** When we strike at the membrane (or skin) of a drum, it vibrates to produce sound [see Figure 2(*b*)]. Thus, **sound is produced when the stretched membrane (or skin) of a drum vibrates**. If we switch on a transistor radio, the thin cone of its speaker vibrates and produces sound. We can feel these vibrations if we place our fingers gently on the cone of the speaker. So, in a radio or television, the thin cone of the speaker vibrates 'backwards and forwards' rapidly and produces sound. Thus, **the sound of a radio (or television) is produced by the vibrations of the cone of speaker.**

From the above discussion we conclude that : **Sound can be produced by the following methods** :

(*i*) **by vibrating strings (as in a *sitar*),**

(*ii*) **by vibrating air columns (as in a flute),**

(*iii*) **by vibrating membranes (as in *tabla*),** and

(*iv*) **by vibrating plates (as in a bicycle bell).**

We will now discuss how sound reaches from a sound producing object to our ears. This is called transmission of sound or propagation of sound.

Propagation of Sound

Sound is produced by the vibrations of an object. When an object vibrates back and forth in air, then the molecules of air close to this object also start vibrating back and forth with the same frequency. These vibrating air molecules pass on their motion to the next layer of air molecules due to which they also start vibrating back and forth. This process goes on and on. And ultimately, all the air molecules around the sound producing object start vibrating back and forth (just like the vibrating object). When the vibrating air molecules fall on our ears, the ears feel these vibrations as sound. Thus, **when an object vibrates (and makes sound), then the air around it also starts vibrating in exactly the same way and carries sound to our ears through the vibrations of its molecules.** And we say that a sound wave travels from the sound producing object to our ears, through the air.

SOUND PRODUCED BY HUMANS

The human beings produce sound by using the voice box which is called 'larynx'. Voice box (or larynx) is situated in our throat at the top of the wind-pipe (or trachea). **The human voice box (or larynx) contains two ligaments known as vocal cords.** The vocal cords are a kind of strings. **Sound is produced by the vibrations of vocal cords** (see Figure 3). The vocal cords are attached to muscles which change the tension (stretching) in the cords and the distance between the cords.

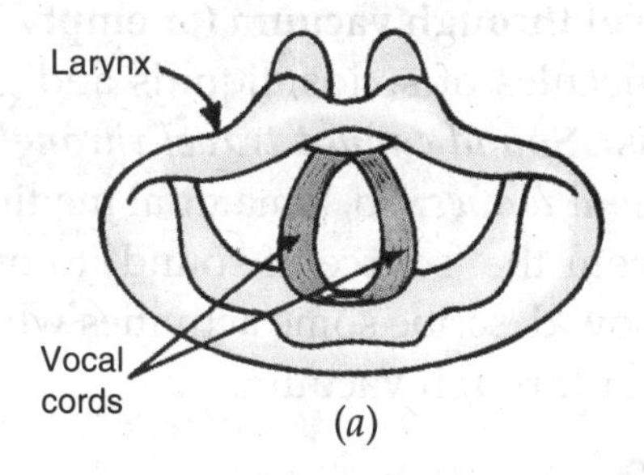

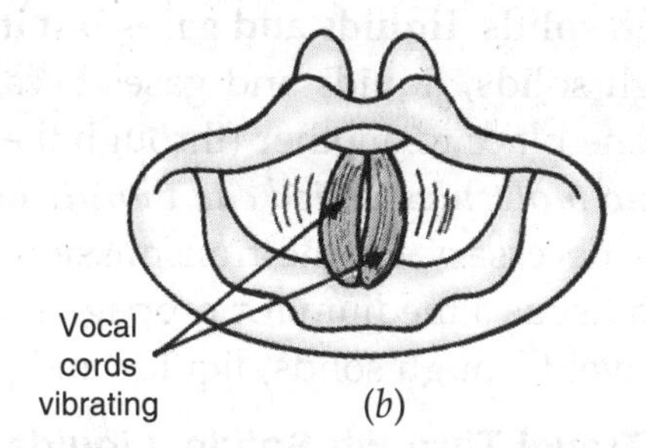

Figure 3. Sound is produced when our vocal cords vibrate.

(*i*) Normally, the muscles of vocal cords are completely relaxed due to which the vocal cords are separated and loose so that air from the lungs passes through them without producing any sound [see Figure 3(*a*)]. Thus, when we are not talking (or singing), the two vocal cords are far apart with a lot of gap between them.

(*ii*) When we want to speak, the muscles of vocal cords contract due to which the two vocal cords become stretched and close together leaving only a narrow slit between them [see Figure 3(*b*)]. **The lungs pass a current of air between the two vocal cords. This air makes the vocal cords vibrate. And the vibrating vocal cords produce sound.** Thus, when we talk or sing (or make any other sound), we actually make our vocal cords vibrate. And vibrations of vocal cords by expelled air produce vocal sounds.

ACTIVITY

We can demonstrate the working of vocal cords to produce sound as follows : Take two rubber strips of the same size. Place these two rubber strips one above the other. Hold the two ends of the rubber strips in your hands and stretch them tight. Keep the stretched rubber strips in front of your mouth and blow air through the thin gap between them (see Figure 4). As the air blows through the stretched rubber strips, a sound is produced. This sound is produced by the vibrations of stretched rubber strips when air rushes through the thin gap between them. Our vocal cords produce sound in a similar way.

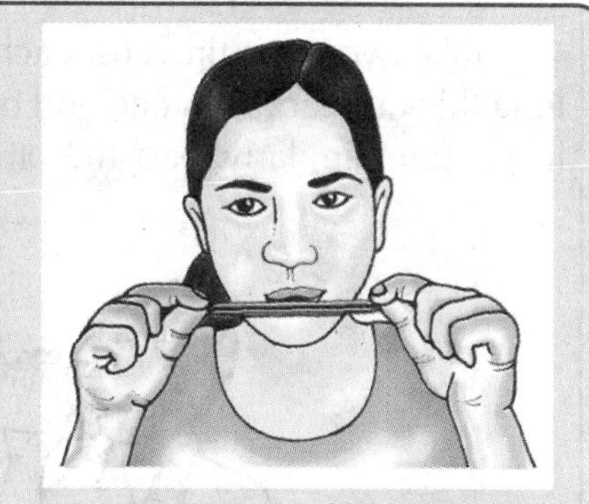

Figure 4. Activity to demonstrate the working of vocal cords to produce sound.

When we talk or sing, then the frequency of sound produced by us changes continuously. The changes in frequency of sound while talking or singing are brought about by the action of muscles attached to the vocal cords in the voice box. **When the muscles attached to vocal cords contract and stretch, the vocal cords become tight and thin, and a sound of high frequency is produced.** On the other hand, **when the muscles relax, the vocal cords become loose and thick, and a sound of low frequency is produced.**

The vocal cords of a man are about 20 mm long. The vocal cords of a woman are about 5 mm shorter than man. **Due to the shorter vocal cords, the frequency (or pitch) of a woman's voice is higher than that of a man.** Small children have very short vocal cords due to which the frequency (or pitch) of their voice is very high. This is why their voice is shrill. So, it is due to the different frequencies (or different pitch) caused by the different lengths of their vocal cords that the voices of men, women and children are different.

SOUND NEEDS A MEDIUM FOR PROPAGATION

The substance through which sound travels is called medium. The medium can be a solid substance, a liquid or a gas. And transmission of sound is called propagation of sound. So, **by saying that sound needs a medium for propagation, we mean that sound needs a solid, liquid or gas for transmission**. In other words, sound needs a material medium like solid, liquid or gas to travel and be heard. ***Sound* can**

travel through solids, liquids and gases but it cannot travel through vacuum (or empty space). Sound can travel through solids, liquids and gases because the molecules of solids, liquids and gases carry sound waves from one place to another (through their vibrations). ***Sound cannot travel through vacuum because vacuum has no molecules which can vibrate and carry sound waves.*** So, a material medium like air, water, wood, etc., is necessary for the transmission of sound from the 'source of sound' to our 'ears'. In other words, sound needs a medium for propagation. We will now describe some activities which will show that sound can travel through solids, liquids and gases, but not through vacuum.

Sound Can Travel Through Solids, Liquids and Gases

ACTIVITIES

Let us press our ear on to one side of a wooden bench and ask a friend to tap or scratch the other end of the bench lightly. We will hear the sound of tapping or scratching through the wooden bench quite loudly. This means that sound can travel through wood, which is a solid. Now take a metre scale made of metal and hold its one end close to your ear very carefully. Ask your friend to scratch the other end of metre scale lightly. You will be able to hear the sound of scratching through the metallic metre scale quite loudly (though other persons around you cannot hear the same sound of scratching). This means that sound can travel through a metal, which is a solid. In general, we can say that **sound travels through solid substances.** We can also make a toy telephone as follows to show that sound travels through solids.

Take two open tin-cans, each having a small hole at the centre of its bottom. Also take about 20 metres long thick thread. Pass one end of the thread into the hole of one tin-can and hold it inside the can by tying it to a pin. Similarly, pass the other end of thread into the hole of second tin-can and tie it to another pin.

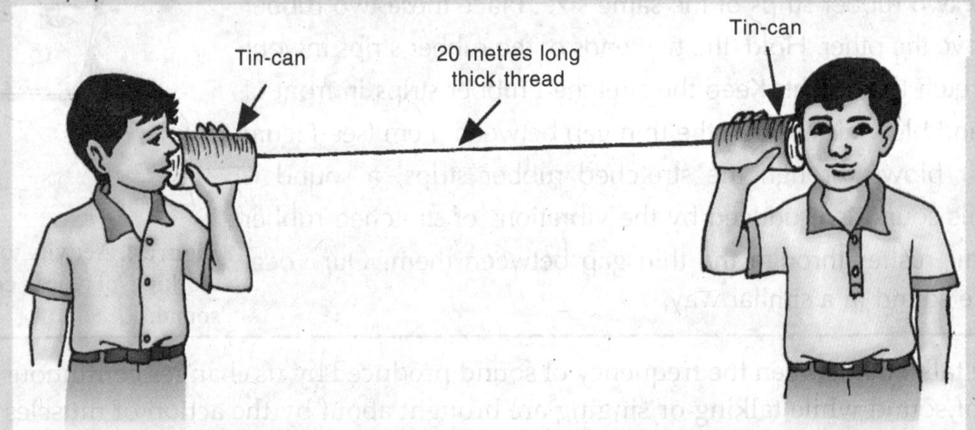

Figure 5. A toy telephone (Here sound travels through the thread, which is a solid substance).

The two tin-cans connected by the thread are now held by two children and taken as far as possible so that the thread gets stretched tightly (as shown in Figure 5).

Now, if one child speaks into one tin-can, he can be heard by the child at the other end who puts his ear to the other tin-can. For example, in Figure 5, the child at the left end is speaking into the toy telephone and the child at the right end (who is 20 metres away) can hear his sound clearly. In this case of toy telephone, sound made by the child while speaking, travels through the thread, which is a solid substance.

The above activities show that sound can travel through solids like wood, metal and thread. In fact, **sound can travel through all the solid substances like metals (iron, steel, etc.), wood, bricks, stone and glass, etc.** We will now describe an activity to show that sound can also travel through liquids. We will use water as liquid in the following activity.

ACTIVITY

Place a squeaking toy (sound making toy) in a polythene bag and immerse it in a bucket full of water. Now, if we put our ear to the side of this bucket and press the toy, we can hear the sound of squeaking toy clearly. In this case the sound of squeaking toy comes to our ear through water contained in the bucket. This shows that sound can travel through water, which is a liquid. In general, we can say that **sound can travel through liquids.** Dolphins and whales which live in the sea can communicate (or talk) with one another under water because sound travels through sea water (which is a liquid).

The following observations will show that sound can also travel through gases. When the telephone bell rings in our home, we can hear its sound even from a distance. In this case, **the sound of ringing telephone bell travels to us through the air in the room, which is a gas (or rather a mixture of gases).** When we talk to a person standing near us, then the sound of our talk travels to the other person through the air around us. The sounds of radio, television, motor cars, buses, trains, aeroplanes, and the chirping of birds, all travel through the air and reach our ears. In fact, **most of the sounds which we hear in our everyday life, reach us through the air.** All the above observations show that sound can travel through air, which is a gas. In general, we can say that **sound can travel through gases.**

From the above discussion we conclude that sound can travel through solids, liquids and gases. We will now describe an activity to show that sound cannot travel through vacuum. The word 'vacuum' means 'empty space'. Even air is not present in vacuum. Thus, **when there is no air in something, we say there is vacuum.** A vacuum can be created in a glass jar by removing all the air from it with the help of a suction pump called vacuum pump.

Sound Cannot Travel Through Vacuum

A material medium (like air) is necessary for the transmission of sound. Sound cannot travel through vacuum (or empty space). This can be shown by the following activity.

ACTIVITY

1. A ringing electric bell is placed inside an airtight glass jar containing air as shown in Figure 6(*a*). We can hear the sound of ringing bell clearly. Thus, when air is present in the glass jar, sound can travel through it and reach our ears.

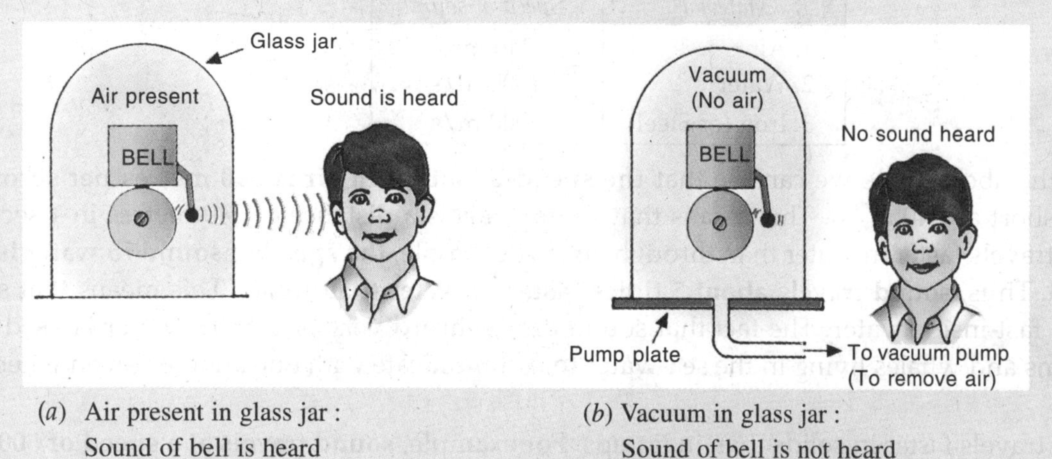

(*a*) Air present in glass jar : Sound of bell is heard

(*b*) Vacuum in glass jar : Sound of bell is not heard

Figure 6. Activity to show that sound cannot travel through vacuum.

2. The glass jar containing ringing bell is placed over the plate of a vacuum pump [see Figure 6(*b*)]. Air is gradually removed from the glass jar by switching on the vacuum pump. As more and more air is removed from the glass jar, the sound of ringing bell becomes fainter and fainter. And when all the air is removed from the glass jar, no sound can be heard at all (though we can still see the clapper striking the bell) [see Figure 6(*b*)]. Thus, when vacuum is created in the glass jar, then the sound of ringing bell placed inside it cannot be heard. This shows that **sound cannot travel through vacuum** (and reach our ears).

3. If air is now put back into glass jar, the sound of ringing bell can be heard again. This shows that **air is necessary for the sound to travel from the ringing bell to our ears.**

We can explain the above observations as follows : When clapper hits the bell, the bell vibrates (and makes sound). The vibrating bell makes the nearby air molecules to vibrate back and forth. These vibrating air molecules make the next layer of air molecules to vibrate, and so on. In this way, ultimately all the air molecules around the ringing bell start vibrating back and forth. When these vibrating air molecules fall on our ears, we can hear the sound of ringing bell. If, however, there is no air between the ringing bell and the ear, then the vibrations of the ringing bell cannot reach our ears and hence we cannot hear the sound of ringing bell. So, **when there is vacuum in glass jar, there are no air molecules to carry sound vibrations.**

The Case of Moon (or Outer Space)

The moon has no air or atmosphere at all. It is all vacuum (or empty space) on the surface of moon. **Sound cannot be heard on the surface of moon because there is no air on the moon to carry the sound waves (or sound vibrations).** So, we cannot talk to one another directly on the moon as we do on earth, even though we may be very close. If an astronaut talks to another astronaut on the moon, he would see the lips moving but no sound will be heard at all. This is because sound cannot travel through the vacuum which exists on the surface of moon. Similarly, there is no air (or any other gas) in outer space to carry sound waves. It is all vacuum in outer space due to which sound cannot be heard in outer space. Thus, the astronauts who land on moon (or walk in outer space) are not able to talk directly to each other. **The astronauts who land on moon (or walk in outer space) talk to each other through wireless sets using radio waves.** This is because radio waves can travel even through vacuum (though sound waves cannot travel through vacuum).

SPEED OF SOUND

Sound takes some time to travel from the sound producing body to our ears. The speed of sound depends on the nature of medium (or material) through which it travels. The speed of sound is different in different materials. In general, **sound travels slowest in gases, faster in liquids and fastest in solids.** The speeds of sound in air (a gas), water (a liquid) and iron or steel (a solid) are given below :

Material	*Speed of sound*
1. Air	340 m/s
2. Water	1500 m/s
3. Iron (or Steel)	5000 m/s

From the above table we can see that **the speed of sound in air is 340 metres per second** (which is written in short as 340 m/s). This means that sound travels a distance of 340 metres in 1 second through air. Sound travels faster in water than through air. For example, **the speed of sound in water is 1500 metres per second.** Thus, **sound travels about 5 times faster in water than in air.** This means that sound can be heard very fast inside water. The fact that sound can be heard very fast inside water is used by creatures like dolphins and whales living in the sea-water to communicate with one another (even when they are far away).

Sound travels faster in solids than in liquids. For example, **sound travels at a speed of 5000 metres per second through iron (or steel).** This is more than 3 times the speed of sound in water. Here is an interesting consequence of the very high speed of sound in iron or steel. If a train is very far away from us, we cannot hear the sound of approaching train through the air. But if we put our ear to the railway line made of steel, then we can hear the sound of the coming train easily even if it is quite far away. This is due to the fact that sound travels much more faster through the railway line made of steel than through air. In fact, **sound travels about 15 times faster in steel than in air.** So, if we want to hear a train approaching from far away, it is more convenient to put the ear to the railway track (railway line) because the sound of train travels much more faster through solid railway track made of steel than through air. Since wood is a solid and air is a gas, so **sound travels faster in wood than in air.** Again, wood is a solid and water is a liquid, so **sound travels faster in wood than in water.** We will now compare the speeds of sound and light.

Sound Travels Slower Than Light

The speed of sound in air is about 340 m/s and the speed of light in air is 300,000,000 m/s. This means that **sound travels at a slow speed but light travels much, much faster than sound**. In fact, the speed of light is very great as compared to the speed of sound. So, though sound may take a few seconds to travel a distance of a few hundred metres, light will take practically no time to reach a distance of even a few kilometres. We will now give an observation from our everyday life which is based on the low speed of sound in air but very high speed of light. It is a common observation that in the rainy season, the flash of lightning is seen first and the sound of thunder is heard a little later (though both are produced at the same time in clouds) (see Figure 7). **It is due to the very high speed of light that we see the flash of lightning first and it is due to comparatively low speed of sound that the thunder is heard a little later**.We will now describe the organ of hearing sound called 'ear'.

Figure 7. Light travels much faster than sound. Due to this, the flash of lightning is seen first and the sound of thunder is heard a little later.

WE HEAR THROUGH OUR EARS

The ears are the sense organs which help us in hearing sound. So, we hear through our ears. We will now describe the construction and working of a human ear briefly. A highly simplified diagram of human ear is shown in Figure 8.

The shape of the outer part of ear (which we see outside the head) is like a funnel. The outer part of ear is called 'pinna' and it is attached to about 2 to 3 centimetre long passage called 'ear canal'. At the end of ear canal a thin, elastic and circular membrane called 'eardrum' is stretched tightly (see Figure 8). There are three small and delicate bones called hammer, anvil and stirrup in the middle part of the ear which are linked to one another. One end of hammer touches the eardrum and its other end is connected to second

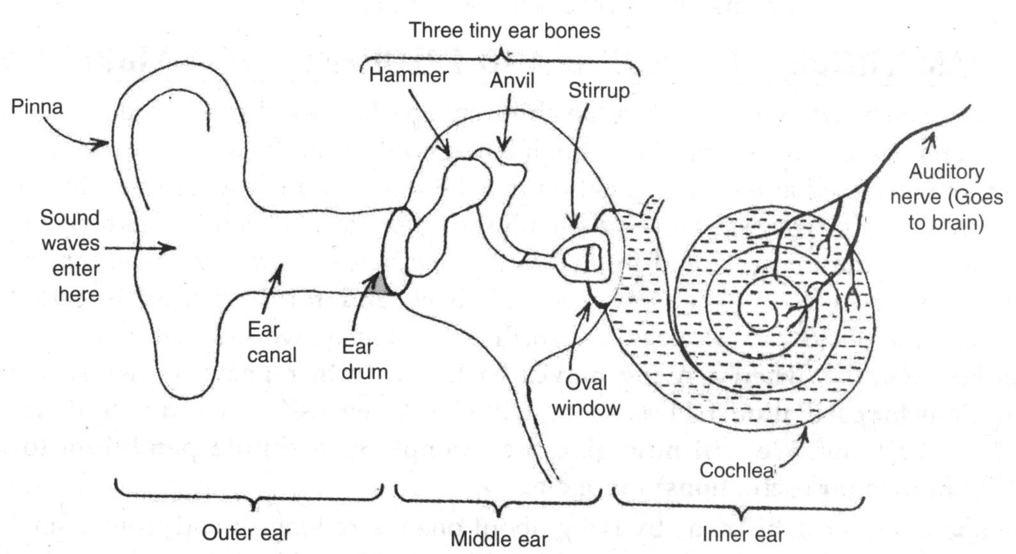

Figure 8. Structure of human ear.

bone anvil. The other end of anvil is connected to third bone called stirrup. And the free end of stirrup touches the membrane over the oval window (see Figure 8). The inner part of ear has a coiled tube called 'cochlea'. One end of cochlea is connected to middle part of ear through the elastic membrane over the oval window. Cochlea is filled with a liquid. The liquid present in cochlea contains nerve cells which are sensitive to sound. The other end of cochlea is connected to auditory nerve which goes into the brain (see Figure 8). Please note that **the three tiny bones in the middle part of ear act as a system of levers and amplify sound vibrations coming from the eardrum before passing them on to the inner part of the ear (cochlea).**

We will now describe the working of ear. The sound waves (coming from a sound producing body) are collected by the pinna of outer part of ear. These sound waves pass through the ear canal and fall on the eardrum. When the sound waves fall on the eardrum, the eardrum starts vibrating back and forth rapidly. The vibrating eardrum causes a small bone hammer to vibrate. From hammer, vibrations are passed on to second bone 'anvil' and then to the third bone 'stirrup'. The vibrating stirrup strikes on the membrane of oval window and passes the amplified vibrations to the liquid in cochlea. Due to this, liquid in cochlea begins to vibrate. The vibrating liquid of cochlea sets up electrical impulses in the nerve cells present in it. These electrical impulses are carried by auditory nerve to the brain (see Figure 8). The brain interprets these electrical impulses as sound and we get the sensation of hearing.

ACTIVITY

We can perform an activity to demonstrate the working of eardrum as follows : Take a plastic tin can and cut its both ends. Stretch a piece of thin rubber sheet (from a burst balloon) across one end of the plastic can and fasten it tightly with a rubber band. Hold the plastic can vertically in your hand with the 'rubber sheet covered end' at the top. Put four or five small grains of a cereal (like rice) on the stretched rubber sheet. Keeping the plastic can vertical, ask your friend to shout 'hurray, hurray', upwards from the open end (lower end) of the plastic can by bringing his mouth below it. We will observe that when the sounds of hurray, hurray fall on the stretched rubber sheet from below, the rice grains placed over it start jumping up and down. The up and down movement of rice grains placed on the stretched rubber sheet tells us that when sound waves fall on it from below the stretched rubber sheet starts vibrating (up and down). This is how the eardrum in our ear works.

We should not put anything (like pin, pencil or pen, etc.) inside our ears. This is because they can tear the eardrum. The tearing of eardrum can make a person deaf. Our ears are very delicate organs. We should take proper care of our ears and protect them from damage.

AMPLITUDE, TIME-PERIOD AND FREQUENCY OF A VIBRATION

We have all seen a swing (or *jhoola*) in the children's park. If we displace the lower end of a swing to one side and then release it, the swing starts moving 'backwards and forwards' repeatedly. Though the top end of swing remains fixed at the same position but the lower end of swing (on which we sit), keeps on moving backwards and forwards, again and again (The alternate 'backwards and forwards' motion is also called 'back and forth' motion or 'to-and-fro' motion). *The motion of a swing is actually an example of vibrations or oscillations.* We can now say that: **A repeated 'back and forth' motion is called vibrations (or oscillations).** When an object moves back and forth continuously, we say that it is making vibrations (or oscillations). For example, **when a swing moves back and forth repeatedly, we say that the swing is making vibrations (or oscillations).** Please note that whether we use the word vibrations or oscillations, it will mean the same thing. **We will now give the example of a simple pendulum to understand the meaning of vibrations (or oscillations) more clearly.**

A simple pendulum can be made by tying about one metre long thread from a small metal ball and suspending it from a height as shown in Figure 9(*a*). The small metal ball of pendulum is called bob. When the pendulum is at rest (not vibrating), then its bob is in the normal position or central position *A* [see Figure 9(*a*)].

If we displace the bob of pendulum to the left side to position *B* and then release it, it will start vibrating (or oscillating) like a swing between positions *B* and *C* [see Figure 9(*b*)]. The bob first goes from position *B* to position *C*, and then comes back to *B*. It again goes from position *B* to position *C*, and then comes back to *B*. This motion of the pendulum bob is repeated again and again. We say that the

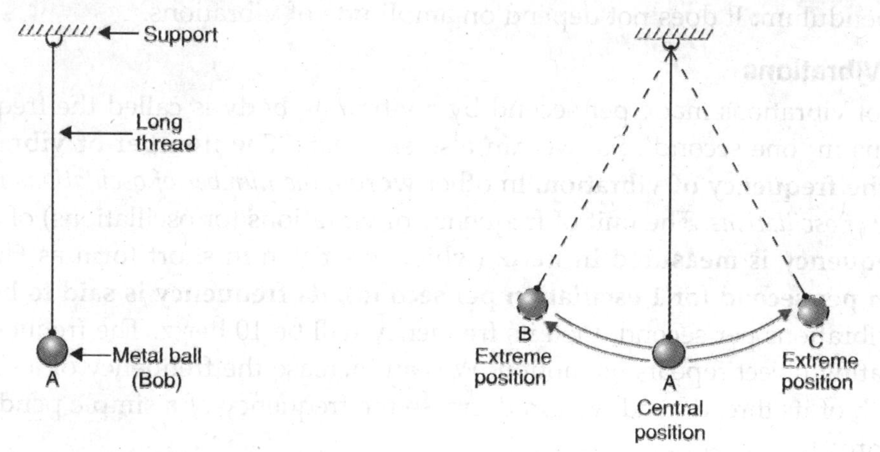

(*a*) A simple pendulum (at rest)

(*b*) A vibrating (or oscillating) simple pendulum

Figure 9.

pendulum bob is vibrating (or oscillating) between positions *B* and *C*. In Figure 9(*b*), the position *A* of bob is called central position and the positions *B* and *C* are called the extreme positions of the bob. When the pendulum bob goes from one extreme position *B* to the other extreme position *C*, and then comes back to *B*, we say that it completes one vibration (or one oscillation). **Every vibration (or oscillation) has three characteristics : amplitude, time-period, and frequency.** These are discussed below.

1. Amplitude of Vibrations

When a simple pendulum vibrates, its bob goes to equal distances on either side of its central position. For example, in Figure 9(*b*), the pendulum bob goes to equal distances *AB* and *AC* from its central position. *The maximum distance to which the bob of a vibrating pendulum goes from its central position is called amplitude of vibrations (or amplitude of oscillations).* In Figure 9(*b*), the distance *AB* is the amplitude of vibration of this simple pendulum. Since the distance *AB* is equal to distance *AC*, so we can also say that distance *AC* is the amplitude of vibration of this pendulum. From this discussion we conclude that : **The maximum displacement of a vibrating object from its central position is called the amplitude of vibrations.** In other words, *the maximum displacement of an oscillating object from its central position is called amplitude of oscillations.* The amplitude actually tells us how far the vibrating object is displaced from its central position. We can increase the amplitude of vibrations of a simple pendulum by raising the height from which the pendulum bob is initially released. Similarly, we can decrease the amplitude of vibrations by releasing the pendulum bob from a smaller height.

2. Time-Period of Vibrations

One complete to-and-fro movement of the pendulum bob is called one vibration (or one oscillation). *The time taken by pendulum bob to complete one vibration (or one oscillation) is called the time-period of pendulum.* For example, in Figure 9(*b*), the time taken by the pendulum bob to travel from position *B* to position *C*, and back to *B*, will be the time-period of this pendulum. In general, we can say that : **The time taken by a vibrating object to complete one vibration is called its time-period.** In other words, *the time taken by an oscillating object to complete one oscillation is called its time-period.* The unit of measuring time-period is 'second'.

The time taken by one vibration (or one oscillation) of a simple pendulum is very short and hence cannot be measured accurately. So, to find the time taken by one vibration (or time-period), we measure the time taken by a large number of vibrations. Dividing the 'total time' by the 'total number of vibrations', we get the time for one vibration (or time-period) of the pendulum. For example, we can measure the time for, say 20 vibrations of the pendulum by using a stop watch. Now, dividing this 'time' by '20' will give us the time taken by one vibration. That is, it will give us the time-period (of vibration) of pendulum. **For a given pendulum, the time-period is the same every time.** The time-period of a pendulum depends only on the length of pendulum. It does not depend on amplitude of vibrations.

3. Frequency of Vibrations

The number of vibrations made per second by a vibrating body is called the frequency of vibration. 'Per second' means in 'one second'. So, we can also say that: **The number of vibrations made in one second is called the frequency of vibration.** In other words, *the number of oscillations made in one second is called the frequency of oscillations.* The unit of frequency of vibrations (or oscillations) of a vibrating object is hertz. That is, **frequency is measured in hertz** (which is written in short form as Hz). **When an object makes 1 vibration per second (or 1 oscillation per second), its frequency is said to be 1 hertz.** And if an object makes 10 vibrations per second, then its frequency will be 10 hertz. The frequency actually tells us how fast the vibrating object repeats its motion. We can increase the frequency of a simple pendulum by reducing the length of its thread. And we can decrease the frequency of a simple pendulum by increasing the length of its thread.

To find the frequency of a simple pendulum, we measure the time taken by the pendulum to make a large number of vibrations. Dividing the 'number of vibrations' by the 'time taken' we get the number of vibrations made in one second. This will give us the frequency of the pendulum. For example, we can measure the time for, say 20 vibrations of the pendulum by using a stop watch. Now, dividing '20' by the 'time taken' will give us the number of vibrations made in one second. This will be the frequency of the pendulum. The calculation of frequency of a pendulum will become more clear from the following example.

Sample Problem. A pendulum makes 15 oscillations in 5 seconds. What is the frequency of the pendulum ?

Solution. The number of oscillations made by a pendulum in 1 second is called its frequency. Now :

In 5 seconds, pendulum makes = 15 oscillations

So, In 1 second, pendulum makes $= \frac{15}{5}$ oscillations

$= 3$ oscillations

Thus, the frequency of this pendulum is 3 oscillations per second or 3 hertz.

Please note that the frequency of vibrations (or oscillations) of a simple pendulum is very low. So, a vibrating simple pendulum produces a sound having very low frequency. And **the very low frequency sound produced by a vibrating simple pendulum cannot be heard by our ears.** An object must vibrate at a frequency of at least 20 hertz to be able to produce audible sound (which can be heard by our ears). We will now give the relation between time-period and frequency.

Relation between Time-Period and Frequency

We have just studied that 'time-period is the time required to make 1 vibration' and 'frequency is the number of vibrations made in 1 second'. This means that **time-period is equal to the reciprocal (or inverse) of frequency.** That is :

$$\text{Time-period} = \frac{1}{\text{Frequency}}$$

This is the relation between the 'Time-period' and 'Frequency' of vibrations (or oscillations). We will now use this relation to solve a numerical problem.

Sample Problem. What is the time-period of a pendulum which is vibrating with a frequency of 10 hertz ?

Solution. We know that: $\text{Time-period} = \frac{1}{\text{Frequency}}$

$= \frac{1}{10}$

= 0.1 second

Thus, the time-period of this pendulum is 0.1 second.

CHARACTERISTICS OF SOUND : LOUDNESS, PITCH AND QUALITY

A sound has three characteristic properties by which it can be recognised. These are loudness, pitch and quality. Two musical sounds may differ from one another in one or more of these properties. We will now discuss the three characteristics of sound in somewhat detail. Let us discuss loudness first.

1. Loudness

Sounds are produced by vibrating objects. If more energy is supplied to an object by plucking it or hitting it more strongly, then the object will vibrate with a greater amplitude and produce a louder sound. Thus, **the loudness of sound depends on the amplitude of vibrations of the vibrating object.** *Greater the amplitude of vibrations, louder the sound will be.* For example, when a *sitar* string is plucked lightly, then it vibrates with a small amplitude and produces a faint sound (or feeble sound). On the other hand, when a *sitar* string is plucked hard, then it vibrates with a large amplitude and produces a very loud sound. This has been shown in Figure 10. In Figure 10(*a*), the *sitar* string is vibrating with a small amplitude, so it

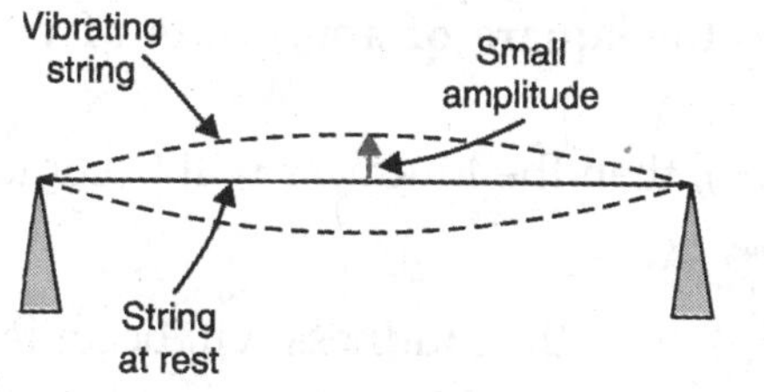

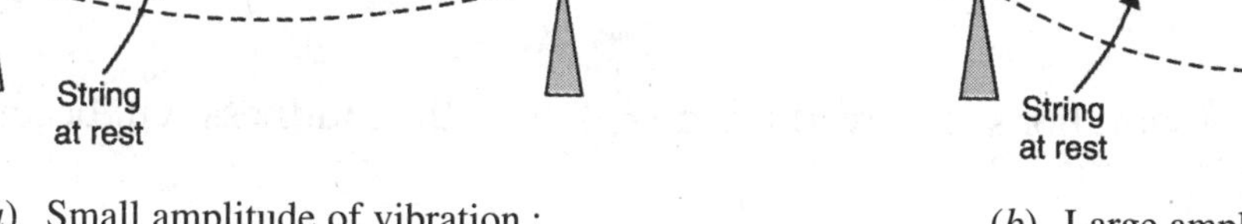

(*a*) Small amplitude of vibration : Faint sound

(*b*) Large amplitude of vibration : Loud sound

Figure 10. Loudness of sound depends on the amplitude of vibrations of a *sitar* string.

produces a faint sound. On the other hand, in Figure 10(*b*), the *sitar* string is vibrating with a large amplitude, so it produces a loud sound.

ACTIVITY

We can demonstrate the dependence of loudness of sound on the amplitude of vibrations of the sound producing object by performing an activity as follows : Take a stainless steel tumbler and a stainless steel

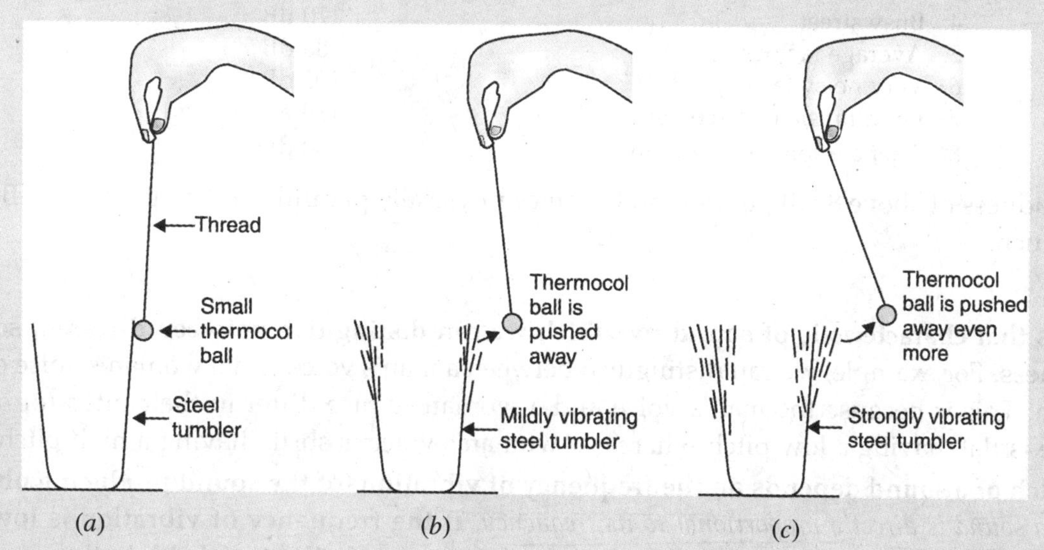

Figure 11. Sound making steel tumbler vibrates (back and forth) and pushes away the thermocol ball. The displacement of thermocol ball is a measure of the amplitude of vibrations of the tumbler.

spoon. Also tie a small thermocol ball to a thread and suspend this thermocol ball touching the rim of steel tumbler as shown in Figure 11(*a*).

(*a*) Strike the steel tumbler gently at the rim with a spoon (to make it vibrate). We will observe that the tumbler produces a feeble sound and, at the same time, the suspended thermocol ball is pushed away to a small distance by the vibrations of the tumbler [see Figure 11(*b*)]. Since the thermocol ball is pushed away by a small distance, this means that the tumbler is vibrating with a small amplitude. Thus, *when the amplitude of vibrations of steel tumbler is small, the sound produced is feeble (less loud).*

(*b*) Now strike the steel tumbler hard at the rim with the spoon. We will observe that the tumbler produces a very loud sound and, at the same time, the suspended thermocol ball is pushed away to a large distance by the vibrations of the tumbler [see Figure 11(*c*)]. Since the thermocol ball is pushed away by a large distance, this means that the tumbler is vibrating with a large amplitude. Thus, *when the amplitude of vibrations of steel tumbler is large, the sound produced is very loud.*

From the above discussion we conclude that the loudness of sound depends on the amplitude of vibrations of sound producing objects. When the amplitude of vibrations is large, the sound produced is loud. On the other hand, when the amplitude of vibrations is small, the sound produced is feeble (or faint). Actually, **the loudness of sound is directly proportional to the square of amplitude of vibrations (of sound producing object)**. This means that :

(*i*) If the amplitude of vibrations is **doubled** (made 2 times), then the loudness will become **four times** [because $(2)^2 = 4$].

(*ii*) And if the amplitude of vibrations is **halved** (made $\frac{1}{2}$), then the loudness will become **one-fourth** [because $\left(\frac{1}{2}\right)^2 = \frac{1}{4}$].

The loudness of sound is expressed in the unit called decibel. The symbol of decibel is **dB**. The softest sound which humans can hear is said to have a loudness of 0 dB (zero decibel). The loudness of sounds coming from some of the common sources of sound around us are given below :

Sound	*Loudness*
1. Normal breathing	10 dB
2. Whispering	30 dB
3. Normal conversation	60 dB
4. Busy street	70 dB
5. Average factory	80 dB
6. Very noisy factory	100 dB
7. Loud music in disco	110 dB
8. A jet aeroplane taking off	130 dB

At a loudness of above 80 dB, the sound becomes physically painful. And at about 140 dB level, sound hurts too much.

2. Pitch

Pitch is that characteristic of sound by which we can distinguish between different sounds of the same loudness. For example, we can distinguish between a man's voice and a woman's voice even without seeing them. This is because the man's voice and a woman's voice differ in their pitch (or shrillness). A man's voice is flat, having a low pitch whereas a woman's voice is shrill, having a high pitch.

The pitch of a sound depends on the frequency of vibration (of the sound producing object). In fact, *the pitch of a sound is directly proportional to its frequency.* **If the frequency of vibration is low, the sound produced has a low pitch.** On the other hand, **if the frequency of vibration is high, the sound produced has a high pitch.** A man's voice has a low frequency, so it has a low pitch. A woman's voice has high frequency, so it has a high pitch. The voice of a small baby has a higher frequency than the voice of even a woman, so the pitch of a baby's voice is higher than that of a woman. Thus, the sounds of low frequency are said to have a low pitch while the sounds of high frequency are said to have a high pitch.

A sound having high frequency (or high pitch) is said to be shrill. The voice of a woman has a higher frequency (or higher pitch) than that of a man due to which **the voice of a woman is shriller than that of a man**. The voice of man having low frequency (or low pitch) is said to be deep (or flat). The voice of a small baby has a higher frequency (or higher pitch) than that of a woman due to which **the voice of a small baby is even more shrill than that of a woman**. It is the frequency of vibration of the sound producing object which determines the pitch (or shrillness) of a sound. **As the frequency of vibration of an object increases, the pitch (or shrillness) of sound produced by it also increases**. When we go from a man to a woman and then a small baby, the frequency of vibration of their vocal cords increases due to which the pitch (or shrillness) of their voice increases. **When we switch on a table fan, we can hear the increase in the pitch of sound produced as the speed of rotation of fan (or frequency of rotation of fan) increases.**

The membrane of a drum vibrates with a low frequency, therefore, a drum produces a low-pitched sound (see Figure 12). On the other hand, the air in a whistle vibrates with high frequency due to which

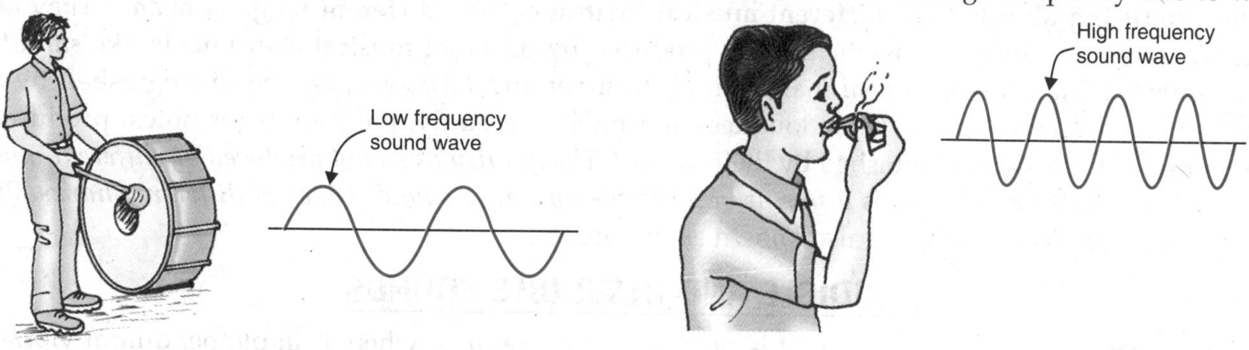

Figure 12. The beating of drum produces a low-pitched sound (having low frequency).

Figure 13. The blowing of whistle produces a high-pitched sound (having high frequency).

the whistle produces a sound having a higher pitch (than that of a drum) (see Figure 13). The sound produced by a whistle is said to be shrill. A bird (say, a crow) makes a high-pitched sound (having a high frequency) whereas a lion makes a low-pitched roar (having a low frequency). The roar of a lion is, however, very loud whereas the sound produced by a bird is comparatively quite feeble.

We can show the relation between the frequency and pitch of a sound producing object (or vibrating object) by performing a simple activity as follows.

ACTIVITY

We take a metal ruler and place it near the edge of a table in such a way that a large part of the ruler projects out of the table top [see Figure 14(*a*)]. The other end of ruler is pressed firmly on the table with our hand. We now press the free end of ruler downwards with our other hand and then let it go. The free end of ruler starts vibrating and produces a low pitch sound.

Metal ruler

Vibrating length of ruler

(*a*)

Metal ruler

Vibrating length of ruler

(*b*)

Figure 14. A vibrating ruler.

Let us now decrease the vibrating length of the ruler which is projecting out of the table [see Figure 14(*b*)]. We again press the free end of the ruler and then let it go. The ruler now starts vibrating faster and produces a high pitch sound. Thus, when we decrease the length of the vibrating ruler, it vibrates faster and the pitch of sound produced by it becomes higher. In fact, shorter the projection of vibrating ruler, higher is the pitch of sound produced by it. Actually, as we go on decreasing the length of the vibrating part of the ruler, the frequency of its vibrations goes on increasing. And this increase in frequency of vibrations, leads to the production of higher pitched sounds.

3. Quality

We can distinguish between the sounds (or notes) produced by a *sitar* and a flute even without seeing these musical instruments. This is because the sounds produced by a *sitar* and a flute differ in quality. Similarly, we can distinguish between the sounds (or notes) produced by singers like Mohammad Rafi and Mukesh by listening to their songs on radio, even without seeing them. This is also because the sounds produced by Mohammad Rafi and Mukesh differ in quality. We can now say that : **Quality is that characteristic of sound which enables us to distinguish between the sounds produced by different sound producing objects (like different musical instruments or different singers) even if they are of same loudness and pitch**. Thus, the sounds produced by different musical instruments like *sitar*, flute, piano, violin, guitar, *veena, shehnai,* harmonium, trumpet and *tabla*, etc., can be distinguished by their quality (even if they are of the same loudness and pitch). Similarly, the sounds (or notes) produced by different singers can be distinguished by their quality. ***The quality of sound produced by different musical instruments or different singers is different because they produce sound waves of different shapes.*** Please note that the 'quality' of sound is also known as 'timbre'.

AUDIBLE AND INAUDIBLE SOUNDS

Whenever an object vibrates, sound is produced. For example, when a simple pendulum vibrates, a sound is produced. But the sound produced by a vibrating simple pendulum is not heard by us. This is because the frequency of vibration of a simple pendulum is very low. It may be 2 or 3 hertz only. And our ears cannot hear sounds of such low frequencies. We say that it is an inaudible sound (which cannot be heard by us). From this example we conclude that **all the vibrating objects do not produce audible sound.** *An object must vibrate at the rate of at least 20 times per second to be able to produce audible sound.* In other words, an object must vibrate with a frequency of at least 20 hertz to be able to produce audible sound. **The sounds having too low frequencies which cannot be heard by human ear are called infrasonic sounds (or just infrasonics)**. Thus, **the sounds of frequencies less than 20 hertz are called infrasonics.** The infrasonic sounds cannot be heard by human beings. Rhinoceros can produce infrasonic sounds having frequencies less than 20 Hz. They can also hear infrasonic sounds.

If the frequency of a sound is less than 20 hertz, it cannot be heard by us; and if the frequency of a sound is more than 20,000 hertz, even then it cannot be heard by us. The human ear can hear sounds which have frequencies between 20 hertz and 20,000 hertz. Thus, **the range of audible frequencies of sound for human hearing is from 20 hertz to 20,000 hertz**. This means that the lower limit of frequency of human hearing (or audible sound) is 20 hertz and the upper limit of frequency of human hearing (or audible sound) is 20,000 hertz. The upper limit of frequency of human hearing 20,000 hertz can also be written as 20 kilohertz, that is, 20 kHz.

The sounds having too high frequencies which cannot be heard by human ear are called ultrasonic sounds (or just ultrasonics). Thus, **the sounds of frequencies greater than 20,000 hertz are called ultrasonics.** The ultrasonic sounds cannot be heard by human beings. For example, a sound having a very high frequency of 40,000 hertz is ultrasonic sound which cannot be heard by human beings.

The human beings can neither produce ultrasonic sound nor can they hear ultrasonic sound. But there are some animals which can produce ultrasonic sounds as well as hear ultrasonic sounds. For example, bat is an animal which screams at a very high frequency, much beyond the limit of our hearing. In other words, **bat produces ultrasonic sound during screaming** (see Figure 15). We cannot hear the screams of a bat because its screams consist of ultrasonic sound having a frequency of more than 20,000 hertz (which is beyond our limit of hearing). **The bats can also hear ultrasonic sounds.** Some other animals like **dogs, monkeys, deer and leopards can also hear ultrasonic sounds** (which cannot be heard by human beings). Since dogs can hear ultrasonic sounds, therefore, some dog owners use special high frequency whistles which only dogs can hear. The crime-branch police also uses special high frequency whistles which produce ultrasonic sounds, to give commands to their dogs.

Figure 15. Bats can produce ultrasonic sounds as well as hear ultrasonic sounds.

The sounds having too high frequency (greater than 20,000 Hz) which cannot be heard by human beings is also called just 'ultrasound'. For example, a sound of frequency 100,000 hertz is an ultrasound. Due to its very high frequency, ultrasound has a greater penetrating power than ordinary sound. The ultrasound is reflected just like ordinary sound waves and produces echoes. But the echoes produced by ultrasound cannot be heard by our ears, they can only be detected by special equipment. These days, ultrasound is used for a large number of purposes. **Some of the important uses of ultrasound** are given below :

(*i*) Ultrasound is used **as a diagnostic tool in medical science** to investigate inside of the human body.

(*ii*) Ultrasound is used **to study the growth of foetus (developing baby)** inside the mother's womb (see Figure 16).

(*iii*) Ultrasound is used **in the treatment of muscular pain and a disease called arthritis** (which is inflammation of joints).

(*iv*) Ultrasound is used **to measure the depth of sea (or ocean).** It is also used **to locate under-water objects** like shipwrecks, submarines and shoals of fish, etc.

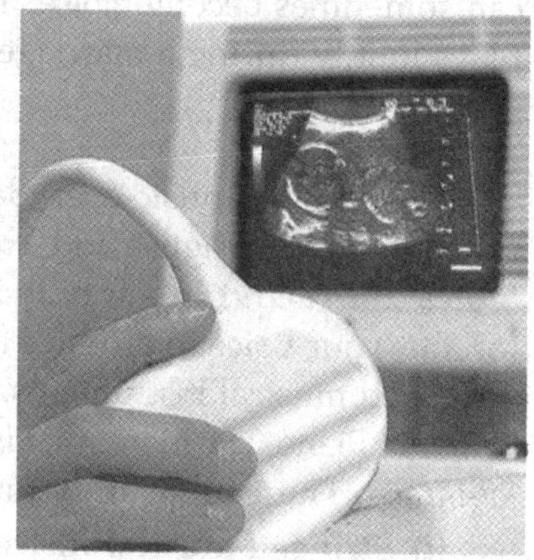

Figure 16. An ultrasound scan of a foetus in the womb.

Please note that the ultrasound equipment works at sound frequencies higher than 20,000 Hz. Another point to be noted is that the speed of infrasonic sound waves as well as ultrasonic sound waves is the same as that of audible sound waves. We will now discuss noise and music.

NOISE AND MUSIC

We hear different types of sounds in our daily life. Some of these sounds are unpleasant to hear whereas other sounds are pleasant to hear. **The unpleasant sounds around us are called noise.** Noise is produced by the irregular vibrations of the sound producing source (see Figure 17). Some of the examples of noise are as follows : Running of mixer and grinder in the kitchen produces noise ; Blowing of horns of

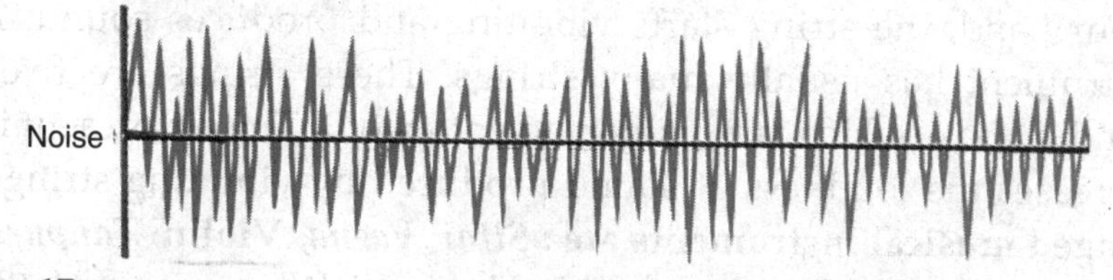

Figure 17. Noise (unpleasant sound) consists of irregular and spiky sound waves.

motor vehicles (like cars, buses and trucks, etc.) causes noise ; Bursting of crackers produces noise ; Barking of dogs produces noise ; Shattering of glass produces noise ; Taking off, landing and flying of aeroplanes causes noise. The various sounds coming from a construction site (where a building is under construction) is also noise. If in a classroom, all the students start talking together loudly, the sounds thus produced will be called noise. We do not enjoy hearing the unpleasant sounds of noise. A noisy sound causes discomfort to the ears.

The sounds which are pleasant to hear are called musical sounds (or music). Musical sounds (or music) are produced by the regular vibrations of the sound producing source (see Figure 18). Some of the examples of musical sounds (or music) are as follows : The strings of a *sitar* make regular vibrations,

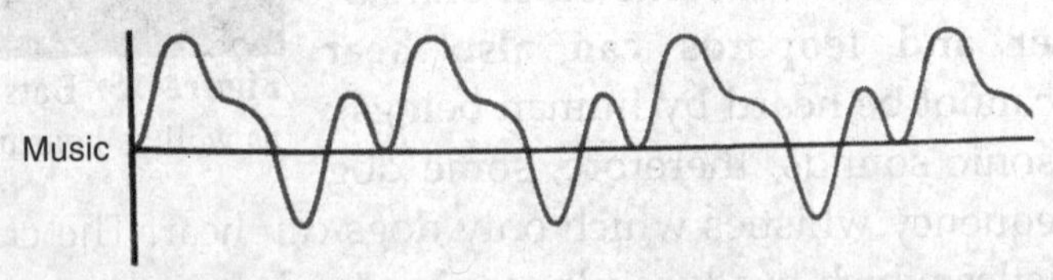

Figure 18. Music (pleasant sound) consists of regular, gently curving sound waves.

so a sitar produces muscial sound. The sound produced by a harmonium is also a musical sound. In fact, all the musical instruments (like *sitar*, *veena*, guitar, violin, piano, flute, *shehnai*, *tabla*, *dholak* and *mridangam*, etc.) produce musical sounds. The speakers of radio, stereo-systems and television also produce musical sounds (or music). When a person sings a melodious song, he (or she) also produces musical sounds. We enjoy the musical sounds produced by the musical instruments or singers because they give pleasant sensation in our ears. **If, however, a musical sound becomes too loud, it would become noise.** Thus, music can sometimes become noise. This happens when very loud music is played at a disco and when a band plays loudly during a marriage procession.

MUSICAL INSTRUMENTS

We all hear music everyday from radio and television. The various sounds produced in music are called 'notes' in English and '*swara*' in Hindi. **The arrangement of sounds of different frequencies called 'notes' (or *swara*) in a way that is pleasant to hear, is called music.** When we sing a song, we modulate (or alter) our voice to produce these musical notes or '*swara*'. The instruments which make musical sounds are called musical instruments. Some of the common musical instruments are : *Sitar, Veena,* Violin, Guitar, *Tanpura,* Piano, Harmonium, *Shehnai,* Flute (*Bansuri*), *Nadaswaram, Tabla, Mridangam,* Cymbals (*Manjira*) and *Jal-tarang.* **There are mainly four types of musical instruments :**

(*i*) Stringed musical instruments,
(*ii*) Wind musical instruments,
(*iii*) Membrane musical instruments, and
(*iv*) Plate type musical instruments

We will now discuss these four types of musical instruments in somewhat detail, one by one. Let us start with the stringed musical instruments.

1. Stringed musical instruments produce musical sounds by the vibrations of stretched strings (or stretched wires). A string is a piece of thin wire. In a stringed musical instrument, thin metal strings (thin metal wires) are fixed tightly between two points. When the stretched string of a musical instrument is plucked or bowed with the fingers of our hand, the string starts vibrating and produces sound. A stringed musical instrument has usually many strings. These strings are fixed tightly on a large sounding box (which is usually made of wood). The air present in the sounding box increases the loudness of sound produced by vibrating strings. **The examples of stringed musical instruments are : *Sitar, Veena,* Violin, *Tanpura, Santoor,* Guitar, Piano and *Ektara*.** *Sitar* is shown in Figure 19. When we pluck the string of a musical instrument like *sitar*, the sound that we hear is not only that of the string. In fact, the whole musical instrument is forced to vibrate and it is the sound of the vibrations of the whole body of the musical instrument that we hear.

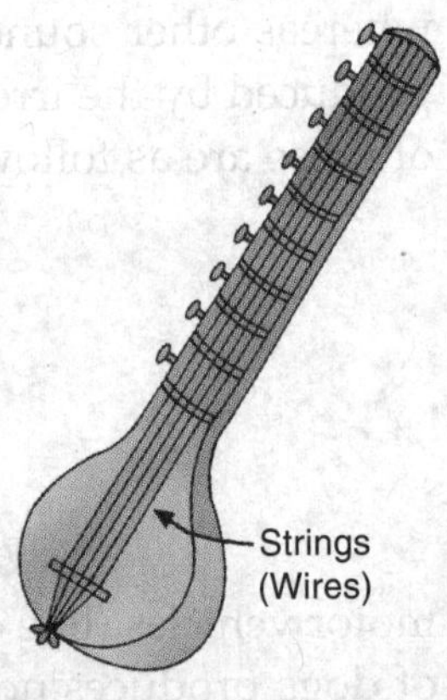

Figure 19. *Sitar.*

2. Wind musical instruments produce musical sounds by the vibrations of air columns inside them. Blowing air is called wind. In a wind musical instrument, a column of air enclosed in a wooden tube (or metal tube) vibrates and produces musical sound. We have to pump air (usually from our mouth) into the wind instrument to make it work and produce sounds. **Some of the wind musical instruments are : *Shehnai*, Flute (*Bansuri*), *Nadaswaram* and Trumpet.** Harmonium is a keyboard wind musical instrument. A wind musical instrument called flute is shown in Figure 20.

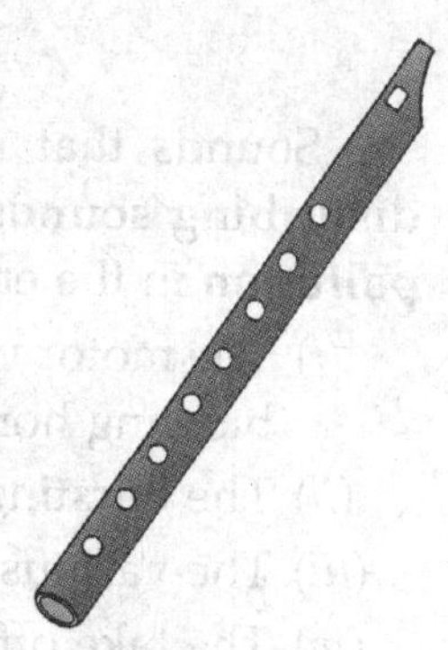

Figure 20. Flute (*Bansuri*).

3. Membrane is a thin sheet of skin. **Membrane type musical instruments produce sounds by the vibrations of thin stretched membranes (or stretched skins).** In a membrane type musical instrument, a thin membrane fixed tightly over a hollow wooden drum vibrates and produces sound. We have to strike the stretched membrane of the instrument with our hands or with sticks to make it vibrate and produce sound. **The examples of membrane type musical instruments are** : ***Mridangam*, *Tabla*, *Dholak*, Drum and *Dhapli*.** A membrane type musical instrument called *mridangam* is shown in Figure 21. Please note that when we strike the membranes of *mridangam*, the sound that we hear is not only that of the membranes but of the whole body of the *mridangam* (which is forced to vibrate). *Mridangam*, *tabla* and *dholak*, etc., are actually rhythm instruments (or *taal* instruments) which are usually used alongwith other musical instruments.

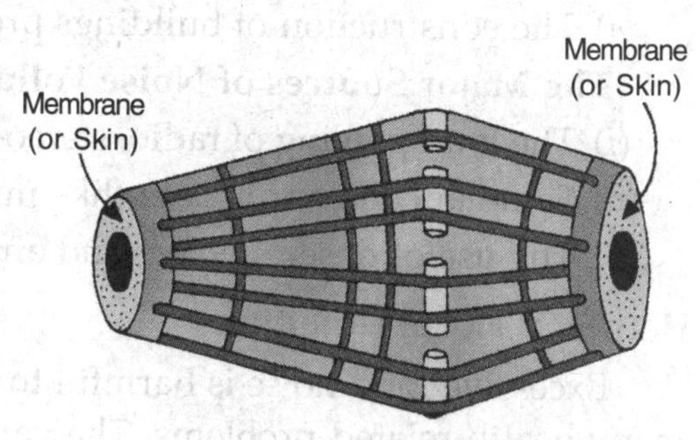

Figure 21. *Mridangam.*

4. Plate type musical instruments produce musical sounds by the vibrations of thick plates (or objects made of plate). The plate type musical instruments are simply beaten (or struck) to produce musical sounds. **The cymbals (*manjira*) is a plate type musical instrument** (see Figure 22). The cymbals consist of two concave brass plates. When the two metal plates of cymbals are struck together, they make a ringing musical sound. **The bell used in performing *pooja* or that in temples is also a plate type musical instrument**. The bell is a hollow metal vessel which emits musical sound when struck with a hammer fixed inside it. **Noot (or *matka*) is also a plate type musical instrument** which makes sound when struck. *Ghatam* and *kartal* are also plate type musical instruments. *Jal-tarang* is also a kind of plate type musical instrument.

Figure 22. Cymbals (*Manjira*).

The '*Jal-tarang*' instrument consists of a number of cups containing different amounts of water. When the cups containing water are struck with two sticks in a proper way, then musical sounds are produced (see Figure 23). The frequency (or pitch) of the sound produced in *jal-tarang* is adjusted by putting the appropriate amount of water in each cup. The cup containing minimum water produces the sound of lowest frequency (or lowest pitch). As the amount of water in the cups goes on increasing, the frequency (or pitch) of the sound produced also goes on increasing. So, the cup having maximum amount of water will produce sound of highest frequency (or highest pitch).

Figure 23. The '*jal-tarang*' musical instrument.

NOISE POLLUTION

Sounds that are loud and unnecessary are called noise. **The presence of loud, unwanted and disturbing sounds in our environment is called noise pollution.** Some of the **major sources of noise pollution** in the environment around us are as follows :

(*i*) The motor vehicles (like cars, buses and trucks, etc.) running on the road produce noise pollution by blowing horns and sounds of their engines.

(*ii*) The bursting of crackers on various social and religious occasions produces noise pollution.

(*iii*) The various machines in factories make loud sounds and cause noise pollution.

(*iv*) The take off, landings, and flying of aeroplanes produces noise pollution.

(*v*) The playing of loudspeakers and bands at marriages and other social functions causes noise pollution.

(*vi*) The construction of buildings produces a lot of noise pollution in the surroundings.

The Major Sources of Noise Pollution in the Homes are the Following :

(*i*) The loud playing of radio, stereo-systems and televisions (at high volume) produces noise pollution.

(*ii*) Some kitchen appliances (like mixer and grinder) cause noise pollution.

(*iii*) The use of desert coolers and air conditioners produces noise pollution.

Harms of Noise Pollution

Excessive loud noise is harmful to us. The presence of excessive noise in the surroundings may cause many health-related problems. The various harms of noise pollution (or loud noise) are as follows :

(*i*) Loud noise can cause great harm to our ears. Constant loud noise reduces the hearing power of our ears. Loud noise can even damage the ears permanently and cause deafness.

(*ii*) Loud noise can cause a person to lose concentration in his work or studies.

(*iii*) Loud noise can cause an ailment called hypertension (high blood pressure).

(*iv*) Loud noise can cause irritation and headache.

(*v*) Loud noise during night-time disturbs our sleep. Continued lack of sleep is bad for health.

Measures to Control Noise Pollution

Noise pollution should be controlled to prevent its harmful effects on human beings and other living things. We can control the noise pollution to some extent by taking the following measures :

(*i*) We should not play radio, stereo-systems and television too loudly.

(*ii*) The horns of motor vehicles should not be blown unnecessarily.

(*iii*) The bursting of crackers should be avoided.

(*iv*) The noise-making factories and airports should be shifted away from the residential areas of the city.

(*v*) Loudspeakers should be played at low volume during marriages and other social functions. No loudspeakers should be allowed to be used late in the night.

(*vi*) Trees should be planted along the roads and around buildings to reduce the noise pollution from the roads and other activities from reaching the residents of the area.

Suppose our parents are going to buy a house. The property dealer says that two houses are available for sale : one house on the roadside and another house three lanes away from the road. We should suggest our parents to buy the house which is three lanes away from the busy road. This is because being away from the busy road will reduce noise pollution caused by heavy traffic on the road. This will make us live comfortably in the house.

Hearing Impairment (Damaged Hearing)

Hearing impairment means that a person has damaged hearing ability and cannot hear properly. There are many reasons why some people have hearing problems. **Some people are born with poor hearing.** *The* *people with very poor hearing are said to be deaf*. Thus, deafness is the total hearing impairment. You might be surprised to know that most deaf people can still hear some sounds. Deaf people often use a sign language with their hands to communicate with others effectively.

Figure 24. Evelyn Glennie. A famous percussion player who is totally deaf.

Partial hearing loss is generally due to an illness, ear infection, injury or old age. Partial hearing loss (or partial hearing impairment) can also be caused by **noise pollution.** A person having partial hearing loss can hear sounds properly by using 'hearing aid'. *Hearing aid is a small sound amplifying device worn on the ear by a partially deaf person (so as to hear properly)*. There are many instances in the world where even totally deaf persons have made great achievements in various fields and lived normal lives. For example, Evelyn Glennie is a famous percussion player who herself cannot hear sounds at all (see Figure 24). She plays with different orchestras in concert halls all over the world. This is amazing because she became severely deaf when she was just 12 years old. She says that she can feel the vibrations of the music through her body. We are now in a position to **answer the following questions :**

Very Short Answer Type Questions

1. How is sound produced ?
2. What should an object do to produce sound ?
3. How does a sound making object differ from one that is silent ?
4. Name the part which vibrates to produce sound in the following :
   (*a*) Drums (*b*) *Sitar* (*c*) Flute
5. What brings the sound of a ringing telephone bell to our ears ?
6. What is the length of vocal cords in a man ?
7. Out of a man and a woman :
   (*a*) who has shorter vocal cords ?
   (*b*) who produces sound of higher pitch ?
8. Give any four sources of sound in a market place.
9. Name the sound producing organ in humans.
10. Which part of our body vibrates when we speak ?
11. What does the working of a toy telephone tell us about sound ?
12. Name one solid, one liquid and one gas through which sound can travel.
13. Which of the following cannot transmit sound ?
    Water, Vacuum, Aluminium, Oxygen gas
14. Is the speed of sound more in water or in steel ?
15. Where would sound travel faster — in wood or in water ?
16. In which medium sound travels faster : air or iron ?
17. In which medium sound travels fastest : air, water or steel ?
18. Out of solids, liquids and gases :
    (*a*) in which medium sound travels slowest ?
    (*b*) in which medium sound travels fastest ?
19. What is the speed of sound in air ?
20. Which of the following is the speed of sound in water and which in steel ?
    (*a*) 5000 m/s (*b*) 1500 m/s (*c*) 340 m/s
21. Name the organs of hearing in our body.
22. Name that part of ear which vibrates when outside sound falls on it.
23. Name the three tiny bones present in the middle part of ear.
24. What is the function of three tiny bones in the ear ?
25. Name the nerve which carries electrical impulses from the cochlea of ear to the brain.
26. What is the name of passage in outer ear which carries sound waves to the eardrum ?
27. Name the quantity whose unit is 'hertz'.

28. What is the relation between 'time-period' and 'frequency' of an oscillating body ?

29. Name three characteristics which are used to describe oscillations (or vibrations).

30. What is the scientific name for the following ?
The number of vibrations made per second.

31. What name is given to the maximum displacement of a vibrating body from its central position ?

32. If 125 oscillations are produced in 5 seconds, what is the frequency in hertz ?

33. How does loudness depend on the amplitude of vibrations ?

34. By how much will the loudness of a sound change when the amplitude of vibrations is : (*a*) doubled ? (*b*) halved ?

35. Name the unit used to measure the loudness of sound. Also write its symbol.

36. What is the loudness of a normal conversation in decibels ?

37. On what factor does the pitch of a sound depend ?

38. How is pitch related to frequency ?

39. Name the characteristic of sound which enables us to distinguish between a man's voice and a woman's voice even without seeing them.

40. Arrange the following sounds in the order of increasing frequencies (keeping the sound of lowest frequency first) :
(*i*) Baby's voice (*ii*) Man's voice (*iii*) Woman's voice

41. Which produces sound of a higher pitch : a drum or a whistle ?

42. Name the characteristic of sound which depends on : (*a*) amplitude (*b*) frequency.

43. Name the characteristic of sound which can distinguish between the 'notes' (musical sounds) played on a flute and *a sitar* (both the notes having the same pitch and loudness).

44. Write the full form of dB.

45. What is the name of very high frequency sounds which cannot be heard by the human ear ?

46. Why do we not hear the screams of a bat ?

47. Which of the following frequency of sound can be heard by a dog but not by a man ?
(*a*) 50,000 hertz (*b*) 15,000 hertz.

48. Name the substance which vibrates in a flute to produce sound.

49. State whether the following statements are true or false :
(*a*) Sound cannot travel in vacuum.
(*b*) The number of oscillations per second of a vibrating object is called its time-period.
(*c*) If the amplitude of vibrations is large, sound is feeble.
(*d*) The lower the frequency of vibration, the higher is the pitch.
(*e*) If the amplitude of vibrations is doubled, the loudness of sound also gets doubled.
(*f*) When the amplitude of vibrations is halved, the loudness of sound becomes one-fourth.
(*g*) Unwanted or unpleasant sound is termed as music.
(*h*) Noise pollution may cause partial hearing impairment.

50. Fill in the following blanks with suitable words :
(*a*) Sounds are produced by .... objects.
(*b*) The human voice box is called ....
(*c*) Sound cannot travel in ....
(*d*) A set of three tiny.......in the middle part of ear passes on sound vibrations from the eardrum to the liquid in cochlea.
(*e*) The unit of frequency is ....
(*f*) The time taken by an object to complete one oscillation is called..............
(*g*) The shrillness of a sound is determined by the ....of vibration.
(*h*) Unpleasant sound is called..............
(*i*) Sound which is pleasing to the ear is called..........sound.
(*j*) A person having partial hearing loss can hear properly by wearing a device called hearing ....on the ear.

Short Answer Type Questions

51. (*a*) What is the name of the strings in the human voice box which vibrate to produce sound ?
(*b*) What makes these strings vibrate ?

52. Describe how sound is produced by the human voice box (or larynx).
53. What is the frequency of the sound produced when the vocal cords are : (*a*) tight and thin ? (*b*) loose and thick ?
54. Why are the voices of men, women and children different ?
55. If you want to hear a train approaching from far away, why is it more convenient to put the ear to the track ?
56. State one observation from everyday life which shows that sound travels much more slower than light.
57. Explain why, the flash of lightning is seen first but the sound of thunder is heard a little later (though lightning and thunder take place in the sky at the same time and same distance from us).
58. Name the object (or part) which vibrates to produce sound in the following musical instruments :
(*a*) *Sitar* (*b*) *Dholak* (*c*) Flute (*d*) Cymbals (*e*) *Veena* (*f*) *Tabla*
59. Name one musical instrument each in which the sound is produced :
(*a*) by vibrating a stretched string. (*b*) by vibrating air enclosed in a tube.
(*c*) by vibrating a stretched membrane (*d*) by vibrating metal plates.
60. Give two examples of each of the following :
(*a*) stringed musical instruments. (*b*) wind musical instruments.
(*c*) membrane musical instruments. (*d*) plate type musical instruments.
61. Which of the sounds having the following frequencies can be heard by the human beings and which cannot ?
(*a*) 6 hertz (*b*) 5000 hertz (*c*) 10000 hertz (*d*) 35000 hertz (*e*) 18 kHz
62. (*a*) What is the upper limit of frequency of human hearing ?
(*b*) What is the lower limit of frequency of human hearing ?
(*c*) Name one animal which can produce ultrasonic sounds.
(*d*) Name two animals which can hear ultrasonic sounds.
63. (*a*) What is a vibration (or an oscillation) ? Define 'amplitude' of vibration of an object.
(*b*) What is the frequency of a vibrating body whose time-period is 0.05 second ?
64. (*a*) State two methods of producing sound.
(*b*) How does sound from a sound producing body travel through air to reach our ears ?
65. (*a*) Why a sound cannot be heard on the moon ?
(*b*) How do astronauts talk to one another on the surface of moon and why ?
66. (*a*) What is meant by the (*a*) 'pitch' of sound, and (*b*) 'quality' of sound ?
(*b*) What is ultrasound ? State two uses of ultrasound.
67. (*a*) What differences will you hear in a sound if there is an increase in (*i*) amplitude, and (*ii*) frequency ?
(*b*) Calculate the time period of a pendulum which is vibrating with a frequency of 10 hertz.
68. (*a*) How can you show that a sounding *tabla* is vibrating ?
(*b*) On what factor does the loudness of a sound depend ?
69. When we put our ear to a railway line, we can hear the sound of an approaching train even when the train is far off but its sound cannot be heard through the air. Why ?
70. Why sound cannot travel through vacuum (or through outer space) ?
71. (*a*) What type of pollution is caused by the working of mixer and grinder in the kitchen ?
(*b*) Why should we not put a pin or pencil in our ears ?
72. Name any two common musical instruments and identify their vibrating parts.
73. What is the difference between noise and music ? Can music become noise sometimes ?
74. Draw a labelled diagram of larynx and explain its functions.
75. (*a*) Give two causes of noise pollution from the homes.
(*b*) What are the usual causes of the partial hearing loss suffered by a person ?

Long Answer Type Questions

76. How can you show that sound cannot travel through a vacuum ? Draw a labelled diagram of the apparatus used.
77. (*a*) What is meant by the 'time-period' of a vibrating object ? State its unit.
(*b*) Define 'frequency' of a vibrating object. Name the unit in which frequency is measured.

(*c*) A pendulum oscillates 40 times in 4 seconds. Calculate its (*i*) time-period, and (*ii*) frequency. Can we hear the sound produced by the oscillations of this pendulum ? Give reason for your answer.

78. Draw a neat and labelled diagram of the human ear. Explain its working.

79. (*a*) What is noise ? Give two examples of sounds which are considered noise.
(*b*) What is a musical sound ? Give two examples of musical sounds.

80. (*a*) What is meant by noise pollution ? Mention some of the sources of noise pollution in your surroundings.
(*b*) Explain how, noise pollution (or excessive loud noise) is harmful to human beings.
(*c*) State the various measures which can be taken to control (or reduce) noise pollution in our surroundings.
(*d*) What can be done along the roads to reduce noise pollution caused by traffic from reaching the residents of the area ?

Multiple Choice Questions (MCQs)

81. Voice of which of the following is likely to have the minimum frequency ?
(*a*) baby girl (*b*) baby boy (*c*) a man (*d*) a woman

82. Sound can travel through :
(*a*) gases only (*b*) solids only (*c*) liquids only (*d*) solids, liquids and gases

83. Which of the following vibrates when a musical note is produced by the cymbals in an orchestra ?
(*a*) stretched strings (*b*) stretched membranes
(*c*) metal plates (*d*) air columns

84. A musical instrument is producing a continuous note. This note cannot be heard by a person having a normal hearing range. This note must then be passing through :
(*a*) water (*b*) wax (*c*) vacuum (*d*) empty vessel

85. Which of the following sound frequencies can be heard by a woman having a normal hearing range ?
A. 25000 Hz B. 15 kHz C. 40000 Hz D. 25 Hz
(*a*) A and B (*b*) B and C (*c*) B and D (*d*) only D

86. When we change a feeble sound to a loud sound, we increase its :
(*a*) frequency (*b*) amplitude (*c*) speed (*d*) timbre

87. Before playing the orchestra in a musical concert, a *sitarist* tries to adjust the tension and pluck the strings suitably. By doing so he is adjusting :
(*a*) intensity of sound only
(*b*) amplitude of sound only
(*c*) frequency of the *sitar* string with the frequency of other musical instruments
(*d*) loudness of sound

88. A key of mechanical piano is first struck gently and then struck again but much harder this time. In the second case :
(*a*) sound will be louder but pitch will not be different
(*b*) sound will be louder and the pitch will also be higher
(*c*) sound will be louder but pitch will be lower
(*d*) both loudness and pitch will remain unaffected

89. One of the following can hear infrasound. This one is :
(*a*) dog (*b*) bat (*c*) rhinoceros (*d*) humans

90. The speed of highly penetrating ultrasonic waves is :
(*a*) lower than those of audible sound waves (*b*) higher than those of audible sound waves
(*c*) much higher than those of audible sound waves (*d*) same as those of audible sound waves

91. The ultrasound waves can penetrate into matter to a large extent because they have :
(*a*) very high speed (*b*) very high frequency (*c*) very high quality (*d*) very high amplitude

92. The frequencies of four sound waves are given below. Which of these sound waves can be used to measure the depth of sea ?
(*a*) 15,000 Hz (*b*) 10 kHz (*c*) 50 kHz (*d*) 10,000 Hz

93. Which of the following frequency of sound can be generated by a vibrating simple pendulum as well as by the vibrating vocal cords of a rhinoceros ?
(*a*) 5 kHz (*b*) 25 Hz (*c*) 10 Hz (*d*) 15,000 Hz

94. Which of the following are used to study the growth of foetus inside the mother's womb ?
(*a*) radio waves (*b*) X-rays (*c*) infrared waves (*d*) sound waves

95. We can distinguish between the musical sounds produced by different singers on the basis of the characteristic of sound called :
(*a*) frequency (*b*) timbre (*c*) pitch (*d*) loudness

96. The maximum speed of vibrations which produce audible sound will be in :
(*a*) dry air (*b*) sea water (*c*) ground glass (*d*) human blood

97. The sound waves travel fastest :
(*a*) in solids (*b*) in liquids (*c*) in gases (*d*) in vacuum

98. The speeds of sound in four different media are given below. Which of the following is the most likely speed in m/s with which the two under water whales in a sea talk to each other when separated by a large distance ?
(*a*) 340 (*b*) 5170 (*c*) 1280 (*d*) 1530

99. The velocities of sound waves in four media P, Q, R and S are 18,000 km/h, 900 km/h, 0 km/h, and 1200 km/h respectively. Which medium could be a liquid substance ?
(*a*) P (*b*) Q (*c*) R (*d*) S

100. Which of the following modes is utilised in the production of sound by humans ?
(*a*) vibrating membranes (*b*) vibrating plates (*c*) vibrating strings (*d*) vibrating air columns

Questions Based on High Order Thinking Skills (HOTS)

101. Three different vibrating objects produce three types of sounds X, Y and Z. The sounds X and Y cannot be heard by a man having normal range of hearing but sound Z can be heard easily. The sound X can be heard by a bat whereas the sound Y can be heard by a rhinoceros. What type of sounds are X, Y and Z ?

102. Your parents are going to buy a house. They have been offered one house on the roadside and another house three lanes away from the roadside. Which house would you suggest your parents should buy ? Explain your answer.

103. The sound from an insect is produced when it vibrates its wings at an average rate of 500 vibrations per second :
(*a*) What is the time-period of the vibrations ?
(*b*) What is the frequency of the vibrations in hertz ?
(*c*) Can we hear this sound ? Why or why not ?

104. There are three small bones in the middle ear.
(*a*) Name the three bones.
(*b*) Which of these bones is in touch with oval window ?
(*c*) Which of these bones is in touch with the ear drum ?
(*d*) Which bone is in touch with the other two bones ?

105. Explain why, if we strike a steel tumbler with a metal spoon lightly, we hear a feeble sound but if we hit the tumbler hard, a loud sound is heard.

ANSWERS

2. Vibrate **5.** Air **13.** Vacuum **30.** Frequency **31.** Amplitude **32.** 25 hertz **34.** (*a*) Becomes four times (*b*) Becomes one-fourth **37.** Frequency **39.** Pitch **40.** Man's voice ; Woman's voice ; Baby's voice **41.** A whistle **43.** Quality (or Timbre) **47.** 50,000 Hz **48.** Air **49.** (*a*) True (*b*) False (*c*) False (*d*) False (*e*) False (*f*) True (*g*) False (*h*) True **50.** (*a*) vibrating (*b*) larynx (*c*) vacuum (*d*) bones (*e*) hertz (Hz) (*f*) time-period (*g*) frequency (*h*) noise (*i*) musical (*j*) aid **53.** (*a*) High (*b*) Low **61.** Can be heard : 5000 hertz, 10000 hertz, 18 kHz; Cannot be heard : 6 hertz, 35000 hertz **63.** (*b*) 20 Hz **67.** (*a*) (*i*) Sound will become louder (*ii*) Sound will become high-pitched (*b*) 0.1 s **71.** (*a*) Noise pollution **77.** (*c*) (*i*) 0.1 s (*ii*) 10 Hz ; No (because it is an infrasonic sound of frequency 10 hertz which is below our hearing range) **81.** (*c*) **82.** (*d*) **83.** (*c*) **84.** (*c*) **85.** (*c*) **86.** (*b*) **87.** (*c*) **88.** (*a*) **89.** (*c*) **90.** (*d*) **91.** (*b*) **92.** (*c*) **93.** (*c*) **94.** (*d*) **95.** (*b*) **96.** (*c*) **97.** (*a*) **98.** (*d*) **99.** (*d*) **100.** (*c*) **101.** X : Ultrasonic sound ; Y : Infrasonic sound ; Z : Audible sound **102.** I would suggest my parents to buy the house which is three lanes away from the busy road. This is because being away from the busy road will reduce noise pollution caused by heavy traffic on the road **103.** (*a*) 0.002 s (*b*) 500 hertz (*c*) Yes ; Because its frequency is within the hearing range of humans **104.** (*a*) Hammer, Anvil, Stirrup (*b*) Strirrup (*c*) Hammer (*d*) Anvil **105.** Amplitude of vibrations small, so feeble sound ; Amplitude of vibrations large, so loud sound

CHAPTER

14

Chemical Effects of Electric Current

We have learnt in earlier classes that some solid materials allow electric current to pass through them whereas others do not allow electric current to pass through them. **The materials which allow electric current to pass through them easily are called good conductors of electricity.** On the other hand, **those materials which do not allow electric current to pass through them easily are called poor conductors of electricity (or non-conductors of electricity).** For example, the metals such as copper and aluminium allow electricity to pass through them easily (or conduct electricity), so they are good conductors of electricity. On the other hand, the materials such as rubber, plastic and wood do not allow electric current to pass through them (do not conduct electricity), so they are poor conductors of electricity (or non-conductors of electricity). In our earlier Classes, we have tested the electrical conductivity of solid materials (like metals, etc.). In this Class we will test the electrical conductivity of liquids.

Do Liquids Conduct Electricity

Just as some of the solids conduct electricity, in the same way, *some of the liquids also conduct electricity*. **The liquids that conduct electricity are solutions of acids, bases and salts in water**. For example, a solution of sulphuric acid, hydrochloric acid, or any other acid, in water conducts electricity. Vinegar contains acetic acid and lemon juice contains citric acid. Vinegar and lemon juice also conduct electricity (because they are solutions of acids). Similarly, a solution of sodium hydroxide, potassium hydroxide, or any other soluble base in water conducts electricity. And a solution of copper sulphate, common salt (sodium chloride), or any other salt in water also conducts electricity. ***There are, however, some important differences in the conduction of electricity by solids (such as metals) and liquids (such as solutions of acids, bases and salts)***. These differences are given below :

***(i)* In solids (like metals), electricity is carried by electrons but in liquids, electricity is carried by ions (positively charged ions and negatively charged ions)**. For example, in a solid like copper metal, electricity is carried by electrons but in a liquid like copper sulphate solution ($CuSO_4$ solution), electricity is carried by copper ions (Cu^{2+}) and sulphate ions (SO_4^{2-}).

(ii) **When electricity is passed through a solid, then no chemical change takes place but when electricity (or electric current) is passed through a liquid, then a chemical change takes place.** For example, when electricity is passed through a copper wire, no chemical change takes place in it but when electricity (or electric current) is passed through acidified water, then a chemical change takes place in which water is decomposed into hydrogen and oxygen gases.

From the above discussion we conclude that some liquids conduct electric current (or electricity). **The liquids which conduct electricity are called conducting liquids** (or conducting solutions). *When electric current (or electricity) is passed through conducting liquids, then chemical changes take place.* **The chemical changes which take place in conducting liquids on passing electric current through them are called chemical effects of electric current.** In this Chapter, we will study the chemical effects of electric current in detail. Before we go further, we should know the meaning of some new terms like electrolyte, electrodes (anode and cathode), and electrolytic cell which will help us in understanding the chemical effects of electric current. These are described below.

A liquid (or solution of a substance) which can conduct electricity is called an electrolyte. In other words, *a conducting liquid is called an electrolyte* (A conducting liquid means a liquid which conducts electricity). A solution of copper sulphate salt in water is a liquid. Copper sulphate solution conducts electricity, therefore, copper sulphate solution is an electrolyte. A conducting liquid or electrolyte contains ions (positively charged ions and negatively charged ions). The flow of these ions conducts electricity through the conducting liquid or electrolyte. **The solutions of acids, bases and salts in water are electrolytes.**

Electrolytes are of two types : strong electrolytes and weak electrolytes. **A strong electrolyte is a liquid (or solution) which conducts electricity very well.** A strong electrolyte is a very good conductor of electricity because it contains a lot of ions in it. Some of the examples of strong electrolytes are : Sulphuric acid solution, Hydrochloric acid solution, Nitric acid solution, Sodium hydroxide solution, Potassium hydroxide solution, Common salt solution (Sodium chloride solution), Copper sulphate solution and Silver nitrate solution. **A weak electrolyte is a liquid (or solution) which conducts electricity to a lesser extent.** A weak electrolyte is a weak conductor of electricity because it contains lesser number of ions. Some of the examples of weak electrolytes are : Vinegar (acetic acid solution), Lemon juice (citric acid solution), Carbonic acid solution, Ammonium chloride solution, Ordinary water (Tap water) and Rain water. **A solid electrical conductor through which an electric current enters or leaves something like a dry cell or an electrolytic cell, is called an electrode.** Electrodes are carbon rods or metal rods depending upon where they are being used. Electrodes are of two types : anode and cathode. The electrode which is connected to the positive terminal of the battery gets positively charged. **The positively charged electrode is called anode.** The electrode which is connected to the negative terminal of the battery gets negatively charged. **The negatively charged electrode is called cathode.**

An arrangement having two electrodes kept in a conducting liquid (or electrolyte) in a vessel is called an electrolytic cell. For example, if we keep two carbon electrodes in a beaker containing acidified water, it will be an electrolytic cell. An electrolytic cell is used for carrying out chemical reactions (or chemical changes) by passing an electric current through the conducting liquid (or electrolyte).

The conducting liquids (or electrolytes) are not as good conductors of electricity as metals are. So, in performing the activities on conduction of electricity through liquids, a single electric cell is not sufficient (because it gives only a small amount of electric current). We have to use a battery made from a number of electric cells (or dry cells) joined together to study the conduction of electricity through liquids. This battery will supply a larger amount of electric current. Thus, a battery (of cells) is used to pass an electric current through the electrolyte taken in the electrolytic cell during the study of chemical effects of electric current.

To Test Whether a Liquid Conducts Electricity or Not

We will now describe an activity to test whether a given liquid (or solution) conducts electricity or not. In other words, we will now describe an activity to test whether a given liquid allows electric current to pass through it or not. We will test two liquids for the conduction of electricity, dilute hydrochloric acid solution and sugar solution, one by one.

ACTIVITY

Take a small beaker. Fix two iron nails on a rubber cork about 1 cm apart and place this cork in the beaker as shown in Figure 1(*a*). The two iron nails will act as the two electrodes. Connect the two nails to the two terminals of a battery by including a torch bulb and a switch in the circuit. Pour a solution of dilute hydrochloric acid in the beaker carefully. Now pass electric current through the hydrochloric acid solution by closing the switch. We will observe that as soon as we switch on the current, the bulb starts glowing [see Figure 1(*a*)]. The bulb can glow only if the hydrochloric acid solution taken in the beaker conducts electricity (making the circuit complete). So, ***the glowing of bulb in this case tells us that hydrochloric acid solution conducts electricity.*** In other words, hydrochloric acid solution is a conducting liquid (or electrolyte). *Since the bulb glows brightly, we also conclude that hydrochloric acid solution is a very good conductor of electricity* (or electric current). If we repeat this activity by taking sulphuric acid solution, sodium hydroxide solution, common salt solution (sodium chloride solution), copper sulphate solution, vinegar or lemon juice in the beaker, the bulb glows again (though it may not glow equally bright in all the cases). This shows that all these solutions are conducting liquids (or electrolytes).

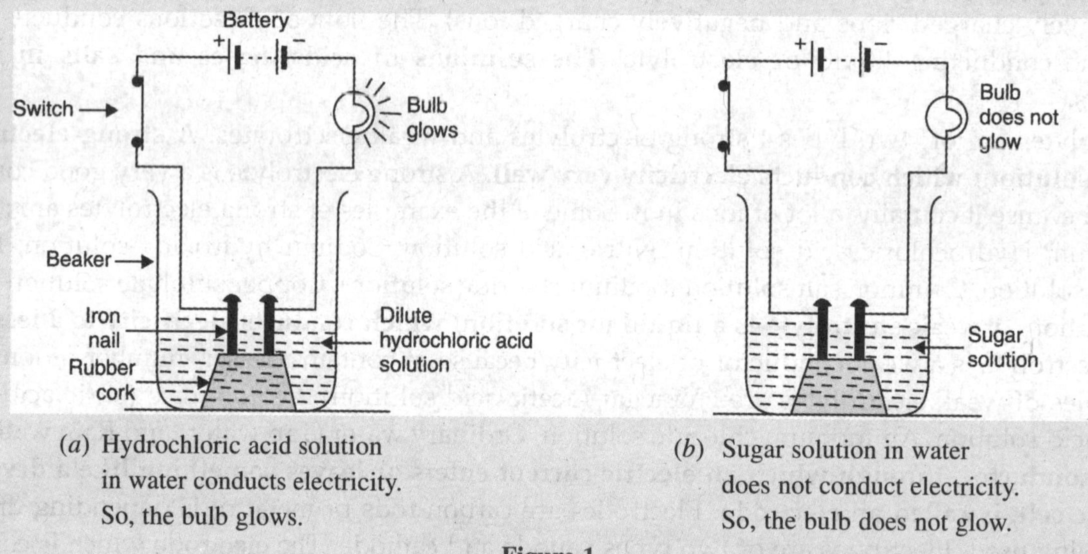

(*a*) Hydrochloric acid solution in water conducts electricity. So, the bulb glows.

(*b*) Sugar solution in water does not conduct electricity. So, the bulb does not glow.

Figure 1.

Let us now take sugar solution in the beaker and switch on the electric current by closing the switch. We will observe that the bulb does not glow in this case [see Figure 1(*b*)]. This shows that ***sugar solution does not conduct electricity.*** In other words, sugar solution is not a conducting liquid. Sugar solution is a poor conductor (or non-conductor) of electricity. If we repeat this activity by taking glucose solution, distilled water, alcohol solution, milk, vegetable oil and honey in the beaker, one by one, we will find that the bulb does not glow at all. This means that glucose solution, distilled water, alcohol solution, milk, vegetable oil and honey are all poor conductors (or non-conductors) of electricity. They are not conducting liquids. They are non-electrolytes.

Weak Conductors of Electricity

We have just studied that when an electric current is passed through hydrochloric acid solution, the **bulb glows brightly** indicating that the hydrochloric acid solution is a very good conductor of electricity [see Figure 2(*a*)]. This is not so in the case of vinegar (which is acetic acid solution) or lemon juice (which is citric acid solution). When an electric current is passed through vinegar (or lemon juice) taken in a beaker, then the **bulb glows very dimly** (even when a large battery is applied) [see Figure 2(*b*)]. **The very dim glowing of bulb indicates that though vinegar and lemon juice conduct electricity but they are**

(*a*) Hydrochloric acid solution is a very good conductor of electricity, so the bulb glows brightly

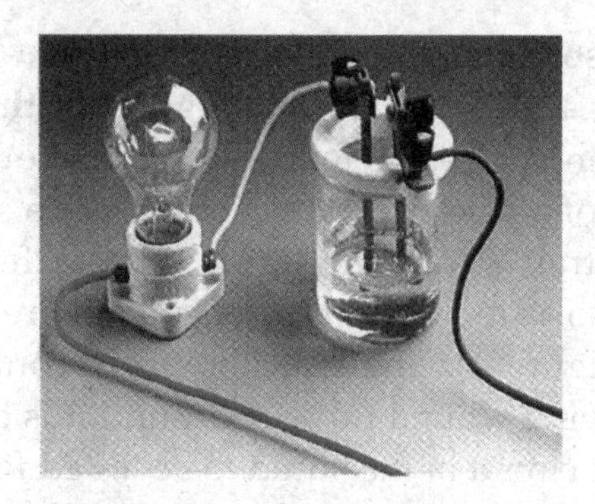

(*b*) Vinegar (or lemon juice) is a weak conductor of electricity, so the bulb glows very dimly

Figure 2. Testing the conduction of electricity through (*a*) hydrochloric acid solution, and (*b*) vinegar (or lemon juice).

weak conductors of electricity. Please note that if a large battery is not used to pass current through vinegar solution (or lemon juice), then the bulb may not glow at all. And if we take ordinary water (tap water) in the beaker and pass electricity through it, then the bulb will not glow at all even if a large battery is applied. This is because ordinary water (or tap water) is a very weak conductor of electricity. Thus, **in some situations, even if the liquid is somewhat conducting, the bulb may not glow**. There are, however, some methods (other than the glowing of bulb) to test the electrical conductivity of weak conducting liquids. These are described below.

Detection of Weak Current Flowing Through a Liquid

The current flowing in a circuit is usually detected by using a small electric bulb (such as a torch bulb). When an electric current flows through a bulb then due to the heating effect of current, the filament of bulb gets heated to a high temperature, becomes white hot and glows to produce light. When the current flowing through a circuit is large, the heating effect is large due to which the bulb glows brightly. And when the current flowing through the circuit is small, the heating effect is also small, due to which the bulb glows dimly. **If, however, the current flowing through the circuit is too weak, then the heating effect produced by current in the filament is too little, due to which the filament does not get heated sufficiently and hence the bulb does not glow at all.** The current in the circuit may be weak if the electrical conductivity of the liquid we are testing is very low. So, if we test a liquid having very low electrical conductivity by using a torch bulb in the circuit, then the bulb may not glow at all (due to too weak current flowing through it). So, ***a torch bulb cannot be used to detect weak electric current flowing through a liquid in a circuit.***

The weak electric current flowing through liquids (having low electrical conductivity) can be detected in two ways :

(*i*) by using a LED (Light-Emitting Diode), and

(*ii*) by using a compass (surrounded by turns of circuit wire).

We can use LED (light-emitting diode) in place of a torch bulb for detecting weak electric current passing through a liquid. LED is a semi-conductor device which glows even when a very weak current passes through it. There are two wires (called leads) attached to an LED (see Figure 3). One lead is slightly longer than the other. While connecting LED in the circuit (in place of bulb), the longer lead is always connected to the positive terminal of the battery and the shorter lead is connected to the negative terminal of the battery. Please note that LED is connected in a circuit (for detecting an electric current flowing through it) where a torch bulb is normally connected.

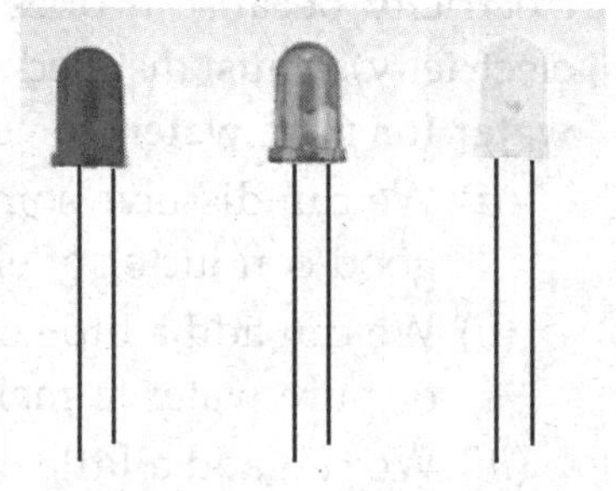

Figure 3. Three LEDs (Light-Emitting Diodes)

We can also detect weak electric current flowing through a liquid by using a compass (by making use of the magnetic effect of current). This can be done as follows : We take out the cardboard tray from the inside of a discarded matchbox. Place a small compass inside this cardboard tray. Wrap an electric wire a few times around the cardboard tray so as to make a type of coil of wire around the compass (as shown in Figure 4). The matchbox tray containing the compass inside it and having wound up wire around it is connected in place of torch bulb in the circuit of the liquid to be tested for conductivity. **Even if a weak electric current flows through the liquid in the circuit, the magnetic needle of compass will show deflection.** This is because even a weak electric current flowing through a wire produces a magnetic field around it. And this magnetic field of electric current acts on the magnetic needle of the compass and deflects it from its usual north-south position. So, **if a compass surrounded by wound up electric wire of a circuit including a liquid in it shows deflection, it will mean that the liquid conducts electricity (or the liquid is a conductor of electricity).** In order to detect the electrical conductivity of a liquid, the two ends of wire (wound around the compass) are connected at a place in the circuit where the two terminals of a torch bulb are normally connected. We can test the electrical conductivity of liquids such as vinegar, lemon juice, tap water, rainwater, sea water, distilled water, milk, vegetable oil and honey with the help of a compass surrounded by wire of circuit.

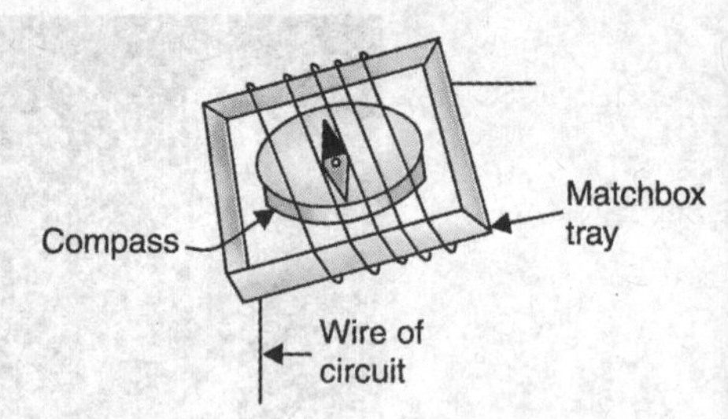

Figure 4. A compass kept in the matchbox tray and surrounded by the turns of electric wire of circuit.

(*a*) When liquids such as vinegar, lemon juice, tap water, rainwater and sea water are included in the circuit, a deflection in the magnetic needle of compass is observed. This shows that vinegar, lemon juice, tap water, rainwater and sea water are conductors of electricity.

(*b*) When liquids like distilled water, milk, vegetable oil and honey are included in the circuit, no deflection is observed in the magnetic needle of compass. This shows that the liquids such as distilled water, milk, vegetable oil and honey are poor conductors of electricity (or non-conductors of electricity).

From the above discussion we conclude that some liquids are good conductors of electricity whereas other liquids are poor conductors of electricity (or non-conductors of electricity). **Most of the liquids that conduct electricity are solutions of acids, bases and salts**. This is because the solutions of acids, bases and salts contain charged particles (called ions) which conduct electricity through these solutions.

The Case of Distilled Water, Tap Water, Sea Water and Rainwater

If we take some distilled water in a beaker and pass electricity through it, we will find that there is no deflection in the magnetic needle of compass (included in the circuit in a suitable way). This shows that **distilled water does not conduct electricity**. In other words, distilled water is a poor conductor of electricity (or a non-conductor of electricity). Let us now dissolve a pinch of common salt (sodium chloride) in distilled water. When salt is dissolved in distilled water, we get salt solution. If we test this salt solution by passing current through it, we will find that the magnetic needle of compass shows deflection. Thus, **distilled water becomes a good conductor of electricity on dissolving a little of salt in it**. In other words, *salt solution is a good conductor of electricity*. Distilled water is a poor conductor (or non-conductor) of electricity because it does not contain any dissolved salts in it (which can provide it ions to conduct electricity). Thus, distilled water is pure water and it does not conduct electricity. **We can make distilled water (or pure water) to conduct electricity in the following ways :**

(*i*) We can dissolve some common salt (or any other salt) in distilled water or pure water to make it a good conductor of electricity.

(*ii*) We can add a little of acid (such as dilute sulphuric acid, lemon juice or vinegar) in distilled water or pure water to make it a good conductor of electricity.

(*iii*) We can add a little of a base (such as sodium hydroxide or potassium hydroxide) in distilled water or pure water to make it a good conductor of electricity.

The water that we get from sources such as taps, handpumps, wells and ponds, etc., is not pure. The water of taps, handpumps, wells and ponds contains small amounts of several salts which are naturally present in it. So, water from all these sources is a conductor of electricity. For example, **tap water is a conductor of electricity** because it contains small amounts of various salts dissolved in it. These salts come naturally to the tap water because tap water comes from the rivers, lakes or tube-wells (which contain dissolved salts). The small amounts of mineral salts present naturally in tap water are beneficial for human health but these salts make tap water a conductor of electricity.

We should never operate an electric switch or touch any working electrical appliance with wet hands. It is dangerous to operate an electric switch or touch a working electrical appliance with wet hands because the tap water present on wet hands is a conductor of electricity due to which it may conduct electric current from the electric switch (or electrical appliance) to our hand and give us an electric shock. Similarly, we should never handle an electrical appliance while standing barefooted on a wet floor.

In case of a fire, before the firemen use big water hoses (flexible water pipes) to throw water on a burning house (or building), they usually cut off the electricity supply of that area. The electric supply is cut off to prevent electrocution of firemen who are busy in fire-fighting operations. Ordinary water is a conductor of electricity, so if the electricity supply is not cut off and firemen come in contact with wet electric switches, electric wires and other electrical appliances, they may get electrocuted.

Tap water is drinking water. **Drinking water contains small amounts of dissolved salts in it but sea water contains a large amount of dissolved salts in it.** Since the sea water contains more salts, therefore, the electrical conductivity of sea water is much more than that of drinking water. So, if we test drinking water and sea water for their electrical conductivity by passing electric current through them, one by one, we will find that the deflection of magnetic needle of compass is much more in the case of sea water (than in the case of drinking water). This is because due to the presence of a large amount of dissolved salts in it, **sea water is a much better conductor of electricity than drinking water** (which contains only a small amount of dissolved salts in it).

Rainwater is said to be pure water. But when rainwater falls to the earth through the atmosphere, it dissolves an acidic gas carbon dioxide from the air and forms a weak acid called carbonic acid. The rainwater may also dissolve other acidic gases such as sulphur dioxide and nitrogen oxides (which are present in polluted air) to form small amounts of other acids such as sulphuric acid and nitric acid. **Due to the presence of small amounts of acids in it, rainwater becomes a conductor of electricity.** So, if we pass electric current through a sample of rainwater, we will see a deflection in the magnetic needle of compass. This will show that **rainwater is a conductor of electricity**. Since rainwater is a conductor of electricity, **it is not safe for an electrician to carry out electrical repairs in the outdoor area during heavy downpour (heavy rain).** The rainwater being a conductor of electricity may cause electrocution of the electrician. We will now answer a question taken from the NCERT science book.

Sample Problem. The bulb does not glow in the experimental set-up shown in Figure 5. List the possible reasons. **(NCERT Book Question)**

Answer. It is possible that :

(*i*) the connections of wires in the circuit may be loose.

(*ii*) the bulb may be fused.

(*iii*) the battery may be dead (all used up).

(*iv*) the liquid may be a very weak conductor of electricity.

(*v*) the liquid may be a poor conductor (non-conductor) of electricity.

+ −
Liquid

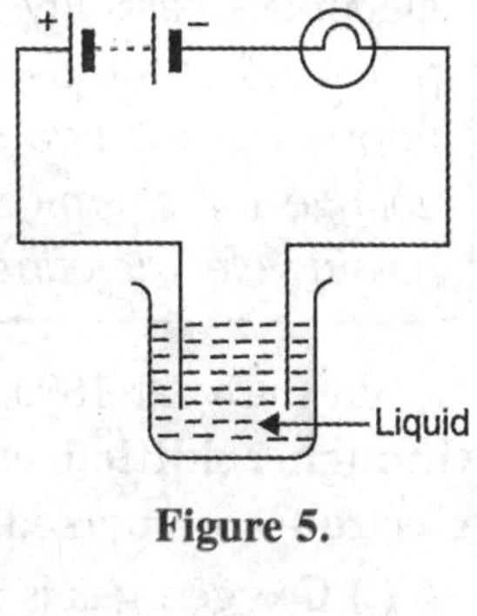

Figure 5.

CHEMICAL EFFECTS OF ELECTRIC CURRENT

Electric current can bring about chemical changes, so it is said to have a chemical effect. For example, when electric current is passed through acidified water by using carbon electrodes, then a chemical reaction takes place to form hydrogen gas and oxygen gas. This chemical reaction can be written as :

$$\text{Water} \xrightarrow[\text{(Chemical reaction)}]{\text{Electric current}} \text{Hydrogen} + \text{Oxygen}$$

Water is a chemical compound, and hydrogen and oxygen are elements. So, in this reaction, a chemical compound 'water' has been decomposed into two elements, hydrogen and oxygen, by the action of electric current. So, this reaction is an example of the chemical effect of electric current. The breaking up of water into hydrogen and oxygen is actually a 'chemical decomposition' reaction caused by passing an electric current through acidified water (which is a conducting liquid).

The chemical decomposition produced by passing an electric current through a conducting liquid is called electrolysis. The decomposition of acidified water into hydrogen and oxygen by passing an electric current (or electricity) is an example of electrolysis. Please note that pure water is not a good conductor of electricity. Water is made an electricity 'conducting liquid' by the addition of a small amount of acid to it. The water containing a little of acid is called 'acidified water'. We use acidified water for carrying out the electrolysis of water. We can also add some base (such as sodium hydroxide) to make water a good conductor for carrying out its electrolysis. Water can be decomposed into hydrogen and oxygen gases by electrolysis. This can be demonstrated as follows.

To Demonstrate the Chemical Effect of Electric Current

We will now describe an activity to demonstrate the chemical effect of electric current.

ACTIVITY

Take out two carbon rods from two discarded dry cells carefully. Clean their metal caps with sand paper. Wrap copper wires around the metal caps of the carbon rods and join them to a battery through a switch (see Figure 6). We call the two carbon rods 'carbon electrodes' or just 'electrodes'. Take 250 mL of water in a beaker. Add a few drops of dilute sulphuric acid to water to make it more conducting. Now, immerse the carbon rods (or carbon electrodes) in acidified water in the beaker (see Figure 6). Make sure that the metal caps of the carbon rods are above the level of water in the beaker. Pass electric current through acidified water in the beaker by closing the switch. Wait for 4 to 5 minutes and observe the two carbon electrodes carefully. We will see that the bubbles of gases are produced at the two carbon electrodes. *The formation of gas bubbles at the two carbon electrodes shows that a chemical change (or chemical reaction) has taken place in water on passing electric current through it.*

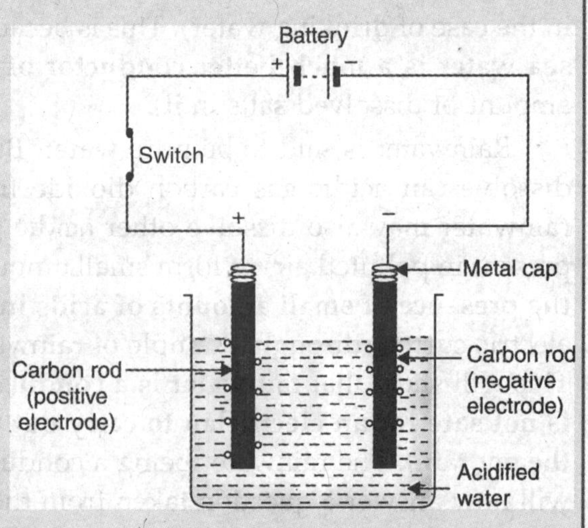

Figure 6. Experimental set-up to demonstrate the chemical effect of electric current by passing current through acidified water (Electrolysis of water).

In the year 1800, a British chemist, William Nicholson, had shown that **if electric current is passed through acidified water, then bubbles of oxygen gas and hydrogen gas are produced at the two electrodes immersed in it.**

(*i*) Oxygen gas is formed at the positive electrode (anode) which is connected to the positive terminal of the battery, and

(*ii*) Hydrogen gas is formed at the negative electrode (cathode) which is connected to the negative terminal of the battery.

We have just described the chemical effect of electric current on water (or rather acidified water). An electric current can also produce a chemical effect on some other substances such as acids, bases, salt solutions and certain molten compounds (melted compounds). When an electric current is passed through these liquids (or molten compounds), chemical reactions take place. We call these chemical reactions 'electrolysis'. From this discussion we conclude that **when an electric current flows through a conducting solution, it causes a chemical reaction (or chemical change).** The chemical reactions brought about by an electric current may produce one (or more) of the following effects :

(*i*) bubbles of a gas (or gases) may be formed on the electrodes.

(*ii*) deposits of metals may form on electrodes.

(*iii*) changes in colour of solutions may occur.

These are some of the chemical effects of the electric current. **The chemical effect produced by an electric current depends on the nature of conducting solution (through which it is passed), and on the nature of electrodes used for passing the electric current.** For example, bubbles of gases are formed when an electric current is passed through acidified water ; deposit of metal is formed when electric current is passed through copper sulphate solution (during electroplating); and a change of colour occurs when electric current is passed through a cut potato.

ACTIVITY TO DEMONSTRATE THE CHANGE IN COLOUR CAUSED BY THE CHEMICAL EFFECT OF ELECTRIC CURRENT

Cut a potato into two halves. Take one piece of cut potato and insert two iron nails into it a little distance apart from one another (see Figure 7). The iron nails are the two electrodes in this case. Connect the two terminals of a battery to the two iron nails by including a compass and a switch in the circuit as shown in Figure 7. Pass the electric current through cut potato piece by closing the switch. *We will observe a deflection in the compass needle showing that potato conducts electricity to some extent.* Let us continue to pass electric current through potato piece for about half an hour. *We will notice a greenish-blue spot on the cut surface of potato around the iron nail which is connected to the positive terminal of the battery* (see Figure 7). There is, however, no coloured spot around the other nail which is connected to the negative terminal of the battery. The formation of a greenish-blue spot around the positive electrode inserted in the surface of a cut potato shows that the chemical effect of current can bring about change in the colour of a conducting solution (A fresh potato contains solution of many substances dissolved in water).

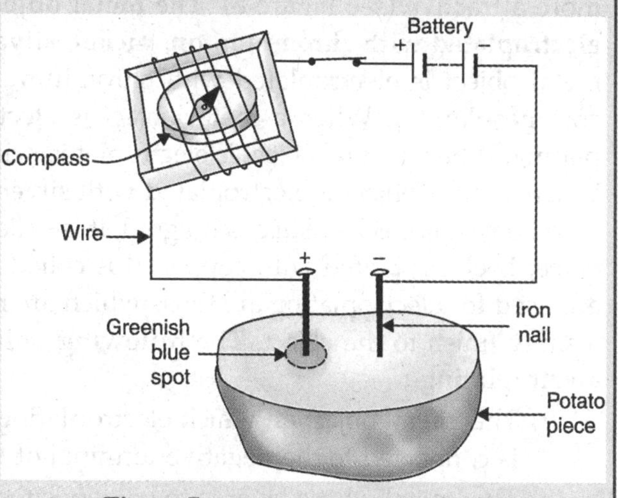

Figure 7. Experimental set up to test the electrical conductivity of potato.

Thus, *an electric current produces a chemical effect in potato leading to a change in colour.*

The fact that a greenish-blue spot on potato surface is always formed around the electrode connected to the positive terminal of a battery can be used to identify the positive terminal of a battery which is concealed in a box (and whose terminals cannot be seen from outside). From the above activity we conclude that **the fresh fruits and vegetables conduct electricity to some extent due to the presence of various salt solutions in them.** Some fruits and vegetables also contain acid solutions in them.

APPLICATIONS OF THE CHEMICAL EFFECT OF ELECTRIC CURRENT

The chemical effect of electric current is used in industries (or factories) for the following purposes :

(*i*) Electroplating metals,
(*ii*) Purification of metals,
(*iii*) Production of certain metals from the ores,
(*iv*) Production of chemical compounds, and
(*v*) Decomposing chemical compounds.

We will now describe all these applications of the chemical effect of electric current in somewhat detail, one by one. Let us start with electroplating. Please note that in the term 'electroplating', the word 'electro' stands for 'electric current' and 'plating' means 'the act of covering a metal object with a thin layer or coating of a different metal'. Electroplating is one of the most common applications of the chemical effect of electric current.

ELECTROPLATING

The process of depositing a thin layer of a desired metal over a 'metal object' with the help of electric current is called electroplating. The purpose of electroplating is to protect the metal objects from corrosion (or rusting) or to make the metal objects look more attractive. In other words, electroplating is done (*i*) for protection against corrosion (or rusting), and (*ii*) for decorative purposes. For example, bathroom taps made of iron (or steel) are electroplated with chromium metal to prevent their corrosion (or rusting). And at the same time, chromium plating gives a shining appearance to the bathroom taps due to which they look more attractive (see Figure 8). **The metal objects (or metal articles) are usually electroplated with chromium, tin, nickel, silver, gold or copper metals.** When a metal object is electroplated with chromium, it is called chromium plating (or chrome plating). When a metal object is electroplated with tin, it is called tin plating. When a metal object is electroplated with nickel, it is called nickel plating. When a metal object is electroplated with silver, it is called silver plating. When a metal object is electroplated with gold, it is called gold plating. And when a metal object is electroplated with copper, it is called copper plating. The metals which are used for electroplating are those which are resistant to corrosion and also give a shiny finish to the object. **The following points should be remembered while electroplating :**

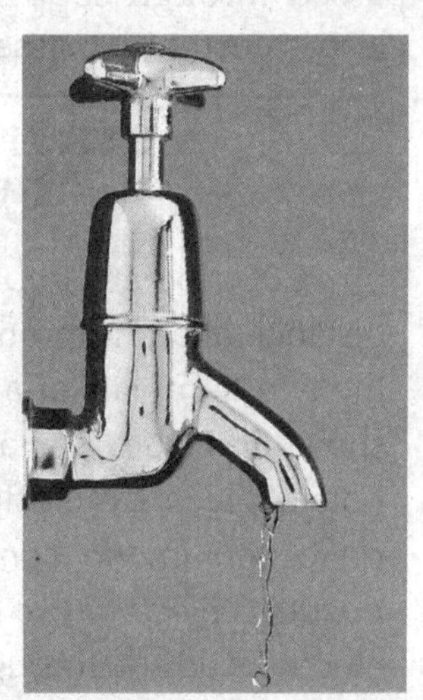

Figure 8. This tap has been electroplated with chromium (or chromium plated).

(*i*) The 'metal object on which electroplating is to be done' is made the negative electrode (cathode) : It is connected to the negative terminal of the battery.

(*ii*) The 'metal whose layer is to be deposited' is made the positive electrode (anode) : It is connected to the positive terminal of the battery.

(*iii*) A water soluble salt of the 'metal to be deposited' is taken as the electrolyte (The electrolyte contains the metal to be deposited in the form of a soluble salt).

We will now describe the process of electroplating by taking the example of copper plating. We will electroplate a key made of iron metal with a thin layer of copper metal (Any other object made of iron can also be used in place of key). **For electroplating an iron object with copper metal (or copper plating) :**

(*a*) The iron object is made negative electrode (cathode). This means that the iron object is connected to the negative terminal of the battery.

(*b*) A copper plate is made positive electrode (anode). This means that a copper plate is connected to the positive terminal of the battery.

(*c*) Copper sulphate solution ($CuSO_4$ solution) is taken as electrolyte (It contains copper metal in dissolved form as copper ions, Cu^{2+}).

ACTIVITY FOR ELECTROPLATING (COPPER PLATING)

We will electroplate an iron object in the form of an iron door key with copper metal. In other words, we will describe the copper plating of an iron key. The experimental set-up for copper plating an iron key is shown in Figure 9. Take 250 mL of distilled water in a clean beaker. Dissolve two teaspoonfuls of copper sulphate in it. This will give us a blue coloured copper sulphate solution. Add a few drops of dilute sulphuric acid to copper sulphate solution to make it more conducting. Take a copper plate of about 10 cm × 4 cm size and a door key made of iron. Clean the surfaces of copper plate and iron key by rubbing with sand paper. Then wash them with water and dry them.

(*i*) Immerse the cleaned copper plate in copper sulphate solution in the beaker. Connect the copper plate to the positive terminal of a battery through a switch (see Figure 9). This copper plate becomes the positive electrode (or anode).

(*ii*) Immerse the cleaned iron key also in copper sulphate solution at a small distance from the copper plate. Connect the negative terminal of the battery to the iron key (see Figure 9). This iron key becomes the negative electrode (or cathode).

(*iii*) Switch on the electric current by closing the switch. Allow the current to pass for about 15 minutes.

(*iv*) Now remove the copper plate and iron key from the copper sulphate solution and look at them carefully. We will find that the copper plate has dissolved a little and the iron key has got a reddish layer of copper metal all over its surface. Thus, the iron key has been electroplated with copper. That is, the iron key has been copper plated. The copper sulphate solution remains unchanged.

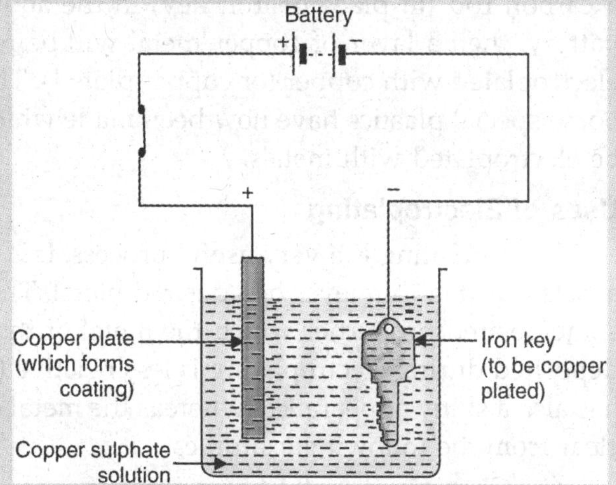

Figure 9. Experimental set-up for electroplating (Here copper plating an iron key).

The process of electroplating copper on an iron key can be explained in the most simple way as follows : The copper sulphate solution (taken as electrolyte) has copper metal in the dissolved form. In fact, the copper sulphate solution consists of free positively charged copper ions (Cu^{2+}) and negatively charged sulphate ions (SO_4^{2-}). When electric current is passed through copper sulphate solution, then the following changes take place :

(*a*) The dissolved copper metal present in copper sulphate solution as positively charged copper ions (Cu^{2+}) gets attracted to the negatively charged electrode 'iron key'. The positively charged copper ions (Cu^{2+}) lose their positive charge on coming in contact with negatively charged iron key and form copper atoms (Cu). These copper atoms deposit on the iron key to form a thin layer (or coating) of copper metal all over the surface of iron key. In this way, copper metal in the electrolyte comes out of the solution and forms a thin layer on the iron key (which is the cathode here).

(*b*) The copper metal of positively charged copper plate electrode dissolves by forming positively charged copper ions, Cu^{2+} (This happens because copper atoms lose their negative charge to positively charged electrode). The copper ions thus formed go into the copper sulphate solution. In this way, the loss of copper ions from copper sulphate solution (utilised in copper plating) is made up and the process continues. Since the copper ions are taken out from the solution at the negative electrode (iron key) but put into solution at the positive electrode (copper plate), therefore, the concentration of copper sulphate solution (or electrolyte) remains constant.

From the above discussion we conclude that **during copper plating of an iron key, copper metal is transferred from the copper plate to the iron key through the copper sulphate solution (with the help of electric current).** In other words, during electroplating with copper, copper metal gets transferred from positive electrode to the negative electrode (the negative electrode being the metal object to be electroplated).

Carbon (in the form of graphite) is a non-metal which is a good conductor of electricity. So, if we take a carbon rod (in place of iron key) in the above activity and connect it to the negative terminal of the battery, then a layer of copper metal will be formed on the carbon rod. Thus, a carbon rod can also be electroplated with copper (or copper plated). This is because carbon rod is a good conductor of electricity. Some special plastics have now been made which are good conductors of electricity. Such plastics can also be electroplated with metals.

Uses of Electroplating

Electroplating is a very useful process. It is widely used in industry for coating metal objects (or metal articles) with a thin layer of a desired metal. The metal 'which is deposited' in the form of a thin layer has some desired properties which the metal of the object does not possess. For example, the metal which is deposited during electroplating is less reactive (than the metal of the object) and hence resists corrosion. It has also a shiny appearance (whereas the metal of the object has a dull appearance). This will become more clear from the following examples.

(*i*) Chromium metal has a shiny appearance and it does not corrode (it does not rust). Chromium metal is, however, quite expensive and hence it is not economical to make whole object out of chromium. So, the object (or article) is made of a cheaper metal (like iron or steel) and only a thin coating of chromium metal is deposited all over its surface by electroplating (chromium plating). After chromium plating, it looks as if the whole iron (or steel) object is made of the chromium metal. For example, chromium plating is done on many objects made of iron metal (or steel) such as bicycle handlebars, bicycle bells, wheel rims, bathroom fittings (taps, etc.), LPG stoves, motor cycle parts, and many, many other objects (see Figure 10). Metallic car bumpers and car grills are also chromium plated (see picture on top of page 258). We know that an iron (or steel) object is not very attractive to look at. Moreover, iron (and steel) objects corrode or rust gradually. So, a thin coating (or layer) of chromium metal on iron and steel objects (deposited by electroplating) makes the iron and steel objects look shiny and attractive, as well as protects them from corrosion (or rusting). If, however, the thin layer of chromium plating on an object is accidently scratched, then the shiny coating of chromium comes off and the iron or steel surface beneath it gets exposed. Rusting of iron and steel objects can then take place. From the above discussion we conclude that **electroplating is used to cover iron and steel objects with a thin layer of chromium metal. This chromium layer gives an attractive, shiny surface and also protects iron and steel objects from rusting.**

Figure 10. Some electroplated objects (chromium plated objects).

(*ii*) Tin metal has a shiny appearance, it does not corrode and it is non-poisonous. It is less reactive than iron. Tin 'cans' used for storing food are made by electroplating tin metal on to iron. Due to tin plating over the surface of iron, the food does not come in contact with iron and is protected from getting spoilt. In fact, *the less reactive and shiny metals (like chromium, tin and nickel) are electroplated on more reactive and dull looking metals (like iron and steel) to protect them from corrosion and give them an attractive finish.*

(*iii*) Electroplating is used to give objects made of a cheap metal a coating of a more expensive metal to make them look more attractive. For example, less expensive metals are electroplated with more expensive metals like silver and gold to make jewellery (or ornaments). These ornaments have the appearance of silver or gold but they are much less expensive (than ornaments made of pure silver or gold).

So, many times, the ornaments worn by women may appear to be made of gold but they are not really of gold. They are gold plated ornaments. However, with repeated use, the gold coating of these electroplated ornaments wears off revealing the cheaper metal silver (or some other metal) beneath it. These ornaments have then to be gold plated again. Similarly, aluminium objects look more attractive when electroplated with nickel.

The conducting solutions (or electrolytes) used in electroplating contain various type of salts which may be poisonous. The conducting solutions are also usually acidic in nature. So, in electroplating factories, the disposal of used conducting solutions (or electrolytes) is a major problem. **The conducting solutions (or electrolytes) used in electroplating process are polluting wastes and hence should be disposed of in a proper way so as to protect the environment**. There are specific guidelines issued by the Government for the safe disposal of electrolyte wastes produced by the electroplating factories.

PURIFICATION OF METALS

The chemical effect of electric current is used in the purification of impure metals (which are extracted from their naturally occurring compounds called ores). In the purification of an impure metal by using the chemical effect of current (or electrolysis) :

(*a*) A thick rod of impure metal is made positive electrode (or anode) : It is connected to the positive terminal of the battery.

(*b*) A thin strip of pure metal is made negative electrode (or cathode): It is connected to the negative terminal of the battery.

(*c*) A water soluble salt of the metal to be purified is taken as electrolyte.

On passing electric current, the metal dissolves from the impure anode and goes into electrolyte solution. The metal present in dissolved form in electrolyte gets deposited on the cathode in the pure form. The impurities are left behind in the electrolyte solution. The metals like copper, zinc and aluminium, etc., are purified by the process of electrolysis by using the chemical effect of electric current.

ACTIVITY TO PURIFY IMPURE COPPER METAL

Take 250 mL of distilled water in a clean beaker. Dissolve two teaspoonfuls of copper sulphate in it. Add a few drops of dilute sulphuric acid to copper sulphate solution. A thick rod of impure copper metal is made positive electrode (anode) by connecting it to the positive terminal of the battery (see Figure 11). A thin plate of pure copper metal is made negative electrode (cathode) by connecting it to the negative terminal of the battery (see Figure 11). Switch on the electric current by closing the switch. Allow the current to pass for about half an hour. It will be observed that the impure copper rod goes on becoming thinner and thinner whereas the pure copper plate goes on becoming thicker and thicker. This is because the impure copper metal of 'anode' goes on dissolving in copper sulphate solution whereas the pure metal from copper sulphate solution goes on depositing on copper plate 'cathode'. Impurities present in impure rod of copper fall to the bottom of the beaker. *Please note that the process of purification of impure copper is like electroplating 'copper metal' on 'copper metal'.* This is because the copper metal of impure copper rod (anode) gets deposited on the pure copper plate (cathode).

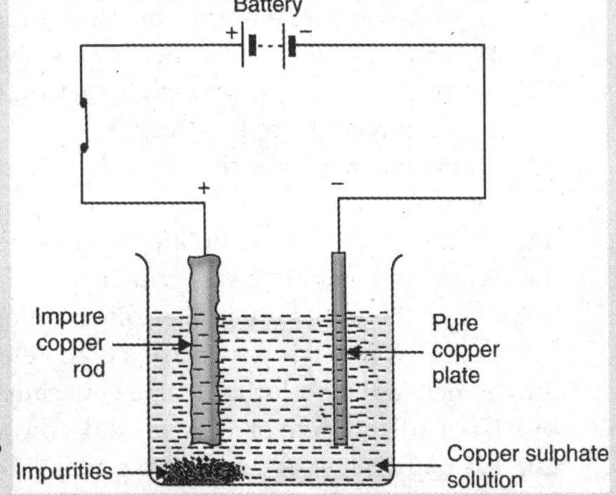

Figure 11. Experimental set-up to purify impure copper metal.

Production of Metals

The chemical effect of electric current is used in the production (or extraction) of certain metals from their naturally occurring compounds called ores. Actually, the chemically reactive metals are produced by this method. For example, the reactive metals such as sodium, aluminium and magnesium, etc., are produced by passing electric current (or electricity) through their compounds in molten state (melted state). Sodium metal is produced by the electrolysis of molten sodium chloride whereas aluminium metal is produced by the electrolysis of molten aluminium oxide. In all the cases, ***metal is produced at the negatively charged electrode (or cathode).*** This is because the positively charged ions of metals present in their molten compounds are attracted by the negatively charged (or oppositely charged) electrode.

Production of Compounds

The chemical effect of electric current (or electrolysis) is used in the production of various chemical compounds. For example, sodium hydroxide (or caustic soda) is produced by the electrolysis of an aqueous solution (water solution) of sodium chloride.

Decomposition of Compounds

The chemical effect of electric current (or electrolysis) is used to decompose various chemical compounds into their elements. For example, water can be decomposed by passing electric current (or electrolysis) into two elements : hydrogen and oxygen. We are now in a position to **answer the following questions :**

Very Short Answer Type Questions

1. Do liquids also conduct electricity ?
2. Name two liquids which conduct electricity and two liquids which do not conduct electricity.
3. Name a device which glows even when a weak electric current passes through it.
4. Write the full form of LED.
5. How would you classify lemon juice – a good conductor or a poor conductor of electricity ?
6. Vinegar is a sour liquid. State whether vinegar will conduct electricity or not.
7. What effect does an electric current produce when flowing through a conducting liquid (or conducting solution) ?
8. When electric current is passed through acidified water, then hydrogen and oxygen are formed. What type of effect of current is illustrated by this statement ?
9. Acidified water is electrolysed by using carbon electrodes. What is produced at : (*a*) positive carbon electrode ? (*b*) negative carbon electrode ?
10. Give one example of the chemical effect of electric current.
11. What should be done to decompose water into hydrogen and oxygen ?
12. Name the process in which a coating of one metal can be deposited on the surface of another metal by using current from a battery.
13. Name the metal which is usually electroplated on car parts such as bumpers and bicycle handlebars made of steel.
14. Which metal is electroplated on iron for making 'cans' used for storing food ?
15. Name two metal objects which have a coating of another metal.
16. Name the most common application of the chemical effect of electric current.
17. Name two metals which are usually electroplated on cheaper metals for making jewellery (or ornaments).
18. Which is the polluting waste generated by electroplating factories ?
19. Give a list of five objects around you which are electroplated.
20. Name two metals which are purified by using the chemical effect of current (or electrolysis).
21. Name two metals which are produced (or extracted) by using the chemical effect of electric current.
22. Name one chemical compound which is produced by using the chemical effect of electric current.
23. Name one compound which is decomposed into hydrogen and oxygen by using the chemical effect of electric current.
24. State whether the following statements are true or false :
    (*a*) Rainwater is a non-conductor of electricity.

(*b*) A piece of fresh potato does not conduct electricity at all.

25. Fill in the following blanks with suitable words :

(*a*) Most liquids that conduct electricity are solutions of ...., ... and ....

(*b*) LED glows even when a ....electric current passes through it.

(*c*) The passage of an electric current through a conducting solution causes............effect.

(*d*) When electric current is passed through acidified water, then ....and..........are formed.

(*e*) The process of depositing a layer of any desired metal on another metal by means of electricity is called ....

(*f*) If you pass current through copper sulphate solution, copper gets deposited on the plate connected to the ....terminal of the battery.

Short Answer Type Questions

26. Which of the following liquids conduct electricity and which do not conduct electricity ?
Lemon juice, Milk, Vinegar, Common salt solution, Sulphuric acid solution, Sugar solution, Distilled water, Honey, Sea water, Rainwater.

27. Why is it dangerous to touch a working electrical appliance with wet hands ?

28. What is the advantage of using LED in testing the electrical conductivity of liquids ?

29. Which effect of electric current is utilised for detecting the flow of current through a solution :

(*a*) when a torch bulb is used ?

(*b*) when a compass is used ?

30. What happens to the needle of a compass kept nearby when electric current is switched on in a wire ? Why does this happen ?

31. Explain why, distilled water does not conduct electricity but tap water conducts some electricity.

32. Distilled water does not conduct electricity. What substances can be added to distilled water in small amounts to make it a good conductor of electricity ? Why ?

33. Which of the two is a better conductor of electricity : drinking water (tap water) or sea water ? Give reason for your answer.

34. Why does a brand new bicycle have shining handlebar and wheel rims ? What will happen if these are accidently scratched ?

35. Is it safe for the electrician to carry out electrical repairs outdoors during heavy downpour ? Explain.

36. When the free ends of a conductivity tester (made by using a battery connected to a wire wound around a compass) are dipped into a solution, the magnetic needle shows deflection. Can you give the reason for this deflection.

37. A beaker contains an acidified copper sulphate solution. A copper plate and a carbon rod are kept in this copper sulphate solution. The copper plate is connected to the positive terminal of a battery whereas the carbon rod is connected to the negative terminal of the battery. What will you observe when an electric current is passed through this set-up for a considerable time ?

38. Does pure water conduct electricity ? If not, what can we do to make it conducting ?

39. In case of a fire, before the firemen use the water hoses to throw water to douse fire, they shut off the electricity supply for the area. Explain why this is done.

40. (*a*) Which effect of electric current is utilised when a thin layer of chromium metal is deposited on an iron tap ? What is this process known as ?

(*b*) For electroplating copper on an iron object, which terminal of the battery (positive or negative) is connected to the iron object ? Also name the electrolyte you will use for this purpose.

Long Answer Type Questions

41. (*a*) What is meant by the chemical effect of electric current ? Explain with the help of an example.

(*b*) Name any two applications of the chemical effect of electric current.

(*c*) What is electrolysis ? Explain why, in the electrolysis of water, 'acidified water' is used.

42. (*a*) Name three types of substances in which an electric current can produce a chemical effect.

(*b*) State some of the characteristics of chemical changes brought about by the chemical effect of electric current.

(*c*) Why does an electric bulb glow when a current passes through it ?

43. (*a*) What is meant by electroplating ? What is the purpose of electroplating ?
(*b*) Which properties of chromium metal make it suitable for electroplating it on car bumpers, bath taps and bicycle handlebars, etc., made of iron ?

44. A strip of impure copper metal is given to you. Describe briefly how you will purify it by using the chemical effect of electric current. Draw a labelled diagram of the experimental set up used for this purpose.

45. With the help of a labelled diagram, describe briefly how an iron key can be electroplated with copper.

Multiple Choice Questions (MCQs)

46. In an activity to check the conduction of electricity through two liquids labelled *A* and *B* by using a bulb, it is observed that the bulb glows brightly for liquid *A* while it glows very dimly for liquid *B*.
(*a*) Liquid *A* is a better conductor than liquid *B*.
(*b*) Liquid *B* is a better conductor than liquid *A*.
(*c*) Both liquids are equally conducting.
(*d*) Conducting properties of liquids cannot be compared in this manner.

47. Which of the following does not conduct electricity ?
(*a*) Vinegar solution (*b*) Sugar solution (*c*) Lemon juice solution (*d*) Caustic soda solution

48. Which of the following metals should be electroplated on a tiffin box made of steel ?
(*a*) copper (*b*) chromium (*c*) silver (*d*) tin

49. Which of the following metals are produced by the electrolysis of their molten compounds ?
A. Copper B. Silver C. Aluminium D. Sodium
(*a*) A and B (*b*) B and C (*c*) C and D (*d*) only D

50. Which of the following types of energy can be used to decompose water into its elements ?
(*a*) heat energy (*b*) light energy (*c*) chemical energy (*d*) electrical energy

51. Which of the following compounds is manufactured by using the chemical effect of electric current ?
(*a*) ammonium hydroxide (*b*) sodium carbonate
(*c*) magnesium hydroxide (*d*) sodium hydroxide

52. Which of the following objects should not be chrome-plated ?
(*a*) car bumper (*b*) gas stove (*c*) frying pan (*d*) bicycle bell

53. Which of the following is a weak electrolyte ?
(*a*) carbonic acid (*b*) sodium hydroxide (*c*) copper sulphate (*d*) nitric acid

54. Electrolytes conduct electricity due to the movement of :
(*a*) electrodes (*b*) atoms (*c*) electrons (*d*) ions

55. Non-metals are generally non-conductors of electricity. The non-metal whose one of the forms can be used to make electrodes in electrolysis experiments is :
(*a*) iodine (*b*) carbon (*c*) silicon (*d*) phosphorus

56. The decomposition produced by passing current through a conducting liquid is called :
(*a*) dialysis (*b*) hydrolysis (*c*) electrolysis (*d*) electroplating

57. Which of the following is not an application of the chemical effect of electric current ?
(*a*) electroplating of metals (*b*) purification of metals
(*c*) decomposition of elements (*d*) decomposition of compounds

58. One of the following is not used for electroplating metal articles. This one is :
(*a*) nickel (*b*) chromium (*c*) sodium (*d*) silver

59. An arrangement having two carbon rods kept in a conducting liquid in a vessel is known as :
(*a*) rechargeable cell (*b*) storage cell (*c*) biological cell (*d*) electrolytic cell

60. Which of the following effects is not produced by the chemical reactions brought about by an electric current ?
(*a*) bubbles of gases on electrodes (*b*) deposits of metals on electrodes
(*c*) change in colour of solution (*d*) formation of a precipitate

61. Which of the following can be electroplated with chromium ?
A. Bakelite B. Graphite C. Steel D. Teflon
(*a*) A and B (*b*) B and C (*c*) only C (*d*) B and D

62. In order to obtain a coating of silver metal on a flower vase made of copper, the electrolyte has to be :
(*a*) silver nitrate solution (*b*) copper nitrate solution
(*c*) sodium nitrate solution (*d*) copper sulphate solution

63. The process of purification of an impure metal is like :
(*a*) electroplating a metal on the same type of metal
(*b*) electroplating a metal on another type of metal
(*c*) producing a chemical compound by electrolysis
(*d*) decomposing a chemical compound by electrolysis.

64. The device which can be used to detect very small current flowing in an electric circuit is :
(*a*) LEAD (*b*) dB (*c*) MCB (*d*) LED

65. If plus sign (+) denotes the positive electrode and minus sign (–) denotes the negative electrode, then which of the following statement is correct for an iron spoon to be copper-plated ?
(*a*) Iron spoon (+), copper plate (–), Iron sulphate electrolyte
(*b*) Iron spoon (–), copper plate (+), Iron sulphate electrolyte
(*c*) Copper plate (–), Iron spoon (+), Copper sulphate electrolyte
(*d*) Copper plate (+), Iron spoon (–), Copper sulphate electrolyte

Questions Based on High Order Thinking Skills (HOTS)

66. A student staying in a coastal region tests the drinking water and also the sea water with a circuit in which a part of the connecting wire is wound around a matchbox containing compass. He finds that the compass needle deflects more in the case of sea water. Can you explain the reason ?

67. When an electric current is passed through a cut potato for a considerable time, a coloured spot is formed around one of the electrodes :
(*a*) What is the colour of the spot ?
(*b*) Around which electrode (positive or negative electrode) the coloured spot is formed ?
(*c*) Which effect of electric current is involved in this case ?

68. In the process of purification of copper metal, a thin plate of pure copper and a thick rod of impure copper are used as electrodes, and a metal salt solution is used as an electrolyte :
(*a*) Which electrode is connected to the positive terminal of the battery ?
(*b*) Which electrode is connected to the negative terminal of the battery ?
(*c*) Which metal salt solution is taken as electrolyte ?

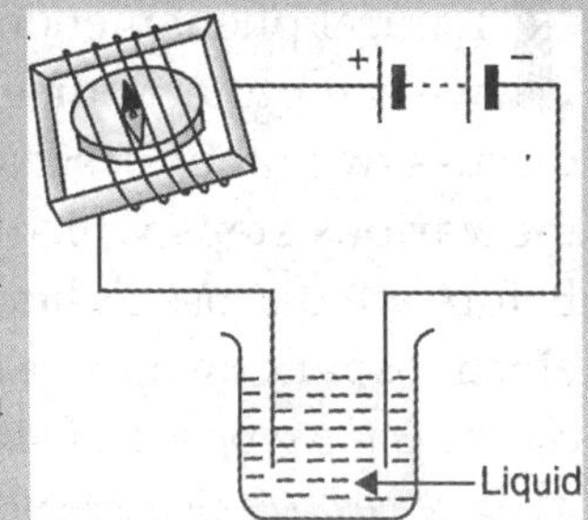

69. A student had heard that rainwater is as good as distilled water. So, he collected some rainwater in a clean glass beaker and tested it. To his surprise he found that the compass needle showed deflection. What could be the reason ?

70. Name three liquids which when tested in the manner shown in figure given alongside may cause the magnetic needle of compass to deflect.

ANSWERS

5. A good conductor of electricity **6.** Vinegar will conduct electricity **8.** Chemical effect **9.** (*a*) Oxygen gas (*b*) Hydrogen gas **12.** Electroplating **15.** Bathroom taps ; Bicycle handlebars **16.** Electroplating **18.** Electrolytes **24.** (*a*) False (*b*) False **25.** (*a*) acids ; bases ; salts (*b*) weak (*c*) chemical (*d*) hydrogen ; oxygen (*e*) electroplating (*f*) negative **29.** (*a*) Heating effect of current (*b*) Magnetic effect of current **36.** The solution conducts electric current. This electric current produces magnetic effect around the wire and deflects magnetic needle **37.** A red-brown layer of copper metal will be deposited on the carbon rod connected to the negative terminal of the battery. The carbon rod will get copper-plated **40.** (*a*) Chemical effect ; Electroplating (*b*) Negative terminal of battery ; Copper sulphate **46.** (*a*) **47.** (*b*) **48.** (*d*) **49.** (*c*) **50.** (*d*) **51.** (*d*) **52.** (*c*) **53.** (*a*) **54.** (*d*) **55.** (*b*) **56.** (*c*) **57.** (*c*) **58.** (*c*) **59.** (*d*) **60.** (*d*) **61.** (*b*) **62.** (*a*) **63.** (*a*) **64.** (*d*) **65.** (*d*) **66.** Sea water contains a large amount of dissolved salts in it. Due to the presence of a large amount of dissolved salts in it, sea water is a much better conductor of electricity than drinking water (which contains only a small amount of dissolved salts in it). The greater electric current passing through sea water produces stronger magnetic field in wire and hence deflects the compass needle more **67.** (*a*) Greenish blue (*b*) Around positive electrode (*c*) Chemical effect **68.** (*a*) Thick rod of impure copper (*b*) Thin plate of pure copper (*c*) Copper sulphate solution **69.** The rainwater falling through the atmosphere dissolves some acidic gases such as carbon dioxide, sulphur dioxide and nitrogen oxides and hence becomes acidic. This acidic rainwater conducts electric current which produces magnetic field and deflects the compass needle **70.** Vinegar ; Lemon juice ; Common salt solution.

Some Natural Phenomena

In Class VII we have studied about winds, storms and cyclones. We learnt that *cyclones* are a destructive natural phenomenon which can cause a lot of damage to human life and property. We also learnt that we can protect ourselves from cyclones to some extent by taking timely action. In this Chapter we will discuss two other destructive natural phenomena called *lightning* and *earthquakes*. We will also learn about the various steps which can be taken to minimise the destruction caused by lightning and earthquakes. Before we discuss lightning, we should first learn about 'electric charges'. This is because **lightning is an electric spark on a huge scale** which is caused by the accumulation of electric charges in the clouds. *Electric charge is the property of matter which is responsible for electrical phenomena*. Electric charge exists in two forms : *positive electric charge* and *negative electric charge*. Let us discuss electric charge in detail.

Electric Charge

About 2500 years ago, a Greek scientist called Thales observed that when a material known as 'amber' was rubbed with a silk cloth, it started attracting tiny feathers. Thales said that amber acquires electric charge (or electricity) on rubbing with silk. And the force of this electric charge attracts the tiny feathers. We can also show the existence of electric charges (or electricity) by performing some simple activities as follows.

ACTIVITY

We rub a plastic comb with our dry hair and bring it near tiny pieces of paper. We will find that the comb attracts the tiny pieces of paper towards itself. Due to this attraction, the tiny pieces of paper move towards the comb and stick to it (see Figure 1). Actually, **when we rub the plastic comb with our dry hair, the plastic comb gets electric charge due to friction. The electrically charged comb then exerts an electric force on the tiny pieces of paper and attracts them.** We can also take a plastic scale

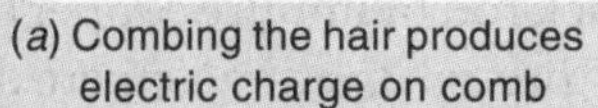

(*a*) Combing the hair produces electric charge on comb

(*b*) Electrically charged comb then attracts tiny pieces of paper

Figure 1.

(in place of plastic comb). We will find that a plastic scale rubbed in dry hair attracts tiny pieces of paper. Similarly, if we rub one end of a pen made of plastic with a sheet of paper (or even with our dry hair) for about one minute and then bring it near tiny pieces of paper, it attracts the tiny pieces of paper towards itself. Due to this attraction, the tiny pieces of paper stick to the pen. Here also the plastic pen gets electric charges on rubbing with a sheet of paper (or dry hair). The electrically charged pen then exerts an electric force on the tiny pieces of paper and attracts them.

A yet another example is that of a balloon. If we take an inflated rubber balloon, rub it carefully with a piece of woollen cloth and then touch it with a wall, the balloon sticks to the wall. In this case, when we rub the rubber balloon with a woollen cloth, the rubber balloon gets electric charges due to friction. This electrically charged balloon then exerts an electric force of attraction on the wall due to which it sticks on the wall. The electric charges which appear on a plastic comb or plastic scale (rubbed with dry hair), on plastic pen (rubbed with a sheet of paper) and on a rubber balloon (rubbed with a piece of woollen cloth) are the *static electric charges* (or stationary electric charges) which remain *bound* to the surface of an object and do not move. The static electric charges are also known as charges at rest. In this Chapter we will be dealing with static electric charges. We will now discuss uncharged objects and charged objects. Please note that whether we use the word 'object' or 'body' in these discussions it will mean the same thing.

Uncharged and Charged Objects

An object having no electric charge on it is called an uncharged object. An uncharged object does not have any effect on other objects. **An object having electric charge on it is called a charged object**. A charged object attracts other uncharged objects. This point will become clear from the following example. If we take a glass rod and bring it near some tiny pieces of paper, it will not have any effect on them. If, however, the **glass rod** is first rubbed with a piece of **silk cloth** and then brought near the tiny pieces of paper, then the glass rod attracts the tiny pieces of paper towards itself. These observations can be explained by saying that initially the glass rod is electrically neutral or uncharged (having no electric charge), so it has no effect on the tiny pieces of paper. But **when the glass rod is rubbed with silk cloth, then it gets electric charge**. The electrically charged glass rod exerts a force on the tiny pieces of paper and hence attracts them. From this example we find that a glass rod rubbed with silk acquires the ability to attract small, uncharged pieces of paper. The objects showing this effect (of attracting other objects) are said to be electrically charged or just charged. ***The process of giving electric charge to an object is called charging the object.***

Charging an Object by Rubbing (or Friction)

The simplest method of charging an object is to rub it with another suitable object (such as silk cloth, woollen cloth, hair, paper or polythene, etc.). When an object is rubbed with another object, then there is friction between them. This friction charges the object. **The charging of an object by rubbing it with another object is called charging by friction**. The charging of a glass rod by rubbing it with a silk cloth is an example of charging by friction. Here are some more examples of charging by friction. When a plastic comb is rubbed with dry hair, the plastic comb acquires an electric charge due to friction. The plastic comb gets electrically charged. The electrically charged comb then attracts tiny pieces of paper. Similarly, when a ballpoint pen refill is rubbed vigorously with a piece of polythene, the refill acquires electric charge by friction. The refill is said to be electrically charged. The charged refill can attract tiny pieces of paper kept near it. And when an inflated rubber balloon is rubbed with a piece of woollen cloth, it gets charged due to friction. **The electric charges generated by rubbing (or friction) are static electric charges. These electric charges remain bound on the surface of the charged object**. They *do not* move by themselves.

Please note that **all the insulator objects (like glass rod, plastic comb, plastic scale, plastic straw, ballpen refill and rubber ballon, etc.) can be charged by rubbing while held in hand**. This is because being insulators (non-conductors), they *do not* conduct electric charges produced on their surface through our hand and body into the earth. On the other hand, **conducting objects made of metals (like a steel**

spoon) cannot be charged by rubbing while held in hand. This is because as soon as a conductor (like a steel spoon) gets charged by rubbing with another material, the electric charges produced on its surface flow through our hand and body into the earth. And the conductor object (like a steel spoon) remains uncharged. A metal object (like a steel spoon) can be charged by rubbing only when held by an insulating material like polythene, etc. Copper, aluminium and iron are also metals. So, a copper rod, an aluminium rod or an iron rod also cannot be charged by friction, if held by hand. Please note that ***our skin is a conductor of electricity which allows electric charges to move through it.*** But since the electric charges on insulators remain bound to their surface, they cannot reach the skin of our hand.

Friction Charges Both the Objects Which are Rubbed Together

When two objects are rubbed together, then both the objects get charged by friction (but with *opposite* charges). For example, when a glass rod is rubbed with a silk cloth, then both, the glass rod as well as the silk cloth get charged. The charged glass rod can attract tiny pieces of paper and the charged silk cloth can also attract tiny pieces of paper. The electric charges acquired by glass rod and silk cloth are, however, *opposite* in nature. ***As a convention, the electric charge acquired by a glass rod (rubbed with silk) is called positive charge.*** So, in this case, the glass rod acquires a *positive* electric charge whereas the silk cloth gets a *negative* electric charge. Similarly, when we rub a ballpoint pen refill with a piece of polythene, then both, the refill as well as polythene get charged but with opposite charges. The ballpoint pen refill acquires a *positive* charge whereas polythene gets the *negative* charge. Please note that **all the objects made of clear plastic called acrylic plastic (such as ballpen refill and clear plastic scale, etc.) get positive charge on rubbing.** On the other hand, **all the objects made of polythene plastic (such as polythene sheet and polythene plastic comb, etc.) get negative charge on rubbing.**

When we rub a plastic comb in dry hair, then both, the comb and hair get charged but with opposite charges. In this case, the plastic comb gets *negative* charge whereas the hair get *positive* charge. Again when we rub an inflated rubber balloon with a piece of woollen cloth, then both, the balloon as well as woollen cloth get electrically charged but with opposite charges. The rubber balloon gets *negative* electric charge whereas the woollen cloth acquires *positive* electric charge. In fact, when any two objects are rubbed together, then both the objects get charged but with *opposite* charges. If one object gets *positive* charge then the other object will get *negative* charge. Now, **which object will get positive charge and which object will get negative charge, depends on the nature of materials of which the two objects are made.**

How Rubbing Charges Various Objects

All the objects (like glass rod, silk cloth, rubber balloon, woollen cloth, plastic comb, hair, ballpoint pen refill, polythene, etc.) are made up of tiny particles called atoms. All the atoms contain two types of electric charges inside them : positive electric charges called protons and negative electric charges called electrons. **In an uncharged object, the number of positively charged particles (protons) and negatively charged particles (electrons) in the atoms are equal.** The equal number of positive and negative electric charges balance each other and make the object electrically neutral (having no over-all charge).

The positively charged protons are held strongly inside the nucleus of atoms, so protons cannot be transferred from the atoms of one object to another object during rubbing. But the negatively charged electrons are held loosely in the atoms (away from the nucleus), therefore, some of the electrons can be transferred from the atoms of one object to another object by rubbing.

(i) **The object which loses negatively charged electrons during rubbing, acquires a positive electric charge** (because then the number of positive protons in it becomes more than the number of negative electrons).

(ii) **The object which gains negatively charged electrons during rubbing, acquires a negative electric charge** (because then the number of negative electrons in it becomes more than the number of positive protons).

Whether an object will lose electrons or gain electrons during rubbing depends on the nature of material of the object. The materials like glass, woollen cloth, hair and ballpen refill lose electrons more easily and hence get positively charged on rubbing. On the other hand, the materials like silk, rubber balloon, plastic comb and polythene gain electrons more easily and hence get negatively charged during rubbing.

Types of Electric Charges and Their Interactions

It has been shown by experiments that **there are two types of electric charges : positive charges and negative charges.** A positive charge *repels* another positive charge, but a positive charge *attracts* a negative charge. Similarly, a negative charge *repels* another negative charge, but a negative charge *attracts* a positive charge. We will now describe some activities to show the existence of two types of electric charges and their interactions.

ACTIVITY 1

(*i*) Take two rubber balloons and inflate them. Hang the two inflated balloons with long threads in such a way that though they are closeby but they do not touch each other [see Figure 2(*a*)]. Rub both the balloons with a woollen cloth and release them. We will see that the two balloons move apart as if they are pushing away each other (or repelling each other) [see Figure 2(*b*)]. In this activity, we have brought close together two balloons which are made of the same material (rubber) and rubbed them with the same material (woollen cloth), so the two balloons must have acquired the same type of electric charges or similar electric charges. **Since two similarly charged balloons repel each other, we conclude that similar charges repel each other.** The same type of electric charges or similar electric charges are also called like electric charges. So, we can also say that **like charges repel**. Actually, when the two balloons are rubbed with a woollen cloth, they acquire *negative* electric charges. *The two negatively charged balloons hung near each other exert a force of repulsion on each other and hence move apart* (*away from each other*) [see Figure 2(*b*)].

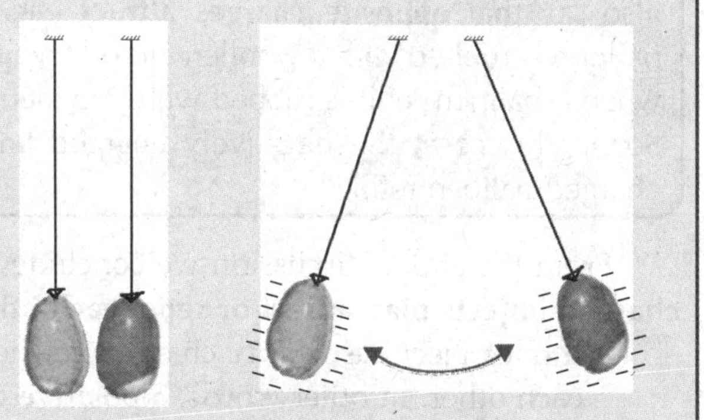

(*a*) Two uncharged balloons hung close to each other

(*b*) Two balloons having like charges (negative charges) repel each other

Figure 2.

ACTIVITY 2

Let us now take two used ballpoint pen refills. Rub one refill with a piece of polythene to charge it. Place this charged refill in a glass beaker carefully (see Figure 3). Now rub the other ballpoint pen refill also with a piece of polythene to charge it. Bring this charged refill near the first charged refill placed in the beaker. We will see that the charged refill placed in the beaker moves away from the charged refill held in our hand as if they are repelling each other (see Figure 3). In this case, both the ballpoint pen refills are made of the same material (acrylic), and both of them have been rubbed with the same material (polythene), so their electric charges should also be of the same type (or similar). Now, **since two similarly charged ballpoint pen refills repel each other, we conclude that similar charges repel each other**. In other words, **like charges repel**. Actually, when the two ballpen refills are rubbed with polythene, they acquire *positive* electric charges. *The two positively charged ballpen refills brought near each other exert a force of repulsion on each other and hence move away from each other* (see Figure 3).

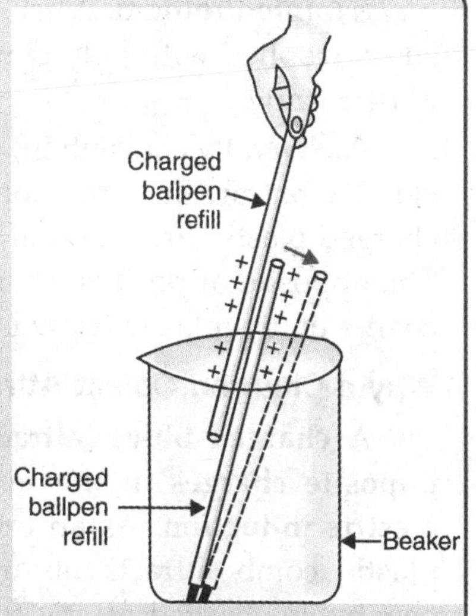

Figure 3. Two ballpoint pen refills having like charges (positive charges) repel each other.

ACTIVITY 3

Rub a ballpoint pen refill with polythene and place this charged refill in a beaker as before. Take a balloon and also charge it by rubbing with a piece of woollen cloth. Now, hold the charged balloon in your hand and bring its charged end near the charged refill placed in the beaker. We will see that the charged refill moves towards the charged balloon as if it is being attracted by the charged balloon (see Figure 4). In this case the ballpen refill and balloon are made of two different materials (acrylic and rubber), and they have been rubbed with different materials (polythene and woollen cloth), so their electric charges must be different (or unlike). Now, **since a charged balloon and a charged ballpen refill having unlike charges (or different charges) attract each other, we conclude that unlike charges attract each other**. Unlike charges are also called opposite charges, so we can also say that **opposite charges attract each other**. Actually, when a balloon is rubbed with a woollen cloth, it acquires a *negative* charge and when a ballpen refill is rubbed with polythene, it gets a *positive* charge. So, in this case, the negatively charged balloon attracts a positively charged ballpen refill.

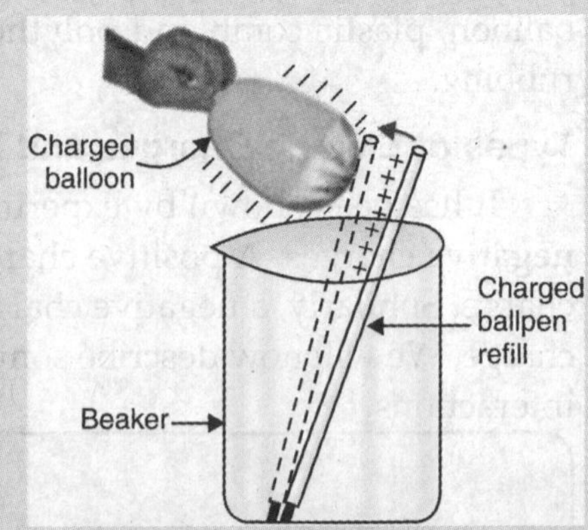

Figure 4. The balloon and ballpoint pen refill having unlike charges (negative and positive charges) attract each other.

From the above discussion we conclude that **depending on the nature of their electric charges, two charged objects may attract or repel each other.**

(*i*) If one object has *positive* charge and the other object has *negative* charge, then the two objects *attract* each other. In other words, **a positive charge and a negative charge attract each other.**

(*ii*) If the two objects have *positive* charges, they *repel* each other. In other words, **two positive charges repel each other.**

(*iii*) If the two objects have *negative* charges, they *repel* each other. In other words, **two negative charges repel each other.**

Many times when we take off woollen or synthetic clothes (like polyester and nylon clothes), our body hair stand erect on their ends. This is because rubbing (or friction) while taking off these clothes charges the body hair with the same kind of electric charge. ***Due to their like charges, the body hair repel one another. This repulsion makes the body hair stand erect.*** We will now answer a question taken from the NCERT book.

Sample Problem. When a charged glass rod (rubbed with silk cloth) is brought near a charged plastic straw (rubbed with polythene), there is attraction between the two. What is the nature of charge on the plastic straw ? **(NCERT Book Question)**

Answer. By convention, the charge present on a glass rod (rubbed with silk cloth) is said to be positive (+). There can be attraction between a charged glass rod and a charged plastic straw only if their charges are opposite to each other. The opposite of positive charge is negative charge. So, the nature of charge on the plastic straw is negative (–).

Why a Charged Object Attracts an Uncharged Object

A charged object attracts an uncharged object by producing opposite charges in the nearer end of the uncharged object by electric induction. As an example, we will explain how a charged plastic comb attracts an uncharged piece of paper. Suppose a negatively charged plastic comb is held over a small piece of paper (see Figure 5). The negatively charged plastic comb produces opposite charges (positive charges) in the top end of paper (which is nearer to

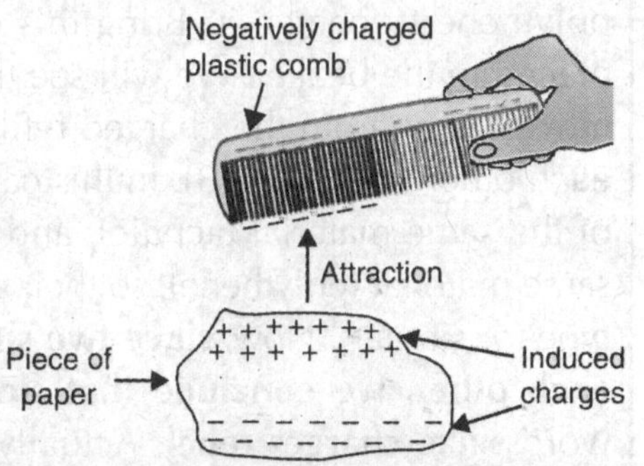

Figure 5. A negatively charged plastic comb attracting a piece of paper by inducing positive charges in its nearer end.

it) by electric induction (see Figure 5). This is because negative charge of plastic comb repels the electrons from the top side of paper to its bottom side, leaving the top of paper positively charged. The plastic comb has negative charge and the top of paper has now positive charge. The attraction between opposite charges (negative charge on plastic comb and positive charge on top of paper) results in the paper being attracted by the plastic comb. Please note that even a positively charged glass rod will attract the uncharged piece of paper by inducing opposite charges (negative charges) in the top end of the paper.

So far we have discussed the charging of various objects. We will now describe how the electric charge on an object is detected (even if the electric charge is very weak). **The electric charge on an object can be detected by using an instrument called electroscope**. We will now discuss the electroscope.

ELECTROSCOPE

The electroscope is a device for detecting electric charge on an object. By using an electroscope, we can tell whether an object is electrically charged or not. We will now describe the construction and working of a simple electroscope.

ACTIVITY

Take an empty jam bottle (which is a glass bottle). Cut a piece of cardboard slightly bigger than the mouth of jam bottle. Make a thin hole in the middle of cardboard piece. Now take a metal paper clip (steel paper clip) and open it up in such a way that it forms a shape of 5 (five) with a hook on the lower side. Insert this opened up metal clip in the hole made in cardboard piece. Take a thin strip of aluminium foil about 5 centimetres long and 1 centimetre wide. This strip of aluminium foil is folded in the middle and suspended from the hook on the lower side of the opened up metal paper clip.

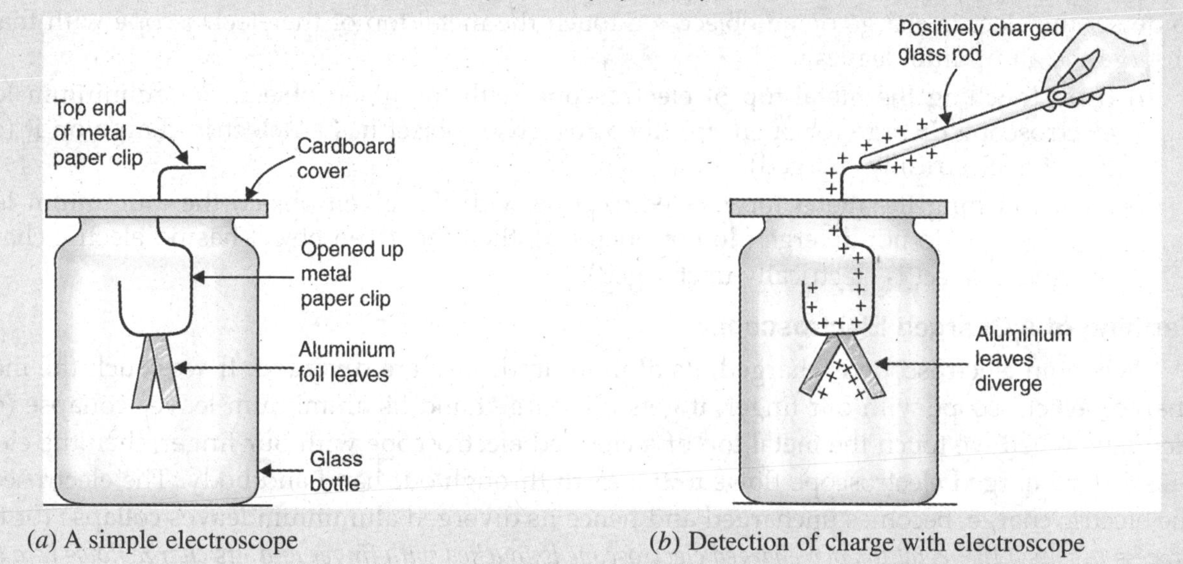

(*a*) A simple electroscope (*b*) Detection of charge with electroscope

Figure 6.

Place the cardboard piece carrying the metal clip and folded aluminium foil strip on the mouth of jam bottle in such a way that the hook carrying the folded aluminium foil is inside the bottle [as shown in Figure 6(*a*)]. If we look at the folded aluminium foil strip kept on the metal hook, it appears like two thin leaves of aluminium foil attached to the metal hook. At this stage, the two aluminium leaves are uncharged and lie close to each other [see Figure 6(*a*)]. Our electroscope is ready.

We will now describe the working of electroscope. Charge a glass rod by rubbing its one end with a piece of silk cloth. Touch the charged glass rod with the top end of metal clip [see Figure 6(*b*)]. We will see that the two aluminium leaves move away from each other [see Figure 6(*b*)]. We say that the aluminium leaves diverge (or open up). We will now explain why the aluminium leaves diverge (or open up) when touched with a charged glass rod.

When we touch the top end of metal clip with the positively charged glass rod, then some of its positive charge is transferred to the top end of metal clip. Now, since the metal clip is a good conductor of electricity, it conducts the positive electric charge to the two aluminium leaves held on its other end. In this way, the two aluminium leaves get charged with the same kind of electric charge—positive charge. We know that similar charges (or like charges) repel each other. So, **the two aluminium leaves having similar charges (positive charges) repel each other due to which they move apart or diverge** [see Figure 6(*b*)].

Let us now charge a plastic comb with negative electric charge by rubbing it with dry hair. When the top end of metal clip of the electroscope shown in Figure 6(*a*) is touched with the negatively charged plastic comb, even then the aluminium leaves of the electroscope diverge (or open up). **In this case, the two aluminium leaves get negatively charged and hence repel each other.** This repulsion causes the aluminium leaves to diverge.

From the above activity we conclude that **whether an object is positively charged or negatively charged, it will cause the aluminium leaves of the electroscope to diverge when touched with the metal top of the electroscope**. Please note that greater the amount of electric charge on an object, greater the aluminium leaves will diverge. If, however, an object is uncharged (having no electric charge) then the aluminium leaves will not diverge at all. In the above activity, the electric charge of a charged glass rod (or charged plastic comb) has been transferred to the aluminium leaves through a metal paper clip. So, in general we can say that ***an electric charge can be transferred from a charged object to another object through a metal conductor.***

Detection of Charge With Electroscope

We will now describe how the charge on an object can be detected by using an electroscope. In order to detect the electric charge on an object, we touch the metal top of the electroscope with that object and observe the aluminium leaves :

(*i*) If on touching the metal top of electroscope with the given object, the aluminium leaves of the electroscope diverge (or open up), then the given object has an electric charge on it (or the given object is electrically charged).

(*ii*) If on touching the metal top of electroscope with the given object, the aluminium leaves of the electroscope do not diverge (do not open up), then the given object has no electric charge on it (or the given object is electrically uncharged).

Earthing of a Charged Electroscope

When an electroscope is charged, its aluminium leaves are diverged. If we touch the metal top of a charged electroscope with our finger, it gets discharged and its aluminium leaves collapse (or fold up). Actually, when we touch the metal top of a charged electroscope with our finger, then the electric charge present on charged electroscope flows to the earth through our hand and body. The electroscope loses all the electric charge, becomes uncharged and hence its diverged aluminium leaves collapse (or fold up). *The process in which the metal top of a charged electroscope is touched with finger and its charge flows into earth through our hand and body, is called earthing*. In general we can say that : **The process of transferring an electric charge from a charged object to the earth is called earthing.** Earthing can also be done by connecting the metal top of the charged electroscope to the earth directly by means of a metal wire (called earth wire). In that case the electric charge of electroscope will flow to the earth through the metal wire.

The electric current in the wiring of houses and other buildings is due to the flow of electric charges (called electrons) through them. **Earthing is provided in the wiring of houses and other buildings to protect us from electric shocks** which may occur due to any leakage of electric current from the body of an electrical appliance. Metal wires (called earth wires) are used for this purpose.

Electric Discharge : Production of Sparks

Air is a non-conductor of electricity. So, normally air does not conduct electric charges (or electric current). If, however, the amount of electric charges on two oppositely charged surfaces is very large, then the air between them conducts electricity in the form of a spark (a spark is a flash of light which is seen for a very short time). Actually, the electric energy heats the air to such a high temperature that it glows. This glow of air is seen as a spark or flash. Thus, if the amount of opposite electric charges on two surfaces is very large, the insulation of air between them breaks and it allows electric charges to pass through it.

The passage of electric current in air due to movement of electric charges is called electric discharge. During electric discharge, the positive and negative electric charges cancel out each other and an **electric spark** and a **crackling sound** are produced. In nature, electric discharge within a cloud during thunderstorm produces huge electric sparks known as lightning alongwith a loud sound called thunder. And electric discharge between a thundercloud and the earth also produces lightning followed by thunder. **We will now explain the process of lightning in terms of the production of electric charges by friction in the clouds in the sky followed by electric discharge.**

LIGHTNING

During rainy season, many times we see a flash of light in the clouds in the sky which is followed by a loud sound called thunder (*garjan*). **The bright flash of light which we see in the clouds is called lightning**. Lightning is an electric discharge in the atmosphere between oppositely charged clouds (or between charged cloud and the earth). Lightning is actually a great electric spark in the sky. Lightning is produced by the electric charges in the sky. The electric nature of lightning was established by a scientist named Benjamin Franklin. We will now describe how a storm cloud gets electrically charged and then produces lightning.

A cloud is a visible mass of condensed water vapours floating in the atmosphere, high above the ground. A heavy, dark, rain cloud is called storm cloud. When a storm cloud develops in the sky, strong winds move upwards through the cloud and make the water drops present in the cloud to rub against one another. This rubbing together of water drops produces extremely large electric charges in the cloud due to friction. The *small* water drops acquire a *positive* charge and, being lighter, move to the upper part of the cloud with rising wind. On the other hand, the *larger* water drops acquire a *negative* charge and, being heavier, come down in the lower part of the cloud. In this way, **the top of the cloud becomes positively charged whereas the bottom of the cloud becomes negatively charged** (see Figure 7).

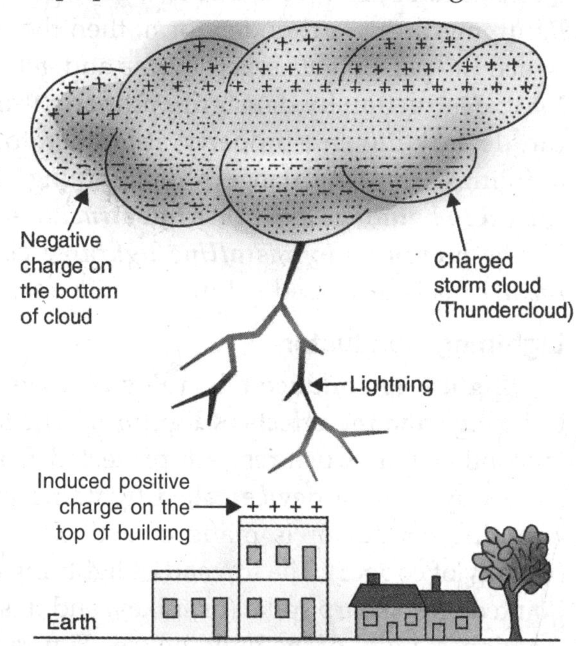

Figure 7. A view of lightning striking a building.

When the amount of opposite electric charges on the top and bottom of a storm cloud becomes extremely large, then electric charges start flowing with high speed through the air between them. When the positive and negative charges of a cloud meet, they produce an intense spark of electricity between the cloud in the sky. We see this electric spark as a flash of lightning is the sky. The electric sparks of lightning heat the nearby air in the sky to very high temperatures. Due to this heat, the air in the sky expands rapidly and produces a loud sound which we call thunder.

Lightning usually occurs within a cloud in the sky. It is called 'sheet lightning'. Sometimes, however, **lightning also occurs between a cloud and the earth (or tall objects of the earth). It is then called 'fork lightning'**. This is described below.

Storm clouds carry electric charges. Now, if a storm cloud having negative charges at its bottom passes over a tall building, it induces positive charges on the roof of the building (see Figure 7). When the electric charges on the bottom of the cloud become extremely large, then these tremendous electric charges present on the bottom of the charged cloud suddenly flow to the roof of the building and we see a flash of lightning coming towards the building. We say that lightning has struck the building (see Figure 7). Thus, lightning strikes the earth or its tall structures when electric charges flow between the cloud and the earth through a tall building, a tree or any other object.

Lightning strikes are more frequent in the hilly areas because in such areas clouds are comparatively closer to the ground than in the planes. In the planes, lightning usually strikes tall structures like tall buildings, factory chimneys, radio and TV transmission towers or big trees. This is because all these tall objects are closer to the charged clouds than the ground.

When we take off woollen or synthetic clothes (like polyester or nylon clothes), sometimes we hear a crackling sound. And if it is dark (as during night), we can even see tiny sparks. This happens as follows : When we take off a woollen (or synthetic) sweater, it rubs against our shirt. The rubbing together of sweater and shirt produces opposite electric charges on them. The discharge of these electric charges produces tiny sparks of light as well as crackling sound. In 1752 an American scientist, Benjamin Franklin, showed that the tiny sparks of light observed while putting off clothes and the lightning in the sky are essentially the same phenomenon (of electric discharge). Sometimes we also see sparks on an electric pole when the naked electric wires fixed on the tall poles become loose and touch one another. Such sparks are quite common when a strong wind is blowing and shaking the over-head electric wires too much. Sometimes we also see electric sparks in our home when a plug is loose in its socket and a current is switched on.

Dangers of Lightning

A flash of lightning carries a lot of electric energy. When lightning strikes a building, its tremendous electric energy can set the building on fire or cause serious damage to its structure. When lightning strikes a tree, it can burn up the tree and damage it by its enormous electric energy. And when a person is hit by lightning during a thunderstorm, then the electric energy passes through the body of the person due to which the person gets severe burns and gets killed. Thus, **when lightning strikes the earth, it can cause a lot of destruction by damaging property (buildings, etc.), trees and killing people.** Since lightning strikes can destroy life and property, it is, therefore, necessary to take measures to protect ourselves and our buildings from the dangers of lightning. ***The damage caused to buildings and other tall structures by lightning can be prevented by installing lightning conductors on them.*** This is discussed below.

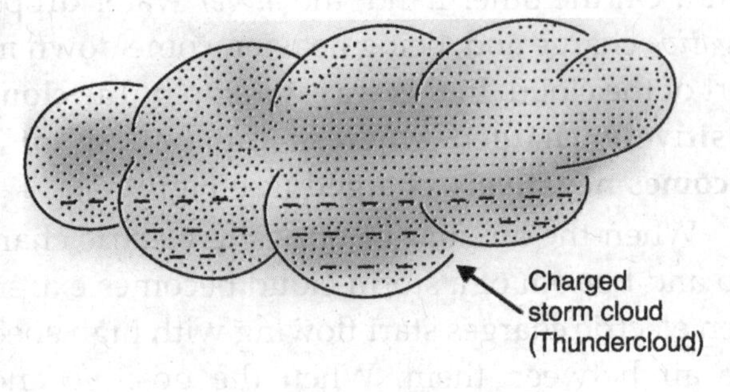

Lightning Conductor

Lightning conductor is a device used to protect a building from the effects of lightning. The tall buildings (and other tall structures) are protected from lightning strikes by using a device called lightning conductor. A lightning conductor is made of a thick strip of metal (usually of copper). The top end of lightning conductor is pointed like a sharp spike (or spikes) and it is fixed above the highest point of the building (see Figure 8). From the top of the building, the thick metal strip runs along the outer wall of the building to the ground. The lower end of metal strip is joined to a metal plate and buried deep in the ground near the base of the building (see Figure 8).

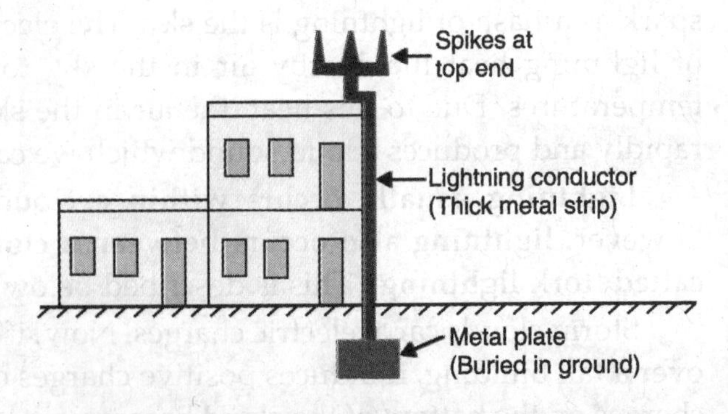

Figure 8. Diagram to show the use of a lightning conductor to protect a tall building from lightning.

If lightning strikes, it will hit the top of the lightning conductor rather than the building. The electric energy of lightning passes through the metal strip and gets discharged safely into the ground through the buried metal plate. Since no electric energy produced by lightning passes through the building, no damage is caused to it. Thus, lightning can be discharged harmlessly into the ground (or earth) through the lightning conductor fitted on tall buildings and other tall structures like factory chimneys, radio and TV transmission towers and monuments like *Qutab Minar*. We know that metals are good conductors of electric charges (or electricity). So, **a lightning conductor made of a metal works by conducting the electric energy of lightning into the earth**. We can see the lightning conductors fixed on many tall buildings, factory chimneys, etc. Lightning conductor was invented by Benjamin Franklin.

If a tall building is not protected with a lightning conductor, then the tremendous electric energy produced during lightning would pass through the walls of the building, causing damage to the material of the walls and making the walls unsafe. It can even set the building on fire. A lightning conductor protects a tall building against lightning by providing its electric energy an easy path to go to the ground. Thus, **the function of a lightning conductor is to conduct any lightning strikes safely to the earth (without causing any damage to the building).**

Measures to Protect Ourselves from Lightning

We can take the following measures (or steps) to protect ourselves from lightning strikes during a thunderstorm :

(*i*) No open space is safe during lightning and thunderstorm. A house (or any other building) is a safe place during lightning. So, if we are in an open space (such as a park, playground, or road, etc.), we should rush to a safer place like a house or some other building nearby on hearing the thunder and observing the lightning in the sky.

(*ii*) Open vehicles like motorbikes, scooters, tractors, and construction machinery (like earth movers, etc.) are not safe during lightning and thunderstorm. So, we should leave such open vehicles during lightning and take shelter inside a house (or some other building). If, however, we are travelling by a covered vehicle like a car (or bus) when thunderstorm and lightning occur, then we are safe inside the car (or bus) with windows and doors of the vehicle closed.

(*iii*) If a person is in open space when thunderstorm and lightning begin, and there is no suitable shelter available nearby, then the following precautions should be taken for protection from lightning :

(*a*) When in open space, a person should never stand under a tree to take shelter during a thunderstorm because there is danger of lightning striking the tree and burning it up. This lightning can also pass through the body of the person standing under the tree and kill him. If, however, a person is in a forest, he should take shelter under a short tree because a short tree is less likely to be hit by lightning. On the other hand, a tall tree (being nearer to the thunderclouds) is more likely to be hit by lightning.

(*b*) When in open space, a person should not lie on the ground during the thunderstorm and lightning. A person should squat low on the ground during lightning. The person should place his hands on his knees with his head between the hands (as shown in Figure 9). This position will make the person the shortest object around which is unlikely to be hit by lightning. The person should also stay away from electric poles, telephone poles and other metal objects during lightning.

Figure 9. The safe position during lightning (when a person is in open space with no shelter nearby).

(*iv*) We should avoid raising an umbrella over our head during lightning. This is because lightning may strike the top end of the metal rod of umbrella (held high over the head) and harm us.

(*v*) The TV antennas and dish antennas fixed on tall buildings are especially prone to lightning strikes. We should, therefore, switch off our TV sets during frequent lightning otherwise TV sets may get burnt.

(*vi*) Lightning can strike metal pipes (such as water pipes) fixed in buildings. So, during a thunderstorm when lightning is taking place, we should avoid touching the metal pipes fixed in a house or building.

EARTHQUAKES

An earthquake is a sudden shaking (or trembling) of the earth which lasts for a very short time. An earthquake is caused by the violent movements of rocks deep inside the earth's crust. Earthquakes occur all the time all over the earth. Most of these earthquakes are so mild that they are not even noticed (or felt) by us. Major earthquakes are much less frequent but they are very dangerous. **Earthquakes can cause immense damage to houses, other buildings, bridges, dams and people, etc. A lot of people get killed when they get buried under the debris of collapsed houses and other buildings during an earthquake. Earthquakes can also cause floods, landslides, and tsunamis** (read as sunamis).

Figure 10. Buildings destroyed by the Gujarat earthquake of January 2001.

A major earthquake can cause damage to life and property on a large scale. A major earthquake occurred in India on 26th January 2001 in Bhuj district of Gujarat (see Figure 10). Another major earthquake occurred on 8th October 2005 in Uri and Tangdhar towns of North Kashmir. A great loss of human life and property (houses and other buildings, etc.) occurred in both these earthquakes. A major tsunami (caused by an earthquake) occurred in the Indian Ocean on 26th December 2004. All the coastal areas around the Indian Ocean suffered huge loss of life and property during this tsunami.

Earthquakes are a destructive natural phenomena. The other destructive natural phenomena such as cyclones and lightning can be predicted in advance to some extent so that we get some time to take measures to protect ourselves and minimise the damage to life and property. This is not so in the case of earthquakes. **Earthquake is a destructive natural phenomenon which cannot be predicted in advance**. Nobody can tell when and where an earthquake will occur. This unpredictable nature of earthquakes makes them even more dangerous. In order to understand why earthquakes occur, we should first understand the inner structure of the earth. This is described below.

Structure of Earth

The inside of earth is made up of three main layers : Core, Mantle and Crust (see Figure 11).

***(i)* CORE.** The innermost part of the earth is called its core (see Figure 11). The core of earth is made up mostly of iron. The core of earth is extremely hot. Most of the earth's core (called outer core) is liquid (molten iron) whereas the inner part of core (called inner core) is under such high pressure that it is solid (solid iron).

***(ii)* MANTLE.** The central region of earth (between the core and crust) is called mantle (see Figure 11). Mantle is the middle layer of the earth. Mantle is mostly made of dense, solid rocks. Some of the mantle is, however, a mixture of solid rocks and hot molten rocks (liquid rocks) like the lava from a volcano. Heat coming from the core of earth warms the mantle. This heating sets up huge convection currents in the mantle. The giant convection currents occurring in the mantle can make the mantle move very, very slowly.

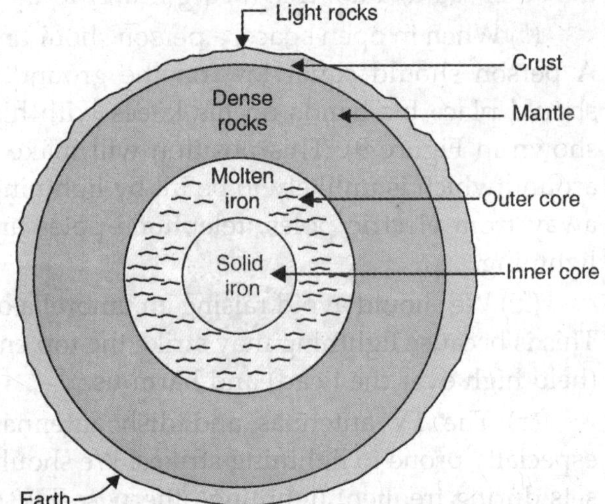

Figure 11. The structure of earth.

***(iii)* CRUST.** The outermost layer of earth is called crust (see Figure 11). The crust of earth is made of comparatively lighter rocks than that of mantle. The crust of earth is thicker where there is land (or continents). The crust of earth under the oceans is thinner. **The whole**

crust of earth is not in one piece. The crust of earth is made of many huge pieces of flat rocks (rather like tiles on a bathroom floor). Each piece of the earth's crust is called a plate. Thus, **the crust of earth is divided into many plates**. The plates of earth on which continents exist are called *continental plates* whereas those plates of earth on which oceans exist are called *oceanic plates*. The plates of crust are in fact very, very large fragments of earth's crust. The solid plates which make up the earth's crust are floating on the partially molten rocks of mantle beneath (see Figure 12). **Due to convection currents taking place in the mantle, the plates of earth's crust are moving around very, very slowly** (see Figure 12). ***The reason the earthquakes occur is that the earth's crust is made of a number of plates which are able to move.*** Earthquakes occur mostly at the edges of moving plates of the crust (or boundaries of the moving plates of the earth).

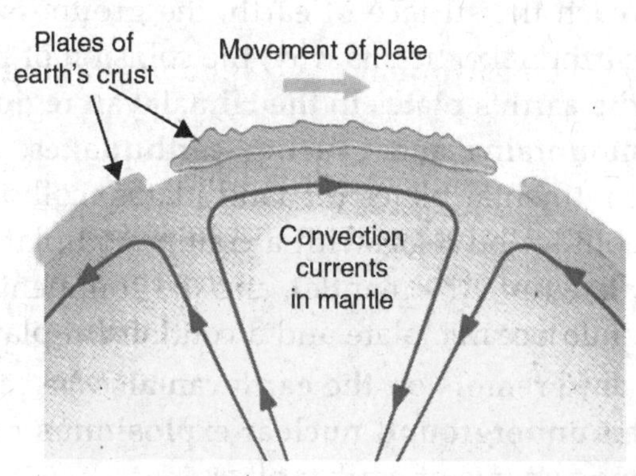

Figure 12. Earth's crust is made up of many huge plates of solid rocks. Convection currents taking place in the earth's mantle move the plates of earth's crust very, very slowly.

Why Do Earthquakes Occur

The earthquakes occur when the moving plates of the earth's crust :

(*i*) slide past one another, and

(*ii*) collide with one another.

We will describe both these cases in detail, one by one

1. The plates of earth's crust are made up of huge rocks having rough edges. Due to their highly rough edges, the movement of two crust plates relative to one another is not smooth. When the two huge plates of earth's crust slide past one another, they rub against one another ferociously and the rocks on their edges get entangled. Due to entanglement of the rocks at their edges, the two crust plates stop moving for some time. During this time, the plates are still pushing against one another and trying to move *but they are not moving* (due to entanglement of their rocky edges). This builds up pressure between the two plates of crust. When sufficient pressure has been built up between the two crust plates, the entangled rocks of two plates break open with a big jolt (see Figure 13). **When the entangled rocks of the two crust plates break open suddenly with a big jolt, the earthquake occurs releasing a tremendous amount of energy**. This sudden release of tremendous amount of energy produces shock waves (or seismic waves) which make the earth shake. After the earthquake, the plates of earth's crust start moving again and continue to move until they get entangled again.

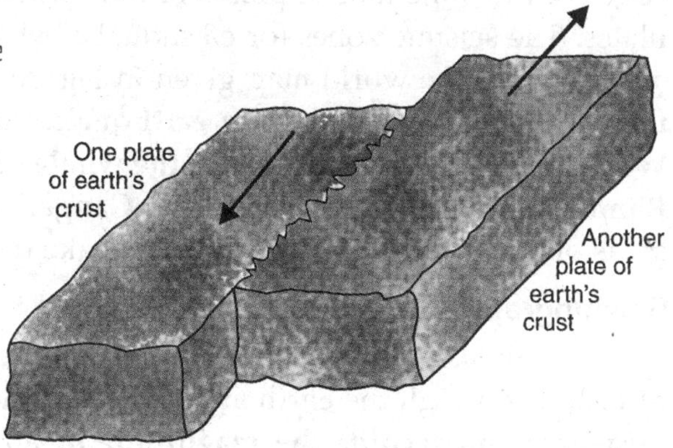

Figure 13. An earthquake is caused when two rocky plates of earth's crust moving in opposite directions slide past each other, rub ferociously, get entangled and then break open suddenly with a jolt.

2. When the two plates of the earth's crust moving in opposite directions collide with each other head on, then the ends of these crust plates buckle and fold forming new mountains and causing earthquakes (see Figure 14). A tremendous amount of energy is released when the two huge plates of the earth's crust collide with each other. This tremendous amount of energy sends shock waves (called seismic waves) throughout the earth. When these shock waves

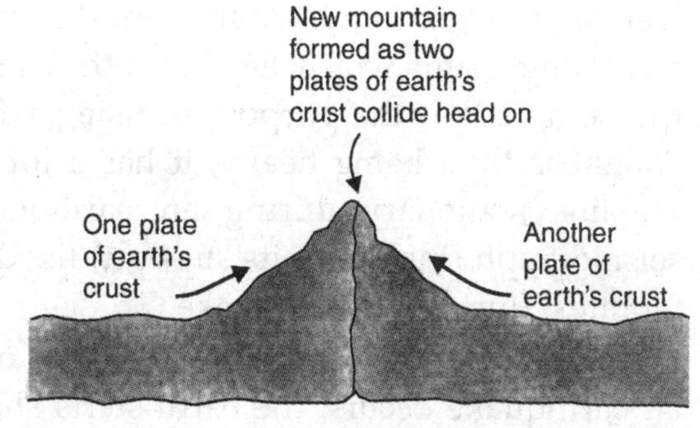

Figure 14. An earthquake is also caused when the two rocky plates of earth's crust moving in opposite directions collide head on, buckle and fold raising new mountains.

reach the surface of earth, the ground starts shaking violently. This shaking of the ground is felt as an earthquake. In this way, the collision of two moving plates of earth's crust also causes earthquake. In fact, **the earth's plates in the Himalayan region push against one another (or collide with one another) raising mountains and causing earthquakes**. Actually, what we have described above happens when two continental plates (or land plates) collide head on with each other. If, however, a moving oceanic plate collides head on with a continental plate (land plate), then the oceanic plate goes under the continental plate (into the mantle), and the continental plate buckles and folds to form a new mountain. This collision of an oceanic plate and a continental plate also causes an earthquake.

Tremors on the earth can also be caused when a volcano erupts or a big meteorite hits the earth or an underground nuclear explosion is carried out. Most of the earthquakes are, however, caused by the movements of earth's plates.

Seismic Zones (or Fault Zones)

Since earthquakes are caused by the movements of earth's plates, the boundaries of the plates are the weak zones where earthquakes are most likely to occur. **The weak zones of earth's crust (which are more prone to earthquakes) are called 'seismic zones' or 'fault zones'**. Most major earthquakes occur in the well defined 'seismic zones' called earthquake belts. The seismic zones mark the edges of the huge mobile pieces of the earth's crust called plates. The seismic zones (or earthquake belt) of the earth are shown in the world map given in Figure 15. **In India, the areas most threatened by earthquakes are Kashmir, Western and Central Himalayas, the whole of North-East, Rann of Kutch, Rajasthan and Indo-Gangetic Plane.** Some areas of South India also fall in earthquake danger zone.

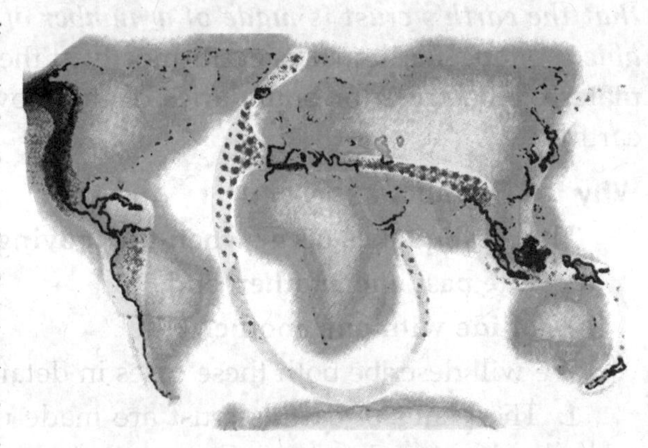

Figure 15. The encircling belt (shown in orange colour) in this figure indicates the seismic zones of the world where most of the earthquakes occur.

Seismograph

Earthquake tremors produce shock waves (called seismic waves). The shock waves travel in all directions through the Earth and also reach the surface of the earth. **Seismograph is an instrument which measures and records the magnitude of an earthquake in terms of the shock waves it produces**. A seismograph detects the shock waves produced by an earthquake and also records the shock waves on the paper in the form of a graph. Seismograph is also known as seismometer. A seismograph consists of a heavy weight (like a heavy metal ball) suspended from a support with the help of a strong wire (see Figure 16). The support is attached to the base of seismograph. And the base of seismograph is fixed rigidly to a solid rock on the surface of earth. A pen is attached to the lower end of the hanging heavy weight. This pen can trace lines on a graph paper wound around a rotating drum which lies beneath it (as shown in Figure 16). Please note that the purpose of using a freely suspended heavy weight is that, being heavy, it has a lot of inertia and hence it remains stationary during an earthquake when the rest of seismograph (including its support) fixed to the earth shakes (or vibrates) during an earthquake.

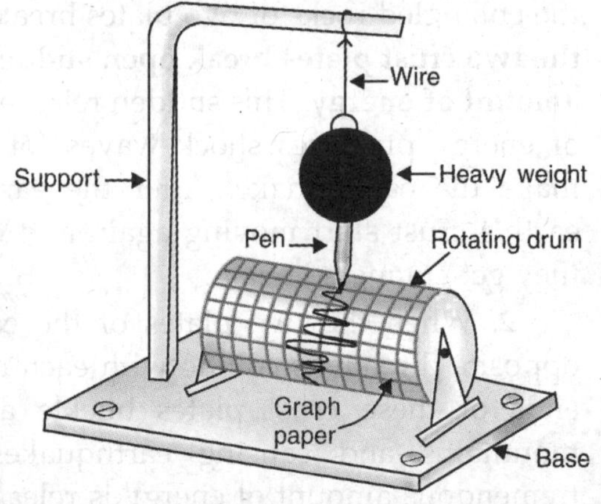

Figure 16. A seismograph.

We will now describe the working of a seismograph. When an earthquake occurs, the earth starts shaking due to which the base of seismograph fixed to the earth also starts shaking. But the freely suspended heavy weight (and the pen attached to it) do not shake during the earthquake, they remain stationary. Since the graph paper on the rotating drum shakes with the shaking earth, the pen attached to suspended weight records the vibrations produced by earthquake on the graph paper which moves under it (see Figure 16). Actually, the pen traces the relative movement between the shaking earth and the stationary, suspended heavy weight. This trace on the graph paper records the earthquake. A trace produced by a seismograph during an earthquake is shown in Figure 17. The seismograph record is also known as seismogram. A seismograph record (or seismogram) shows seismic waves or earthquake waves recorded on a graph paper by a seismograph. It also shows the duration (or time) for which the earthquake lasts, in seconds.

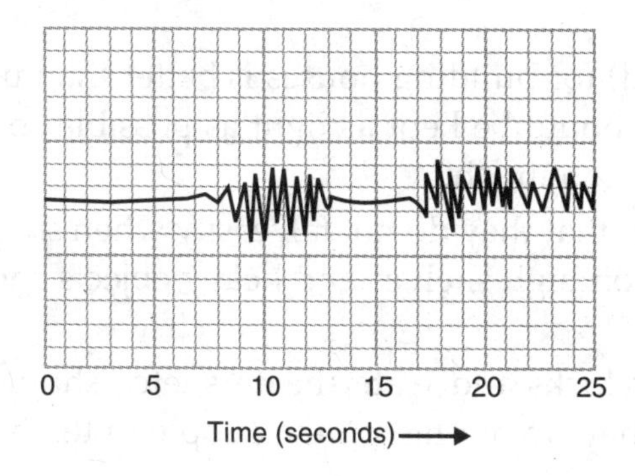

Figure 17. A seismograph record (or seismogram).

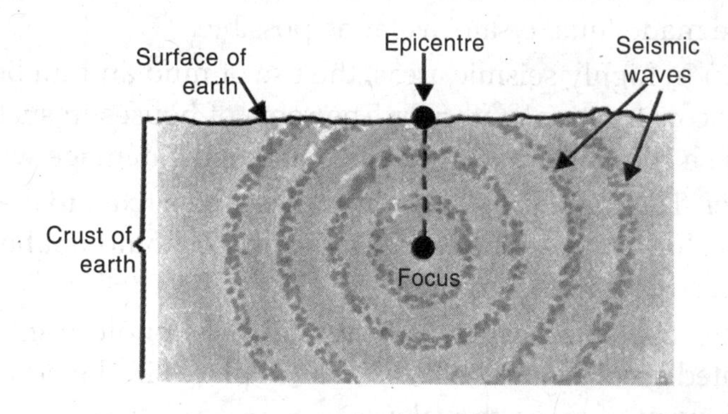

Figure 18. A map of the earthquake.

Earthquake recording laboratories (called seismic stations) have been established all around the world. By studying the seismograph records of an earthquake produced at various seismic stations, scientists can construct a complete map of an earthquake giving its focus, epicentre, magnitude and duration, etc. (see Figure 18). **The place inside the earth's crust where the earthquake is generated, is called 'focus' of the earthquake** (see Figure 18). The focus of an earthquake is deep underground. *The seismic waves (or earthquake waves) spread from the focus and travel through the earth in all directions.* Focus is actually the 'source'of an earthquake. **The point on earth's surface directly above the focus is called epicentre** (see Figure 18).

Richter Scale

The magnitude (or intensity) of an earthquake is expressed on the Richter Scale. The Richter Scale is a series of numbers from 1 to 12 used to express the magnitude (or size) of an earthquake. Just like the Decibel Scale for measuring the loudness of sound, the Richter Scale is not a linear scale. This means that an earthquake of magnitude 2 on Richter Scale is not two times as strong as an earthquake of magnitude 1 on this scale. In fact, an earthquake of magnitude 2 is ten times as strong as an earthquake of magnitude 1. Similarly, an earthquake of magnitude 3 on Richter Scale is 100 times as strong as an earthquake of magnitude 1 on the same scale. And an earthquake of magnitude 6 on the Richter Scale is 1000 times stronger than an earthquake of magnitude 4. In other words, an earthquake of magnitude 6 on Richter Scale has 1000 times more destructive energy than an earthquake of magnitude 4.

An earthquake of magnitude 1.5 on the Richter Scale is the smallest earthquake that can be felt by us. An earthquake of magnitude 4.5 on the Richter Scale causes some damage but not much damage. **The earthquakes having magnitudes higher than 7 on the Richter Scale are really destructive earthquakes**. Both, Bhuj and Kashmir earthquakes had magnitudes greater than 7.5 on the Richter Scale. The earthquakes which measure 8.5 or more on Richter Scale are devastating. During these earthquakes, the affected part of earth's surface shakes violently. Such earthquakes can destroy entire cities and villages causing a great loss of life and property.

Protection Against Earthquakes

Earthquakes are highly destructive and they cannot be predicted. So, it is necessary that we take precautions to protect ourselves all the time (especially if we live in a seismic zone of the country). Some of the important precautions which can be taken by the people living in seismic zones (or earthquake prone areas) for protection against earthquakes are as follows :

(*i*) **All the houses and other buildings in seismic zones should be designed and constructed in such a way that they can withstand major earthquake tremors.** It is possible to do so by using modern building technology which is available these days. In our country, the Central Building Research Institute at Roorkee has developed knowhow to make quake-proof houses. The people should consult qualified architects and structural engineers while constructing houses and other buildings in earthquake prone areas so that they can be made 'quake safe' as far as possible.

(*ii*) In highly seismic areas, the use of mud and timber (wood) for building houses is better than using heavy construction materials. The roofs of houses in such areas should be kept as light as possible so that in case a roof falls during an earthquake, the damage will not be too much.

(*iii*) The cupboards and shelves should be fixed to the walls so that they do not fall easily when shaking occurs during an earthquake. Glass bottles should not be placed on high shelves and heavy objects should be placed low to the ground.

(*iv*) The objects such as heavy mirrors, photo frames, wall clocks and water-heaters, etc., should be mounted securely on the walls at such places in the house that they do not fall on the people in the house in the event of an earthquake.

(*v*) All buildings (especially tall buildings) should have fire-fighting equipment in working order because some buildings may catch fire during an earthquake (due to electric short circuits).

In case an earthquake occurs, we should take the following precautions to protect ourselves depending on whether we are at home or outdoors at that moment.

If we are at home when an earthquake occurs, then :

(*i*) We should take shelter under a sturdy table or a kitchen counter and stay there till the shaking due to earthquake stops. All this while, we should cover our head with hands.

(*ii*) We should stay away from tall and heavy objects (like steel almirahs, cabinets, book racks and refrigerators, etc.,) so that if they topple, they may not fall on us. We should also stay away from glass windows that may shatter due to vibrations.

(*iii*) If we are in bed, we should just move on the bed to be close to a wall but not get up from the bed. We should protect our head with a pillow.

If we are outdoors when an earthquake occurs, then :

(*i*) We should stay at a clear spot away from buildings, trees and over-head power lines, etc. We should also sit on the ground (so that we may not fall down due to shaking of ground).

(*ii*) If we are in a car (or bus), we should not come out of it. The car (or bus) should be driven slowly to a clear spot away from buildings, trees and over-head electric wires, etc. We should not come out of the vehicle till the tremors stop.

We are now in a position to **answer the following questions :**

Very Short Answer Type Questions

1. What are the two kinds of electric charges ?
2. What kind of electric charge is acquired :
   (*a*) by a glass rod rubbed with silk cloth ?
   (*b*) by a plastic comb rubbed with dry hair ?
3. What type of electric charge is acquired by a rubber balloon when rubbed with a woollen cloth ?
4. A negatively charged object attracts another charged object placed near it. What is the nature of charge on the other object ?
5. A positively charged object repels another charged object kept close to it. What is the nature of charge on the other object ?
6. A negatively charged object repels another charged object held near it. What is the nature of charge on the other object ?

7. A glass rod is rubbed with a silk cloth. What type of charge is acquired by (*a*) silk cloth, (*b*) glass rod ?
8. An inflated rubber balloon and a woollen cloth are rubbed together. What type of charge is acquired by (*a*) woollen cloth, and (*b*) rubber balloon ?
9. Name the device to detect electric charge on a body.
10. When an object is touched with the metal top of an electroscope, its aluminium leaves diverge. What conclusion do you get from this observation ?
11. What name is given to the flash of light which occurs in the sky during the rainy season ?
12. Why should a person not stand under a tree during a thunderstorm ?
13. Name the scientist who showed that lightning is electric in nature.
14. Name the device which is used to protect a tall building from lightning.
15. What name is given to the phenomenon in which the earth shakes suddenly for a very short time ?
16. Name one destructive natural phenomenon which cannot be predicted in advance.
17. List three states in India where earthquakes are more likely to occur.
18. Name the instrument used to measure and record an earthquake.
19. What was the magnitude of Bhuj and Kashmir earthquakes on the Richter Scale ?
20. Name the scale on which the magnitude (or intensity) of an earthquake is expressed.
21. For what purpose is Richter Scale used ?
22. Name two events (other than earthquakes) which can cause tremors on the earth.
23. In the context of an earthquake, which one is deep under the ground : focus or epicentre ?
24. State whether the following statements are true or false :
    (*a*) Like charges attract each other.
    (*b*) A charged glass rod attracts a charged plastic straw.
    (*c*) Lightning conductor cannot protect a building from lightning.
    (*d*) Earthquakes can be predicted in advance.
    (*e*) An earthquake of magnitude 2 on Richter Scale is ten times as strong as an earthquake of magnitude 1 on the same Scale.
    (*f*) The plates of earth's crust are continuously moving.
    (*g*) An earthquake is measured and recorded by using an instrument called electrocardiograph.
25. Fill in the following blanks with suitable words :
    (*a*) Like charges............ ; unlike charges...........
    (*b*) Rubbing glass with silk makes a............charge on the glass.
    (*c*) Combing your hair makes a...............charge on the comb.
    (*d*) The negatively charged particles which are transferred from one object to another during charging by friction are called...........
    (*e*) The charging of an object by rubbing it with another object is called charging by............
    (*f*) In an electroscope, the aluminium leaves diverge because like charges........
    (*g*) ....is provided in buildings to protect us from electric shocks due to any leakage of electric current.
    (*h*) Lightning is nothing but an.............spark.
    (*i*) Each fragment of earth's crust is called a..........

Short Answer Type Questions

26. Why does a plastic comb rubbed with dry hair attract tiny pieces of paper ?
27. How will you charge a glass rod by the method of friction ?
28. How will you charge an inflated rubber balloon by the method of friction ?
29. How will you charge a plastic comb (plastic scale or plastic pen) by the method of friction ?
30. How will you charge a ballpoint pen refill by the method of friction ?
31. What will you observe when the metal top of an electroscope is touched with a glass rod which has been rubbed with silk cloth ? Give reason for your answer.
32. What will you observe when the metal top of an electroscope is touched with a plastic comb rubbed in dry hair ? Give reason for your answer.
33. What happens when we touch the metal top of a charged electroscope with our finger ? What is this process known as ?

34. Explain why, a charged body loses its charge when we touch it with our hand.

35. What happens when the two plates of earth's crust moving in opposite directions slide past one another ?

36. What happens when two moving plates of earth's crust collide head on with each other ?

37. How will you find out whether an object is charged or not ?

38. Explain why, it might be dangerous to raise an umbrella over our head in a thunderstorm.

39. A person is in open space during a thunderstorm with no shelter (not even a tree) available nearby. Describe the safe position which he should take to protect himself from lightning. Why is this position considered safe ?

40. Suggest three measures to protect ourselves from lightning.

41. Explain why, sometimes when we take off the woollen sweater or a polyester shirt in a dark room, we can see tiny sparks of light and hear a crackling sound.

42. (*a*) Name the material of which a lightning conductor is made.
(*b*) What is the shape of the top end of a lightning conductor ?
(*c*) Where is the upper end of the lightning conductor fixed in a building ?
(*d*) Where is the lower end of the lightning conductor fixed and how ?

43. What precautions would you take to protect yourself during an earthquake if you are inside the house ?

44. What precautions would you take to protect yourself during an earthquake if you are outdoors ?

45. State any two precautions which should be observed by people living in seismic zones for protection against earthquakes.

Long Answer Type Questions

46. What is an electroscope ? Draw a labelled diagram of an electroscope and explain its working.

47. (*a*) What is lightning ? How is lightning produced between clouds in the sky ?
(*b*) Why does lightning usually strike tall buildings ?
(*c*) What damage can be done when lightning strikes on the earth ?

48. (*a*) How does a lightning conductor protect a tall building ? Name the scientist who invented the lightning conductor.
(*b*) Why are lightning strikes more frequent in hilly areas ?

49. (*a*) What is an earthquake ? What are the two main situations in which earthquakes occur ?
(*b*) Define (*i*) focus, and (*ii*) epicentre, of an earthquake.
(*c*) What are the various effects of an earthquake ?

50. (*a*) Name the three layers of earth. Draw a labelled diagram to show the structure of earth.
(*b*) What is a seismograph ? Draw a labelled diagram of a seismograph.

Multiple Choice Questions (MCQs)

51. Which of the following cannot be charged by friction, if held by hand ?
(*a*) a plastic scale (*b*) a copper rod (*c*) an inflated balloon (*d*) a woollen cloth

52. When a glass rod is rubbed with a piece of silk cloth, then :
(*a*) the glass rod and silk cloth both acquire positive charge.
(*b*) the glass rod becomes positively charged while the silk cloth has a negative charge.
(*c*) the glass rod and silk cloth both acquire negative charge.
(*d*) the glass rod becomes negatively charged while the silk cloth has a positive charge.

53. Which of the following are transferred from one object to another when these two objects are charged by friction ?
(*a*) atoms (*b*) protons (*c*) neutrons (*d*) electrons

54. The electric nature of lightning was established by a scientist named :
(*a*) Isaac Newton (*b*) Robert Hooke (*c*) Benjamin Franklin (*d*) Thales

55. A plastic comb is rubbed with dry hair whereas a glass rod is rubbed with a piece of silk cloth. Which of these will get negatively charged ?
A. Plastic comb B. Glass rod C. Dry hair D. Silk cloth
(*a*) A and B (*b*) B and C (*c*) A and D (*d*) B and D

56. The magnitude of an earthquake is measured on :
(*a*) Celsius scale (*b*) Kelvin scale (*c*) Decibel scale (*d*) Richter scale
57. An earthquake of magnitude 2 on Richter scale is :
(*a*) two times as strong as an earthquake of magnitude 1
(*b*) four times as strong as an earthquake of magnitude 1
(*c*) ten times as strong as an earthquake of magnitude 1
(*d*) hundred times as strong as an earthquake of magnitude 1
58. The epicentre of an earthquake is :
(*a*) deep under the crust of earth (*b*) in the mantle of earth
(*c*) on the surface of earth (*d*) in the core of earth
59. The waves generated by the earthquake tremors are called :
(*a*) ultrasonic waves (*b*) rhythmic waves (*c*) systemic waves (*d*) seismic waves
60. When an object gets negatively charged by the process of friction, then :
(*a*) the object loses some electrons (*b*) the object loses some protons
(*c*) the object gains some electrons (*d*) the object gains some protons
61. The device used for detecting charge (positive or negative) on an object is called :
(*a*) stethoscope (*b*) telescope (*c*) kaleidoscope (*d*) electroscope
62. A charged object attracts an uncharged object by producing opposite charges in the nearer end of the uncharged object by the process of :
(*a*) electric potential (*b*) electric induction
(*c*) friction (*d*) electromagnetic induction
63. A lightning conductor is a device which transfers :
(*a*) electric energy (*b*) light energy
(*c*) solar energy (*d*) photoelectric energy
64. When a plastic comb is rubbed with dry hair, the hair get positively charged by friction. In this porcess :
(*a*) the hair lose some positive protons
(*b*) the hair gain some positive protons
(*c*) the hair lose some negative electrons
(*d*) the hair gain some negative electrons
65. Which of the following part of the earth is made up of molten iron ?
(*a*) mantle (*b*) inner core (*c*) outer core (*d*) crust
66. Which of the following area of India is not the most threatened by earthquakes ?
(*a*) North-East (*b*) Kashmir (*c*) West Bengal (*d*) Rajasthan
67. The place inside the earth's crust where the earthquake is generated is called :
(*a*) seismic zone of the earth (*b*) epicentre of the earthquake
(*c*) fault zone of the earth (*d*) focus of the earthquake
68. An inflated rubber balloon is rubbed with a woollen cloth whereas a ballpoint pen refill is rubbed with a polythene bag. Which of these will get positively charged ?
A. Inflated rubber balloon B. Woollen cloth C. Ballpoint pen refill D. Polythene bag
(*a*) A and B (*b*) B and C (*c*) A and D (*d*) B and D
69. Lightning can even burn up a tree. Lightning contains a tremendous amount of :
(*a*) heat energy (*b*) electric energy (*c*) chemical energy (*d*) nuclear energy
70. The tremendous electric charges in the atmosphere which produce sheet lightning in the clouds are produced by the process of :
(*a*) friction (*b*) induction (*c*) conduction (*d*) convection

Questions Based on High Order Thinking Skills (HOTS)

71. Explain why, a charged balloon is repelled by another charged balloon whereas an uncharged balloon is attracted by a charged balloon. **(NCERT Book Question)**
72. Explain why, a glass rod can be charged by rubbing when held by hand but an iron rod cannot be charged by rubbing, if held by hand.
73. A glass rod is rubbed with a silk cloth and an inflated rubber balloon is rubbed with a woollen cloth.

Now, out of glass rod, silk cloth, rubber balloon and woollen cloth :
(*a*) which two objects acquire negative charge ?
(*b*) which two objects acquire positive charge ?

74. What will you observe when the metal top of an electroscope is touched with :
(*a*) a positively charged object ?
(*b*) a negatively charged object ?

75. An earthquake measures 3 on Richter Scale :
(*a*) Would it be recorded by a seismograph ?
(*b*) Is it likely to cause much damage ?

ANSWERS

4. Positive charge **5.** Positive charge **6.** Negative charge **10.** The object has an electric charge on it (It is electrically charged) **16.** Earthquakes **23.** Focus **24.** (*a*) False (*b*) True (*c*) False (*d*) False (*e*) True (*f*) True (*g*) False **25.** (*a*) repel ; attract (*b*) positive (*c*) negative (*d*) electrons (*e*) friction (*f*) repel (*g*) Earthing (*h*) electric (*i*) plate **51.** (*b*) **52.** (*b*) **53.** (*d*) **54.** (*c*) **55.** (*c*) **56.** (*d*) **57.** (*c*) **58.** (*c*) **59.** (*d*) **60.** (*c*) **61.** (*d*) **62.** (*b*) **63.** (*a*) **64.** (*c*) **65.** (*c*) **66.** (*c*) **67.** (*d*) **68.** (*b*) **69.** (*b*) **70.** (*a*) **71.** A charged balloon repels another charged balloon because like charges repel ; A charged balloon attracts an uncharged balloon by producing opposite charges in the nearer end of uncharged balloon by electric induction. **72.** A glass rod can be charged by rubbing when held in hand because glass rod is an insulator which does not conduct electric charges produced on its surface through our hand and body into the earth ; An iron rod cannot be charged by rubbing when held in hand because iron rod is a conductor due to which as soon as it gets charged by rubbing, the electric charges produced on its surface flow through our hand and body into the earth and it remains uncharged **73.** (*a*) Silk cloth and Rubber balloon acquire negative charge (*b*) Glass rod and Woollen cloth acquire positive charge **74.** (*a*) The aluminium leaves of the electroscope diverge (or open up) (*b*) The aluminium leaves of the electroscope diverge (or open up) **75.** (*a*) Yes (*b*) No

CHAPTER

16

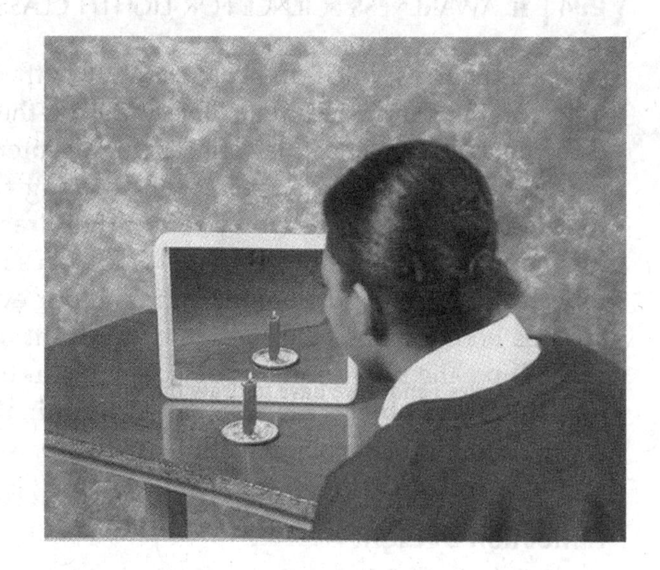

Light

Light is a form of energy. **Light is needed to see things around us.** We are able to see the beautiful world around us because of light. We can read a book, see pictures in a magazine and watch TV and movies due to the existence of light. And it is light which makes us see our image in a looking mirror. We detect light with our eyes.

What Makes Things Visible

Though we see various things (or objects) around us with our eyes but eyes alone cannot see any object. For example, we cannot see objects in a dark room or in the darkness of night even when our eyes are wide open. We need a source of light to make the objects (or things) visible. So, as soon as light from a torch (an electric bulb or a tube-light) falls on the object, we are able to see it clearly even in a dark room or in the darkness of night. *It is only when light coming from an object enters our eyes that we see that object.* This light may have been emitted by the object itself or may have been reflected by the object. Thus, **it is light which makes things visible to us. Light enables us to see things from which it comes or from which it is reflected.**

Luminous Objects and Non-Luminous Objects

There are two types of objects around us : luminous objects and non-luminous objects. **The objects which emit their own light are called luminous objects**. The luminous objects are, in fact, *the sources of light*. Luminous objects produce their own light and then emit this light. **The sun, other stars, lighted electric bulb, glowing tube-light, torch, fire, and flame of a burning candle, are all luminous objects**. A luminous object can be seen because the light given out by it enters our eyes. For example, we can see the sun because the light given out by sun (or light emitted by sun) enters our eyes. Luminous objects are very small in number.

All the objects cannot give out their own light. **The objects which do not emit their own light are called non-luminous objects**. Actually, the non-luminous objects cannot make their own light. Since non-luminous objects cannot produce light, therefore, they cannot emit their own light. **The moon, earth, other planets, table, chair, book, trees, plants, flowers, human beings, fan, bed, mirror, diamond, walls, floor and roads, are some of the examples of non-luminous objects**. In fact, *most of the objects around us are non-*

luminous objects (which do not have light of their own). The non-luminous objects can be seen only when light coming from a luminous object falls on them. This light is reflected by the non-luminous object in all directions. And when this reflected light enters our eyes, we can see the non-luminous object. This is because to us the light appears to be coming from the non-luminous object. Thus, **we can see the non-luminous objects because they reflect light (received from a luminous object) into our eyes**. For example, the moon is a non-luminous object which does not have its own light. We can see the moon because moon reflects light (received from the sun) into our eyes. Thus, moon is a reflector of sunlight. Similarly, we can see this book because the sunlight (bulb-light or tube-light) falling on it is reflected by the book into our eyes. Thus, the non-luminous objects shine in the light of luminous objects and become visible to us. **The non-luminous objects are also called illuminated objects** (because they get illuminated or lighted up by the light of luminous objects falling on them). ***Most of the objects around us (being non-luminous) are seen by the reflected light.***

Reflection of Light

When light falls on the surface of an object, the object sends this light back. **The process of sending back light rays which fall on the surface of an object, is called reflection of light.** The reflection of light is studied by using a plane mirror. A plane mirror reflects almost all the light which falls on it. This means that a plane mirror changes the direction of light which falls on it. We will now study the direction in which the light falling on a plane mirror is reflected.

In order to study the reflection of light, we need an apparatus which can produce a thin beam of light. We use an apparatus called 'ray-box' to produce a thin beam of light in science activities. A ray-box has a light bulb inside it and there is a narrow slit in front of the box (see Figure 1). When the light bulb is switched on, a very thin beam of light (or a narrow beam of light) comes out of the narrow slit of the ray-box. This narrow beam of light is then used to study the reflection of light from a plane mirror. Thus, **the ray-box acts as a source of light in the 'reflection of light' activities.** The thin beam of light produced by a ray-box is visible on a white sheet of paper, so its path on paper can be traced by using a pencil. Please note that though a thin beam of light is made up of several rays of light but for the sake of simplicity and convenience, a thin beam of light is considered to be a ray of light. The 'ray-box' is also known as 'ray-streak apparatus'.

ACTIVITY TO STUDY THE REFLECTION OF LIGHT FROM A PLANE MIRROR

We take a plane mirror strip *MM'* and place it sideways on a white sheet of paper so that its reflecting surface (shining surface) is towards the left side (see Figure 1). Mark the position of mirror on the sheet of paper with a pencil.

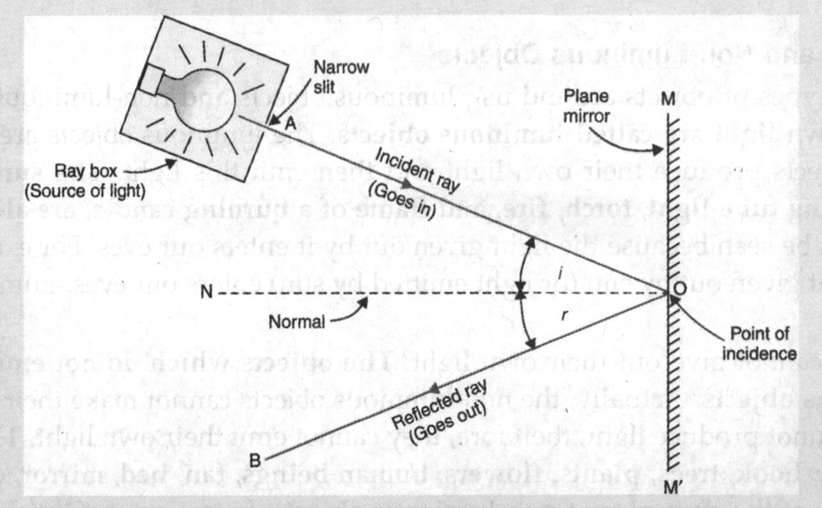

Figure 1. Arrangement to study the reflection of light from a plane mirror.

Keep the ray-box at position *A* in front of the plane mirror (see Figure 1). By opening the slit of ray-box, shine a narrow beam of light *AO* on the plane mirror. We will see that the beam of light *AO* strikes the mirror surface at point *O*, it gets reflected and then goes in another direction *OB* (see Figure 1). We mark the point *O* on the sheet of paper and also trace the paths of rays of light *AO* and *OB* on the sheet of paper by using a scale and pencil. At point *O* we draw a line *ON* perpendicular to the surface of the mirror *MM'*.

Let us measure the angles *AON* and *NOB*. We will find that the angle *AON* is equal to the angle *NOB*. Now, the angle *AON* is the angle of incidence and the angle *NOB* is the angle of reflection, so this activity shows that the angle of reflection is equal to the angle of incidence. In this activity, the incident ray *AO*, the reflected ray *OB* and the normal *ON*, all lie in the plane of paper. They neither come up out of paper nor go down into paper. This shows that the incident ray, the reflected ray, and the normal (perpendicular) at the point of incidence, all lie in the same plane.

Before we go further and study the laws of reflection of light, it is necessary to understand some important terms connected with the reflection of light clearly. These terms are : Incident ray, Point of incidence, Reflected ray, Normal, Angle of incidence and Angle of reflection. These are described below :

1. The ray of light which falls on the mirror surface is called incident ray. In Figure 1, the ray of light *AO* coming from the ray-box falls on the mirror surface, therefore, *AO* is the incident ray. The incident ray tells us the direction in which the light from a source falls on the mirror. The incident ray always goes towards the mirror.

2. The point at which the incident ray strikes the mirror is called the point of incidence. In Figure 1, the incident ray *AO* strikes the mirror (or touches the mirror) at point *O*, therefore, *O* is the point of incidence. The point of incidence tells us where exactly light falls on the mirror surface.

3. When the incident ray falls on a mirror, the mirror sends it back in another direction. And we say that the mirror has reflected the ray of light. **The ray of light which is sent back by the mirror is called the reflected ray.** In Figure 1, the ray of light *OB* is sent back by the mirror, therefore, *OB* is the reflected ray. The reflected ray tells us the direction in which the light goes after reflection from the mirror. The reflected ray always goes away from the mirror. *Please note that there can be only one reflected ray for a given single incident ray falling on a plane mirror.* This is because the same ray of light is called incident ray before it strikes the mirror and becomes reflected ray after it rebounds from the mirror.

4. The 'normal' is a line drawn at right angles to the mirror surface at the point of incidence. In other words, **the 'normal' is a line which is perpendicular to the mirror surface at the point of incidence.** In Figure 1, the dotted line *ON* is the normal to the mirror surface *MM'* at the point of incidence *O*. We usually represent 'normal' to the mirror by a dotted line to distinguish it from the incident ray and the reflected ray. Please note that 'normal' is just a line which is perpendicular to the mirror surface, and it should not be called 'normal ray'. The 'normal' is an imaginary line which is drawn on paper for the sake of convenience in understanding the laws of reflection. Please note that ***the 'normal' lies exactly in-between the incident ray and the reflected ray.***

5. The angle between incident ray and normal is called the angle of incidence. In Figure 1, *AO* is the incident ray and *NO* is the normal. So, the angle *AON* is the angle of incidence. The angle of incidence is represented by the letter i (i = incidence). Please note that the angle of incidence is made by the incident ray with the normal to the mirror surface and not with the mirror surface itself.

6. The angle between reflected ray and normal is called the angle of reflection. In Figure 1, the reflected ray is *OB* and the normal is *NO*. So, the angle *BON* is the angle of reflection (We can also say that the angle *NOB* is the angle of reflection). The angle of reflection is represented by the letter r (r = reflection).

Laws of Reflection of Light

When a ray of light falls on a plane mirror, it gets reflected (see Figure 2). The reflection of light from a plane mirror takes place according to two laws which are known as laws of reflection of light. The laws of reflection of light are as follows :

1. According to the first law of reflection : **The incident ray, the reflected ray, and the normal (at the point of incidence), all lie in the same plane.** In Figure 2, the incident ray *AO*, the reflected ray *OB* and the normal *ON*, all lie in the same plane, the plane of paper. They are neither coming up out of the paper; nor going down into the paper.

2. According to the second law of reflection : **The angle of reflection is always equal to the angle of incidence.** If the angle of incidence is *i* and the angle of reflection is *r*, then :

$$\angle i = \angle r$$

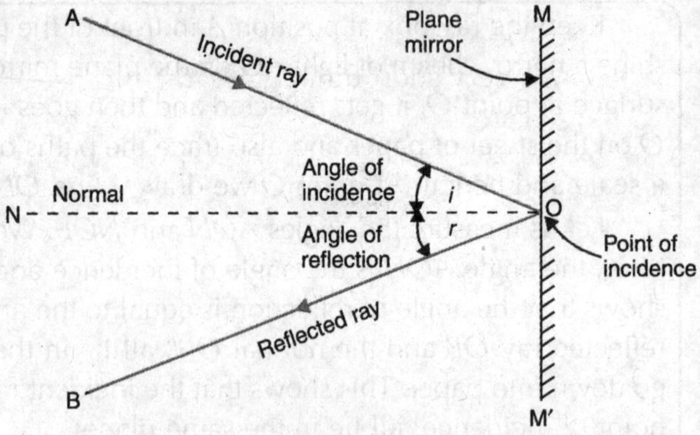

Figure 2.

In Figure 2, if we measure the angle of reflection *NOB*, we will find that it is exactly equal to the angle of incidence *AON*.

The second law of reflection will become more clear from the following examples : The second law of reflection says that the angle of reflection is always equal to the angle of incidence. This means that if the angle of incidence for a ray of light is 35°, then the angle of reflection will also be 35° (because they have to be equal). This is shown in Figure 3. In Figure 3, the angle of incidence *AON* is 35°, so the angle of reflection *NOB* is also 35°.

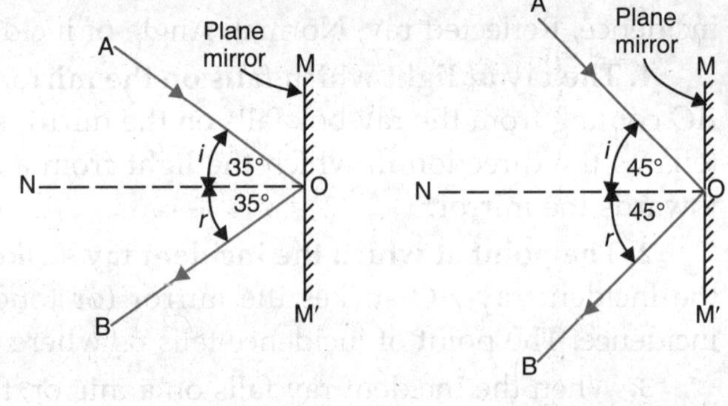

Figure 3. Figure 4.

If we change the angle of incidence, the angle of reflection will also change accordingly. The new angle of reflection will also be equal to the new angle of incidence. For example, if the angle of incidence is changed to 45°, then the angle of reflection will also change and become 45° (see Figure 4). Now, when the angle of incidence is 45°, then the angle of reflection is also 45°. So, in this case the reflected ray is at an angle of 45° + 45° = 90° to the incident ray. From this we conclude that if the reflected ray is at an angle of 90° to the incident ray, then the angle of incidence will be half of 90°, that is, $\frac{90°}{2} = 45°$ (see Figure 4) .

We will now describe **what happens when a ray of light falls normally (or perpendicularly) on the surface of a plane mirror.** When a ray of light is incident normally (or perpendicularly) on a plane mirror, it means that it is travelling along the 'normal' to the mirror surface (see Figure 5). The angle of incidence for such a ray of light is zero. Since the angle of incidence is zero, so according to the second law of reflection, the angle of reflection should also be zero. This means that the reflected ray will also travel back from the mirror along the normal (see Figure 5). Thus, **a ray of light which is incident normally (or perpendicularly) on a mirror is reflected back along the same path.** This is because the angle of incidence for such a ray of light is 0° and the angle of reflection is also 0°. Thus, if the incident ray goes to a mirror along normal, the reflected ray will also travel back along normal. In this case the same line represents incident ray, normal and the reflected ray (see Figure 5).

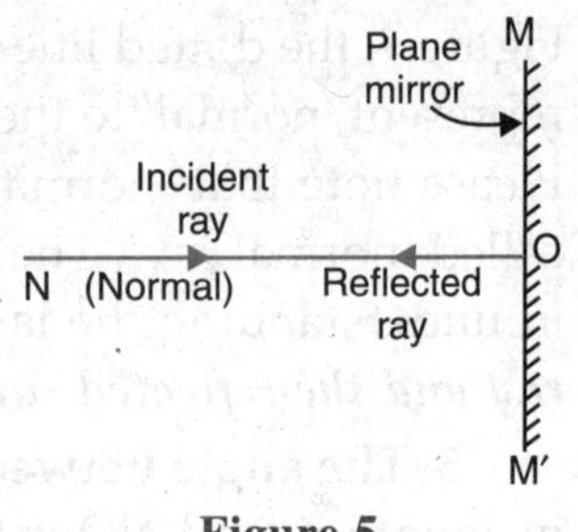

Figure 5.

Please note that **whenever light is reflected, laws of reflection are obeyed.** As we will see after a while, we can find out the nature and position of an image formed by a plane mirror by using the laws of reflection of light. We will now answer some questions based on the laws of reflection of light.

Sample Problem 1. An incident ray makes an angle of 35° with the surface of a plane mirror. What is the angle of reflection ?

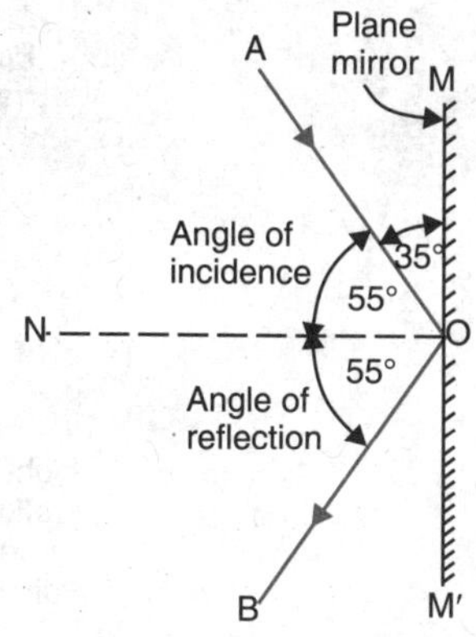

Solution. In order to find out the angle of reflection, we should first know the angle of incidence. In this case, the incident ray makes an angle of 35° with the surface of the mirror (see Figure 6), so the angle of incidence is not 35°. The angle of incidence is the angle between incident ray and normal. So, in this case, the angle of incidence will be 90° – 35°= 55°. Since the angle of incidence is 55 degrees, therefore, the angle of reflection is also 55 degrees. This is shown clearly in Figure 6.

Plane mirror
A
M
Angle of incidence
35°
55°
N
O
55°
Angle of reflection
B
M′

Figure 6.

Sample Problem 2. Two plane mirrors *PQ* and *QR* are kept at right angles to each other as shown in Figure 7. A ray of light *AB* is incident on the mirror *PQ* at an angle of 30° as shown in Figure 7. Draw the path of the reflected ray from the second mirror *QR* and find the angle of reflection for the mirror *QR*.

(NCERT Book Question)

Solution. (*i*) When the ray of light *AB* is incident on plane mirror *PQ* making an angle of incidence *ABN* of 30°, it will be reflected from the mirror *PQ* making an equal angle of reflection of 30° with the normal *BN*. So, we draw a line *BC* making an angle of 30° with the normal *BN* (see Figure 8). The line *BC* will be reflected ray of light and the angle *NBC* (of 30°) will be the angle of reflection for the mirror *PQ* (see Figure 8).

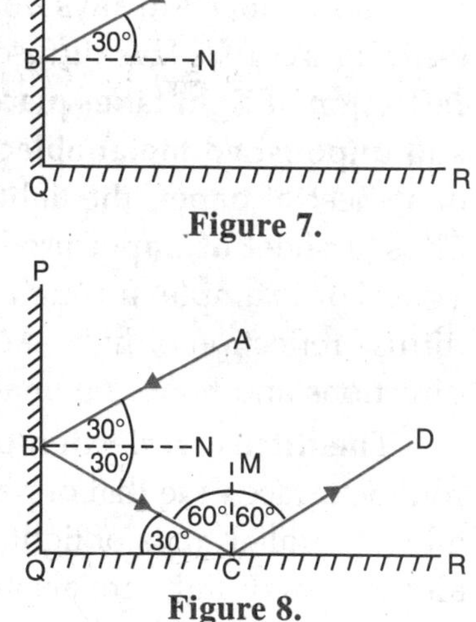

Figure 7.

(*ii*) The reflected ray *BC* of mirror *PQ* meets the second mirror *QR* at point C making an angle of 30° with the surface of mirror *QR* (This is because angle *NBC* and angle *BCQ* are alternate angles and hence equal) (see Figure 8). The reflected ray *BC* of mirror *PQ* becomes incident ray *BC* for the mirror *QR*. The angle of incidence for ray *BC* on mirror *QR* will be 90° – 30° = 60°. Since the angle of incidence for ray of light *BC* on mirror *QR* is 60°, therefore, the reflected ray *CD* for mirror *QR* will also make an equal angle of reflection of 60° (as shown in Figure 8).

Figure 8.

REGULAR REFLECTION AND DIFFUSE REFLECTION OF LIGHT

In regular reflection, a parallel beam of incident light is reflected as a parallel beam in one direction. In this case, parallel incident rays remain parallel even after reflection and go only in one direction (see Figure 9). **Regular reflection of light occurs from smooth surfaces like that of a plane mirror (or highly polished metal surfaces).** For example, when a parallel beam of light falls on the smooth surface of a plane mirror, it is reflected as a parallel beam in only one direction as shown in Figure 9. Thus, **a plane mirror produces regular reflection of light.** Images are formed by regular reflection of light. For example, a smooth surface (like that of a plane mirror) produces a clear image of an object due to regular reflection of light. A highly polished metal surface and a still water surface also produce regular reflection of light and form images. This is why we can see our face in a polished metal object as well as in the still water surface of a pond or lake. A polished wooden table and a marble floor with water spread over it are very smooth and hence produce regular reflection of light.

The regular reflection of light from a smooth surface can be explained as follows : All the particles of a smooth surface (like a plane mirror) are facing in *one direction*. Due to this the angle of incidence for all the parallel rays of light falling on a smooth surface is the same and hence the angle of reflection for all the rays of light is also the same . *Since the angle of incidence and the angle of reflection are the same (or equal), a beam of parallel rays of light falling on a smooth surface is reflected as a beam of parallel light rays in one direction only* (see Figure 9).

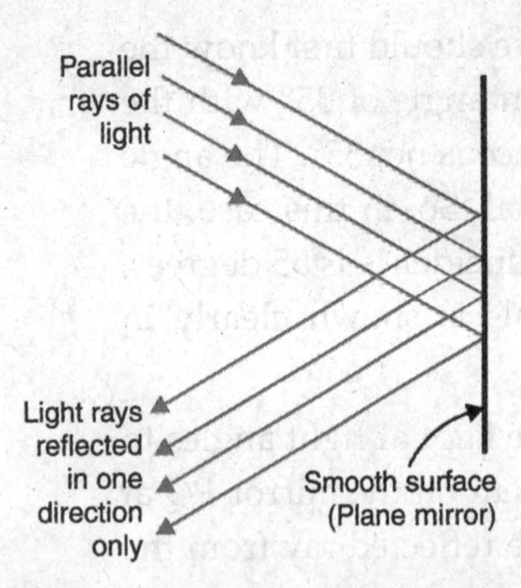

Figure 9. Regular reflection : Incident light is reflected in only one direction.

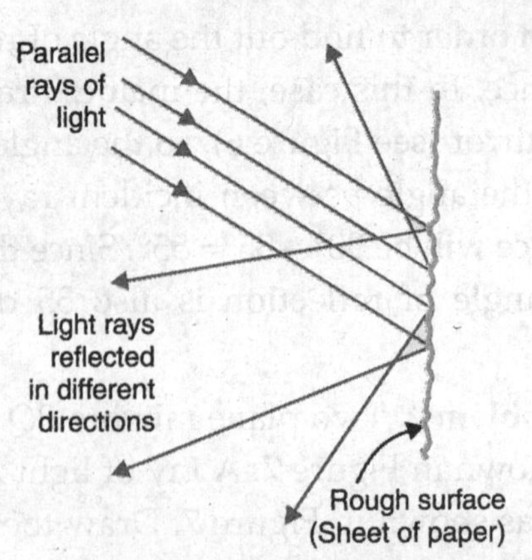

Figure 10. Diffuse reflection : Incident light is reflected in different directions.

In diffuse reflection, a parallel beam of incident light is reflected in different directions. In this case, the parallel incident rays do not remain parallel after reflection, they are scattered in different directions (see Figure 10). The diffuse reflection is also known as *irregular reflection* or scattering. **The diffuse reflection of light takes place from rough surfaces like that of paper, cardboard, chalk, table, chair, walls and unpolished metal objects.** For example, when a parallel beam of light rays falls on the rough surface of a sheet of paper, the light is scattered by making reflected rays in different directions (see Figure 10). Thus, **a sheet of paper produces diffuse reflection of light.** No image is formed in diffuse reflection of light. For example, a rough surface (like that of paper) does not produce an image of the object due to diffuse reflection of light. Actually, the light rays falling on the rough surface of paper are scattered in all directions and hence no image is formed.

The diffuse reflection of light from a rough surface can be explained as follows : The particles of a rough surface (like that of paper) are all facing in *different directions*. Due to this, the angles of incidence for all the parallel rays of light falling on a rough surface are different and hence the angles of reflection for all the rays of light are also different. *Since the angles of incidence and the angles of reflection are different, the parallel rays of light falling on a rough surface go in different directions* (see Figure 10). Please note that **the diffuse reflection of light is not due to the failure of the laws of reflection.** Diffuse reflection is caused by the roughness (or irregularities) in the reflecting surface of an object (like paper or cardboard, etc.). The laws of reflection are valid at each point even on the rough surface of an object.

The surfaces of most of the objects are rough (or uneven) to some extent. So, **most of the objects around us cause diffuse reflection of light and scatter the light falling on them in all directions.** In fact, we can see these objects only because they scatter light rays falling on them in all directions. For example, a book lying on a table can be seen from all parts of the room due to diffuse reflection of light from its surface. The surface of book, being rough, scatters the incident light in all parts of a room. Hence the book can be seen from all parts of the room. A cinema screen has a rough surface and causes diffuse reflection of light falling on it. The cinema screen receives light from a film projector and scatters it in all directions in the cinema hall so that people sitting anywhere in the hall can see the picture focused on the screen. We will now study the formation of an image by a plane mirror.

FORMATION OF IMAGE IN A PLANE MIRROR

When an object is placed in front of a plane mirror, its image is formed in the plane mirror. We can locate the image of an object in a plane mirror by using the laws of reflection of light. This can be done as follows.

Suppose a small object *O* (say, a pin-head) is placed in front of a plane mirror *MM'* (see Figure 11). An image of this point object will be formed in the plane mirror. We can find out the position of image of this object in the plane mirror by drawing a ray-diagram as follows :

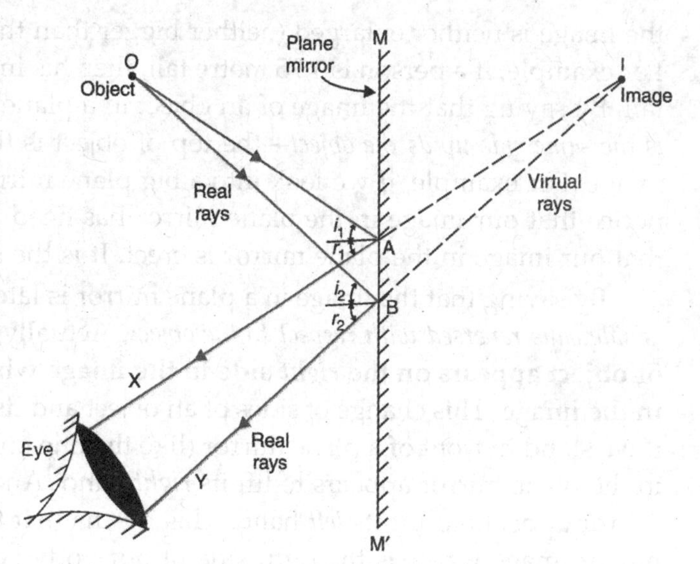

Figure 11. Formation of image of a point object in a plane mirror.

1. We take two diverging rays *OA* and *OB* coming from the object *O*. The rays *OA* and *OB* are incident rays which strike the mirror at points *A* and *B* making different angles of incidence.

2. The first incident ray *OA* is reflected by the mirror at point *A* and goes in the direction *AX* after reflection (making an angle of reflection r_1 equal to the angle of incidence i_1). So, *AX* is the first reflected ray.

3. The second incident ray *OB* is reflected by the mirror at point *B* and it goes in the direction *BY* after reflection (making an angle of reflection r_2 equal to the angle of incidence i_2). Thus, *BY* is the second reflected ray.

4. The two reflected rays *AX* and *BY* are diverging rays (going apart) which cannot meet in front of the mirror to form a real image. So, we produce the reflected rays *AX* and *BY* backwards (behind the mirror) by dotted lines. On producing backwards, the reflected rays meet at point *I* behind the mirror (see Figure 11).

5. When we look into the plane mirror from an angle (as shown in Figure 11), the reflected rays *AX* and *BY* enter our eye. The reflected rays appear to be coming from point *I* behind the mirror. So, point *I* is the image of object *O*.

Since the reflected rays of light do not actually meet at the image point *I* but only appear to do so, we say that a virtual image of object *O* is formed at *I*. Such a virtual image cannot be obtained on a screen.

Characteristics of Image Formed by a Plane Mirror

(*i*) The image formed by a plane mirror is *virtual* (or unreal).

(*ii*) The image formed by a plane mirror is *behind the mirror.*

(*iii*) The image formed in a plane mirror is the *same distance behind the mirror as the object is in front of it.*

(*iv*) The image formed in a plane mirror is of the *same size as the object.*

(*v*) The image formed by a plane mirror is *erect.*

(*vi*) The image in a plane mirror is *laterally inverted.*

We have studied the characteristics of images formed by a plane mirror in detail in Class VII. Here we will explain the various characteristics of an image formed by a plane mirror briefly. By saying that the image formed by a plane mirror is virtual, we mean that *the image formed by a plane mirror cannot be obtained on a screen.* It can be seen only by looking into the plane mirror. For example, when we look into a plane mirror, we see the image of our face. Our image in the plane mirror is virtual (or unreal). Our image seen in the plane mirror cannot be formed on a screen placed behind the plane mirror (where the image appears to be). In fact, when we look into a plane mirror (like the one on a dressing table), we find that our image in the plane mirror is formed *behind the plane mirror.*

By saying that the image in a plane mirror is the same distance behind the mirror as the object is in front of it, we mean that *if a person is standing at a distance of 1 metre in front of a plane mirror, then his image will be formed at the same distance of 1 metre behind the plane mirror.* Now, since the person is 1 metre in front of plane mirror and his image is 1 metre behind the plane mirror, therefore, the distance between the person and his image will be 1 metre + 1 metre = 2 metres. In other words, if a person is 1 metre in front of a plane mirror, then he will seem to be 2 metres away from his image.

When we say that the image in a plane mirror is of the same size as the object, we mean that *the dimensions of image (such as length, breadth and height) are exactly the same as that of the object.* In other words, the image is neither enlarged (neither bigger than the object) nor diminished (nor smaller than the object). For example, if a person is 1.75 metre tall, then his image in the plane mirror will also be exactly 1.75 metre tall. By saying that the image of an object in a plane mirror is erect, we mean that *the image in plane mirror is the same side up as the object*—the top of object is the top of image and bottom of object is the bottom of image. For example, if we look into a big plane mirror, we will see the image of our whole body. We will notice that our image in the plane mirror has head on the top and feet at the bottom just like us. We say that our image in the plane mirror is erect. It is the same side up as we are.

By saying that the image in a plane mirror is laterally inverted, we mean that *the image in a plane mirror is sideways reversed with respect to the object*. Actually, **in an image formed by a plane mirror, the *left* side of object appears on the *right* side in the image whereas the *right* side of object appears on the *left* side in the image.** This change of sides of an object and its mirror image is called **lateral inversion.** For example, if we stand in front of a plane mirror (like the one on a dressing table) and lift our *left* hand, then our image in the plane mirror appears to lift its *right* hand. And if we lift our *right* hand, then our image in the plane mirror appears to lift its *left* hand. This means that the left side of our body becomes the right side in the mirror image whereas the right side of our body becomes left side in the mirror image. We say that our image in the plane mirror is laterally inverted (or sideways reversed). The change in the sides of an object and its mirror image is due to the phenomenon of lateral inversion.

REFLECTED LIGHT CAN BE REFLECTED AGAIN

If the rays of light reflected by a plane mirror are incident on another plane mirror, then the reflected rays are reflected again. In this case, the reflected rays of the first plane mirror become incident rays for the second plane mirror. The further reflection from second plane mirror also takes place according to the laws of reflection of light. ***An optical instrument (or device) in which reflected light is reflected again is a periscope.*** We will now describe a periscope.

A periscope is a long, tubular device through which a person can see objects that are out of the direct line of sight. A periscope gives us a higher view than normal. For example, by using a periscope, we can see the objects on the other side of a high wall which cannot be seen by us directly. The periscope makes use of two plane mirrors to see over the top of things. Actually, ***a periscope works on the reflection of light from two plane mirrors arranged parallel to one another.*** A periscope consists of a long tube T having two plane mirrors M_1 and M_2 fitted at its two ends (as shown in Figure 12). The two plane mirrors are fitted in such a way that they are parallel to one another and their reflecting surfaces face each other. Each plane mirror, however, makes an angle of 45° with the side of the tube (see Figure 12). There are two holes in the periscope tube : one hole is in front of the top mirror M_1 and the other hole is in front of the bottom mirror M_2.

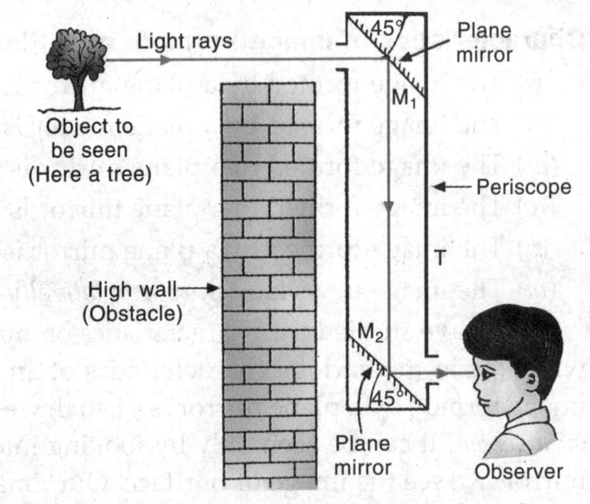

Figure 12. Diagram to show the working of a periscope.

We will now describe the working of a periscope. Suppose there is a tree behind a high wall which we cannot see directly (see Figure 12). We can, however, see this tree by using a periscope. This can be done as follows : The upper hole of periscope is turned towards the object to be seen (here a tree) so that the top mirror M_1 faces the object. We look into the periscope from the bottom hole in front of lower mirror M_2. The light rays coming from the tree fall on the plane mirror M_1. Mirror M_1 reflects these rays of light downwards, towards the second mirror M_2 (see Figure 12). The mirror M_2 then reflects the reflected rays of light towards the eye of the person looking into the periscope through the lower hole. Since the light rays coming from the tree enter the eye of the person (or observer), it is possible to see the image of tree through the periscope (even though the tree cannot be seen directly).

In a periscope, the top mirror reflects light and the bottom mirror reflects the 'reflected light'. So, the working of a periscope demonstrates that reflected light can be reflected again. The working of a periscope also explains how reflection from two plane mirrors enables us to see objects which are not visible directly.

Some of the Uses of Periscopes are Given Below :

(*i*) A periscope is used to see over the heads of a crowd (say, as in a football match).

(*ii*) A periscope is used by soldiers sitting in a trench (or bunker) to observe the enemy activities outside (over the ground).

(*iii*) A periscope is used by a navy officer sitting in a submarine to see ships over the surface of water in the sea (even though the submarine itself may be submerged under water).

The fact that reflected light can be reflected again enables a person to see the hair cut at the back of his head at a hair dresser's shop. This happens as follows : After giving the hair cut to a person, the hair dresser holds a small plane mirror behind the head of the person. The light coming from hair at the back of head is reflected by this small mirror on to a big mirror which is in the front of the person. This big mirror reflects the 'reflected light' again due to which the person can see the image of the back hair of his head showing how the hair have been cut at the back side of his head.

Multiple Images

We know that a plane mirror forms only a single image of an object placed in front of it. We will now describe what happens when an object is placed between two plane mirrors which are inclined at an angle to each other. **When two plane mirrors are kept inclined at an angle, they can form multiple images of an object.** This is because the image of object formed in one plane mirror acts as object for the other plane mirror. It has been found that if two plane mirrors are inclined at an angle x, then the number of images formed in them is given by the formula :

$$\text{No. of images formed} = \frac{360°}{x} - 1$$

By using this formula, we can calculate the number of images formed (or seen) when two plane mirrors are inclined at angles of 180°, 120°, 90°, 60°, 45° and 0°, respectively. As an example, let us calculate the number of images formed in two plane mirrors inclined at an angle of 90°. In this case the angle between two plane mirrors (x) is 90°. So,

$$\begin{aligned}\text{No. of images formed} &= \frac{360°}{90°} - 1\\ &= 4 - 1\\ &= 3\end{aligned}$$

Thus, **two plane mirrors inclined at an angle of 90° form *three* images of an object placed between them.**

When the two plane mirrors are inclined at an angle of 90°, they are said to be at *right angles* to each other. So, we can also say that **the two plane mirrors arranged at right angles to each other form *three* images of an object placed between them.** Thus, if we take two plane mirrors, set them at right angles to each other (with their edges touching), and place a coin in-between these mirrors, then we will see three images of the coin in the two plane mirrors.

We can also do the calculations to find the number of images of an object formed when the two plane mirrors are inclined at angles of 180°, 120°, 60°, 45° and 0°. The results of these calculations are given below:

Angle between two plane mirrors	*No. of images formed*
180°	1
120°	2
90°	3
60°	5
45°	7
0°	Infinite

From the above table we can see that **as the angle between the two plane mirrors decreases, the number of images formed increases.** And when the angle between two plane mirrors becomes 0°, that is, when the two plane mirrors become parallel to each other, then an infinite number of images are formed.

Thus, **if an object is placed between two parallel plane mirrors facing each other, then theoretically, an infinite number of images (very, very large number of images) should be formed**. In reality, only a limited number of images are seen. This is because some light is absorbed by the mirrors at each successive reflection due to which many images are so faint that they cannot be seen by us. From this discussion we conclude that : **We can see a large number of images of ourselves if we stand before two plane mirrors hanging on opposite walls.** This is because our image formed in one mirror acts as object for the other mirror, and this process goes on and on, resulting in a large number of images. We can, however, not see all the images because they become fainter and fainter with the increasing number of reflections of light.

The fact that multiple images (or a number of images) of an object are formed by plane mirrors which are kept inclined at an angle to one another is used in kaleidoscope to make numerous beautiful patterns which are liked by children as well as by adults. Let us discuss the kaleidoscope now.

Kaleidoscope

The kaleidoscope is an instrument or toy containing inclined plane mirrors which produce multiple reflections of coloured glass pieces (or coloured plastic pieces) and create beautiful patterns. The kaleidoscope consists of three long and narrow strips of plane mirrors inclined at 60° to one another forming a hollow prism, and fitted into a cardboard tube. One end of the cardboard tube is closed by an opaque disc (cardboard disc) having a small hole at its centre. The other end of cardboard tube is closed with two circular discs of glass : the inner disc being of transparent glass (clear glass) and the outer disc of ground glass (translucent glass). A number of small pieces of different coloured glass (or plastic) and having different shapes are kept between the two glass discs (which can move around freely in the space between the two glass discs).

Figure 13. A pattern produced by kaleidoscope.

When we hold the kaleidoscope tube towards light and look inside it through the small hole, we see beautiful patterns of coloured glass (see Figure 13). Actually, **the coloured glass pieces act as objects and the inclined plane mirrors form multiple images of these glass pieces by repeated reflections, which look like beautiful patterns (or designs).** If we turn the kaleidoscope tube slightly, the glass pieces will rearrange and produce a new pattern. A kaleidoscope produces hundreds of ever-changing coloured patterns (or designs). *An interesting feature of a kaleidoscope is that we can never see the same pattern again*. Every time a new pattern is formed. Kaleidoscopes are used by **designers** of wall papers and fabrics, as well as by **artists** to get ideas for new patterns.

ACTIVITY

We can make a kaleidoscope ourselves as follows : Take three rectangular strips of plane mirror each about 15 cm long and 4 cm wide. Join the three plane mirror strips lengthwise by using adhesive tape so as to form a hollow prism (see Figure14). The reflecting surfaces (shiny surfaces) of the three mirror strips are kept facing one another. Fix the hollow prism formed by joining three mirror strips in a cardboard tube which is somewhat longer than the mirror strips. Close one end of the cardboard tube with a cardboard disc having a small hole at the centre (see Figure 14). At the other end of cardboard tube, fix a transparent glass disc touching the hollow prism of plane mirror strips. Place several small pieces of coloured glass (like broken pieces of different coloured glass bangles) on the transparent glass disc and then fix a ground glass disc to close the end of cardboard tube. Enough space should be left between the transparent glass disc and ground glass disc to allow the coloured glass pieces (placed between them) to move around freely when the cardboard tube is rotated. The kaleidoscope is now ready. When we peep in through the hole on the front side and rotate the cardboard tube gently, we will be able to see a variety of beautiful patterns (or designs) formed by the multiple reflections of coloured glass pieces.

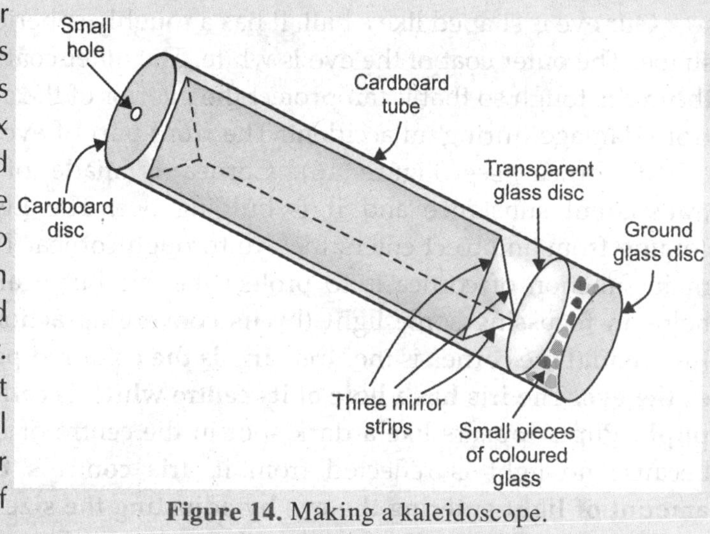

Figure 14. Making a kaleidoscope.

Sunlight—White or Coloured

The sunlight is referred to as white light. **The white sunlight actually consists of seven colours (mixed together)**. The fact that white sunlight consists of lights of seven different colours can be shown by using a glass prism as follows.

ACTIVITY

Take a glass prism and place it on a table in a darkened room. Place a white cardboard screen at some distance behind the prism. Allow a thin beam of sunlight (coming through a tiny hole in the window) to fall on the prism. (see Figure 15). We will see that the beam of white sunlight splits on entering the glass prism and forms a broad patch of seven colours (called spectrum) on the white screen placed on the other side of prism (see Figure 15). **The splitting up of white light into seven colours on passing through a transparent medium like a glass prism is called dispersion of light**. The formation of spectrum (band of seven colours) shows that white sunlight is made up of seven colours. The seven colours of the spectrum of white light are : Violet, Indigo, Blue, Green, Yellow, Orange and Red. It is, however, usually not possible to distinguish all the seven colours of the spectrum easily due to overlapping of various colours.

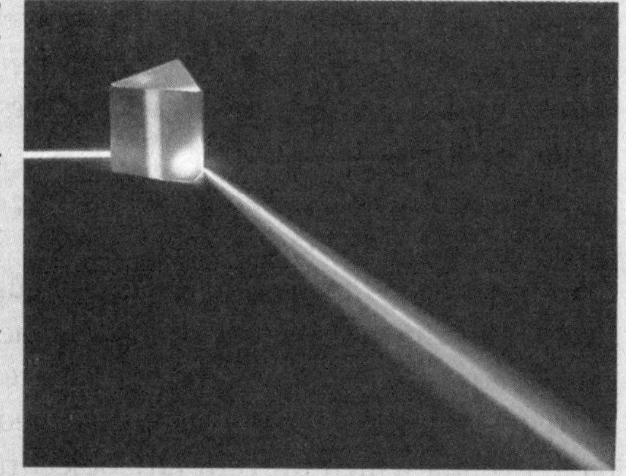

Figure 15. Dispersion of light.

Rainbow in the sky is a natural phenomenon showing the dispersion of sunlight. Rainbow is produced by the dispersion of sunlight by tiny rain drops suspended in the atmosphere (which act as tiny prisms made of water). The formation of rainbow also tells us that sunlight consists of seven colours.

THE HUMAN EYE

Eye is one of our most important sense organs. The eye enables us to see the various objects around us. **The main parts of the human eye are : Cornea, Iris, Pupil, Ciliary muscles, Eye lens** (which is a flexible convex lens), **Retina and Optic nerve**. All these parts of the eye are shown in the simplified diagram of human eye given in Figure 16. We will now describe the construction and working of the human eye.

Construction of the Eye

Our eye is shaped like a ball. It has a roughly spherical shape. The outer coat of the eye is white. The outer coat of the eye is tough so that it can protect the interior of the eye from damage during an accident. The front part of eye is called cornea (see Figure 16). Cornea is made of a transparent substance and it is bulging out. The light coming from an object enters the eye through cornea. The main function of cornea is to protect the eye but it also helps in focussing some light (by its converging action). Just behind the cornea is the 'Iris'. **Iris is the coloured part of the eye. The iris has a hole at its centre which is called pupil**. Pupil appears like a dark spot in the centre of iris because no light is reflected from it. **Iris controls the amount of light entering the eye by adjusting the size of pupil**. The iris is actually that part of the eye which gives the eye its distinctive colour. When we say that a person has green eyes, we mean that the colour of iris in his eyes is green.

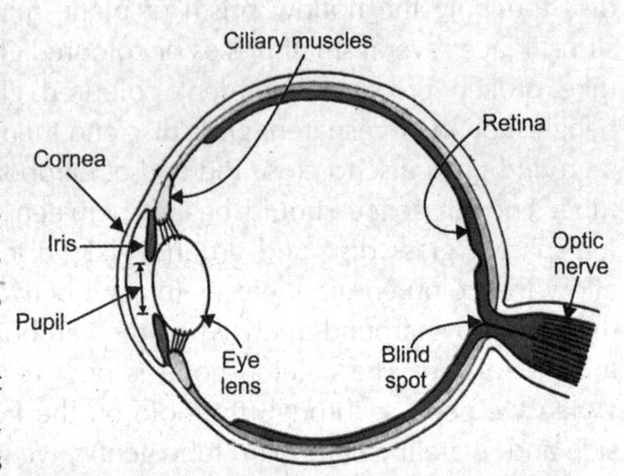

Figure 16. Structure of the human eye.

Behind pupil is the eye-lens (see Figure 16). **The eye-lens is a convex lens** made of a transparent and flexible material like jelly. The eye-lens is actually a living lens because it is made up of transparent living cells which allow light to pass through them. **The eye-lens is held in position by ciliary muscles** whose one side is attached to the eye-lens and the other side is attached to the eye-ball. **The ciliary muscles can change the curvature of eye-lens and make it thin or thick according to the need of the eye.** In other words, the ciliary muscles can change the focal length (and hence converging power) of the eye-lens according to the requirements of the eye to see distant objects or nearby objects. Thus, the focal length (and hence converging power) of the eye-lens is controlled by ciliary muscles. Please note the difference between convex eye-lens and the ordinary convex lens made of glass : *The eye-lens is a flexible convex lens whose thickness and hence focal length (or converging power) can be changed by the action of ciliary muscles. On the other hand, a glass convex lens has a fixed thickness due to which its focal length (or converging power) is also fixed, and cannot be changed.*

The retina is a screen on which the image is formed in the eye. The retina is behind the eye-lens, at the back part of the eye (see Figure 16). **The eye-lens focusses the image of an object on the retina.** The retina is attached to optic nerve (see Figure 16). **The optic nerve carries the image formed on retina to the brain** in the form of electrical signals. The space between cornea and eye-lens is filled with a viscous liquid called 'aqueous humour'. And the space between eye-lens and retina is filled with another liquid called 'vitreous humour'. We will discuss aqueous humour and vitreous humour in higher classes. The eyes also have eyelids which prevent any object from entering the eyes. The eyelids also shut out light when not required. We will now describe the working of human eye.

Working of the Eye

When we look at an object, then light rays coming from the object enter the pupil of the eye and fall on the eye-lens. **The eye-lens is a convex lens, so it converges the light rays and produces a real and**

inverted image of the object on the retina. The retina has a large number of light sensitive cells. When the image of the object falls on the retina, then the light sensitive cells generate electrical signals. The retina sends these electrical signals to the brain through the optic nerve and we are able to see the object. Please note that although the image of an object formed on the retina is inverted, but our brain interprets this image as that of an erect object.

The Function of Iris and Pupil

The iris automatically adjusts the size of pupil according to the intensity of light received by the eye from the surroundings. If the amount of light around us is very high (as during the day light), then iris contracts the pupil (makes the pupil small) and hence reduces the amount of light entering the eye (see Figure 17). On the other hand, if the amount of light around us is small (as in a dark room or during the night) then iris expands the pupil (makes the pupil large) so that more light can enter the eye (see Figure 18). This is because we need to allow more light to go into the eyes when the outside light is dim so as to see properly.

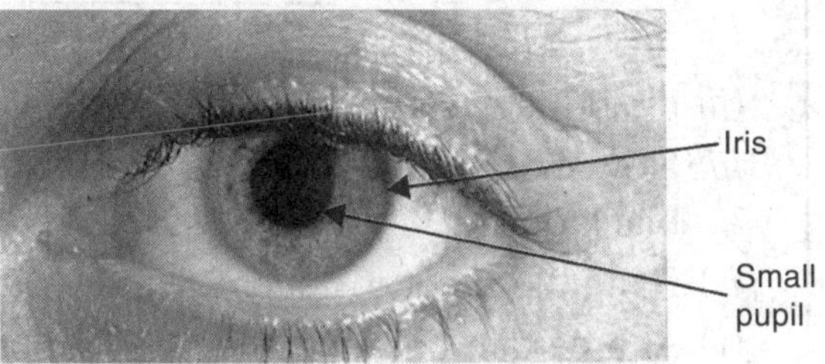

Figure 17. When outside light is bright, pupil becomes small, so that less light goes into the eye.

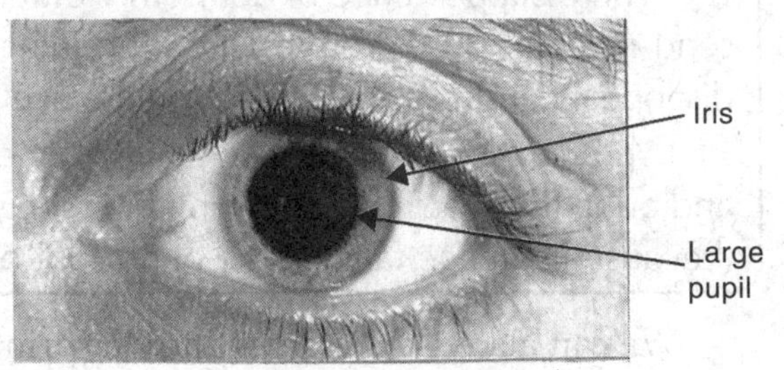

Figure 18. When outside light is dim, pupil becomes large, so that more light enters the eye.

It should be noted that **the adjustment of the size of pupil takes some time**. For example, when we go from a bright light to a darkened cinema hall, at first we cannot see our surroundings clearly. After a short time our vision improves, and we can see the persons sitting around us. This is due to the fact that in bright sunlight the pupil of our eye is small. So, when we enter the cinema hall, very little light enters our eye and we cannot see properly. After a short time, the pupil of our eye expands and becomes large. More light then enters our eye and we can see clearly. On the other hand, if we go from a dark room into bright sunlight or switch on a bright lamp, then we feel the glare in our eyes. This is due to the fact that in a dark room, the pupil of our eye is large. So, when we go out in bright sunlight or switch on a bright lamp, a large amount of light enters our eyes and we feel the glare. Gradually, the pupil of our eye contracts. Less light then enters our eye and we can see clearly.

Rods and Cones

The retina of our eye has a large number of light-sensitive cells. There are two kinds of light-sensitive cells on the retina : *rods* and *cones*.

(*i*) Rods are the rod-shaped cells present in the retina of an eye which are sensitive to dim light. Rods are the most important for vision in dim light (as during the night). Rod cells of the retina, however, do not provide information about the colour of the object.

(*ii*) Cones are the cone-shaped cells present in the retina of an eye which are sensitive to bright light (or normal light). **Cones also cause the sensation of colour of objects in our eyes**. Thus, cones present in the retina of the eye are responsible for colour vision. Cones, however, do not function in dim light.

Blind Spot

At the junction of optic nerve and retina in the eye, there are no light sensitive cells (no rods or cones) due to which no vision is possible at that spot. This is called blind spot. Thus, **blind spot is a small area of the retina insensitive to light where the optic nerve leaves the eye** (see Figure 16). When the image of an object is formed at the blind spot in the eye, it cannot be seen by the eye. Blind spot is not sensitive to light because there are no light-sensitive cells like rods or cones in this region. **The existence of blind spot in the eye can be demonstrated as follows.**

ACTIVITY

(*i*) Take a drawing sheet and draw a thick cross and a circular dot on this sheet of paper about 7 cm apart (as shown in Figure 19).

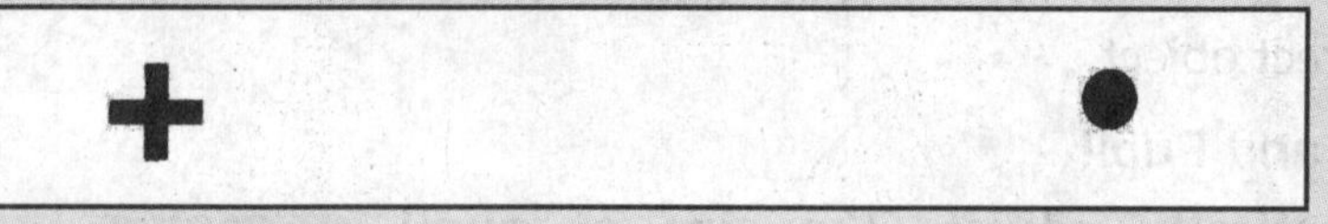

Figure 19.

(*ii*) Hold the sheet of paper at an arm's length from your eyes.

(*iii*) Now close your left eye and look at the cross on the paper with your right eye. You will also see the black dot.

(*iv*) Bring the sheet of paper towards you *slowly*.

(*v*) At a certain distance, the dot disappears (because its image has fallen on the blind spot of your right eye).

The disappearance of dot from view shows that there is a point on the retina of the eye which cannot send message to the brain when the image of dot falls on it. This point is the blind spot. The black dot disappears and cannot be seen by the eye because its image falls on the blind spot of the right eye.

You can repeat the above activity by closing the right eye, keep looking at the dot mark with left eye, and bring the sheet of paper from arm's length towards you *slowly*. At a certain distance, the cross disappears (because its image falls on the blind spot of your left eye).

We can also use the cross and dot printed on this page of the book to show the existence of blind spot in the eyes. All that we have to do is to hold this book at arm's length and bring it slowly towards us in the way described above.

Persistence of Vision

If we look at a bright object and then close our eyes, the image of the object will remain in our eyes for a very short duration. Thus, the image formed on the retina of the eye does not fade away immediately. **The image of an object seen by our eyes persists (or remains) on the retina for about $\frac{1}{16}$th of a second even after the object has disappeared from our view**. This happens because the stimulated light-sensitive cells of the retina take a little time to come back to their original state. **The ability of an eye to continue to see the image of an object for a very short duration even after the object has disappeared from view, is called persistence of vision**. It is due to the phenomenon of persistence of vision that we are able to see movie pictures in a cinema hall. This is because ***if the still pictures of a moving object are flashed on our eyes at a rate faster than 16 pictures per second then (due to persistence of vision), the eyes perceive this object as moving.*** This will become more clear from the following discussion.

The movies that we see are actually a number of still pictures (in the form of a long film) which are taken in proper sequence with a movie camera. These still pictures are projected on the screen of a cinema hall at the rate of 24 pictures per second (which is faster than 16 pictures per second) with the help of a film projector. Under these conditions, the image of one picture seen on the screen persists on the retina till the image of next picture falls on the screen, and so on. Due to this, the slightly different images of the successive pictures present on the film merge smoothly into one another and give us the feeling of moving images (or movie). In this way we are able to see the moving images (or moving pictures) of actors and actresses on a cinema screen.

ACTIVITY

We can demonstrate the phenomenon of persistence of vision by performing an activity as follows. Take a piece of cardboard and a pencil [see Figure 20(*a*)]. Make the sketch of a bird on one side of the cardboard and the sketch of a cage on its back side [see Figures 20(*b*) and (*c*)]. Make two small, horizontal slits (or cuts) in the piece of cardboard, one near its top and the other near its bottom. Insert the pencil

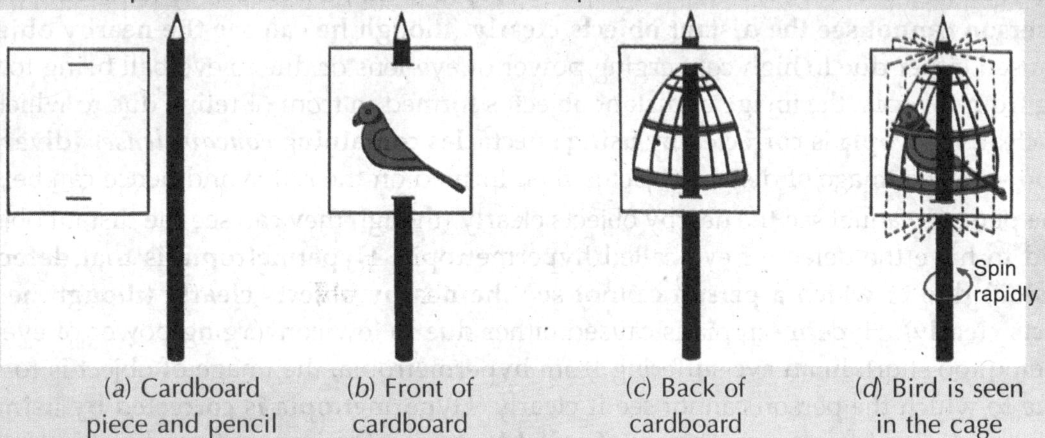

(*a*) Cardboard piece and pencil (*b*) Front of cardboard (*c*) Back of cardboard (*d*) Bird is seen in the cage

Figure 20.

through the two slits on the cardboard as shown in Figure 20(*b*). Hold the lower end of pencil between the palms of your hands. Move your palms forwards and backwards quickly so as to spin the cardboard rapidly. Look at the cardboard when it is spinning rapidly. We will find that when the cardboard spins rapidly, then bird is seen to be in the cage [see Figure 20(*d*)]. This happens due to the phenomenon of persistence of vision. When the cardboard is spinning rapidly, the image of bird persists on our retina for a very short while during which the image of cage also falls on the retina, making the bird appear to be in the cage.

Range of Vision of a Normal Human Eye

The farthest point from the eye at which an object can be seen clearly is known as the "far point" of the eye. The far point of a normal human eye is at **infinity.** This means that the far point of a normal human eye is at a very large distance. **The nearest point up to which the eye can see an object clearly without any strain, is called the "near point" of the eye.** The near point of a normal human eye is at a distance of **25 centimetres** from the eye. The near point of an eye is also known by another name as the least distance of distinct vision. This means that the least distance of distinct vision for a normal human eye is about 25 centimetres. In other words, *the most comfortable distance at which one can read a book with normal eyes is about 25 centimetres from the eyes*. The minimum distance at which the eye can see objects distinctly (or clearly), however, varies with age. From the above discussion we conclude that **the range of vision of a normal human eye is from infinity to about 25 centimetres.** That is, a normal human eye can see the objects clearly which are lying anywhere between infinity to about 25 centimetres.

A normal eye can see the distant objects as well as the nearby objects by focussing the images of distant objects as well as nearby objects on its retina by changing the thickness (or converging power) of its lens. The ciliary muscles of the eye can change the thickness of the soft and flexible eye-lens and hence change the converging power of eye-lens. ***A thin eye-lens has less converging power whereas a thick eye-lens has greater converging power on the rays of light coming from an object.***

(*i*) When the eye is looking at a distant object, the ciliary muscles are *relaxed* due to which the eye-lens is *thin* (or less convex). The thin eye-lens has smaller converging power which is sufficient to converge the *parallel* rays of light coming from a distant object to form its image on the retina.

(*ii*) When the same eye is looking at a nearby object, the ciliary muscles get *stretched* due to which the eye-lens becomes *thick* (or more convex). The thick eye-lens has greater converging power which is required to converge the *diverging* rays of light coming from a nearby object to form its image on the retina.

Defects of the Eye

We will now discuss three common defects of the eye (or defects of vision) called myopia, hypermetropia and cataract, very briefly.

(*i*) Some persons cannot see the distant objects clearly (though they can see the nearby objects clearly). They are said to have a defect of eye called myopia. **Myopia is that defect of eye (or defect of vision) due to which a person cannot see the distant objects clearly (though he can see the nearby objects clearly).** Myopia is caused either due to high converging power of eye-lens or due to eye-ball being too long. In an eye suffering from myopia, the image of distant object is formed in front of retina due to which the person cannot see it clearly. **Myopia is corrected by using spectacles containing *concave lenses* (diverging lenses) of suitable power.** The image of distant object is then formed on the retina and hence can be seen clearly.

(*ii*) Some persons cannot see the nearby objects clearly (though they can see the distant objects clearly). They are said to have the defect of eye called hypermetropia. **Hypermetropia is that defect of eye (or defect of vision) due to which a person cannot see the nearby objects clearly (though he can see the distant objects clearly)**. Hypermetropia is caused either due to low converging power of eye-lens or due to eye-ball being too short. In an eye suffering from hypermetropia, the image of object is formed behind the retina due to which the person cannot see it clearly. **Hypermetropia is corrected by using spectacles containing *convex lenses* (converging lenses) of suitable power.** The image of nearby object is then formed on the retina and hence the person can see it clearly.

(*iii*) A yet another defect of the eye which usually comes in old age is cataract. **The medical condition in which the lens of eye of a person becomes progressively cloudy resulting in blurred vision is called cataract.** Cataract develops when the eye-lens of a person becomes cloudy (or even opaque) due to the formation of a membrane over it. Cataract decreases the vision of the eye gradually. It can even lead to total loss of vision of the eye. The vision of the person can be restored after getting **surgery** done on the eye having cataract. The opaque lens is removed from the eye of the person by surgical operation and a new artificial lens is inserted in its place. **The eye-defect called cataract cannot be corrected by any type of spectacle lenses.**

Care of the Eyes

Our eyes are a wonderful gift of nature. We must take proper care of our eyes and protect them from any kind of injury or damage so that they remain good throughout our whole life. **Some of the precautions to protect our eyes and maintain healthy eyesight are as follows :**

1. Wash your eyes at least twice a day with clean water.

2. Too little light or too much light is bad for eyes. Too little light causes eyestrain and headache. So, **do not read or write in dim light because it puts strain on the eyes and may cause headache**. Too much light (like that from viewing the sun directly, electric welding or laser torch) can injure the retina and damage the eyes. So, **we should never look directly at the sun or other powerful lights**. We should protect our eyes from the glare of bright light.

3. Always read by keeping the book at normal distance for distinct vision. **Do not read by bringing the book too close to your eyes or by keeping it too far from the eyes.**

4. Raise your eyes from time to time while reading, writing or watching television so as to relax the eyes.

5. Protect your eyes from injuries and foreign bodies (like dust particles and insects, etc). If something (like a dust particle or insect, etc.) gets into the eyes, splash the eyes with a lot of clean water. **Do not rub the eyes with hands to prevent injury to the eyes**. If need be, consult a doctor for treatment.

6. If you find difficulty in reading a book or writing on the blackboard or you have to squeeze your eyes to see clearly, get your eyes checked by an eye-specialist doctor immediately. And if doctor recommends, wear spectacles to regain normal eyesight.

7. In case of an injury or any other problem to the eyes, consult an eye-specialist doctor. Self-treatment can be dangerous to the eyes.

8. Vitamin A is essential for keeping the eyes healthy. Deficiency of vitamin A in the diet is responsible for many ailments of the eye including night blindness. **The inability of eyes to see properly in dim light**

(especially at night) is called night blindness. Night blindness is an eye disease which is caused by the lack (or deficiency) of vitamin A in the diet of a person. **We should also include those food items in our diet which contain vitamin A to prevent night blindness and keep our eyes healthy**. The food items which are rich in vitamin A are : Carrots, Cod-liver oil (Fish liver oil), Green vegetables (such as Spinach), Cabbage, Broccoli, Eggs, Milk, Butter, Curd, Cheese and fruits such as Papaya and Mango.

Eyes of Other Animals

Animals have eyes of different shapes and sizes. For example :

(*i*) The eyes of a crab are quite small but they enable the crab to look all around. Due to this a crab can see an enemy even if it comes from behind.

(*ii*) Butterflies have large eyes which appear to be made up of thousands of little eyes. The eyes of a butterfly enable it to see not only in the front and sides but at the back as well.

(*iii*) A night bird (owl) can see very well in the night (in dim light) but not during the day (when there is bright light). This is due to the following reasons :

(*a*) The owl has *big eyes* having large cornea and a *large pupil* to allow more light in the eye (see Figure 21).

(*b*) The owl's eyes have a *large number of rods in the retina* (which are sensitive to dim light) and *only a few cones* (which are sensitive to bright light).

(*iv*) The day light birds (such as kite and eagle) can see very well during the day (in bright light) but not in the night (when there is dim light). The eyes of day light birds (such as kite and eagle) are *small* and have *more cones in retina* (which are sensitive to bright light) but *fewer rods* (which are sensitive to dim light).

Figure 21. An owl has large eyes to see in dim light.

Visually Challenged Persons Can Read and Write

Some persons may lose their eyesight due to an eye injury (received during an accident) or due to a disease. And some persons cannot see at all from birth. **Those persons who are 'unable to see' are known as visually challenged persons.** Such persons develop their other senses more sharply. They try to identify things by touching and by listening to voices more carefully.

The most popular resource for visually challenged persons which can make them read and write is braille. **Braille is a written language for the visually challenged persons in which characters (letters and numbers, etc.) are represented by patterns of raised dots.** The visually challenged persons recognise the words written on a braille sheet by touching the patterns of dots (which are raised slightly to make it easier to touch). The braille was developed by Louis Braille who was himself a visually challenged person. There is a braille code for common languages, mathematics and scientific notations. Many Indian languages can be read and written by using the braille system. Braille writer slate and stylus help the visually challenged persons in taking notes, reading and writing. Type-writer like devices for braille are also available now. Auditory aids such as cassettes, tape recorders, talking books and talking calculators are also very useful for such persons.

There are many visually challenged persons in the world who have great achievements to their credit. Helen Keller, an American lecturer and author is perhaps the most famous and inspiring visually challenged person of the world. Helen Keller lost her eyesight when she was only 18 months old. But because of her resolve and courage she studied up to graduation in a University and wrote a number of books, including 'The Story of My Life'. Some visually challenged Indians have also made many achievements. For example, Ravindra Jain, who is visually challenged by birth, obtained the Sangeet Prabhakar degree from Allahabad University. He is a famous lyricist, singer and music composer. A visually challenged boy Diwakar is a very good singer. We are now in a position to **answer the following questions :**

Very Short Answer Type Questions

1. What is meant by 'incident ray' ?
2. What is meant by 'reflected ray' ?
3. How many reflected rays can there be for a given single incident ray falling on a plane mirror ?
4. What do you understand by the term 'point of incidence' ?
5. What is 'normal' in the reflection of light from a plane mirror ?
6. Define the angle of incidence.
7. Define the angle of reflection.
8. A ray of light is incident on a plane mirror at an angle of 30°. What is the angle of reflection ?
9. An incident ray makes an angle of 75° with the surface of a plane mirror. What will be the angle of reflection ?
10. A ray of light is incident normally (perpendicularly) on a plane mirror. Where will this ray of light go after reflection from the mirror ?
11. What is the angle of incidence when a ray of light is incident normally on a plane mirror ?
12. What is the angle of reflection when a ray of light is incident normally on a plane mirror ?
13. What is the angle of incidence of a ray of light if the reflected ray is at an angle of 90° to the incident ray ?
14. Name the apparatus which is used to obtain a thin beam of light.
15. What type of reflection of light takes place from :
    (*a*) a rough surface ?
    (*b*) a smooth surface ?
16. Which type of reflection of light, regular reflection or diffuse reflection, leads to the formation of images ?
17. What type of reflection of light takes place from :
    (*a*) a cinema screen ?
    (*b*) a plane mirror ?
18. If an object is placed at a distance of 7.5 cm from a plane mirror, how far would it be from its image ?
19. Is the image of an object in a plane mirror virtual or real ?
20. Name the phenomenon responsible for the following effect :
    When we sit in front of a plane mirror and write with our right hand, it appears in the mirror that we are writing with the left hand.
21. Name a device which works on the reflection of reflected light.
22. How are the two plane mirrors in a periscope arranged :
    (*a*) with respect to one another ?
    (*b*) with respect to sides of the tube ?
23. What will be the number of images formed when an object is placed between two parallel plane mirrors facing each other ?
24. Name an instrument or toy which works by producing multiple reflections from three plane mirrors to form beautiful patterns.
25. State one use of kaleidoscope.
26. Name the device used to split white light into seven colours.
27. What happens when a beam of sunlight is passed through a glass prism ?
28. What type of lens (convex or concave) is present in the human eye ?
29. What is the range of vision of a normal human eye ?
30. Name the point inside the human eye where the image is not visible.
31. Name the phenomenon which enables us to see movies in a cinema hall.
32. Name an eye ailment (or eye-disease) caused by the deficiency of vitamin A in the diet.
33. What is the name of transparent front part of an eye ?
34. What is the name of a small opening in the iris of an eye ?
35. Which part of the eye gives it its distinctive colour ?
36. Write the names of the main parts of the human eye.
37. What happens to the size of the pupil of our eye in dim light ?
38. What happens to the size of the pupil of our eye in bright light ?

39. State whether the following statements are true or false :
(a) The moon is an illuminated object.
(b) Diffuse reflection means the failure of the laws of reflection of light.
(c) In a kaleidoscope, a pattern seen once can never be seen again.
40. Fill in the following blanks with suitable words :
(a) The angle of .... equals the angle of reflection.
(b) A person 1 m in front of a plane mirror seems to be......... m away from his image.
(c) If you touch your.........ear with right hand in front of a plane mirror, it will be seen in the mirror that your right ear is touched with..........
(d) The size of pupil becomes...........when you see in dim light.
(e) Night birds have.........cones than rods in their eyes.
(f) The image of an object persists on the retina of an eye for about.........second even after the object has disappeared.
(g) If the still pictures of a moving object are flashed on our eyes at a rate faster than.............pictures per second, the eye perceives the object as moving.
(h) In a movie, the still pictures in proper sequence are projected on the screen usually at the rate of ....pictures per second.

Short Answer Type Questions

41. Suppose you are in a dark room. Can you see objects in the room ? Can you see objects outside the room ? Explain.
42. What makes things visible to us ? Why cannot we see a book which is placed (a) behind a wooden screen, and (b) in a dark room ?
43. We can see the sun because it is glowing. How are we able to see the moon ?
44. Name the two types of reflection of light. Which type of reflection makes us see an object from all directions ?
45. A wall reflects light and a mirror also reflects light. What difference is there in the way they reflect light ?
46. Explain why, a book lying on a table in a room can be seen from all the parts of the room.
47. What is the full form of $\angle i$ and $\angle r$? What is the relation between them ?
48. You see your image in a plane mirror ? State two characteristics of the image so formed.
49. What is a periscope ? How many mirrors are there in a periscope ?
50. State the various uses of a periscope.
51. Explain how, a hair dresser makes you see hair at the back of your head after the hair cut is complete.
52. How many images of an object will be formed when the object is placed between two plane mirrors which are inclined at the following angles to one another ?
(a) 120° (b) 45° (c) 180° (d) 60° (e) 90°
53. Two plane mirrors are set at right angles to each other. A coin is placed in-between these two plane mirrors. How many images of the coin will be seen ?
54. How many images of a candle will be formed if it is placed between two parallel plane mirrors separated by 40 cm ?
55. Explain why, when an object is placed between two plane mirrors inclined at an angle, then multiple images are formed.
56. How can you show that white light (say, sunlight) consists of seven colours ?
57. What information do you get about sunlight from the formation of a rainbow ?
58. What is meant by 'dispersion of light' ? Name a natural phenomenon which is caused by the dispersion of sunlight in the sky.
59. How many plane mirror strips are there in a kaleidoscope ? How are they arranged ?
60. How does eye adjust itself to deal with light of varying intensity ?
61. Explain why, we cannot see our surroundings clearly when we enter a darkened cinema hall from bright sunshine but our vision improves after some time.
62. How does the eye-lens differ from the ordinary convex lens made of glass ?
63. Name the part of the eye :
(a) which controls the amount of light entering the eye.

(*b*) which converges light rays to form the image.
(*c*) on which image is formed.
(*d*) which carries the image to brain.
(*e*) which changes the curvature (or thickness) of eye-lens to focus objects lying at various distances.

64. Name the cells on the retina of an eye :
(*a*) which are sensitive to bright light.
(*b*) which are sensitive to dim light.
(*c*) which produce sensation of colour.

65. What are rods and cones in the retina of an eye ?

66. Name any one defect of the eye. How is it corrected ?

67. What is cataract ? How can the vision of a person having cataract be restored ?

68. What is meant by 'persistence of vision' ?

69. Explain how you can take care of your eyes.

70. What should we do if something like a dust particle or an insect gets into our eye ?

71. Name any five food items (including two fruits) which are rich in vitamin A.

72. Explain why, too little or too much light, both are bad for eyes.

73. Explain why, an owl can see well in the night (but not during the day) whereas an eagle can see well during day (but not at night).

74. (*a*) What is 'blind spot' in the eye ?
(*b*) What is night blindness ? What causes night blindness ?

75. What is lateral inversion ? Explain with the help of an example.

Long Answer Type Questions

76. (*a*) What is meant by a luminous object ? Name two luminous objects.
(*b*) What is meant by a non-luminous object ? Name two non-luminous objects.

77. (*a*) What is the difference between regular reflection and diffuse reflection of light ? Name one object which can produce regular reflection of light and another which produces diffuse reflection of light.
(*b*) Draw diagrams to show regular reflection of light and diffuse reflection of light.
(*c*) Which of the following will cause regular reflection of light and which diffuse reflection of light ?
(*a*) Polished wooden table (*b*) Chalk powder
(*c*) Cardboard (*d*) Mirror
(*e*) Paper (*f*) Marble floor with water spread over it.

78. (*a*) Draw a diagram to show the reflection of light from a plane mirror. Label the following on the diagram :
(*a*) Plane mirror (*b*) Incident ray (*c*) Reflected ray
(*d*) Point of incidence (*e*) Normal (*f*) Angle of incidence
(*g*) Angle of reflection
(*b*) State the laws of reflection of light.

79. (*a*) Draw a labelled diagram showing how a plane mirror forms an image of a point object placed in front of it.
(*b*) State the characteristics of the image formed in a plane mirror.

80. (*a*) Draw a labelled diagram of the human eye. Label the following parts on this diagram :
Cornea, Iris, Pupil, Ciliary muscles, Eye-lens, Retina, Optic nerve, Blind spot.
(*b*) What are the functions of the following parts of the eye ?
(*a*) Iris (*b*) Eye-lens (*c*) Ciliary muscles (*d*) Retina (*e*) Optic nerve

Multiple Choice Questions (MCQs)

81. The angle of reflection is equal to the angle of incidence :
(*a*) always (*b*) sometimes (*c*) under special conditions (*d*) never

82. The image formed by a plane mirror is :
(*a*) virtual, behind the mirror and enlarged.
(*b*) virtual, behind the mirror and of the same size as the object.
(*c*) real, at the surface of the mirror and enlarged.
(*d*) real, behind the mirror and of the same size as the object.

83. The least distance of distinct vision for a young adult with normal vision is about :
(*a*) 25 m (*b*) 2.5 cm (*c*) 25 cm (*d*) 2.5 m

84. The angle between an incident ray and the plane mirror is 30°. The total angle between the incident ray and the reflected ray will be :
(*a*) 30° (*b*) 60° (*c*) 90° (*d*) 120°

85. The image of an object formed by a plane mirror is :
(*a*) virtual (*b*) real (*c*) diminished (*d*) upside-down

86. Which of the following is a non-luminous object ?
(*a*) sun (*b*) star (*c*) moon (*d*) fire

87. A device which works on the reflection of light from two plane mirrors arranged parallel to one another is :
(*a*) electroscope (*b*) kaleiodoscope (*c*) periscope (*d*) stethoscope

88. The number of images formed of an object placed between two plane mirrors inclined at right angles to each other is :
(*a*) two (*b*) five (*c*) one (*d*) three

89. As the angle between two plane mirrors is decreased gradually, the number of images of an object placed between them :
(*a*) increases gradually (*b*) decreases gradually
(*c*) first increases then decreases (*d*) first decreases then increases

90. The deficiency of one of the following in the diet of a person for a considerable time can lead to a disease called night blindness. This one is :
(*a*) vitamin B (*b*) vitamin D (*c*) vitamin A (*d*) vitamin C

91. Which of the following is not a part of the human eye ?
(*a*) retina (*b*) auditory nerve (*c*) optic nerve (*d*) ciliary muscle

92. How does the eye change in order to focus on near or distant objects ?
(*a*) the lens moves in or out (*b*) the retina moves in or out
(*c*) the lens becomes thicker or thinner (*d*) the pupil becomes larger or smaller

93. Which of the following changes occur when you walk out of bright sunshine into a poorly lit room ?
(*a*) the pupil becomes larger (*b*) the lens becomes thicker
(*c*) the ciliary muscle relaxes (*d*) the pupil becomes smaller

94. An incident ray makes an angle of 65° with the surface of a plane mirror. The angle of reflection in this case will be :
(*a*) 65° (*b*) 45° (*c*) 25° (*d*) 35°

95. Which of the following produces diffuse reflection of light ?
(*a*) mirror on a dressing table (*b*) water surface of a pond
(*c*) screen in a cinema hall (*d*) polished wooden table

96. The human eye forms the image of an object at its :
(*a*) cornea (*b*) iris (*c*) pupil (*d*) retina

97. The change in converging power of an eye-lens is caused by the action of :
(*a*) iris (*b*) ciliary muscles (*c*) optic nerve (*d*) retina

98. The size of the pupil of the eye is adjusted by :
(*a*) cornea (*b*) ciliary muscles (*c*) optic nerve (*d*) iris

99. The defect of vision in which the eye-lens of a person gets progressively cloudy resulting in blurred vision is called :
(*a*) myopia (*b*) night blindness (*c*) cataract (*d*) hypermetropia

100. A person cannot see the distant objects clearly (though he can see the nearby objects clearly). He is suffering from the defect of vision called :
(*a*) hypermetropia (*b*) myopia (*c*) night blindness (*d*) cataract

Questions Based on High Order Thinking Skills (HOTS)

101. A man stands 10 m in front of a large plane mirror. How far must he walk before he is 5 m away from his image ?

102. A ray of light strikes a plane mirror XY at an angle of incidence of 65°, is reflected from this plane mirror and then strikes a second plane mirror YZ placed at right angles to the first mirror. What is the angle of reflection for the mirror YZ ?

103. The eye of a person exhibits a phenomenon X due to which it can see the image of an object for a short duration of Y even after the object has disappeared from his view. It is due to the phenomenon X that we are able to see moving Z on a television screen. What are X, Y and Z ?

104. Man A has a defect of vision due to which he cannot see the nearby objects clearly (though he can see the distant objects clearly). On the other hand, man B has a defect of vision due to which he cannot see the distant objects clearly (though he can see the nearby objects clearly). The defect in man A can be corrected by using spectacles containing lenses C whereas the defect in man B can be corrected by using spectacles containing lenses D.
(*a*) Name the defect of vision in man (*i*) A, and (*ii*) B.
(*b*) What type of lenses are (*i*) C, and (*ii*) D ?

105. A student makes a device P by using three long and narrow strips of plane mirrors inclined at 60° to one another which enables him to see beautiful patterns made by pieces of coloured glass bangles. On the other hand, another student makes a device Q by using two plane mirrors arranged parallel to each other which helps him to see a football match clearly even when some very tall persons are sitting in front of him in the ground. What are P and Q ?

ANSWERS

8. 30° **9.** 15° **11.** 0° **12.** 0° **13.** 45° **18.** 15 cm **20.** Lateral inversion **21.** Periscope **24.** Kaleiodoscope **30.** Blind spot **39.** (*a*) True (*b*) False (*c*) True **40.** (*a*) incidence (*b*) 2 (*c*) left ; left hand (*d*) large (*e*) fewer (*f*) $\frac{1}{16}$ (*g*) 16 (*h*) 24 **52.** (*a*) 2 (*b*) 7 (*c*) 1 (*d*) 5 (*e*) 3 **53.** 3 **54.** Infinite number of images **81.** (*a*) **82.** (*b*) **83.** (*c*) **84.** (*d*) **85.** (*a*) **86.** (*c*) **87.** (*c*) **88.** (*d*) **89.** (*a*) **90.** (*c*) **91.** (*b*) **92.** (*c*) **93.** (*a*) **94.** (*c*) **95.** (*c*) **96.** (*d*) **97.** (*b*) **98.** (*d*) **99.** (*c*) **100.** (*b*) **101.** 7.5 m **102.** 25° **103.** X : Persistence of vision ; Y : $\frac{1}{16}$ second ; Z : Pictures **104.** (*a*) (*i*) Hypermetropia (*ii*) Myopia (*b*) (*i*) Convex lenses (*ii*) Concave lenses **105.** P : Kaleiodoscope ; Q : Periscope

CHAPTER 17

Stars and the Solar System

The objects which exist in the sky (or in outer space) are called celestial objects. The stars (including the Sun), the planets (including the Earth), satellites (like the Moon), asteroids, comets and meteoroids are all celestial objects. Celestial objects are also known as heavenly objects (or heavenly bodies). In this Chapter we will learn about the various celestial objects which exist in the vast surrounding space called universe.

The Night Sky

During the daytime, we can see only the Sun in the sky. Because of the glare of sunlight, we cannot see any other celestial objects in the sky during the day. But as soon as the Sun sets in the evening and it becomes dark, we can see many celestial objects in the night sky. **The various celestial objects which we can see easily in the night sky are stars, planets, moon and meteors (or shooting stars).**

On a clear, moonless night, we can see about 3000 stars with naked eye. Many more stars can be seen with the help of a telescope. **An important characteristic of stars is that they appear to twinkle in the sky.** That is, their light increases and decreases continuously. The twinkling of stars is an illusion (false show) caused by the disturbance of star's light by Earth's atmosphere. We can also see some star-like objects in the night sky which do not twinkle. These are the planets which revolve around the Sun. **Planets do not twinkle because they are much more nearer to the Earth than the stars.** Moon is the most prominent object which we can see in the night sky. If we keep on looking at the night sky for a considerable time, we may see a bright streak of light flashing across the sky for a few seconds (and then disappearing). This is a meteor which is commonly known as a shooting star. Let us first discuss the stars.

STARS

Stars are the celestial objects (like the Sun) that are extremely hot and have light of their own. Stars emit heat and light continuously. Stars consist mostly of hydrogen gas. The heat and light of stars is produced by the nuclear fusion reactions taking place inside them all the time. In these fusion reactions, hydrogen present inside the stars is converted into helium, with the release of a tremendous amount of heat and light. Stars are very, very large objects having the shape of a ball. The stars are much, much bigger than our Earth. Many of the stars are even bigger than the Sun. **The stars appear to be small because they are very, very far away from us.**

In our everyday language, we do not call the Sun a star. But in reality, the Sun is also a star. **The Sun is the star which is nearest to the Earth. The Sun looks much bigger and brighter because it is much nearer to us than any other star.** Though many other stars are bigger and brighter than the Sun, but they look small and faint. This is because they are very far away from Earth (as compared to the Sun). We can see the stars only at night. **Though the stars are present in the sky even during the daytime, but we cannot see them during the daytime because of the bright light of the Sun.** There are about 10,000 billion stars in the Universe.

We know that the Sun (which is a star) rises up in the east in the morning, travels in the sky the whole day, and then sets in the west in the evening. So, the Sun appears to move in the sky from east to west direction. Same is the case with all other stars. Thus, **the stars appear to move in the sky from east to west direction. This apparent motion of the stars in the sky from east to west is due to the rotation of Earth from west to east on its axis.** *When the Earth moves (or rotates) on its axis from west to east, then the stars in the sky appear to move in the opposite direction : from east to west.*

There is one star which does not appear to move in the sky. It is called Pole Star. Thus, **the star which appears stationary from the Earth is Pole Star.** The Hindi name of Pole Star is *Dhruva Tara.* The Pole Star always remains in the same position in the sky. Though all other stars appear to move from east to west, the Pole Star remains fixed at the same place in the sky in the North direction. **The Pole Star appears to be stationary and does not change its position with time because it lies on the axis of rotation of Earth (which is fixed in space and does not change with time)** (see Figure 1). Since Pole Star remains fixed (or stationary) in the sky, all other stars appear to revolve around the Pole Star. Please note that the Pole Star is not visible from the southern hemisphere of the Earth.

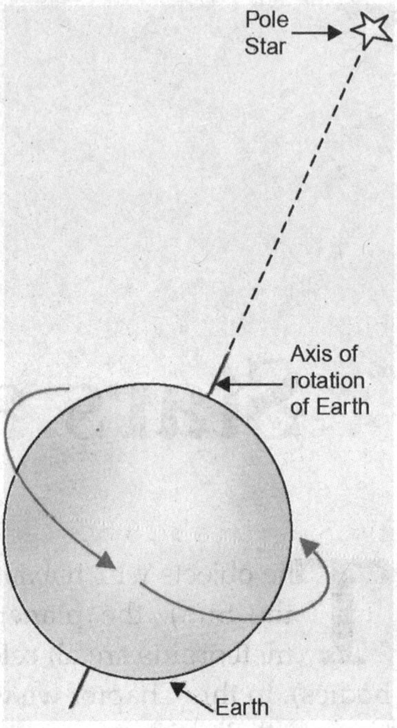

Figure 1. The Pole Star lies on the axis of rotation of the Earth.

ACTIVITY

The fact that the Pole Star does not appear to move and remains stationary in the sky can be shown by performing an activity as follows : Take an umbrella and open it. Make about 10 small stars out of paper. Paste one star at the top position of the central rod of the umbrella and other stars at different places on the umbrella cloth, near the end of each metal 'spoke' (as shown in Figure 2). Now rotate the umbrella by holding its central rod in your hand. Observe the paper stars on the rotating umbrella. We will find that all the stars on the spokes appear to move but there is one star which does not appear to move. The star which does not appear to move is located at the top end of central rod (which is being rotated). In this activity, the central rod of umbrella is like the axis of rotation of Earth and the star fixed at its top end is like the Pole Star.

Figure 2. The paper star at the top end of central rod of umbrella does not appear to move when the umbrella is rotated. It is like the Pole Star.

Before we go further, we should know the meaning of a special unit of distance called 'light year' which is used for measuring the extremely large distances between the various celestial objects (like stars and planets) in this universe. This is described below.

The Unit of Measuring Distances in the Universe : Light Year

On the Earth, we normally express the large distances in the unit of 'kilometre'. The distances in the Universe are so large that 'kilometre' becomes an extremely small and inconvenient unit to express such large distances. For example, the various stars are millions of kilometres away, so 'kilometre' unit is not suitable to express the distances between the stars and other celestial objects. **The distances between the various celestial objects (like the stars and planets) are expressed in the unit of 'light year'.**

One 'light year' is the 'distance travelled by light in one year'. We can calculate the value of *'one light year'* in *'kilometres'* by multiplying the speed of light (which is 3,00,000 km/s) by the number of seconds in one year (which is 3,15,36,000 s). We will find that :

1 light year = 9,460,800,000,000 kilometres

This can be written in short as :

1 light year = 9.46 × 10^{12} kilometres

This shows that 'light year' is an extremely big unit of distance. Please note that year is a unit of time, but light year is not a unit of time. **Light year is a unit of distance.** In addition to light year, **light minute** is also used as a unit of distance in some cases. *One light minute is the distance travelled by light in one minute.*

The distance between the Sun and the Earth is about 150,000,000 kilometres (which is 150 million km). Light from the Sun takes about 8 minutes to reach the Earth. So, **the distance of Sun from the Earth is 8 light minutes.** In other words, we can say that the Sun is about 8 light minutes away from the Earth. **After the Sun, the next nearest star to the Earth is 'Proxima Centauri'.** Proxima Centauri is at a distance of about 40,000,000,000,000 kilometres (which is 40,000,000 million km) from the Earth. Light from the next nearest star, Proxima Centauri, takes about 4.3 years to reach the Earth. So, **the distance of Proxima Centauri star from the Earth is 4.3 light years.** In other words, we can say that Proxima Centauri star is about 4.3 light years away from the Earth. Thus, the light coming from this star which we will see tonight would have left the surface of this star about 4.3 years ago ! Please note that the star nearest to the Earth (after the Sun) is Proxima Centauri and not Alpha Centauri as written in some other books. Some of the stars in the Universe are so far off that light coming from them takes millions of years to reach the Earth. Thus, some of the stars are even millions of light years away from the Earth.

From the above discussion we conclude that the distances between stars and planets are expressed in the unit of 'light year' because these distances are so large that it is inconvenient to express them in the unit of 'kilometres'. It is, however, very convenient to express extremely large distances in the unit of light year. Now, suppose a star is 6 light years away from the Earth. By saying that a star is 6 light years away from the earth, we mean that the distance of this star from the Earth is so much that light from the star takes 6 years to reach the Earth. We will now describe the star groups called constellations.

CONSTELLATIONS

We see thousands of stars on a clear, moonless night. Some of these stars are arranged in groups or patterns. These groups of stars appear to form some recognizable shapes and patterns. For example, one group of stars suggests the outline of a big bear (great bear), another group of stars reminds us of a hunter, and so on. **The group of stars which appears to form some recognizable shape or pattern is known as a constellation.** All the stars of a constellation always remain together. Due to this the shape of a constellation always remains the same. Our ancestors named these 'star groups' or 'constellations' after the objects which they seemed to resemble. About 88 constellations are known at present. **Each constellation has been given a name signifying an animal, a human being or some other object which it appears to resemble.** All the constellations appear to move in the sky from east to west. This is because the Earth (from where we view them), rotates on its axis from west to east. Some of the important constellations are :

(*i*) Ursa Major **(*iii*) Cassiopeia**
(*ii*) Orion **(*iv*) Leo Major**

All these constellations can be easily recognised (or identified) in the night sky. A constellation does not have a small number of stars. A constellation consists of a group of a large number of stars, some of which are bright and can be seen easily with the naked eye whereas others are dim and seen with difficulty. *While describing the various constellations we will give the arrangement of only bright stars in them which can be recognised easily in the night sky.* All the stars which make a constellation are not at the same distance from us. They are just in the same line of sight in the sky. We will now describe the four common constellations, Ursa Major, Orion, Cassiopeia and Leo Major in somewhat detail. Let us start with Ursa Major.

1. Ursa Major Constellation

Ursa Major constellation is one of the most famous constellation which we can see in the night sky. Ursa Major constellation is also known as 'Great Bear', 'Big Bear', 'Big Dipper' or 'Plough'. The Indian name of Ursa Major constellation is '*Saptarishi*'. Though Ursa Major constellation is a group of many stars but seven of its stars are quite bright and can be seen easily. The arrangement of main stars in the Ursa Major constellation is shown in Figure 3. **The Ursa Major constellation consists of seven bright stars which are arranged in a pattern resembling somewhat a big bear** (see Figure 3). The stars marked 1, 2, 3, and 4 are supposed to form the body of the big bear whereas the stars marked 5, 6 and 7 form the tail of the big bear. The head and paws of this bear are formed from some other faint stars which are not shown in Figure 3. It is clear that ***a lot of fertile imagination is required to make out some recognizable shape from a group of stars.*** The 'Ursa Major' is called 'Great Bear' because its stars are arranged in such a way that they seem to form the outline of the body of a 'Great Bear' (or Big Bear) (see Figure 4).

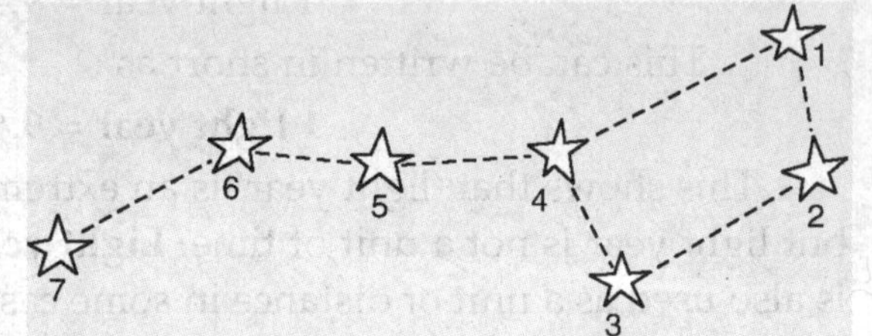

Figure 3. Ursa Major constellation (It is also known as Great Bear, Big Bear, Big Dipper or Plough constellation).

The Ursa Major is also called Big Dipper because it resembles a dipper (a bowl with a handle), which was used to drink water in olden days. The stars marked 1, 2, 3 and 4 (in Figure 3) appear to form the bowl of the dipper whereas the stars marked 5, 6 and 7 seem to form its handle. Please note that Ursa Major constellation also resembles a 'big kite with a tail', a 'big laddle'(a big spoon with a cup-shaped bowl), or a 'big question mark' in the night sky. **The Ursa Major constellation is visible during the summer season** in the early part of the night. It can be seen clearly in the month of April in the northern part of sky at night. The stars marked 1 and 2 in Ursa Major constellation are called pointer stars because the line joining them points to the direction of Pole Star.

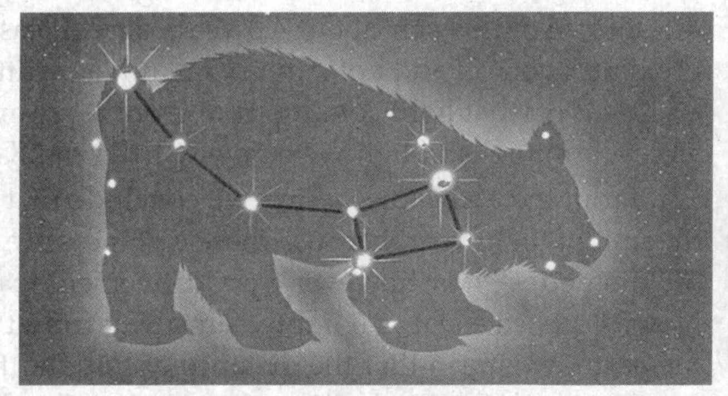

Figure 4. This is how Ursa Major constellation looks like a Big Bear.

We can locate the position of Pole Star in the night sky with the help of Ursa Major constellation. This can be done as follows : Look towards the northern part of the sky on a clear, moonless night during summer (at about 9 pm) and identify the Ursa Major constellation in the sky. Now look at the two pointer stars at the end of the Ursa Major constellation (see Figure 5). Imagine a straight line drawn through the two pointer stars of the Ursa Major constellation. Extend this imaginary line towards the north direction in the sky. This line will lead to a star which is not very bright (see Figure 5). This star is the Pole Star.

Figure 5. Locating the Pole Star in the night sky with the help of Ursa Major constellation.

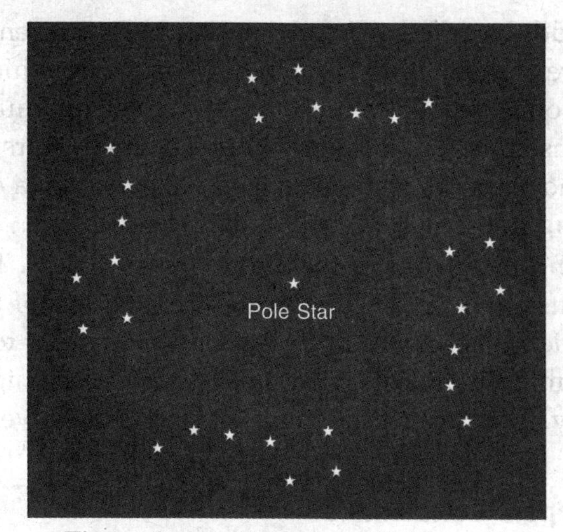

Figure 6. Ursa Major constellation revolves around the Pole Star in the night sky.

If we observe the Ursa Major constellation at different times in the night, we will find that it appears to move from east to west. Since the Pole Star remains fixed in the night sky, therefore, ***the Ursa Major constellation appears to revolve around the Pole Star in the night sky*** (see Figure 6). Please note that some of the northern constellations like Ursa Major are not visible from some points in the southern hemisphere of the Earth.

2. Orion Constellation

Orion is one of the well known and most impressive constellations in the night sky. Orion is also known as 'Hunter'. The Indian name of Orion is *'Mriga'*. There is a greater number of bright stars in Orion than any other constellation. **The Orion constellation consists of seven or eight bright stars** (and several faint stars). Thus, the Orion constellation can be drawn either with seven stars or with eight stars. The Orion constellation drawn with eight main stars is shown in Figure 7. The arrangement of stars in Orion constellation is supposed to resemble a 'hunter' in the kneeling position. The star marked 1 in Figure 7 represents the head of the hunter. The stars marked 2 and 3 in Figure 7 are supposed to form the shoulders of the hunter whereas the stars marked 7 and 8 are imagined to form the feet of the hunter. The three stars

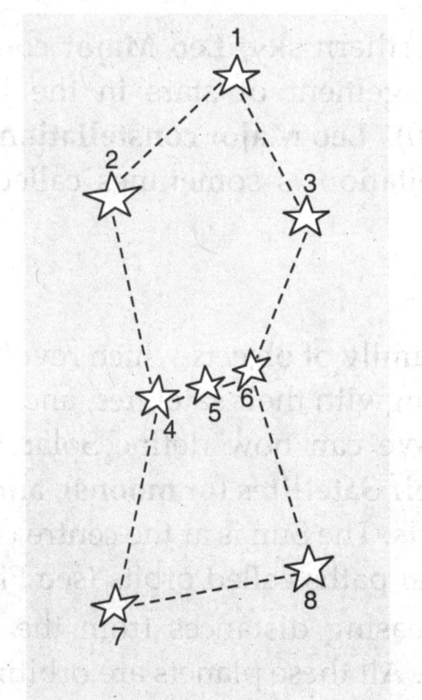

Figure 7. Orion constellation.

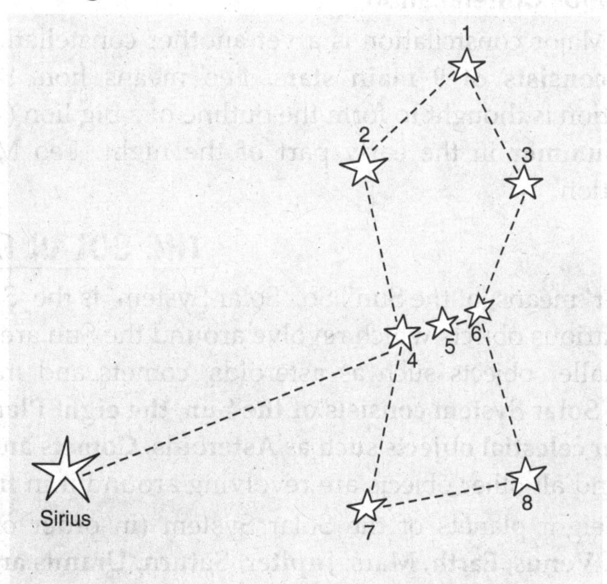

Figure 8. Locating the 'Sirius' Star in the night sky with the help of Orion constellation.

in the middle marked 4, 5 and 6 in Figure 7 represent the belt of the hunter. The arms and weapons of the hunter are formed from some other faint stars which are not shown in Figure 7. Again , a lot of fertile imagination is needed to make out the shape of a hunter from a group of stars in the night sky (When the Orion constellation is drawn with seven main stars, then the star marked 1 at the top in Figure 7 is not shown and the stars marked 2 and 3 are joined by a dotted line). **The Orion constellation is visible in the sky during the winter season** in the late evenings.

The brightest star in the night sky is 'Sirius'. The brightest star called 'Sirius' is located close to the Orion constellation. We can locate the position of Sirius Star in the night sky with the help of Orion constellation. This can be done as follows : In order to locate Sirius, imagine a straight line passing through the three middle stars of Orion constellation in the night sky. Look along this line towards the east direction in the sky. This imaginary line will lead us to a very bright star (see Figure 8). This very bright star is 'Sirius'.

3. Cassiopeia Constellation

Cassiopeia is another prominent constellation in the northern sky. **Cassiopeia constellation consists of 5 main stars.** The 5 main stars of Cassiopeia constellation are arranged to form the shape of distorted letter W or M (depending on the position of Cassiopeia constellation in the sky). Thus, Cassiopeia constellation looks like a distorted letter W or M (see Figure 9). The Cassiopeia constellation is thought to represent an ancient Queen named Cassiopeia seated on a chair. **Cassiopeia constellation is visible during winter** in the early part of the night.

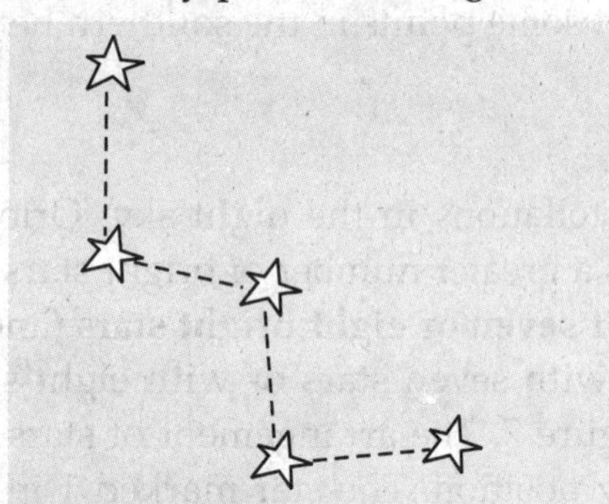

Figure 9. Cassiopeia constellation.

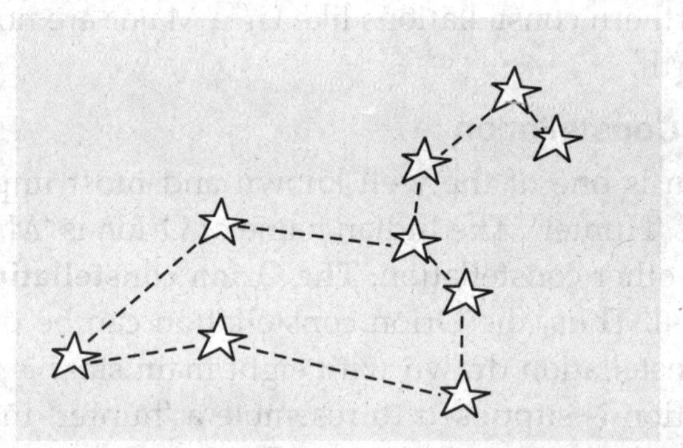

Figure 10. Leo Major constellation.

4. Leo Major Constellation

Leo Major constellation is a yet another constellation in the northern sky. **Leo Major constellation usually consists of 9 main stars.** Leo means lion. So, the arrangement of stars in the Leo Major constellation is thought to form the outline of a big lion (see Figure 10). **Leo Major constellation is visible during summer** in the early part of the night. Leo Major constellation is sometimes called just Leo constellation.

THE SOLAR SYSTEM

'Solar' means 'of the Sun'. So, 'Solar System' is the 'Sun and its family of objects which revolve around it'. The various objects which revolve around the Sun are planets alongwith their satellites, and millions of other smaller objects such as asteroids, comets and meteoroids. We can now define Solar System as follows : **Solar System consists of the Sun, the eight Planets and their Satellites (or moons), and millions of smaller celestial objects such as Asteroids, Comets and Meteoroids**. The Sun is at the centre of the Solar System and all other objects are revolving around it in fixed elliptical paths called orbits (see Figure 11).

The eight planets of the Solar System (in order of their increasing distances from the Sun) are : **Mercury, Venus, Earth, Mars, Jupiter, Saturn, Uranus and Neptune**. All these planets are orbiting around the Sun. All the planets (except Mercury and Venus) have natural satellites (or moons) around them. *Please*

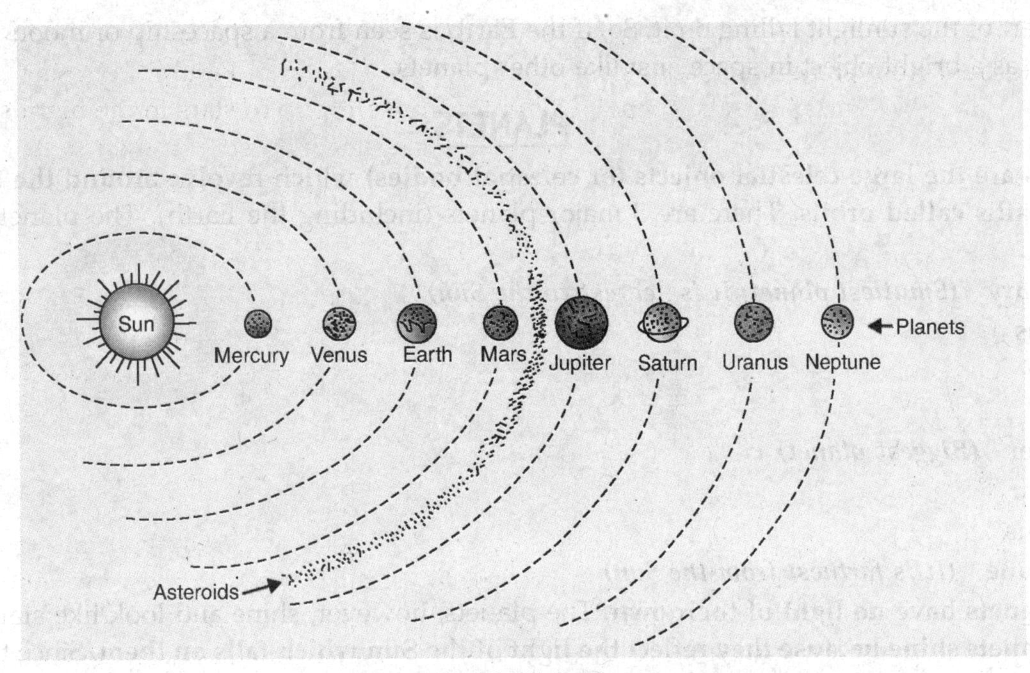

Figure 11. The Solar System.

note that just as planets revolve around the Sun, in the same way, the satellites (or moons) revolve around the planets. When the planets move around the Sun, then their satellites (or moons) also move alongwith them. Asteroids are small rocky bodies which revolve around the Sun between the orbits of the planets Mars and Jupiter (see Figure 11). Comets and meteoroids are also the minor members of the Solar System which revolve around the Sun. The orbits of comets and meteoroids have not been shown in Figure 11 to keep the diagram simple. The Sun is the biggest object in the Solar System. Because of its great size, the Sun has a very powerful gravitational force (or gravitational pull). The gravitational force of the Sun keeps the Solar System together and controls the movements of planets and other members of the Solar System. In other words, **the gravitational pull of the Sun keeps all the planets and other objects revolving around it**. We will now describe all the members of the Solar System in detail, one by one. Let us start with the Sun.

SUN

The Sun is a star (see Figure 12). **It is the star around which the Earth and other planets revolve.** Compared with the millions of other stars, the Sun is a medium-sized star and of average brightness. The Sun appears to be bigger and brighter because it is much more nearer to the Earth than any other star. In fact, Sun is the nearest star to the Earth. **The Sun is a star having a system of planets around it with life on one of its planets called Earth**. The diameter of the Sun is about 100 times the diameter of the Earth, and mass of the Sun is more than a million times the mass of the Earth. The Sun is called '*Surya*' in Hindi.

Figure 12. Sun.

The Sun is an extremely hot object. The temperature at the surface of the Sun is about 6000°C. **The Sun is not a solid object**. It is a sphere of hot gases. **The Sun consists mostly of hydrogen gas**. The nuclear fusion reactions taking place in the centre of the Sun (in which hydrogen is converted into helium), produce a tremendous amount of energy in the form of heat and light. *It is this nuclear energy which makes the Sun shine.* The Sun is continuously emitting huge amounts of heat and light. **The Sun is the main source of heat and light energy for all the planets of the Solar System (including the Earth) and their satellites, etc**. The planets and other objects in the sky reflect a part of the sunlight falling on them due to which they shine and become visible to us. Even our Earth reflects a part of the sunlight falling on it. So, if the Earth is seen from a spaceship or moon, the Earth will also appear as a bright object in space, just like other planets.

PLANETS

Planets are the large celestial objects (or celestial bodies) which revolve around the Sun in closed elliptical paths called orbits. There are 8 major planets (including the Earth). The planets of the Solar System are :

1. **Mercury** *(Smallest planet; It is nearest to the Sun)*
2. **Venus**
3. **Earth**
4. **Mars**
5. **Jupiter** *(Biggest planet)*
6. **Saturn**
7. **Uranus**
8. **Neptune** *(It is farthest from the Sun)*

The planets have no light of their own. The planets, however, shine and look like stars in the night sky. **The planets shine because they reflect the light of the Sun which falls on them**. Since the planets are much nearer to us than the stars, they appear to be big and do not twinkle at night. In fact, **the easiest way to distinguish planets from the stars in the night sky is that the stars appear to twinkle at night but the planets do not twinkle at all**. The planets move around the Sun from west to east, so the relative positions of the planets in the night sky keep changing day by day. In general we can say that **the planets keep changing their positions with respect to stars in the night sky**. This is another characteristic to distinguish planets from the stars. The planets are very small as compared to the Sun or other stars. Every planet has its own fixed path (called orbit) in which it revolves (or moves) around the Sun. Since all the planets revolve in their separate, fixed paths (or fixed orbits), they do not collide with one another while revolving around the Sun.

Out of the 8 planets of the Solar System, 5 planets, Mercury, Venus, Mars, Jupiter and Saturn can be seen easily with the naked eye, so they were known to the ancient astronomers. The 2 planets, Uranus and Neptune are very far off and have been discovered with the help of a telescope. The 8th planet is our own Earth. **A planet may be made of rock and metal, or gas**. The first four planets, Mercury, Venus, Earth and Mars are much nearer to the Sun (than the other four planets). The first four planets are called inner planets. **The four inner planets (Mercury, Venus, Earth and Mars) are made of rocks and have metallic cores**. The four inner planets are comparatively small and dense bodies having solid surfaces like our Earth. Earth is the biggest of the four inner planets. The inner planets have very few natural satellites (or moons) (see Table on the next page).

The planets outside the orbit of Mars are called outer planets. Thus, Jupiter, Saturn, Uranus and Neptune are called outer planets. The outer planets are much farther off from the Sun than the inner planets. The four outer planets are giant planets (very, very big planets). **The four outer planets (Jupiter, Saturn, Uranus and Neptune) are made mainly of hydrogen and helium gases, and not of rock and metal**. They do not have solid surfaces at all. The outer planets have ring systems around them. The outer planets also have a large number of natural satellites (or moons) around them (see Table on the next page).

All the planets revolve around the Sun and also rotate on their axis. The time taken by a planet to complete one revolution around the Sun is called its **period of revolution**. The time taken by different planets to make one revolution around the Sun (or period of revolution) is, however, different. Actually, ***as the distance of a planet from the Sun increases, its period of revolution also increases.*** The time taken by a planet to complete one rotation on its axis, is called its **period of rotation**. The time taken by different planets to rotate once on their axis (or period of rotation), is also different. Some important facts and figures about the planets of the Solar System are given on the next page.

Some Facts and Figures about the Planets

Name of planet	*Diameter of planet*	*Distance from Sun*	*Time taken for one revolution around Sun (Period of revolution)*	*Time taken to turn once on its axis (Period of rotation)*	*Mass of planet compared to Earth taken as 1*	*No. of satellites (or moons)*
1. Mercury	4880 km	58×10^6 km	88 days	58 days	0.055	None
2. Venus	12100 km	108×10^6 km	225 days	243 days	0.8	None
3. Earth	12760 km	150×10^6 km	$365\frac{1}{4}$ days	24 hours	1	1
4. Mars	6780 km	228×10^6 km	687 days	24 hours 37 minutes	0.1	2
5. Jupiter	142800 km	778×10^6 km	$11\frac{3}{4}$ years	9 hours 50 minutes	318	67
6. Saturn	120000 km	1427×10^6 km	$29\frac{1}{2}$ years	10 hours 14 minutes	95	62
7. Uranus	50800 km	2870×10^6 km	84 years	10 hours 49 minutes	15	27
8. Neptune	48600 km	4504×10^6 km	165 years	16 hours 3 minutes	17	14

Please note that till the year 2006 there were 9 planets in the Solar System. The 9th planet was Pluto. Pluto was the farthest planet from the Sun (beyond Neptune). In 2006, the International Astronomical Union (IAU) adopted a new definition of planet. Pluto does not fit this new definition of planet. So, though Pluto is still revolving around the Sun, it is no longer considered a planet of the Solar System. We will learn the new definition of planet in higher classes. The planets are called '*Graha*' in Hindi. We will now describe all the planets in detail, one by one. Let us start with the first planet called Mercury.

1. Mercury

Mercury is the first planet from the Sun (see Figure 13). **Mercury is the planet which is nearest to the Sun.** Since the planet Mercury is closest to the Sun, therefore, it is very hot during the day. **Mercury is the smallest planet of the Solar System**. The planet Mercury has a rocky surface which is covered with craters. Mercury is always close to the Sun and usually hidden by the Sun's glare. Due to this, Mercury is a planet which is quite difficult to see. Mercury planet is, however, visible just before sun-rise or just after sun-set, near the horizon. So, planet Mercury can be seen only at those places where trees and other buildings do not obstruct the view of the horizon. When planet Mercury is visible just before sun-rise in the morning, it is called a 'Morning Star' and when it is visible just after sun-set in the evening, then it is called an 'Evening Star'. **The planet Mercury can be seen either as a Morning Star in the eastern sky just before sun-rise or as an Evening Star in the western sky just after sun-set.**

Figure 13. The planet Mercury.

Planet Mercury can be seen as a 'Morning Star' or as an 'Evening Star' because it lies inside the orbit of the Earth. Please note that actually we cannot use the term 'star' for Mercury because it is a planet and not a star. Mercury is termed as 'Morning Star' or 'Evening Star' because it is a fairly bright object in the sky and appears like a star. In fact, Mercury shines because it reflects light from the Sun.

Planet Mercury shows phases like the Moon. This is due to the fact that Mercury lies inside the Earth's orbit. So, as Mercury revolves around the Sun, its sun-lit surface is visible in varying amounts from the Earth. This produces phases of Mercury. **No life can exist on the planet Mercury** because it is extremely hot and has no water on it. Mercury has also no atmosphere to prevent the deadly ultraviolet radiations of the Sun from reaching its surface. **Mercury has no satellite**. Mercury planet is known as *'Budh Graha'* in Hindi.

2. Venus

Venus is the second planet from the Sun (see Figure 14). ***Venus is the closest planet to the Earth.*** The rotation of Venus on its axis is somewhat unusual. This is because Venus rotates on its axis from east to west (whereas Earth rotates on its axis from west to east). Due to this, on Venus the Sun would rise in the west and set in the east. This is opposite to what happens on the Earth. Venus is a rocky planet. The planet Venus has a dense atmosphere which consists almost entirely of carbon dioxide gas. Venus is the brightest planet in the night sky. In fact, **the planet Venus is the brightest object in the night sky (except the Moon)**. So, we can easily recognise the planet Venus in the night sky by its brightness. The planet Venus appears very bright because its cloudy atmosphere reflects 75 per cent of the light which it receives from the Sun. In fact, Venus reflects more sunlight than any other planet of the Solar System and hence it appears to be the brightest planet.

Figure 14. The planet Venus.

Being quite near to the Sun, the planet Venus is very hot. The planet Venus also gets heated excessively by the trapping of Sun's heat rays by carbon dioxide gas present in its atmosphere (which is called greenhouse effect). The nearness to the Sun alongwith its heating caused by the greenhouse effect makes Venus the hottest planet of the Solar System. Please note that though Mercury is the nearest planet to the Sun, it is not the hottest planet. **The hottest planet is Venus.**

The planet Venus is also called a 'Morning Star' or an 'Evening Star'. **The planet Venus can be seen either as a Morning Star in the eastern sky just before sun-rise or as an Evening Star in the western sky just after sun-set**. Planet Venus can be seen as a 'Morning Star' or as an 'Evening Star' because it lies inside the orbit of the Earth (just like Mercury).

The planet Venus also shows phases like the Moon (see Figure 30 on page 330). This is due to the fact that Venus lies inside the Earth's orbit. So, as Venus revolves around the Sun, its sun-lit surface is presented to the Earth in varying amounts. This produces the phases of Venus. There is no evidence of life on the planet Venus. Life cannot exist on the planet Venus because it is extremely hot, it has no water and there is no sufficient oxygen in its atmosphere. **Venus has no satellite**. Please note that **Mercury and Venus are the only two planets of the Solar System which have no satellites (or moons) revolving around them**. Venus planet is called *'Shukra Graha'* in Hindi.

3. Earth

Earth is the third planet from the Sun. The two planets which lie between the Sun and the Earth are Mercury and Venus. The Earth is spherical in shape. When viewed from the outer space, the Earth appears to be a blue and green ball due to the reflection of sunlight from water and land on its surface (see Figure 15). **Earth is the only planet in the Solar System on which life is known to exist**. The two major factors which are responsible for the existence of life on Earth are : Distance of the Earth from the Sun, and Size of the Earth. The distance of Earth from the Sun is such that it has the correct temperature range for the existence of water and survival of life. The size of the Earth is such that it has sufficient gravitational field to hold on to an atmosphere of many gases (like oxygen and carbon dioxide) which are needed for the evolution and maintenance of life.

Figure 15. The planet Earth.

The various environmental conditions available on Earth which are responsible for the existence and continuation of life on Earth are as follows :

***(i)* The Earth has an atmosphere** (which contains many gases including oxygen and carbon dioxide). The Earth's atmosphere plays an important role in maintaining life on the Earth. For example, **the Earth's atmosphere has sufficient oxygen, the gas we need in order to live**. The Earth's atmosphere also supplies **carbon dioxide** needed for the preparation of food by photosynthesis by the plants.

***(ii)* The Earth has large quantities of water.** In fact, Earth is the only planet to have lots of water. This water helps in the evolution and maintenance of life on Earth. There can be no life without water.

***(iii)* The Earth has a suitable temperature range** for the existence of life. The Earth is neither too hot nor too cold.

***(iv)* The Earth has a protective blanket of ozone layer** high up in the atmosphere. This ozone layer absorbs most of the extremely harmful ultraviolet radiations coming from the Sun and prevents them from reaching the Earth and hence protects the living things on the Earth.

The Earth has two types of motion : (*i*) the Earth rotates on its axis, and (*ii*) the Earth revolves around the Sun. **The Earth rotates (or spins) on an imaginary axis which passes through its North and South Poles** (see Figure 16). **The Earth completes one rotation on its axis in 24 hours which we call one day**. The Earth rotates (or spins) on its axis from west to east. The axis of rotation of Earth is not perpendicular to the plane of the Earth's orbit. **The axis of rotation of Earth is slightly tilted with respect to the plane of its orbit (or path) around the Sun**. In fact, the axis of Earth is tilted at an angle of 23.5 degrees to the perpendicular plane (see Figure 16). The Earth rotates on its axis in the tilted position and it also revolves around the Sun in the same tilted position throughout.

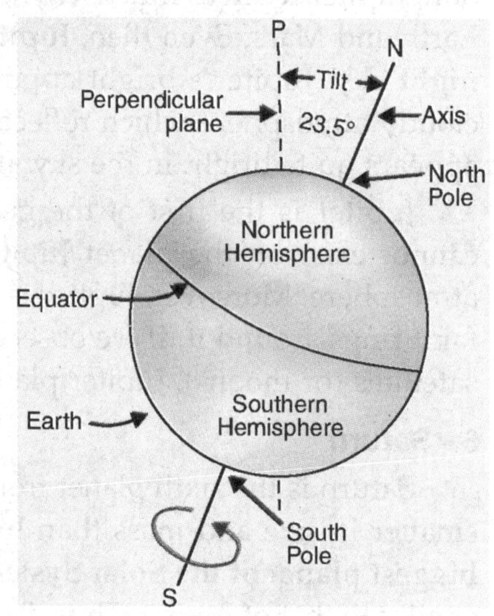

Figure 16. The Earth rotates (or spins) on a tilted axis from west to east.

An important consequence of the rotation of Earth on its axis is that it causes day and night on the Earth. The Earth rotates (or spins) on its axis and also revolves (or moves) around the Sun in an elliptical orbit. **The Earth takes 1 year to complete one revolution around the Sun**. When the Earth moves around the Sun in its orbit, it remains tilted to its orbit throughout. **An important consequence of the motion of tilted Earth around the Sun is that it causes different seasons on the Earth (such as summer, autumn, winter and spring)**. The Earth has one natural satellite called **Moon**. Earth planet is called *'Prithvi Graha'* in Hindi.

4. Mars

Mars is the fourth planet from the Sun. It is the first planet beyond Earth. **Mars is also called the red planet because its surface appears red** (see Figure 17). The red colour of Mars is used to distinguish it from other planets of the Solar System. Mars is visible from the Earth for most part of the year. Please note that Mars (and all other planets beyond Mars) do not appear as 'Morning Stars' or 'Evening Stars', and they also do not show phases like the Moon. This is because they all lie outside the Earth's orbit around Sun. Mars is a small planet having a small mass. Since the planet Mars is very far off from the Sun, so it is quite a cold planet.

Figure 17. The planet Mars.

Of all the planets, Mars is most like the Earth. Mars is a rocky planet. Mars has a thin atmosphere as compared to the Earth. **The thin atmosphere of Mars contains mainly carbon dioxide with small amounts of nitrogen, oxygen, noble gases and water vapour**. The planet Mars appears to have the right ingredients for life. The scientists think that though the atmosphere of Mars contains much less oxygen and water vapour (than that of the Earth), but it could be enough to support some primitive forms of life. Though no evidence of the existence of any form of life on Mars has been found so far but investigations are still going on. **Mars has two natural satellites.** Mars planet is called *'Mangal Graha'* in Hindi.

5. Jupiter

Jupiter is the fifth planet from the Sun (see Figure 18). **Jupiter is the biggest planet of the Solar System.** It is almost twice as large as rest of the planets put together. The mass of Jupiter is also more than the combined mass of all other planets. The diameter of Jupiter is 11 times the diameter of the Earth and its mass is about 318 times that of the Earth. Jupiter is so large that about 1300 Earths can be placed inside this giant planet. Because of its very big size, Jupiter can be seen easily in the night sky. Being very far off from the Sun, Jupiter receives much less heat and light from the Sun as compared to the Earth and Mars. Even then, **Jupiter appears to be a very bright object in the night sky**. Jupiter's bright appearance is due to the fact that it has a thick, cloudy atmosphere which reflects most of the sunlight falling on it. We can easily recognise Jupiter as it appears quite bright in the sky. Jupiter rotates very rapidly on its axis.

Figure 18. The planet Jupiter.

Jupiter is the first of the gas-type planets. **Jupiter is made mainly of hydrogen and helium**. Life cannot exist on the planet Jupiter because it has poisonous gases (like methane and ammonia) in its atmosphere. Moreover, Jupiter is a very cold planet. **Jupiter has 67 satellites (or moons)**. It has also some faint rings around it. If we observe the planet Jupiter through a telescope, we can also see four of its large satellites (or moons). Jupiter planet is called *'Brihaspati Graha'* in Hindi.

6. Saturn

Saturn is the sixth planet from the Sun (see Figure 19). Saturn is somewhat smaller in size and mass than Jupiter. But after Jupiter, **Saturn is the second biggest planet of the Solar System**. The chemical composition of Saturn is very similar to that of Jupiter. Thus, **Saturn is also made up mainly of hydrogen and helium.** *One interesting thing about Saturn is that it is the least dense among all the planets of the Solar System*. The density of Saturn is even less than that of water. Since Saturn is lighter than water, so if we imagine it in a large pool of water, then Saturn will float on water. It will not sink in water.

Figure 19. The planet Saturn.

The most distinguishing feature of Saturn is the system of colourful rings which surround it (see Figure 20). Three distinct sets of rings around Saturn are visible from the Earth. **Saturn is the only planet with a system of well-developed rings encircling it**. The rings of Saturn are made up of tiny particles, all orbiting the Saturn like miniature satellites. The rings of Saturn cannot be seen with naked eyes, they can be observed only with the help of a telescope. **The presence of a well-developed system of rings around Saturn makes it unique in the Solar System**. Being far off from the Sun, Saturn is an extremely cold planet. So, no life can exist on Saturn. **Saturn has 62 satellites (or moons)**. Planet Saturn is called *'Shani Graha'* in Hindi.

Figure 20. Colourful rings of Saturn.

7. Uranus

Uranus is the seventh planet from the Sun (see Figure 21). It can be seen only with the help of a large telescope. In fact, **Uranus was the first planet to have been discovered with the help of a telescope**. Though the diameter of Uranus is almost four times that of the Earth, it appears as a small disc through a telescope. This is because Uranus is very, very far off from the Earth. After Jupiter and Saturn, **Uranus is the third biggest planet of the Solar System**. Like Venus, Uranus also rotates on its axis from east to west. **The most remarkable feature of Uranus is that it has highly tilted axis of rotation**. As a result of the

Figure 21. The planet Uranus.

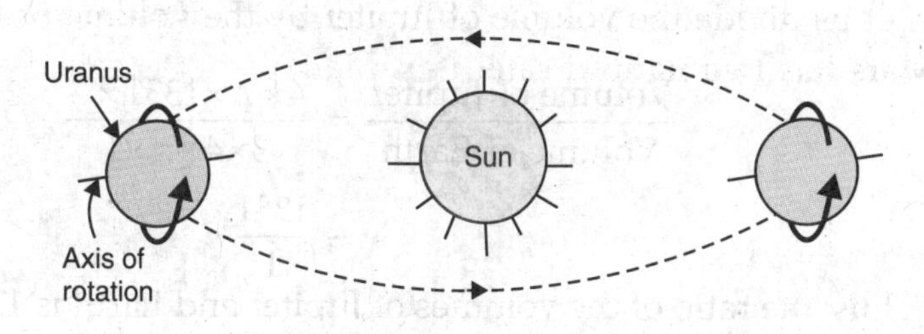

Figure 22. Due to its highly tilted axis of rotation, Uranus appears to roll on its side while orbiting around the Sun.

highly tilted axis of rotation, **Uranus appears to roll on its side while orbiting around the Sun** (see Figure 22). **Uranus is made up mainly of hydrogen and helium**. Uranus is an extremely cold planet. It is also surrounded by an atmosphere of poisonous gases. Due to these reasons, no life can exist on the planet Uranus. **The planet Uranus has 27 satellites (or moons)**. It has also some rings around it. Uranus planet is called *'Indra Graha'* in Hindi.

8. Neptune

Neptune is the eighth planet from the Sun (see Figure 23). It lies beyond Uranus. *Neptune is the outermost planet of the Solar System.* It is the most distant planet from the Sun. Thus, **the planet Neptune is farthest from the Sun**. It is also farthest from the Earth. Neptune is the second planet which was discovered with the help of a telescope. The planet Neptune can be seen as a tiny blue-green speck even by using the most powerful telescope on the Earth. Neptune is made up mainly of liquid and frozen hydrogen and helium gases. Neptune is an extremely cold planet. So, no life can exist on the planet Neptune. **Neptune has 14 satellites (or moons)**. Please note that the number of natural satellites (or moons) around the planets changes over time because the astronomers continue to observe the planets with still better and better telescopes leading to the discovery of new satellites (or moons). Neptune has also some rings around it. Neptune planet is called *'Varun Graha'* in Hindi. We will now solve a problem taken from the NCERT Science book.

Figure 23. The planet Neptune.

Sample Problem. The radius of Jupiter is 11 times the radius of the Earth. Calculate the ratio of the volumes of Jupiter and the Earth. How many Earths can Jupiter accommodate ? **(NCERT Book Question)**

Solution. By saying that the radius of Jupiter is 11 times the radius of the Earth, we mean that if the radius of the Earth is taken to be 1 (in any units), then the radius of Jupiter will be 11 (in the same units). Jupiter and Earth are sphere-shaped objects (like balls). The formula for calculating volume of a sphere is $\frac{4}{3}\pi r^3$ where π (pi) is a constant and r is the radius of the sphere. We will now calculate the volumes of Jupiter and Earth one by one and then find out the ratio of their volumes.

(*i*) Volume of Jupiter $= \frac{4}{3}\pi r^3$

$= \frac{4}{3} \times \pi \times (11)^3$ (for Jupiter, $r = 11$)

$= \frac{4}{3} \times \pi \times 1331$ ... (1)

(*ii*) Volume of Earth $= \frac{4}{3}\pi r^3$

$= \frac{4}{3} \times \pi \times (1)^3$ (for Earth, $r = 1$)

$= \frac{4}{3} \times \pi \times 1$ ... (2)

Let us divide the volume of Jupiter by the volume of Earth to obtain the ratio of their volumes. So,

$$\frac{\text{Volume of Jupiter}}{\text{Volume of Earth}} = \frac{4\times\pi\times1331\times3}{3\times4\times\pi\times1}$$

$$= \frac{1331}{1}$$

Thus, the ratio of the volumes of Jupiter and Earth is 1331 : 1. Since the volume of Jupiter is 1331 times the volume of Earth, therefore, Jupiter can accommodate 1331 Earths.

SATELLITES

Just as a celestial body which revolves around the Sun (or any other star) is called a planet, in the same way, a celestial body which revolves around a planet is called a satellite. Thus : **A satellite is a celestial body that revolves around a planet** (see Figure 24). Earth is a planet. Since the Moon revolves around the

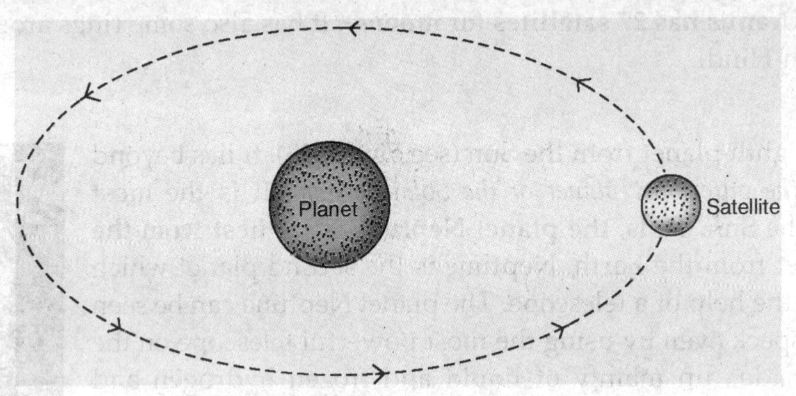

Figure 24. A satellite revolves around a planet.

Earth, therefore, **Moon is a satellite of the Earth.** Please note that the Moon is a natural satellite of the Earth. As we will learn after a while, many artificial satellites (or man-made satellites) are also revolving around the Earth these days. Please note that though we commonly call the Earth's natural satellite as Moon, the natural satellites of all other planets are also called their 'moons'.

Out of the eight planets of the Solar System, the first two planets, Mercury and Venus, do not have satellites. All the remaining six planets have one or more satellites. **The satellites revolve around the planets due to the gravitational pull of the planets**. The satellites have no light of their own. The satellites shine and become visible to us because they reflect the light of the Sun falling on them. We will now describe the Earth's natural satellite called Moon in detail.

Moon

The moon is a natural satellite of the Earth (see Figure 25). The Moon revolves around the Earth on a definite regular path—the Moon's orbit (see Figure 26). The gravitational attraction of the Earth holds the Moon in its orbit. The Earth alongwith Moon, revolves around the Sun (see Figure 26). The Moon also

Figure 25. Moon.

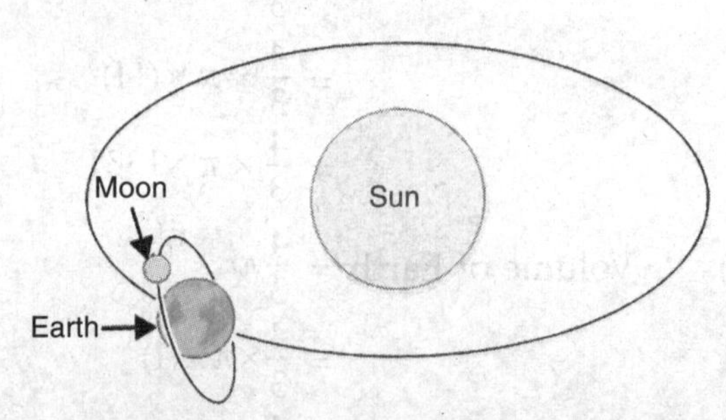

Figure 26. Moon revolves around the Earth, and the Earth alongwith Moon, revolves around the Sun.

rotates on its axis. The Moon is about one-fourth the size of the Earth in diameter but its mass is only about one-eightieth that of the Earth. The distance of Moon from the Earth is 3,84,000 km. **Moon is the closest celestial object to the Earth**. Moon appears to be much bigger than the stars because it is much more nearer to the Earth than the stars. Actually, all the planets and stars are much bigger than the Moon. Moon is a huge ball of rocks. Moon's surface is covered with hard and loose dirt. There are many craters of different

Figure 27. Surface of the Moon. Please note the craters on the Moon's surface.

Figure 28. An astronaut, Edwin Aldrin, on the Moon in 1969.

sizes on the surface of Moon (see Figure 27). The surface of Moon has also a large number of steep and high mountains. We can now say that **the surface of Moon is covered with hard and loose dirt, craters and mountains**. The Moon has no air (or atmosphere). The Moon has no water. **Since there is no air or water on the Moon, therefore, there is no life on the Moon.** On the Moon, days are extremely hot and nights are extremely cold. Man has landed on the Moon. The first man to land on Moon was an American astronaut Neil Armstrong in July 1969. He was followed by another astronaut Edwin Aldrin shortly afterwards (see Figure 28).

The Moon does not produce its own light even then we are able to see it shining in the night sky. **We are able to see the Moon because the sunlight falling on the Moon gets reflected towards the Earth** (see Figure 29). When the sunlight falls on Moon, then a part of this sunlight is reflected by the surface of Moon towards the Earth. When this reflected sunlight enters our eyes on the Earth, to us it appears as if the light is coming from the Moon itself (see Figure 29). Since the sunlight reflected by Moon enters our eyes, we are able to see the Moon. Thus, ***Moon shines with the sunlight reflected by it***. We can, however, see only that part of the Moon from which sunlight is reflected towards us on the Earth. Please note that though the Moon reflects sunlight towards us even during daytime but we cannot see the Moon during daytime due to the glare of the Sun. We can see the Moon only during night time when the Sun is not present in our sky (see Figure 29). The Moon always shows the same face (or same side) to the Earth as it rotates on its axis. This is because the Moon completes one rotation on its axis in the same time which it takes to complete one revolution around the Earth (taking into account the movement of Earth around the Sun). So, we never see the back side of the Moon from the Earth. We will now discuss the **phases of Moon.**

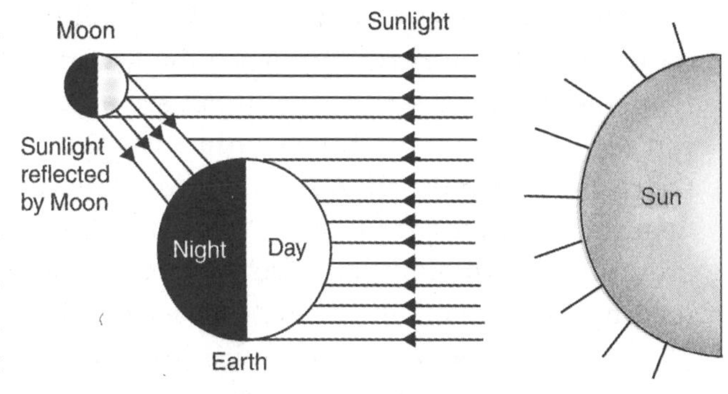

Figure 29. Moon is visible to us because it reflects sunlight towards the Earth.

If we observe the Moon continuously every night for a month, we will find that *there is one day in the month when the Moon cannot be seen in the night* (even when the sky is clear and there are no clouds, etc.). **The day (or rather night) on which the Moon is not visible at all is called the new Moon day** [see Figure 30(*a*)]. New Moon day is called *'Amavasya'* in Hindi. We have a very dark night on this day of the month (because there is no moonlight at all). On the next day, only a small, curve like portion of the Moon appears

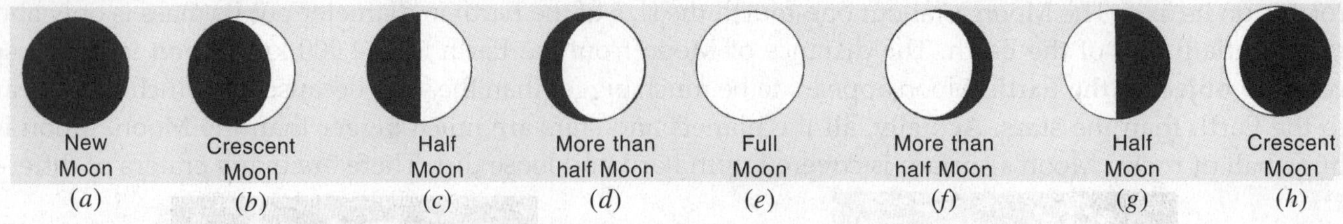

Figure 30. Some of the phases of the Moon (as seen from the Earth).

in the night sky [see Figure 30(*b*)]. This is known as the **Crescent Moon.** Thereafter, every night, the size of the bright, visible part of the Moon appears to become bigger and bigger, giving us many shapes including half Moon and more than half Moon [see Figures 30(*c*) and (*d*)]. After fifteen days (from the new Moon day), we can see the whole bright disc of the Moon in the night sky. So, *there is also one day in a month when the Moon is visible as a perfectly round ball of light in the sky*. **The day (or rather night) on which the whole bright disc of Moon is visible to us on the Earth is called the full Moon day** [see Figure 30(*e*)]. The full Moon day is called '*Purnima*' in Hindi. We have the maximum moonlight during the night on this day of the month. Thereafter, every night, the size of the bright, visible part of the Moon goes on becoming smaller and smaller [see Figures 30(*f*), (*g*) and (*h*)]. And after fifteen days (from the full Moon day), the Moon is not visible again [see Figure 30(*a*)]. We have the new Moon again. This new Moon will again change into full Moon after another fifteen days, and the process is repeated endlessly. From this discussion we conclude that the shape (or appearance) of the visible part of the Moon changes everyday over the whole month. And this change is repeated again and again, month after month. **The different shapes (or appearances) of the bright, visible part of the Moon as seen from the Earth (during a whole month) are called phases of the Moon**. Some of the phases of Moon (as seen from the Earth) are shown in Figure 30. We will now describe why Moon changes its shape everyday. In other words, **we will explain the formation of phases of the Moon.**

Moon has no light of its own. Moon shines and becomes visible to us because it reflects sunlight falling on it towards the Earth. But since the Moon revolves around the Earth and the Earth (alongwith Moon) revolves around the Sun, we cannot see all of the sun-lit surface of Moon from the Earth all the time. We can see only that part of the sun-lit surface of Moon which is towards us (on the Earth). Depending upon the relative positions of the Sun, the Moon and the Earth, we see different amounts (or portions) of the sun-lit surface of Moon from the Earth. So, **as Moon revolves around the Earth once every month and moves around the Sun alongwith Earth, different amounts of its sun-lit surface are turned towards the Earth leading to a change in the appearance of Moon and formation of phases of the Moon.**

From the above discussion it is clear that the phases of Moon occur due to its continuously changing position with respect to the Earth and the Sun. This is shown in Figure 31.

(*i*) When the Moon is on the side of Earth nearest to the Sun (see position 1 in Figure 31), then the side of Moon which is lit by Sun is away from Earth. And the side of Moon which is towards the Earth is in darkness. In this position, Moon appears to be in darkness from Earth and hence cannot be seen. This is called new Moon.

(*ii*) As the Moon moves in its orbit around the Earth from position 1 to 2, we can see a small sun-lit portion of its surface. This is called crescent Moon. As the Moon moves further from position 2 to positions 3 and 4, the sun-lit portion of Moon facing the Earth becomes bigger and bigger giving us half Moon and more than half Moon. This is called the waxing phase (increasing phase) of the Moon.

(*iii*) After fifteen days from new Moon day, the Moon reaches in position 5 which is on the side of Earth farthest from the Sun (see Figure 31). In this position, the whole sun-lit side of the Moon is towards the Earth, and we see the Moon as a full round disc of bright light. This is called full Moon.

(*iv*) As the Moon moves around the Earth further from position 5 to positions 6, 7 and 8, the sun-lit portion of Moon facing the Earth becomes smaller and smaller. This is called the waning phase (decreasing

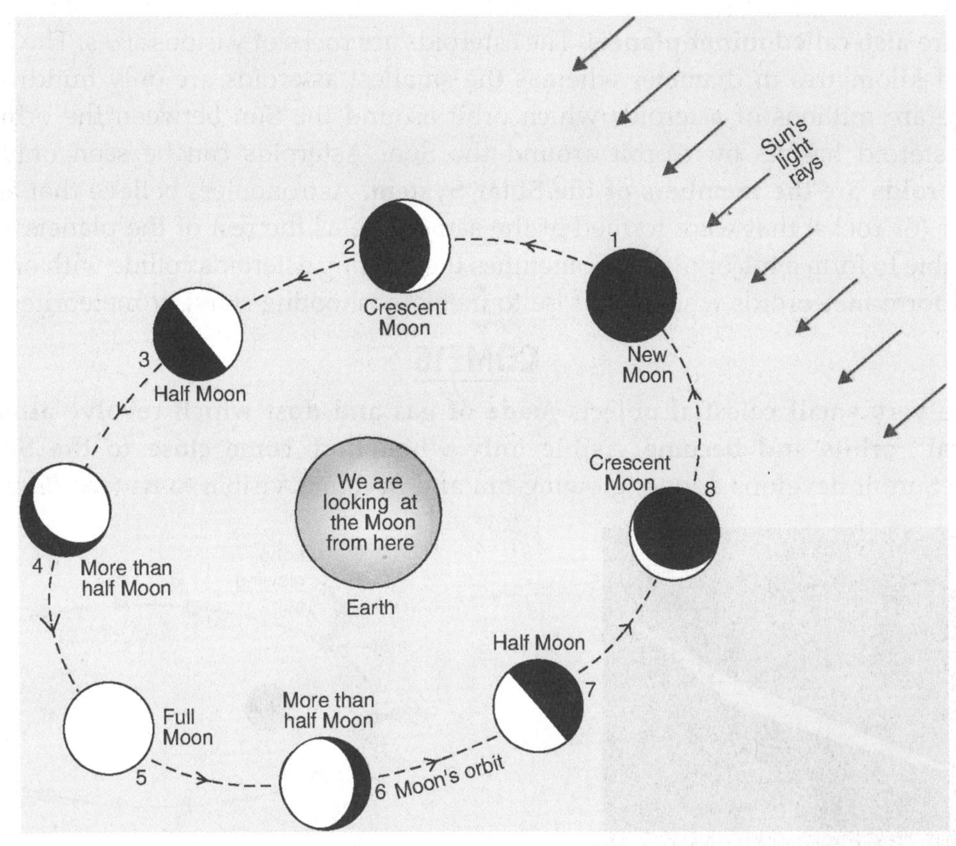

Figure 31. Diagram to show the positions of Moon in its orbit and its corresponding phases.

phase) of the Moon. And ultimately Moon completes the revolution around the Earth and again reaches position 1. So, we have new Moon once again.

Please note that after the new Moon day, the size of sun-lit part of Moon visible from the Earth *increases* everyday. On the other hand, after the full Moon day, the size of sun-lit part of Moon visible from the Earth *decreases* every day. **We can have one 'new Moon' and one 'full Moon' during a month** (which is the time taken by Moon to complete one revolution around the Earth). The time period between one full Moon and the next full Moon is actually 29½ days. In the moon-based calendars (called lunar calendars) this period of 29½ days is taken as a month.

Some Other Members of the Solar System

In addition to the Sun, planets and satellites, the Solar System also includes asteroids, comets and meteoroids. Asteroids, Comets and Meteoroids revolve around the Sun just like the planets but they are much smaller than the planets. The motion of asteroids, comets and meteoroids is also governed by the gravitational force of the Sun. So, asteroids, comets and meteoroids are considered to be the members of the Solar System. We will now describe all these members of the Solar System in detail. Let us start with asteroids.

ASTEROIDS

There is a wide space (or gap) in-between the orbits of planets Mars and Jupiter. A large number of small objects made of rocks revolve around the Sun in the wide space between the orbits of the planets Mars and Jupiter in the Solar System. These are called asteroids. Thus, **asteroids are small celestial objects which revolve around the Sun between the orbits of Mars and Jupiter** (see Figure 32). Asteroids are smaller than the planets.

Figure 32. Asteroids.

Asteroids are also called minor planets. The asteroids are rocks of various sizes. The biggest asteroids are hundreds of kilometres in diameter whereas the smallest asteroids are only hundreds of metres in diameter. There are millions of asteroids which orbit around the Sun between the orbits of Mars and Jupiter. Each asteroid has its own orbit around the Sun. Asteroids can be seen only through large telescopes. **Asteroids are the members of the Solar System.** Astronomers believe that asteroids are the pieces of matter (or rocks) that were formed at the same time as the rest of the planets which somehow could not assemble to form a major planet. Sometimes the orbiting asteroids collide with one another, break into pieces and form meteoroids which give rise to meteors (shooting stars) or meteorites.

COMETS

Comets are very small celestial objects made of gas and dust which revolve around the Sun in highly elliptical orbits and become visible only when they come close to the Sun. As a comet approaches the Sun, it develops a long, glowing tail and becomes visible to us (see Figure 33). A comet

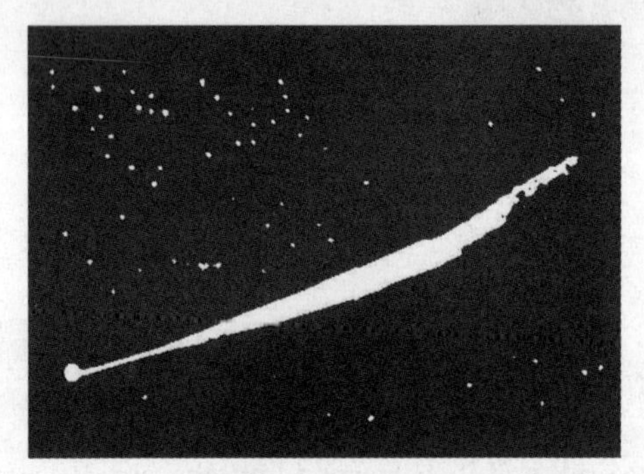

Figure 33. A comet in the night sky.

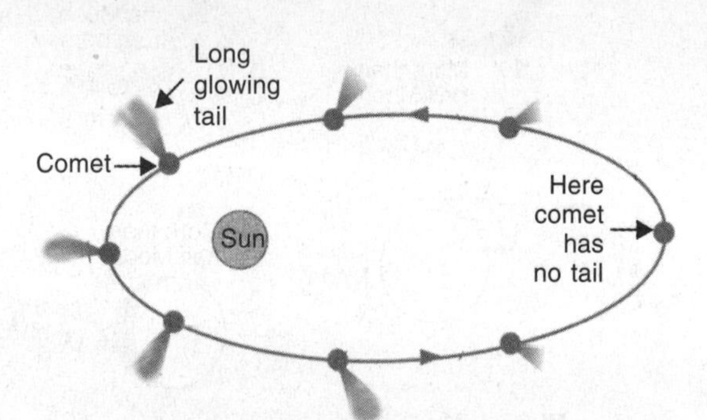

Figure 34. A comet in orbit around the Sun.

becomes visible on coming close to the Sun because the Sun's rays make its gas glow which then spreads to form a tail millions of kilometres long. The head of a comet is small and appears as a bright ball of light but its tail is very, very long. Thus, ***a comet is a collection of gas and dust which appears as a bright ball of light in the sky with a long glowing tail when it comes close to the Sun.*** The tail of a comet always points away from the Sun (see Figure 34). The tail of a comet grows longer as it comes nearer the Sun but it disappears when the comet moves far away from the Sun (see Figure 34). **Comets are smaller than asteroids**. Comets are distinguished from asteroids by the presence of a long glowing tail. Asteroids do not have any glowing tail.

Comets are also the members of our Solar System. Comets revolve around the Sun just like planets. Each comet follows its own orbit around the Sun. The orbits of comets are very, very large and highly elliptical, due to which they remain very far off from the Sun most of the time and can be seen very rarely. **The period of revolution of a comet around the Sun is very, very large**. For example, **Halley's comet has a period of revolution of about 76 years**. That is, Halley's comet is seen after every 76 years. Halley's comet was last seen in 1986. Comets do not last for ever. Each time a comet passes near the Sun, it loses some of its gas and ultimately only the dust particles are left in space. When these dust particles enter into the Earth's atmosphere, they burn up due to heat produced by air resistance and produce meteors (or shooting stars).

Some people think that comets are the messengers of disasters such as wars, epidemics (widespread diseases) and floods, etc. These are, however, myths and superstitions (false beliefs). The appearance of a comet is a natural phenomenon. It has nothing to do with natural or man-made disasters. We should not be afraid of sighting the comets.

METEOROIDS : METEORS AND METEORITES

Meteoroids are celestial objects which range in size from tiny sand grains to big boulders of several hundred tonnes and revolve around the Sun in their orbits. Meteoroids are mainly left behind by the disintegration of asteroids and comets. **Meteoroids are much smaller than asteroids and comets.** Meteoroids are present throughout the Solar System. **Meteoroids are members of the Solar System** because they revolve around the Sun. When (due to some reason), a meteoroid breaks away from its orbit around the Sun and falls towards the Earth, it becomes a meteor. We can now say that *meteoroid is a small body in the Solar System that would become a meteor if it entered the Earth's atmosphere.* It is the visible path of a burning meteoroid that enters the Earth's atmosphere which is called a meteor. We will now describe the meteors in detail.

When the sky is clear and the Moon is not there, many times we see a streak of light in the sky during night which disappears within seconds. It is called a meteor or shooting star. **Meteors are the celestial bodies from the sky which we see as a bright streak of light that flashes for a moment across the sky** (Figure 35). **The meteors are commonly called shooting stars.** When a meteor enters into the atmosphere of Earth with high speed, a lot of heat is produced due to the resistance of air. This heat burns the meteor and the burning meteor is seen in the form of a streak of light shooting down the sky, and it falls on the Earth in the form of dust. A meteor (or shooting star) lasts for a very short time because the tiny rocky pieces (of which it is made), burn and vaporise completely in a few seconds due to the excessive heat produced by atmospheric friction. When the Earth crosses the tail of a comet, swarms of meteors (large number of meteors) are seen. These are known as meteor showers. Some meteor showers occur at regular intervals each year.

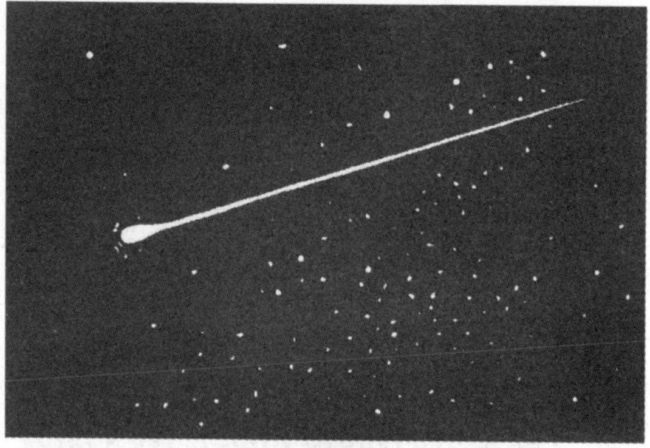

Figure 35. A meteor (shooting star) in the night sky.

A meteor is commonly known as a shooting star though it is not really a star. **A meteor is called a shooting star because, viewed from the Earth, it looks like a streak of starlight shooting across the night sky**. The main difference between a star and a shooting star is that **a star has its own light but a shooting star has no light of its own.** The light of a shooting star is produced when its particles burn on entering the Earth's atmosphere.

Most of the meteors are small and burn up completely on entering the Earth's atmosphere. They fall on the Earth in the form of dust. If, however, a meteor is big, then a part of it may reach the Earth's surface without being burned up in the air. **A meteor which does not burn up completely on entering the Earth's atmosphere and lands on Earth, is known as a meteorite.** Meteorites are sort of stones falling from the sky. The meteorites falling on the Earth from the sky range in size from small pebbles to big blocks many tonnes in weight. More than 3000 meteorites, weighing around a kilogram or more, fall on the Earth each year in different parts of the world (see Figure 36). Meteorites are made of rock (or metal). Meteorites are believed to be the pieces of asteroids or comets which somehow strayed away from their orbit around the Sun and fell to Earth. They are thought to have been formed at the same time as the planets of the Solar System. So, by studying the composition of meteorites, the scientists can get valuable information about the nature of the material of which the various planets (including the Earth) are made.

Figure 36. This meteorite fell in Australia in 1969.

ARTIFICIAL SATELLITES

A man-made space-craft placed in orbit around the Earth is called an artificial satellite. The artificial satellites are also known as man-made satellites. An artificial satellite is shown in Figure 37. An artificial satellite is placed in orbit around the Earth with the help of a launch vehicle called rocket. Rocket carries the artificial satellite from the Earth to a height of a few hundred kilometres (or a few thousand kilometres) above it and gives it a sideways push. This makes the satellite move in an orbit around the Earth. **The motion of artificial satellite around the Earth is maintained by the gravitational pull of the Earth**. In this way, the artificial satellite keeps on revolving around the Earth continuously, without stopping. The artificial satellites revolve around the Earth just like its natural satellite Moon. The main differences between the artificial satellites of the Earth and its natural satellite Moon are as follows :

(*i*) The artificial satellites are much nearer to the Earth than its natural satellite Moon.

(*ii*) The height (or distance) of the artificial satellites from the Earth (and hence their speed around the Earth) can be adjusted according to our needs. This is, however, not possible with the natural satellite of Earth called Moon.

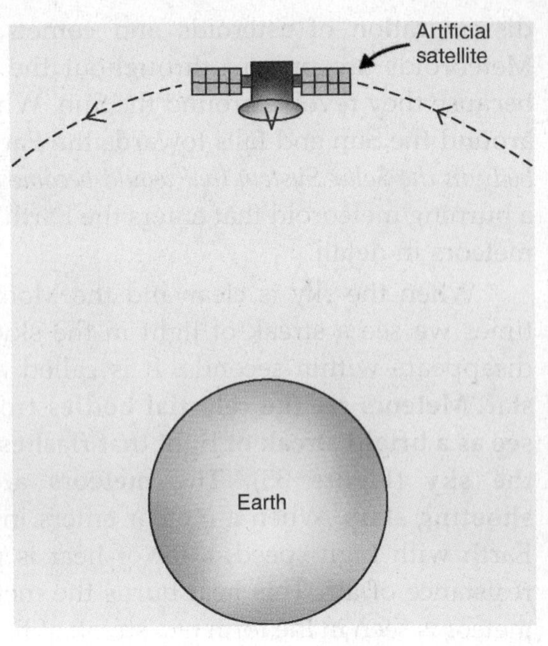

Figure 37. An artificial satellite in orbit around the Earth.

The artificial satellites carry a large variety of equipment and instruments inside them. For example, they may carry sound and picture relaying machines, cameras, infra-red sensors, telescopes and many other type of instruments required for doing different jobs. **The electricity required for running the equipment in an artificial satellite is provided by solar cells.** The solar cells convert sunlight into electricity.

Before we give the various applications (or uses) of artificial satellites, we should know the meaning of the term 'remote sensing'. **The technique of collecting information about an object from a distance (without making a physical contact with the object) is called remote sensing.** Remote sensing satellites can scan the Earth's surface very closely with their cameras and infra-red sensors, even while orbiting high above the Earth. They can see the details on the Earth even up to a fraction of a metre. We will now describe the important applications (or uses) of artificial satellites.

Artificial satellites have many practical applications (or uses) which help our lives in many ways. **The important applications (or uses) of artificial satellites are given below :**

1. **Artificial satellites are used for communications such as long distance transmission of television programmes, radio programmes, telephone calls and internet.** The artificial satellites used for communications purposes are called Communications Satellites.

2. **Artificial satellites are used for weather forecasting (such as rain-fall, snow-fall, etc.) and for giving advance warning of floods and cyclones, etc.** Weather forecasting is done by using artificial satellites called Weather Satellites (or Meteorological Satellites) which are a kind of remote sensing satellites.

3. **Artificial satellites are used for surveying the natural resources of the Earth like minerals, agricultural crops and potential fishing zones in the sea, etc.** This is done by using the artificial satellites called remote sensing satellites.

4. **Artificial satellites are used for spying for military purposes (like observing the movement of enemy troops and military equipment, taking pictures of enemy air-fields and harbours, etc).** Remote sensing satellites are used for spying for military purposes.

5. **Artificial satellites are used to collect information about other planets, stars and galaxies, etc.** The artificial satellites used for studying celestial objects are called 'Astronomy Satellites'.

India has built and launched many artificial satellites. The first artificial satellite launched by India was 'Aryabhatta' (see Figure 38). It was launched in 1975. Some other Indian satellites are : Bhaskara, INSAT, IRS, Rohini, Kalpana-1 and EDUSAT (INSAT stands for Indian National SATellite, IRS stands for Indian Remote-sensing Satellite whereas EDUSAT stands for EDUcation SATellite). The Agency responsible for the development of space science programmes in our country is 'Indian Space Research Organisation' (ISRO). We are now in a position to **answer the following questions :**

Figure 38. Aryabhatta.

Very Short Answer Type Questions

1. Name any two celestial objects which we can see easily in the night sky.
2. Name the star (after the Sun) which is closest to the Earth.
3. Name the star which remains fixed at the same place in the sky in the North.
4. Name the unit which is used to express distances between the various celestial bodies (like stars and planets).
5. Why is the distance between stars and planets expressed in light years ?
6. What do you understand by the statement that a star is 8 light years away from the Earth ?
7. Name the constellation which reminds us of a large laddle or a question mark in the night sky.
8. In which season of the year is the constellation Orion visible in the sky ?
9. In which season of the year is the constellation Ursa Major visible in the sky ?
10. Give two other names of Ursa Major constellation.
11. In which season of the year are the following constellations visible in the night sky ?
    (*a*) Cassiopeia (*b*) Leo Major
12. Name the biggest planet of the Solar System.
13. Name the smallest planet of the Solar System.
14. Which force keeps the members of the Solar System bound to the Sun ?
15. Why does the Sun appear to be bigger and brighter than all other stars ?
16. Which is the main source of heat and light energy for all the members of the Solar System (like planets and satellites, etc.) ?
17. Name a star having a system of planets with life on one of its planets.
18. Name the planet having a well-developed system of rings around it.
19. Name the planets which lie between the Sun and the Earth.
20. How will our Earth look when seen from a space-ship or Moon ?
21. Name the planet (*a*) nearest to the Sun, and (*b*) farthest from the Sun.
22. Name two planets which have been discovered with the help of telescope.
23. Name the hottest planet.
24. Name two planets which show phases like the Moon.
25. Name one planet of the Solar System having life on it.
26. Which characteristics of Mars distinguishes it from other planets ?
27. Which planet is known as the red planet ?
28. In which part of the sky can you find Venus if it is visible as an Evening Star ?
29. Which characteristic of the planet Saturn makes it unique in the Solar System ?
30. Name two planets which can be seen as 'Morning Star' and 'Evening Star'.
31. What name is given to the celestial body which revolves around a planet ?
32. Which two planets have asteroids between them ?
33. Name two objects other than planets which are members of the Solar System ?
34. Name the member of the Solar System which appears in the sky like a bright ball of light with a long, glowing tail.

35. Which celestial body is seen as a bright streak of light coming down the night sky ?
36. Which celestial objects are also called minor planets ?
37. Name one natural and one artificial satellite of the Earth.
38. Name the agency responsible for the development of space science programmes in India.
39. Name the first artificial satellite launched by India.
40. Name the technique of collecting information about the Earth from an orbiting satellite.
41. Write the full name of INSAT.
42. Write the full name of IRS.
43. Name two constellations which are visible in the sky :
    (*a*) in the summer season.
    (*b*) in the winter season.
44. State whether the following statements are true or false :
    (*a*) Constellation Orion can be seen only with a telescope.
    (*b*) Pole Star is a member of the Solar System.
    (*c*) Mercury is the smallest planet of the Solar System.
    (*d*) Uranus is the farthest planet in the Solar System.
    (*e*) There are nine planets in the Solar System.
    (*f*) Comets are members of the Solar System.
    (*g*) INSAT is an artificial satellite.
45. Fill in the following blanks with suitable words :
    (*a*) The stars appear to...............in the sky.
    (*b*) The Sun is a.............whereas Orion is a ....
    (*c*) The group of stars that appears to form a recognisable pattern in the sky is known as............
    (*d*) The brightest star in the night sky is..............
    (*e*) Ursa Major constellation appears to revolve around the ....star in the night sky.
    (*f*) Orion constellation can be used to locate the position of..............star whereas Ursa Major constellation can be used to locate the ....star in the night sky.
    (*g*) The planet which is farthest from the Sun is...........
    (*h*) The planet which appears reddish in colour is............
    (*i*) The small heavenly bodies revolving around the Sun between the orbits of Mars and Jupiter are called ....
    (*j*) Asteroids are found between the orbits of...........and...........
    (*k*) Shooting stars are actually not.............
    (*l*) A celestial body that revolves around a planet is known as........
    (*m*) A meteoroid becomes a...........on entering Earth's atmosphere.
    (*n*) The long distance transmission of television programmes has been made possible with the help of .... satellites.

Short Answer Type Questions

46. What is meant by 'celestial objects' ? Name any three celestial objects.
47. What is a star ? Name the star nearest to the Earth.
48. Why does Pole Star appear to be stationary in the sky ?
49. Do all the stars in the sky move ? Explain.
50. What is a constellation ? Name any two constellations.
51. (*a*) How much time does light take to reach us from the Sun ?
    (*b*) How much time does light take to reach us from the next nearest star 'Proxima Centauri' ?
52. What is Solar System ? Name the different types of celestial objects which are members of the Solar System.
53. What are planets ? How many planets are there in the Solar System ?
54. Name all the planets of the Solar System in the order of their increasing distances from the Sun.
55. What are inner planets and outer planets ? Name them.
56. Why is it difficult to observe the planet Mercury ?
57. State one way in which the planets can be distinguished from the stars in the night sky.

58. What are the various environmental conditions available on Earth which are responsible for the existence and continuation of life on Earth ?
59. State one important consequence of each of the following :
(a) Rotation of Earth on its axis.
(b) Motion of tilted Earth around the Sun.
60. What is a satellite ? Name the natural satellite of the Earth.
61. Moon does not have light of its own. How are we able to see the Moon ?
62. Name the constellation which appears to have the shape of :
(a) a big bear (b) a distorted W or M (c) a hunter (d) a big lion
63. What are asteroids ? Where are they located ?
64. State two differences between the artificial satellites of the Earth and its natural satellite Moon.
65. What is an artificial satellite ? Name any two artificial satellites launched by our country.
66. State five uses of artificial satellites.
67. Define light year. How many kilometres make 1 light year ?
68. What is a comet ? Name the comet which was last seen in 1986 after a period of 76 years.
69. When does a comet become visible to us ?
70. What happens to the tail of a comet when it moves far away from the Sun ?
71. What is a meteor ? What is the other name of a meteor ?
72. What is the difference between a star and a shooting star ?
73. What are meteorites ?
74. What is the difference between a meteor and a meteorite ?
75. What are meteoroids ? Which of two is really a member of the Solar System : Meteoroid or Meteor ?

Long Answer Type Questions

76. (a) What is the number of prominent stars in the Ursa Major ?
(b) Draw a diagram of Ursa Major constellation to show the position of main stars in it.
77. (a) What is the number of prominent stars in the Orion ?
(b) Draw a diagram of the Orion constellation to show the position of prominent stars in it.
78. (a) What is the number of main stars in the Cassiopeia constellation ? Draw a diagram of Cassiopeia constellation to show the position of main stars in it.
(b) What is the number of main stars in Leo Major constellation ? Draw a diagram to show the position of main stars in Leo Major constellation.
79. (a) Explain how you can locate the Pole Star with the help of Ursa Major constellation.
(b) Explain how the position of Sirius Star can be located in the night sky with the help of Orion constellation.
80. (a) What is the difference between a full Moon and a new Moon ? After how many days a full Moon changes into a new Moon ?
(b) What is meant by the phases of the Moon ? What causes the phases of the Moon ?

Multiple Choice Questions (MCQs)

81. Which of the following is not a planet of the Sun ?
(a) Sirius (b) Mercury (c) Saturn (d) Earth
82. Ursa Major is a :
(a) star (b) planet (c) constellation (d) satellite
83. Which of the following is not a member of the Solar System ?
(a) Asteroids (b) Morning Star (c) Satellites (d) Constellations
84. Which of the following is not a planet ?
(a) Mercury (b) Saturn (c) Mars (d) Moon
85. Phases of the Moon occur because :
(a) we can see only that part of the Moon which reflects light towards us.
(b) our distance from the Moon keeps changing.
(c) the shadow of the Earth covers only a part of the Moon's surface.
(d) the thickness of the Moon's atmosphere is not constant.

86. Which of the following is not a member of the Solar System ?
(*a*) an asteroid (*b*) a satellite (*c*) a constellation (*d*) earth

87. After the Sun, the next nearest star to the Earth is :
(*a*) Alpha Centauri (*b*) Pole star (*c*) Sirius (*d*) Proxima Centauri

88. The distances between the various celestial objects are usually expressed in the unit of :
(*a*) kilometres (*b*) light minutes (*c*) light years (*d*) light seconds

89. Which of the following constellations can be seen in the night sky during the winter season ?
A. Orion B. Ursa Major C. Leo Major D. Cassiopeia
(*a*) A and B (*b*) B and C (*c*) A and D (*d*) B and D

90. The brightest star in the night sky called Sirius is located close to one of the following constellations. This constellation is :
(*a*) Great Bear (*b*) Leo Major (*c*) Cassiopeia (*d*) Orion

91. Asteroids revolve around the Sun between the orbits of the planets :
(*a*) Mars and Earth (*b*) Venus and Mars (*c*) Jupiter and Mars (*d*) Mars and Saturn

92. The biggest planet of the solar system is :
(*a*) Mars (*b*) Saturn (*c*) Mercury (*d*) Jupiter

93. The planet with a system of well developed rings encircling it is :
(*a*) Jupiter (*b*) Venus (*c*) Saturn (*d*) Neptune

94. The two pointer stars the line passing through which points to the direction of pole star are a part of the constellation called :
(*a*) Ursa Major (*b*) Orion (*c*) Cassiopeia (*d*) Leo Major

95. The smallest planet of the solar system is :
(*a*) Earth (*b*) Mars (*c*) Mercury (*d*) Saturn

96. Which of the following are non-luminous objects ?
A. Orion B. Morning Star C. Moon D. Pole star
(*a*) A and B (*b*) B and C (*c*) B and D (*d*) only C

97. Which of the following planets show phases like the moon ?
A. Venus B. Mercury C. Jupiter D. Mars
(*a*) A and B (*b*) B and C (*c*) A and C (*d*) C and D

98. Which of the following is the hottest planet in the Solar System ?
(*a*) Earth (*b*) Mercury (*c*) Venus (*d*) Mars

99. Which of the following constellations are visible in the summer season ?
A. Cassiopeia B. Orion C. Leo Major D. Big Bear
(*a*) A and B (*b*) B and C (*c*) A and C (*d*) C and D

100. The agency responsible for the development of space science programmes in India is :
(*a*) INSAT (*b*) ORS (*c*) IRS (*d*) ISRO

Questions Based on High Order Thinking Skills (HOTS)

101. X is a group of stars which is visible during the summer season in the early part of the night. It can be seen clearly in the month of April in the northern part of the sky. It resembles a bowl with a handle. It also resembles a big kite with a tail.
(*a*) What is the general name of groups of stars like X ?
(*b*) Write any two names of X.
(*c*) Is it a part of our Solar System ?
(*d*) How many bright stars are usually observed in X ?
(*e*) Which famous star can be located in the sky with the help of X ?

102. (*a*) In which direction do stars appear to move in the sky ?
(*b*) Why do they appear to move in this direction ?

103. The number of main stars in constellation *A* is 5, in constellation *B* is 7, in constellation *C* can be 7 or 8, whereas in constellation *D* is usually 9. Name the constellations *A*, *B*, *C* and *D*.

104. Which star in the night sky can be located with the help of :
(*a*) Orion constellation ?
(*b*) Ursa Major constellation ?

105. Match items in column *A* with one or more items in column *B* :

A	*B*
(*i*) Inner planets	(*a*) Saturn
(*ii*) Outer planets	(*b*) Pole Star
(*iii*) Constellation	(*c*) Great Bear
(*iv*) Satellite of the Earth	(*d*) Moon
	(*e*) Earth
	(*f*) Orion
	(*g*) Mars

ANSWERS

44. (*a*) False (*b*) False (*c*) True (*d*) False (*e*) False (*f*) True (*g*) True **45.** (*a*) twinkle (*b*) star ; constellation (*c*) constellation (*d*) Sirius (*e*) Pole (*f*) Sirius ; Pole (*g*) Neptune (*h*) Mars (*i*) asteroids (*j*) Mars ; Jupiter (*k*) stars (*l*) satellite (*m*) meteor (*n*) artificial **81.** (*a*) **82.** (*c*) **83.** (*d*) **84.** (*d*) **85.** (*a*) **86.** (*c*) **87.** (*d*) **88.** (*c*) **89.** (*c*) **90.** (*d*) **91.** (*c*) **92.** (*d*) **93.** (*c*) **94.** (*a*) **95.** (*c*) **96.** (*b*) **97.** (*a*) **98.** (*c*) **99.** (*d*) **100.** (*d*) **101.** (*a*) Constellations (*b*) Ursa Major ; Great Bear (*c*) No (*d*) Seven (*e*) Pole star **102.** (*a*) The stars appear to move in the sky from east to west direction (*b*) This is due to the rotation of earth from west to east direction on its axis **103.** A : Cassiopeia ; B : Ursa Major ; C : Orion ; D : Leo Major **104.** (*a*) Sirius (*b*) Pole star **105.** (*i*) *e* ; *g* (*ii*) *a* (*iii*) *c* ; *f* (*iv*) *d*

CHAPTER

18

Pollution of Air and Water

The presence of unusually high concentrations of harmful or poisonous substances in the environment (air, water, etc.) is called pollution. Pollution contaminates the air and water with poisonous substances and makes them impure to such an extent that they become harmful to the human beings, other animals, plants as well as to the non-living things. **An unwanted and harmful substance that contaminates the environment (such as air and water) is called a pollutant.** In most simple words, *a substance that causes pollution is called a pollutant.* The air and water which contain pollutants at levels harmful to humans, other animals, plants and non-living things are said to be polluted. Air and water are both essential for our survival. For example, air is necessary for breathing. We cannot live without air even for a few minutes. Similarly, we cannot live without water for more than a few days (though we can live much longer without food).

Air is a mixture of gases (which is present all around us). **The two main gases present in air are nitrogen and oxygen.** Nitrogen makes up about 78 per cent of air whereas oxygen makes up about 21 per cent of air by volume. **Air also contains small amounts of carbon dioxide, argon and water vapour, etc.** *The mixture of gases containing nitrogen, oxygen, carbon dioxide, argon and water vapour gives us pure air (or clean air)* which is good for us. ***In addition to the normal constituents, the polluted air may also contain harmful substances such as carbon monoxide, sulphur dioxide, nitrogen oxides, smoke and dust, etc.*** When air around us is polluted, we are forced to breathe it to remain alive (though it is harmful to us). Many respiratory problems are caused by breathing in polluted air. In fact, the respiratory problems among children are increasing day by day due to increasing air pollution. Water is the most common liquid which is present around us. Fresh water (usable water) is present in rivers, lakes and ponds. Some fresh water is also present under the ground. Due to increasing water pollution, water-borne diseases are also increasing day by day. In this Chapter we will discuss how air and water get polluted; what are the harmful effects of air and water pollution; and how air and water pollution can be controlled (or minimised).

AIR POLLUTION

The air over large cities is heavily contaminated with harmful gases like sulphur dioxide, nitrogen oxides, carbon monoxide, smoke and dust, etc. **The contamination of air with harmful gases (like sulphur**

dioxide, nitrogen oxides, carbon monoxide), smoke and dust, etc., is called air pollution. The substance whose presence in air makes it impure or contaminated is called an **air pollutant.** *A substance becomes an air pollutant when it is present in air in such concentration which is high enough to have a harmful effect on the living or non-living things.* **The major pollutants which cause air pollution are : Sulphur dioxide, Nitrogen oxides, Carbon monoxide, excess of Carbon dioxide, Chlorofluorocarbons, and Suspended particulate matter** (such as **Dust, Smoke** and **Fly ash**).

Sources of Air Pollution

Most of the air pollution is caused by the burning of fuels such as wood, cow-dung cakes, coal, kerosene, petrol and diesel in homes, motor vehicles (automobiles), factories and thermal power plants, etc. The various sources of air pollution are given below :

(*i*) Smoke emitted from homes by the burning of fuels like wood, cow-dung cakes, kerosene and coal causes air pollution.

(*ii*) Exhaust gases emitted by motor vehicles (automobiles) due to burning of petrol and diesel cause air pollution (see Figure 1). *Motor vehicles are the major cause of air pollution in big cities.*

(*iii*) Smoke emitted by factories and thermal power plants due to burning of coal causes air pollution.

(*iv*) Oil refineries and industries engaged in the production of metals and manufacture of chemicals cause air pollution.

(*v*) Stone crushers, cement factories, asbestos factories and lead processing units cause air pollution.

(*vi*) Use of chlorofluorocarbons in refrigeration, air conditioning and aerosol sprays causes air pollution.

(*vii*) Smoking causes air pollution.

Figure 1. Exhaust gases emitted by motor vehicles cause air pollution. Motor vehicles produce pollutants like *carbon monoxide, carbon dioxide, sulphur dioxide, nitrogen oxides and smoke, etc.*

All the above sources of air pollution are **man-made sources** of air pollution in which pollutants are added to air by various human activities. **Forest fires and volcanic eruptions are the two natural sources of air pollution** which put smoke and dust into the air.

Harmful Effects of Air Pollution

Air pollution produces a large number of bad effects on living and non-living things. Air pollution can cause health problems in human beings. It can kill animals and plants. It can also damage the environment and property (buildings, etc.) We will now give the names, sources and harmful effects of various air pollutants.

(*i*) SULPHUR DIOXIDE. Sulphur dioxide is produced by the **burning of coal** in factories and thermal power plants. Sulphur dioxide is also produced by the **burning of petrol and diesel** in motor vehicles. Actually, the fuels such as coal, petrol and diesel contain some sulphur as impurity which burns to produce sulphur dioxide gas. Oil refineries also emit sulphur dioxide gas into air. **Sulphur dioxide gas in the polluted air causes respiratory problems. If may even cause permanent lung damage. Sulphur dioxide gas in polluted air produces acid rain.** This acid rain damages trees, plants, soil, aquatic animals (like fish), statues, buildings and historical monuments. Sulphur dioxide also contributes to the formation of a deadly air pollutant called **smog.**

(*ii*) NITROGEN OXIDES. Nitrogen oxides are produced by the **burning of fuels like petrol and diesel** in motor vehicles. They are also produced by the **burning of coal** in factories and thermal power plants. Actually, the high temperature produced by the burning of fuels like petrol, diesel and coal makes some of the nitrogen and oxygen of air to combine to form nitrogen oxides. Oil refineries also produce and emit nitrogen oxides into the air. **Nitrogen oxides attack breathing system and lead to lung congestion. They also attack skin.** Just like sulphur dioxide, **nitrogen oxides present in polluted air produce acid rain.** Nitrogen oxides also contribute to the formation of **smog.**

Smog is a deadly air pollutant which is formed by the combination of smoke and fog (The minute water particles suspended in air near the surface of earth during cold weather in winter, is called fog). Smoke contains tiny carbon particles, and harmful gases such as nitrogen oxides and sulphur dioxide, etc. The carbon particles, nitrogen oxides and sulphur dioxide, etc., of smoke combine with the condensed water vapour called fog to form 'smog'. **Smog causes cough and aggravates (makes worse) asthma and other lung diseases, especially in children.**

***(iii)* CARBON MONOXIDE.** Carbon monoxide is produced by the **incomplete combustion of fuels** like wood, coal, kerosene, petrol and diesel in homes, factories and motor vehicles. The **exhaust gases of motor vehicles** (cars, buses and trucks, etc.) contain carbon monoxide which they emit into air (Incomplete combustion of fuels which produces carbon monoxide, takes place in insufficient supply of air). Cigarette smoke also contains carbon monoxide. Carbon monoxide is a very *poisonous* gas. When inhaled, **carbon monoxide combines with the haemoglobin of our blood and reduces the oxygen-carrying capacity of blood.** Due to this, blood is not able to carry sufficient oxygen to our body parts. **This lack of oxygen causes respiratory problems (breathing problems). It causes suffocation.** If too much carbon monoxide is inhaled, it may even cause death.

***(iv)* CARBON DIOXIDE.** Carbon dioxide is produced in excessive amounts by the **burning of large quantities of fuels** such as wood, coal, kerosene, petrol, diesel, LPG, and CNG in homes, factories and motor vehicles. **Though carbon dioxide is a normal constituent of air but excess of carbon dioxide in air is considered a pollutant (because it produces undesirable changes in the environment).** Carbon dioxide is the main greenhouse gas which traps sun's heat in the earth's atmosphere by producing greenhouse effect which leads to global warming. We will study this in detail after a while.

***(v)* CHLORO-FLUORO-CARBONS (CFCs).** Chlorofluorocarbons are the chemical compounds made of **chlorine, fluorine and carbon** elements. They are commonly known as CFCs. Chlorofluorocarbons are used in **refrigeration, air conditioning and aerosol sprays.** So, all these sources release chlorofluorocarbons into the air. *Chlorofluorocarbons are industrially useful gases but they also behave as air pollutants because of their damaging effect on ozone layer* (which exists high up in the atmosphere or upper atmosphere). **Chlorofluorocarbons are depleting the useful ozone layer of the upper atmosphere.** This happens as follows : Chlorofluorocarbons released into the air go up and ultimately reach high into the atmosphere where the protective ozone layer exists. The chlorofluorocarbons react with the ozone gas of ozone layer and destroy it gradually. **Ozone layer prevents the harmful ultraviolet radiations of the sun from reaching the earth.** So, ozone layer protects us from the harmful effects of the ultraviolet radiations of the sun. **The destruction of ozone layer by CFCs will allow the extremely harmful ultraviolet radiations of the sun to reach the earth. These ultraviolet radiations can cause skin cancer, cataract, and destruction of plants, including crops.** In fact, a big hole has already been made by the destruction of ozone gas in the ozone layer over the South Pole of the earth. It is called "ozone hole". The good news is that less harmful chemicals are now being used in place of CFCs.

***(vi)* SUSPENDED PARTICULATE MATTER (SPM).** The finely divided solid or liquid particles suspended in air are called suspended particulate matter. **Some of the examples of suspended particulate matter are : Dust, Smoke and Fly ash.** They remain suspended in air for long periods.

(*a*) Dust consists of tiny particles of earth. Dust is produced by blowing wind, heavy traffic on roads, stone crushers and construction activities. Dust in air spoils our clothes and reduces visibility. **Dust produces allergic reactions in human body and aggravates diseases like bronchitis.** Dust covers the leaves of plants and trees and *prevents* photosynthesis.

(*b*) Smoke is mainly tiny particles of carbon in air. Smoke is produced by the burning of fuels like wood, cow-dung cakes, coal, kerosene, petrol, and diesel in homes, factories, thermal power plants and motor vehicles. Smoke present in air spoils our clothes and blackens the buildings. **Smoke attacks our lungs and causes respiratory diseases.**

(c) **The minute ash particles formed by the burning of coal and carried into air by the gases produced during burning, is called fly ash.** Fly ash is emitted by the chimneys of coal based thermal power plants. **Fly ash particles present in air cause irritation to the eyes, skin, nose, throat and respiratory tract.** Continued breathing in air containing fly ash causes diseases like bronchitis and lung cancer.

The two extremely harmful effects of air pollution on the environment are acid rain and greenhouse effect (or global warming). So, we will now discuss acid rain and greenhouse effect produced by polluted air in detail, one by one. Let us start with acid rain.

ACID RAIN

The burning of fossil fuels like coal and oil in factories, thermal power plants and oil refineries, and petrol and diesel in motor vehicles produce acidic gases like sulphur dioxide and nitrogen oxides which go into air and pollute it. Sulphur dioxide reacts with water vapour present in the atmosphere to form sulphuric acid whereas nitrogen oxides react with water vapour present in the atmosphere to form nitric acid. These acids dissolve in rainwater and fall to the earth in the form of acid rain. Thus, **acid rain is that rain which contains small amounts of acids formed from acidic gases like sulphur dioxide and nitrogen oxides present in polluted air**. Acid rain contains very dilute solutions of sulphuric acid and nitric acid. ***Acid rain causes great damage to living and non-living things.*** The damage caused by acid rain is very, very slow and hence cannot be seen immediately. Acid rain has the following harmful effects :

***(i)* Acid Rain Destroys Forests**. Acid rain damages the forest trees by destroying their leaves [see Figure 2(*a*)]. It causes the leaves of trees to turn yellow and fall off. In the absence of leaves, the roots of trees cannot absorb water from the soil. And due to lack of water the trees die. Acid rain also damages a lot of crop plants every year and causes a big loss to the farmers. Acid rain makes the soil acidic. This acidic soil is not good for the growth of crop plants.

(*a*) Forest trees damaged by acid rain

(*b*) Fish killed due to high acidity of lake water caused by acid rain

(*c*) A statue damaged by acid rain

Figure 2. Harmful effects of acid rain.

***(ii)* Acid Rain Kills Aquatic Animals Such as Fish**. Acid rain causes the water in ponds, lakes and rivers to become much more acidic and unsuitable for the survival of aquatic animals and plants. Due to high acidity of water, the aquatic animals such as fish get killed [see Figure 2(*b*)].

***(iii)* Acid Rain Corrodes the Statues, Buildings and Historical Monuments and Damages Them Slowly.** The statues, buildings and monuments are made of marble or limestone, etc. The acids present in acid rain react with the carbonates present in marble and limestone of a statue, building or monument and corrode it slowly (dissolve it slowly). In this way, acid rain makes the statues, buildings and monuments to crumble away slowly [see Figure 2(*c*)].

The Case of Taj Mahal. The Taj Mahal at Agra (near Delhi) is a beautiful historical monument made of pure, white marble (see Figure 3). The experts have warned that **air pollution around Taj Mahal area is discolouring its white marble and also corroding it slowly**. This poses a threat to the beauty of Taj Mahal.

Figure 3. Taj Mahal monument is being affected by air pollution.

Actually, the Mathura Oil Refinery near Agra as well as the various industries in and around Agra are emitting gaseous pollutants such as sulphur dioxide and nitrogen oxides into the air which cause acid rain. *The acids present in acid rain react with the marble (calcium carbonate) of Taj Mahal monument and corrode it slowly.* The slow corrosion (or eating up) of marble of a monument by acid rain is also known as 'Marble Cancer'. *The suspended particulate matter such as soot particles emitted in the smoke from Mathura Oil Refinery is discolouring the pure white marble of Taj Mahal by turning it yellowish.* The Supreme Court of India has taken several steps to save Taj Mahal from the damage being caused by air pollution (acid rain, etc.). It has ordered all the industries in Agra area to switch over to cleaner fuels like CNG and LPG to reduce air pollution. It has also asked vehicles to be run on CNG or unleaded petrol in the Taj Mahal area.

From the above discussion we conclude that it is not only the living things (such as humans, other land animals, aquatic animals like fish, trees and crop plants, etc.) which are affected by air pollution, even the non-living things (such as soil, statues, buildings and historical monuments) get affected by air pollution.

Before we go further and describe the greenhouse effect, we should know the meaning of 'greenhouse' and how it works. This is described below : **The greenhouse is a structure or building made of glass walls and glass roof in which the plants that need protection from cold weather are grown** (see Figure 4). The glass walls and glass roof of a greenhouse allow the sun's heat rays to go in freely but do not allow the inside heat (reflected by soil, plants and other things in the greenhouse) to go out. In this way, more and more of sun's heat rays are trapped inside the greenhouse due to which the temperature in the greenhouse rises. So, even without an internal supply of heat, the temperature inside a greenhouse becomes higher than that outside. This heat is beneficial for the growth of plants inside the greenhouse (when the outside temperature is very low during winter season). Thus, greenhouse acts as a heat trap (which traps sun's heat energy). **We will now describe why a greenhouse acts as a heat trap.**

Figure 4. A greenhouse. The walls and roof of this greenhouse are made of glass. This greenhouse acts as a heat trap (for sun's heat rays).

The sun is an extremely hot object due to which the heat rays emitted by the sun are of shorter wavelengths. ***The shorter wavelength heat rays coming from the sun can pass through the glass walls and glass roof of a greenhouse and go inside it.*** On the other hand, the inside objects of glasshouse are much less hot (than the sun), so they emit heat rays of longer wavelengths. ***The longer wavelength heat rays emitted by the inside objects of the greenhouse cannot pass through the glass walls and glass roof of a greenhouse and go out.*** Thus, glass is a material which allows sun's heat rays to come in but does not allow heat rays to go out. This is called greenhouse effect. **Due to the presence of a carbon dioxide layer around the earth, our atmosphere acts like the glass roof of an ordinary greenhouse and allows sun's heat rays to be trapped within the earth's atmosphere**. This is why it is called greenhouse effect. Let us discuss it in detail.

GREENHOUSE EFFECT

Carbon dioxide gas present in the atmosphere allows the heat rays of the sun to pass through it and reach the earth but prevents the heat rays reflected from the earth's surface and its objects from passing out of the atmosphere into space. In this way, the sun's heat rays remain trapped in the earth's atmosphere and

17. Name two greenhouse gases ? Which one of them produces the maximum greenhouse effect ?
18. What depletes the ozone layer in the atmosphere ? What are the harmful effects of the depletion of ozone layer on us ?
19. Name one source and one harmful effect of each of the following air pollutants :
(*a*) Sulphur dioxide (*b*) Nitrogen oxides (*c*) Carbon monoxide (*d*) Chlorofluorocarbons (CFCs)
20. Explain why, even clear, transparent and odourless water may not always be safe for drinking.
21. Explain why, hot water released by power plants and industries is considered a pollutant.
22. Why does the increased level of nutrients (or fertilisers) in the lake water affect the survival of aquatic organisms (like fish) ?
23. Explain how, the use of pesticides in agriculture causes water pollution.
24. (*a*) Describe the threat to Taj Mahal monument due to air pollution.
(*b*) State any two ways of controlling air pollution.
25. (*a*) What is potable water ? Name any two methods to make water safe for drinking.
(*b*) State two ways in which you can conserve water at home by preventing its wastage.

Long Answer Type Questions

26. (*a*) What is meant by water pollution ? What are the different ways in which water gets polluted ?
(*b*) State the harmful effects of water pollution.
27. (*a*) What is air ? Write the names of various constituents of air.
(*b*) What is air pollution ? What are the main sources of air pollution ?
28. What is smog ? How is smog formed ? What are its harmful effects ?
29. What is acid rain ? How is acid rain caused ? What are the harmful effects of acid rain ?
30. What is global warming ? What are the likely harmful effects of global warming ?

Multiple Choice Questions (MCQs)

31. Which of the following is not a greenhouse gas ?
(*a*) carbon dioxide (*b*) nitrous oxide (*c*) methane (*d*) nitrogen
32. Which of the following air pollutant reduces the oxygen-carrying capacity of blood to a large extent ?
(*a*) carbon dioxide (*b*) nitrogen monoxide (*c*) carbon monoxide (*d*) sulphur dioxide
33. The constituent of polluted air which contributes in producing acid rain is :
(*a*) nitrogen (*b*) sulphur dioxide (*c*) oxygen (*d*) argon
34. The Kyoto Protocol is associated with one of the following. This one is :
(*a*) reduction in the use of chlorofluorocarbans
(*b*) reduction in the emission of greenhouse gases
(*c*) reduction in the cutting of forest trees
(*d*) reduction in pollution of fresh water sources
35. Which of the following will be reduced in air in a city forest when a lot of dust and fly ash in emitted by a coal-based factory in the vicinity ?
(*a*) nitrogen (*b*) carbon dioxide (*c*) oxygen (*d*) water vapour
36. Which of the following disease cannot be caused by drinking of river water contaminated with untreated sewage ?
(*a*) cholera (*b*) typhoid (*c*) tuberculosis (*d*) diarrhoea
37. Which of the following statement about ozone is correct ?
(*a*) it is essential for breathing (*b*) it absorbs ultraviolet rays
(*c*) its proportion in air is about 3 % (*d*) it is mainly responsible for global warming
38. Drinking water can be made absolutely safe by adding some :
(*a*) aspirin tablets (*b*) iodine tablets (*c*) chlorine tablets (*d*) chlorophyll tablets
39. The excessive use of one of the following in agriculture can cause the death of fish in a pond by oxygen starvation. This one is :
(*a*) fertilisers (*b*) manures (*c*) pesticides (*d*) herbicides
40. Which of the following is usually not a water pollutant ?
(*a*) sewage (*b*) fertiliser (*c*) fly ash (*d*) pesticide

41. Which of the following are used in electric water filters to kill all the harmful micro-organisms present in tap water and make it absolutely safe for drinking ?
(*a*) infrared radiation (*b*) gamma radiation (*c*) visible radiation (*d*) ultraviolet radiation
42. Which of the following is not an air pollutant ?
(*a*) sulphur dioxide (*b*) sewage (*c*) CFCs (*d*) SPM
43. Which of the following air pollutant is capable of preventing photosynthesis in plants ?
(*a*) CFCs (*b*) nitrogen oxides (*c*) dust (*d*) carbon monoxide
44. One of the following does not contribute in producing acid rain. This one is :
(*a*) nitrogen dioxide (*b*) nitrogen monoxide (*c*) carbon monoxide (*d*) sulphur dioxide
45. Which of the following will reach the earth in greater amounts if the amount of chlorofluorocarbons released into the air increases ?
(*a*) infrared rays (*b*) X-rays (*c*) gamma rays (*d*) ultraviolet rays

Questions Based on High Order Thinking Skills (HOTS)

46. The farmers use large amounts of a substance P in the fields to increase the crop yield. The excess of P dissolves in water and runs into a lake. The substance P causes rapid growth of tiny green water plants Q in the lake which cover the whole lake like a green sheet. When the plants Q die, the organisms called R decompose them by utilizing S dissolved in lake water. The amount of dissolved S in water decreases too much due to which the fish living in lake suffocate and die. What are P, Q, R and S ?
47. At many places the wastewater containing human excreta from homes and carried in big underground pipes is dumped into a river as such which pollutes the riverwater.
(*a*) What is the common name of such wastewater ?
(*b*) Name five types of harmful organisms contained in it.
(*c*) Name any five human diseases caused by drinking river water contaminated with such wastewater.
48. Match the items given in column I with one or more items given in column II :

Column I	*Column II*
(*i*) Prevents photosynthesis	(*a*) Sewage dumped in river
(*ii*) Damage ozone layer	(*b*) Excess fertiliser in fields
(*iii*) Produce acid rain	(*c*) Carbon dioxide in air
(*iv*) Kill fish by deoxygenating water	(*d*) Dust in air
(*v*) Causes water borne diseases	(*e*) CFCs
(*vi*) Leads to global warming	(*f*) Sulphur dioxide in air

49. State one way in which the air pollution caused by the burning of fossil fuels in transport and industry can kill the fish living in a lake and one way in which the water pollution caused by an agricultural activity can kill fish living in the same lake.
50. The incomplete combustion of firewood in homes produces a very poisonous gas X. When inhaled, gas X combines with the substance Y present in blood and reduces the capacity of blood to carry gas Z causing respiratory problems and suffocation. What are X, Y and Z ?

ANSWERS

1. Carbon dioxide **2.** Sulphur dioxide ; Nitrogen oxides **4.** Carbon monoxide **6.** Ozone **10.** Oxygen **31.** (*d*) **32.** (*c*) **33.** (*b*) **34.** (*b*) **35.** (*c*) **36.** (*c*) **37.** (*b*) **38.** (*c*) **39.** (*a*) **40.** (*c*) **41.** (*d*) **42.** (*b*) **43.** (*c*) **44.** (*c*) **45.** (*d*) **46.** P : Fertiliser ; Q : Algae ; R : Bacteria ; S : Oxygen **47.** (*a*) Sewage (*b*) Bacteria ; Protozoa ; Fungi ; Viruses ; Parasites (Worms) (*c*) Cholera ; Typhoid ; Diarrhoea ; Dysentery ; Jaundice **48.** (*i*) *d* (*ii*) *e* (*iii*) *c* ; *f* (*iv*) *b* (*v*) *a* (*vi*) *c* **49.** (*i*) Acid rain caused by the burning of fossil fuels makes the lake water too much acidic which kills the fish (*ii*) Some of the fertilisers used in agriculture run into lake water and cause too much growth of algae in lake. Decomposition of dead algae by bacteria by using dissolved oxygen of lake water decreases the amount of oxygen in lake water too much. And fish then die because of oxygen starvation **50.** X : Carbon monoxide ; Y : Haemoglobin ; Z : Oxygen.